February 22–24, 2012
Monterey, California, USA

**Association for
Computing Machinery**

Advancing Computing as a Science & Profession

FPGA'12

Proceedings of the 2012 ACM/SIGDA International Symposium on
Field Programmable Gate Arrays

Sponsored by:
ACM SIGDA

Supported by:
Altera, Microsemi, Microsoft Research, Xilinx, BEEcube, Algo-Logic, and Atomic Rules

With logistical support from:
The Trimberger Family Foundation

**Association for
Computing Machinery**

Advancing Computing as a Science & Profession

The Association for Computing Machinery
2 Penn Plaza, Suite 701
New York, New York 10121-0701

Notice to Past Authors of ACM-Published Articles
ACM intends to create a complete electronic archive of all articles and/or other material previously published by ACM. If you have written a work that has been previously published by ACM in any journal or conference proceedings prior to 1978, or any SIG Newsletter at any time, and you do NOT want this work to appear in the ACM Digital Library, please inform permissions@acm.org, stating the title of the work, the author(s), and where and when published.

ISBN: 978-1-4503-1155-7

Additional copies may be ordered prepaid from:

ACM Order Department
PO Box 30777
New York, NY 10087-0777, USA

Phone: 1-800-342-6626 (USA and Canada)
+1-212-626-0500 (Global)
Fax: +1-212-944-1318
E-mail: acmhelp@acm.org
Hours of Operation: 8:30 am – 4:30 pm ET

ACM Order Number: 420120

Printed in the USA

Foreword

Welcome to the 2012 ACM International Symposium on Field-Programmable Gate Arrays (FPGA). *FPGA* continues to be a premier venue for researchers to present their efforts in a variety of FPGA-related areas: architecture, circuit design, Computer-Aided Design (CAD), FPGA-based computing machines, design studies and applications specially suited to FPGA implementation. In addition to the paper presentations, FPGA provides opportunities for researchers, students and industrial participants to mingle and to personally discuss research.

FPGA'12 is the 20th annual meeting for this symposium. A special collection of papers, *The FPGA-20*, has been compiled to celebrate 20 years of successful FPGA research. A panel of experts chose the papers and Steve Trimberger and André DeHon edited the final contents for this anniversary volume and will present it during the conference.

We received 87 submissions this year and selected 20 of these for presentation as full papers (24% acceptance rate) and these will appear in the proceedings as 10-page papers. An additional 16 submissions were selected as 4-page short-paper presentations; these consist of a 5-minute introductory presentation followed by a poster where delegates can ask detailed questions of the authors. An additional 20 poster-only presentations will allow authors to interact with attendees more directly.

This year's symposium features a pre-conference workshop, "FPGAs in 2032: Challenges and Opportunities in the Next 20 Years," run by Vaughn Betz (University of Toronto) and Lesley Shannon (Simon Fraser University). All attendees are encouraged to attend. A special anniversary banquet will be held at The Monterey Bay Aquarium.

Putting together *FPGA'12* was a team effort. We first thank the authors for providing the content of the program. We are grateful to the program committee who worked very hard in reviewing papers and providing feedback for authors. We also want to thank our sponsors, ACM and ACM SIGDA and our corporate supporters: Algo-Logic, Altera, Atomic Rules, BeeCube, Microsemi, Microsoft Research and Xilinx. We also thank the Trimberger Family Foundation for additional logistical support.

Welcome to FPGA 2012!

Katherine Compton
FPGA'12 General Chair
University of Wisconsin-Madison, USA

Brad Hutchings
FPGA'12 Program Chair
Brigham Young University, USA

Table of Contents

Session 4: Architecture I

Session 5: Applications II

Session 6: Tools and Abstractions

Session 7: Compute Engines and Run-Time Systems

Session 8: Architecture II

Poster Session 1

Poster Session 2

FPGA 2012 Organization

General Chair: Katherine Compton *(University of Wisconsin-Madison)*

Program Chair: Brad Hutchings *(Brigham Young University)*

Publicity Chair: John Lockwood *(Algo-Logic Systems)*

Finance Chair: John Wawrzynek *(University of California, Berkeley)*

Workshop Chairs: Vaughn Betz *(University of Toronto, Canada)*
Lesley Shannon *(Simon Fraser University, Canada)*

Program Committee: Jason Anderson *(University of Toronto, Canada)*
Kia Bazargan *(University of Minnesota, USA)*
Vaughn Betz *(University of Toronto, Canada)*
Chen Chang *(BEEcube, USA)*
Deming Chen *(University of Illinois at Urbana-Champaign, USA)*
Peter Cheung *(Imperial College London, UK)*
Derek Chiou *(University of Texas, Austin, USA)*
Paul Chow *(University of Toronto, Canada)*
Katherine Compton *(University of Wisconsin at Madison, USA)*
Jason Cong *(University of California at Los Angeles, USA)*
George Constantinides *(Imperial College London, UK)*
Carl Ebeling *(University of Washington, USA)*
Jonathan Greene *(Actel, USA)*
Scott Hauck *(University of Washington, USA)*
Lei He *(UCLA, USA)*
Brad Hutchings *(Brigham Young University, USA)*
Mike Hutton *(Altera, USA)*
Ryan Kastner *(University of California Santa Barbara, USA)*
Martin Langhammer *(Altera, UK)*
Miriam Leeser *(Northeastern University, USA)*
Guy Lemieux *(University of British Columbia, Canada)*
Philip Leong *(University of Sydney, Australia)*
David Lewis *(Altera, Canada)*
Mingjie Lin *(University of Central Florida, USA)*
John Lockwood *(Algo-Logic Systems, USA)*
Wayne Luk *(Imperial College London, UK)*
Patrick Lysaght *(Xilinx, USA)*
Stephen Neuendorffer *(Xilinx, USA)*
Jonathan Rose *(University of Toronto, Canada)*
Kyle Rupnow *(Advanced Digital Sciences Center, Singapore)*
Graham Schelle *(Xilinx, USA)*

Additional reviewers (continued):

Andreas Weichslgartner

Ruediger Willenberg

Justin Wong

Jacob Tolar

Tuan

Deepak Unnikrishnan

Tim Vanderhoek

Jasmina Vasiljevic

Kris Vorwerk

Lu Wan

Gene Wu

Xiaojian Yang

Daniel Ziener

Desheng Zheng

Jiefan Zhang

Peng Zhang

Zhiru Zhang

Tobias Ziermann

Yi Zou

FPGA 2012 Sponsors & Supporters

Sponsor:

Supporters:

Also with the support of:

With additional logistics from:

Intra-Masking Dual-Rail Memory on LUT Implementation for Tamper-Resistant AES on FPGA

Anh-Tuan Hoang
Ritsumeikan University, Japan
1-1-1, Noji-Higashi, Kusatsu, Shiga
525-8577, Japan
+81-77-599-4166

anh-tuan@fc.ritsumei.ac.jp

Takeshi Fujino
Ritsumeikan University, Japan
1-1-1, Noji-Higashi, Kusatsu, Shiga
525-8577, Japan
+81-77-561-5150

fujino@se.ritsumei.ac.jp

ABSTRACT

In current countermeasure design trends against differential power analysis (DPA), security at gate level is required in addition to the security algorithm. Several dual-rail pre-charge logics (DPL) have been proposed to achieve this goal. Designs using ASIC can attain this goal owing to its backend design restrictions on placement and routing. However, implementing these designs on field programmable gate arrays (FPGA) without information leakage is still a problem because of the difficulty involved in the restrictions on placement and routing on FPGA.

This paper describes our novel masked dual-rail pre-charged memory approach, called "intra-masking dual-rail memory on LUT," and its implementation on FPGA for tamper-resistant AES. In the proposed design, all unsafe nodes, such as unmasking and masking, and the dual-rail memory and buses are packed into a single LUT. This makes them balanced and independent of the placement and routing tools. The design is independent of the cryptographic algorithm, and hence, it can be applied to available cryptographic standards such as DES or AES as well as future standards. It requires no special placement or route constraints in its implementation. A correlation power analysis (CPA) attack on 1,000,000 traces of AES implementation on FPGA showed that the secret information is well protected against first-order side-channel attacks. Even though the number of LUTs used for memory in this implementation is seven times greater than that of the conventional unprotected single-rail memory table-lookup AES and three times greater than the implementation based on a composite field, it requires a smaller number of LUTs than all other advanced tamper-resistant implementations such as the wave dynamic differential logic, masked dual-rail pre-charge logic, and threshold.

Categories and Subject Descriptors

B.7.1 [**Hardware**]: **Integrated circuits** - *Algorithms implemented in hardware*

General Terms

Design, Security

Keywords

Side-Channel Attack, Differential Power Analysis (DPA), Masking, Dual-Rail Memory, Intra-Masking Dual-Rail Memory on LUT, Tamper Resistance, AES, Field Programmable Gate Array (FPGA).

1. INTRODUCTION

In recent times, Advanced Encryption Standard (AES) has become a widely used encryption standard. It is a round-based symmetric cipher that can be implemented efficiently on all types of platforms. The key of AES [14] is 128, 192, and 256 bits in length, and hence, finding the key using a brute-force attack in an acceptable amount of time is difficult. AES is considered mathematically safe for application in cryptographic devices.

However, side-channel attacks have become a new threat to cryptographic devices such as smart cards [11]. When cryptographic algorithms such as AES and DES are implemented on hardware, they leak some secret information to its side channel, which can be simply measured using an oscilloscope. Side-channel information such as power consumption, timing information, and electromagnetic field can be analyzed in a side-channel attack process to reveal secret internal information (the cryptographic key). Differential power analysis (DPA) introduced by Kocher in 1999 [11] and correlation power analysis (CPA) [1] introduced by Brier in 2004 are the most powerful and commonly known attack techniques. They exploit the correlation between the instantaneous power consumption of a cryptographic device and intermediate data while the device performs encryption operations. Secret information can be extracted from the unprotected cryptographic device using only thousands of traces, as shown in the notes and experiments of Chaudhuri [5] and Mimura [12].

In order to counter these threats, various countermeasure methods have been proposed to prevent side-channel attacks based on DPA. These countermeasures can be roughly categorized into the algorithm-level ([8], [13], [15], [16], [23]) and circuit level ([6], [25], [26]) groups. Countermeasures at the algorithm level include methods such as masking of the processed data ([8], [16], [23]), secret sharing, and threshold implementation ([13], [15]); they depend on cryptographic algorithms and cannot be used when the algorithm is changed. Circuit-level tamper-resistant designs cut off side-channel information leakage at gate level by randomizing power consumption using random switching logic ([18], [26]) or by unifying the power consumption of the cryptographic device using dual-rail logic ([6], [24], [25]); they are currently the focus of many researchers due to their algorithm independency. However, the security of these logics depends on the timing control or balance in the placement and routing process. Hence, they can be applied to ASIC implementation, which requires a complex design process with severe restrictions on placement and routing, but are inappropriate to implement on FPGA, which has automatic placement and routes. Consequently, a series of successful attacking efforts with regard to the imbalance in placement and routing were reported in [12], [21] for circuit-level countermeasure implementations on ASIC and FPGA.

In this paper, we introduce our novel intra-masking dual-rail memory on LUT implementation for AES on FPGA in order to overcome the limitations of other circuit-level countermeasure methods. The processing data on a bus are hidden with a random mask in each round, and the power consumption of a non-linear operation (SBox) is made uniform by complementary memories. Simple memory duplication for the complementary SBox with standard design flow will leak secret information to the side channel due to the imbalance in the routes of the two memories. In order to use the standard design flow supported by the FPGA tool, the two memory elements in a complementary pair must be implemented on the same basic FPGA element, the LUT. Our approach is suitable for all non-linear cryptographic circuits due to its algorithm independency. No specific constraint for placement and routing or non-standard design flow is required during its implementation on FPGA. An experiment on the Side-Channel Attack Standard Evaluation Board GII (SASEBO-GII board) [19] proved the resistance of the method to a first-order CPA attack with 1 million traces. The AES algorithm and the side-channel attack method on it are presented in Section 2. The related work and issues are given in Section 3. Sections 4 and 5 present our original masked dual-rail memory concept, its improvement over intra-masking dual-rail memory on LUT implementations, and evaluations. Finally, the conclusion is given in Section 6.

2. AES ALGORITHM: IMPLEMENTATION AND LEAKAGE INFORMATION TO SIDE-CHANNEL

2.1 AES Algorithm

The AES algorithm is used to encrypt 128-bit data blocks with a key of 128, 192, or 256 bits. In this study, we focused on the AES-128 algorithm, which uses a 128-bit key to encrypt data. The 128-bit data is divided into 16 bytes. The data are arranged in a 4 × 4 matrix called a "state." The algorithm initiates with AddRoundKey processing followed by 10 rounds. Each round consists of four steps: SubBytes, ShiftRows, MixColumns, and AddRoundKey.

SubBytes: SubBytes is a nonlinear substitution step. The 128-bit data are separated into 16 bytes and processed by 16 separate SBoxes, where each byte is replaced by another taken from a substitution box (Rijndael S-Box). This operation provides nonlinearity to the cipher. The SBox is derived from the multiplicative inverse over Galois Field 2^8 ($GF(2^8)$).

ShiftRows: ShiftRows operates on the rows of each state. It cyclically shifts the bytes in each row by a certain offset. The first row is kept unchanged while the second row is cyclically shifted one byte to the left. Similarly, the third and fourth rows are cyclically shifted two and three bytes to the left.

MixColumns: MixColumns mixes all four bytes in the same column of the state together to generate an additional four outputs

(in the byte) for that column. MixColumns guarantees that each original input byte affects all the four output results. ShiftRows and MixColumns together provide diffusion to the cipher.

AddRoundKey: AddRoundKey simply combines the round key (subByte key in each round operation) with the state by an exclusive–or (XOR) operation.

2.2 AES Implementation and Leakage Information

Figure 1 shows a common, non-pipelining implementation of AES on hardware. Ten clocks that correspond to each of the 10 rounds are required. In each round, the SubBytes, ShiftRows, MixColumns, and AddRoundKey transformations for all 16 SBoxes, which correspond to the 16 SBox round keys, are performed in the same clock. A register is used to store the intermediate state of round processing. The power consumption of the implementation at each clock is the overlapped power used by the above four transformations for all 16 SBoxes and SBox keys. This causes difficulties for attackers to expose the secret key compared with software or pipeline implementations, in which the power used by each SBox can be separately measured.

The nonlinear multiplicative inverse computation of the SBox inside the SubBytes transformations can be implemented by looking up a simple table using a Galois field with composite field [20] or using subfield arithmetic [3]. The side-channel leakage information in the 8-bit nonlinear processing of this SBox is the objective of the side-channel attack. The power consumed by each SBox relies on the 0–1 or 1–0 transition inside its gates and wires. These transitions again rely on the data processed by the SBox. The attackers can exploit the input data of each SBox to find the corresponding 8-bit key.

2.3 Side-Channel Attack on AES

The general attack method on AES works according to final round processing, in which the 8-bit ciphertext corresponding to the pursuing SBox is inversely XORed with each hypothetical 8-bit key (among $2^8 = 256$ possibilities) and then processed with the

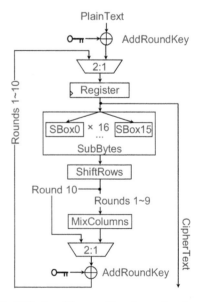

Figure 1. AES algorithm and hardware implementation.

inverse-ShiftRows and inverse-SubBytes transformations to find the hypothetical intermediate value at the register. The Hamming weight (HW) of this data determines the power consumption of the device at the final clock (clock 10), while the Hamming distance (HD) between the values of the register before and after the storage time (rising edge of the clock) affects the power consumption of the implementation at the next clock (clock 11). This power is used to change the value of the register and then change the gate status of SubBytes, ShiftRows, MixColumns, and AddRoundKeys during the processing. The real consumed power and hypothetical HWs (or HDs) mentioned above are used to compute the correlation factors for the hypothesis keys. There are 256 possibilities for the 8-bit key (00_h–FF_h), and hence, there are 256 correlation factors with a constant number of traces. The highest one corresponds to the expected correct key. Sixteen attacks are needed for the whole 128-bit key processed by 16 SBoxes.

3. AVAILABLE TAMPER RESISTANCE: DESIGNS AND PROBLEMS
3.1 Tamper Resistance at Gate Level

3.1.1 Wave Dynamic Differential Logic (WDDL)
Wave dynamic differential logic (WDDL) is a representative of dual-rail pre-charge logic, which makes power consumption uniform at gate level. WDDL gates shown in Figure 2.a were introduced by Tiri in 2004 [27]. The gate consists of a pair of complementary logic-AND and logic-OR to form a pair of complementary outputs. The logic is pre-charged by a zero-spacer (0, 0) at the beginning of each clock cycle by the pre-charge input logic, where (0, 0) means that both output levels of logic-AND and logic-OR are 0s. Consequently, there are only two possible transformations from (0, 0) to (0, 1) and from (0, 0) to (1, 0) inside the WDDL. Hence, there is exactly one transition from 0 to 1 per pair and clock cycle regardless of the logic value carried by the pair. Thus, if a different wire pair maintains symmetry, power consumption remains constant. In contrast, if there are small loading imbalances between the two wires that make up the pair, the amount of energy needed per clock cycle depends on which of the two nodes is switched, and thus, it is correlated with the logical value. Information leakage then occurs.

Figure 2.b shows the implementation of WDDL on FPGA. The two complementary logics in a pair are placed in two LUTs and placed next to each other in a slice. With ideal placement and routing, the two wires that make up a pair are balanced in length, as shown on the right side of Figure 2, and so, the amount of energy needed per clock cycle is uniform. Unfortunately, balancing the capacitive load of the two wires is hard to do using the FPGA placement and routing tool. As a result, there are loading imbalances in the wire pair. Residual information leakage occurs, and Sauvage has reported a successful side-channel attack exploiting this [21].

3.1.2 Masked Dual-Rail Pre-Charge Logic (MDPL)
In dual-rail logic styles such as WDDL, complementary wires need to have the same capacitive load. However, building two perfectly balanced wires is quite challenging in modern LSI design. MDPL combines the ideas of WDDL and random switching logic to randomize the power consumption of cryptographic devices [16]. The data are now represented by masked data and its complement ((data $\oplus$ mask) and /(data $\oplus$ mask)) in an MDPL cell (MDPL-AND gate and MDPL-OR gate), as shown in Figure 3. Computation occurs with two rails in the same manner as the computation in the WDDL method. The benefit of this approach is that complementary wires in MDPL circuits do not need to be balanced. Some LSI implementations using MDPL were performed, and DPA resistances were evaluated. Several researchers including Popp, the inventor of MDPL, claimed that secret information leakage exists due to an early propagation effect and/or power distribution dependence on random numbers in MDPL countermeasures [17], [12].

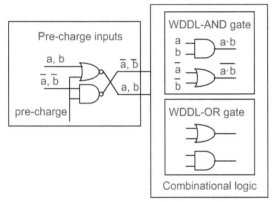

a. WDDL AND/OR logics with pre-charging.

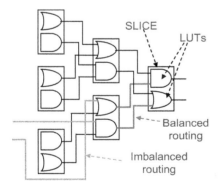

b. WDDL implementation on FPGA.

Figure 2. WDDL logics with pre-charging and implementation on FPGA.

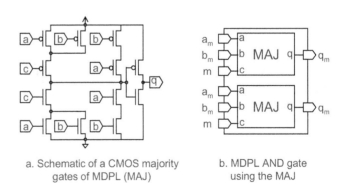

a. Schematic of a CMOS majority gates of MDPL (MAJ)

b. MDPL AND gate using the MAJ

Figure 3. MDPL majority gates and AND gate example.

Furthermore, large implementation area and high power consumption of MDPL are also problems.

3.2 Tamper Resistance in Combination with an Algorithm

Tamper-resistant designs in combination with an algorithm address concerns of masking and data sharing among different computations, such as "perfect mask" suggested by Canright [4], masked composite SBox implementation on FPGA given by Regazzoni [8], or "sharing" introduced by Nikova [15]; this was realized by Moradi [13] with "threshold" implementation. The masking method tries to hide data computed inside the SBox by masking the real data with a round mask. The correct SBox result can be achieved at the end of the computation by the unmasking operation. The sharing method shares several real data points in different parts and makes a pair that is computed together to avoid glitch leakage. The latest threshold implementation effort shares the datapath and key schedule with some mask to hide AES leakage. These methods rely on a cryptographic algorithm to hide secret information, so it is hard to apply these methods to other cases with different algorithms such as DES or future cryptographic standards. The threshold implementation (TI) on FPGA at Yokohama University [10] showed that this method requires a huge FPGA resource.

4. MASKED DUAL-RAIL MEMORY

All of the above methods have some weakness that makes them difficult to apply in future cryptographic device implementations or standard design flows. The WDDL design requires all pairs of complementary wires to be perfectly balanced, which is very hard or even impossible to guarantee. MDPL needs a large implementation area while still allowing some leakage due to an early propagation effect. Tamper resistance in combination with an algorithm cannot be applied to future cryptographic standards with new algorithms. This was the motivation for us to develop a design that is independent of algorithms, easily applicable to standard design flows, and small in hardware size.

In order to overcome the weaknesses of the above methods, we propose a novel implementation for the nonlinear SubBytes operation based on complementary memories with dual-rail addresses and data lines, which we call the masked dual-rail memory method on FPGA. Similar to the balancing problem of WDDL, we must balance the route of each complementary memory element pair, which cannot be achieved by a standard design flow. In order to use the standard design flow on FPGA in its design and implementation, we use the basic element (LUT) of the FPGA as the masked dual-rail memory. The two elements in the complementary pair must be located in the same LUT to maintain the route balance. Intermediate data except that in the SubBytes operation is masked by a random number. This protects the HW and HD of the data from leaking to a side-channel. Moreover, the method is independent of the cryptographic algorithm and thus supports all current standards such as DES and AES as well as future ones.

4.1 Masked Dual-Rail Memory Concept

The concept design of masked dual-rail memory has a complementary memory pair with dual-rail input and output buses. The masked output is selected by a random *new_mask*, as shown in Figure 4. The masked dual-rail memory for the AES SBox

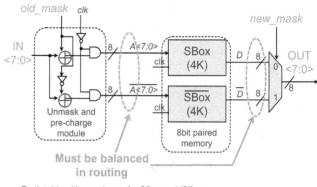

Truth table with pre-charge for SBox and /SBox

old_mask	IN<7:0>	clk	A<7:0>	D	/A<7:0>	/D	
x	xx_h	1	00_h	00_h	00_h	00_h	Pre-charge when clk=1
...	...	...	...	...	...	...	
x	xx_h	1	00_h	00_h	00_h	00_h	
0	00_h	0	00_h	63_h	FF_h	$/63_h = 9C_h$	Evaluate when clk=0
0	01_h	0	01_h	$7C_h$	FE_h	$/7C_h = 83_h$	
...	...	...	...	...	...	...	
0	FE_h	0	FE_h	BB_h	01_h	$/BB_h = 44_h$	
0	FF_h	0	FF_h	16_h	00_h	$/16_h = E9_h$	
			SBox		/SBox		

Figure 4. Masked dual-rail memory concept.

contains two 4-Kbit memories called *SBox* and */SBox*, which are grouped in a pair. The data stored inside the *SBox* memory are the original table lookup values for the SBox in [14]. In contrast, the value inside each element of */SBox* is the complement of that in *SBox* but stored in an inversion sequence. This means that if the value inside address 00_h of *SBox* is 63_h, the complementary address of */SBox*, FF_h, must have the complementary value of 63_h, which is $9C_h$. The unmasking process and pre-charge module are used to pre-charge the dual-rail address buses A and /A by (0, 0), when the clock is high (clk = 1) to guarantee uniform power consumption. It also removes the mask applied to the data at the previous step by XORing the masked AES data with *old-mask* and */old-mask*. It generates the address (or real input data A) for *SBox* and the /address (or the complement of the real input data /A) for */SBox*. Consequently, the complementary values stored inside the two memories appear at the output as *D* and */D*. Finally, *new-mask* is applied to the complementary outputs *D* and */D* to select the final masked data out. The dual-rail output from the memories (*D* and */D*) also requires a pre-charge state for SBox and /SBox (i.e., outputs = 0 if clk = 1). Hence, clk must be considered as one more input address to the memories, which increases the address bus of each memory to 9 bits. The capacity of the memory becomes $2^9 = 512$ elements, which are 8 bits each. Thus, the total size of each memory is 4 Kbits.

Similar to the operations of other dual-rail logic methods, the operation sequence of the masked dual-rail memory starts with a pre-charging time when all gates and pair buses (A, /A, D, and /D) are reset to 0. The evaluation time then starts for setting the dual-rail buses to the corresponding value. Hence, the wire pairs in the pair bus have the transformation of $(0, 0) \rightarrow (0, 1)$ or $(0, 0) \rightarrow (1, 0)$. At any time, exactly one $0 \leftrightarrow 1$ transition occurs in a wire pair. Consequently, if the complementary wires have equivalent capacitive load, power consumption is uniform. In contrast, if there are loading imbalances between the two buses that make up the pair, the amount of energy needed per clock cycle depends on which of the two memories is switched and therefore is correlated with the logical value. Information leakage then occurs.

Figure 5 shows the implementation of AES using the conceptual masked dual-rail memory. The configuration of the AES module is the same as the original one in Figure 1, except that the non-linear SBox modules are replaced by the masked dual-rail memory. Thus, to unmask the SBox input data, masking at the output and mask generator is required. The previous single random mask bit is duplicated to the 8-bit mask with a value of 00_h or FF_h to unmask the intermediate value inside the register. The dual-rail address values inside SBox are generated for the mask with 0 and 1 outputs corresponding with the outputs of SBox or /SBox. A new 1-bit random mask is used to select these two masked outputs to make the masked output of SubByte. The masked output from the dual-rail memory is then processed through ShiftRows, MixColumns, and AddRoundKeys before coming back to the register.

4.2 Straightforward Implementation of Conceptual Masked Dual-Rail Memory and Placement Control on FPGA

In the straightforward implementation method for the masked dual-rail memory on Virtex-5, we separate the 4-Kbit memory shown in Figure 4 into eight 512-bit memories, with each memory in charge of the 1-bit output. The 1-bit output 512-bit memory, which requires LUT9 (9-input LUTs) as the basic memory module, must be separated into smaller parts to fit the FPGA basic element. The complementary memory parts of SBox and /SBox are then grouped together to form complementary LUTs and placed into a single slice. Figure 6 shows the full 8-bit SBox implementation on Virtex-5 FPGA, which is composed of 6-input LUTs (LUT6). Each complementary LUT6 pair can store 32 bits of data for SBox and /SBox together with 32-bit 0's for pre-charging; hence, we need eight pairs to store the 256 bits of data of 1-bit masked dual-rail memory. In total, $8 \times 8 = 64$ LUT6 pairs are required for a single SBox of AES. Hence, $64 \times 16 = 1024$

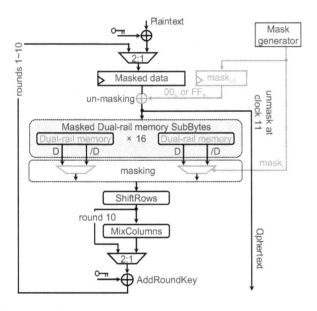

Figure 5. AES implementation using masked dual-rail memory.

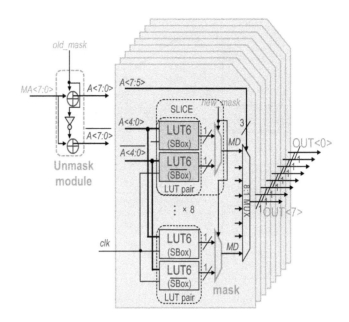

Figure 6. Straightforward masked dual-rail memory SBox implementation on FPGA using LUT pair.

LUT6 pairs, which are equal to 2048 LUT6s, are necessary for the implementation of the 16 SBoxes of the AES device. The unmask module receives the 8-bit masked data ($MA<7:0>$) and XORs with old_mask and /old_mask to create the dual-rail address bus ($A<7:0>$ and $/A<7:0>$) to all the LUT pairs and the final 8:1 multiplexer for the final masked output result (OUT). The masked output (MD) is selected from the outputs of the complementary LUTs using the new_mask signal inside the slice.

The contents of the paired LUTs, which are placed in the same slice as shown in Figure 6, must be formed from the complementary values of SBox and /SBox to balance capacitive load in routing. The memories used as output during the pre-charge time are also separated into the LUTs; hence, half of the content of LUT6 (32 bits) must be used for pre-charge purposes when the clock is high. The SBox and /SBox memory separation and mapping to the LUT6 are shown in Figure 7. The 256-bit SBox data are placed on eight different LUT6 pairs, which have a storable memory size of 32 bits. The values inside /SBox from addresses FF_h to $E0_h$ are complementary to those of SBox from addresses 00_h to $1F_h$. Other pairs are created and grouped in the same manner until the last one, which is formed from the paired addresses ($E0_h$–FF_h) from SBox and ($1F_h$–00_h) from /SBox, is created and grouped. Two LUT6_Ls are used to store the paired memory values to form complementary LUTs. MUX7F in the FPGA slice is used with the new_mask signal as a selector for the masking operation. The combination of two LUT6_L pairs and MUX7F is automatically placed on a single slice by the FPGA placement and routing tool without special manual placement.

4.3 Experiment and Discussion

The experimental conditions with general, inexpensive devices are shown in Table 1. The experiment board (FPGA board) has two FPGA chips: one for the controller (mainly for communication) and the other for cryptographic purposes. 128-bit plaintext is sent from the host PC to the controller FPGA chip

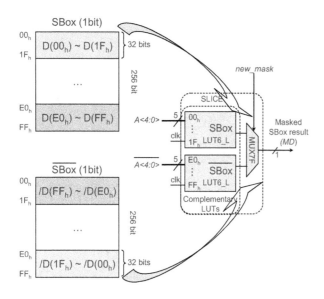

Figure 7. Formation of complementary LUT pair for masked dual-rail memory.

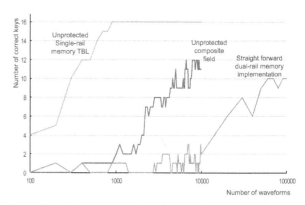

Figure 8. Leakage comparison for straightforward masked dual-rail memory SBox and other implementations on FPGA.

through a USB cable after regular intervals. The data are then sent to a cryptographic FPGA chip for encryption. A LFSR random number generation module is also included into this FPGA chip for round mask generation. The oscilloscope measures the power used by the cryptographic chip during the encryption process and sends it to the host PC. The PC receives this trace before sending a new plaintext to the SASEBO board. Measurement for 100,000 traces is carried out for over 1 h.

Table 1. Experiment conditions

FPGA board	SASEBO-GII [19]
Cryptographic FPGA	Virtex-5 XC5vlx30-1ff324
Measurement	Agilent DSO1024A oscilloscope
Attack method	CPA with 100,000 traces
Measurement time	Over 1 h

Figure 8 shows the attacking result of the conceptual masked dual-rail memory implementation in comparison with other unprotected SBox computation methods, such as conventional single-rail memory table lookup (TLB) and composite field [20]. The results show that the unprotected implementation method easily leaks the most secret information to the side channel after less than 1,000 traces with conventional single-rail memory TBL and 10,000 traces with composite field. In contrast, the conceptual masked dual-rail memory implementation leaks only a small amount of the secret information even after 100,000 traces. This shows that even the straightforward implementation of the masked dual-rail memory is safer than unprotected implementations; however, it still leaks some secret information, and improvement is necessary.

If we look at the design details in Figure 6, we can see that only one unmask module is used, and the dual-rail unmasked addresses are transferred to each complementary memory pair. The automatic placement and routing tool for FPGA may randomly route the two address buses ($A<4:0>$ and $/A<4:0>$) to the pair LUT6 in the slice. Consequently, the lengths of the two buses

may be unbalanced. In addition, the 3-bit selection for the final 8:1 multiplexer ($A<7:5>$) is singly routed from the real address. The power used by these wires relies on the unmasked data that they process. As a result, secret information leaks to the side-channel because of unbalanced routing of the dual-rail address bus to the LUT pair and the single route to the 8:1 MUX shown in Figure 4. Hence, the straightforward implementation shown in Figure 4 must be improved to remove leakages caused by unbalanced routing outside the LUTs.

5. MASKED DUAL-RAIL MEMORY IMPROVEMENT

There are two ways to enhance the security of the conceptual masked dual-rail memory implementation: strictly balancing the wires combining the dual-rail bus to the LUT pair, or masking them until they reach the memory elements (LUT). The first method requires a special placement and routing tool for the FPGA. The second involves redesigning the memory to mount the unmask module into the LUT to form an "intra-masking dual-rail memory on LUT." Here, we explain the application of the second method to improve the conceptual masked dual-rail memory implementation by combining the unmask module for the input and the mask module for the output into the memory to remove the imbalance of the dual-rail capacitive load on the unmasked address. This also secures the input for the final selector shown in Figure 6 ($A<7:5>$) by the masked signal (MA).

5.1 First-Order DPA Attack-Resistant Design

5.1.1 Memory duplication for internal unmasking

In order to internally pack the unmask module shown in Figure 6 into the memory, each memory element must be duplicated into the complementary address, as shown in Figure 9. The idea here is to use two memories: one for the original address and the other for the masked address. The memory whose data are stored at the original address is named "masked by 0." The content of each element in the additional memory (i.e., "masked by 1") is taken from the original but stored in a complementary address; in other words, the value at address 00000_b of the original memory is stored into the address of $/00000_b = 11111_b = 1F_h$ of the second one. The output data from the real address ($mask = 0$) and complementary one ($mask = 1$) have the same value regardless of the mask. old_mask is used to select the output from either the "masked by 0" or "masked by 1" memory. These two memories

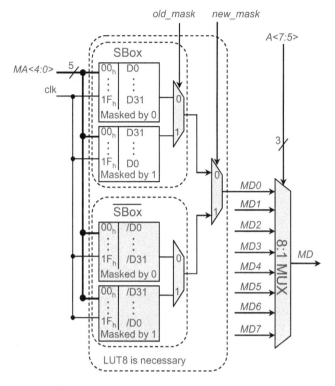

Figure 9. Memory duplication for unmasking.

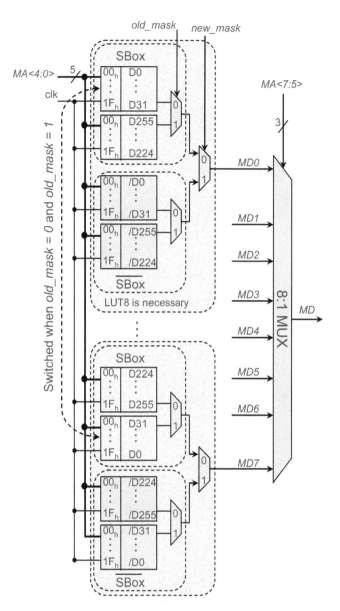

Figure 10. Intra-masking dual-rail memory on LUT.

are then combined into a larger memory, in which *old_mask* is considered as one more address bit in the combinational memory. Figure 9 shows the memory duplication for the unmasking of SBox and /SBox in Figure 6. The combinational memory for both SBox and /SBox on LUT6 now requires an LUT8 (8-input LUT). As in the case of the signals used to select the masked data out (*OUT*) in Figure 6, the select signals in Figure 9 must be the unmasked address $A<7:5>$. This gives an unsafe node to the implementation, where the capacitive load is unbalanced in the dual-rail address bus, and hence, it should be improved.

5.1.2 Intra-masking Dual-Rail Memory on LUT

In order to hide the real address $A<7:5>$ of the final selector in Figure 9, duplicated data must be rearranged among different LUTs, as shown in Figure 10. We named this model "intra-masking dual-rail memory (DRM) on LUT." Instead of grouping the data of the "masked by 0" and "masked by 1" memories into the same LUT8, as shown in Figure 9, we separate the "masked by 1" memory into a different LUT8, which is the last memory (complementary LUT of the first one). The data are now switched between the first LUT8, which corresponds to the address $MA<7:5> = 000_b$ when *old_mask* = 0, or the last LUT8, which corresponds to the address $\overline{MA}<7:5> = 111_b$ when *old_mask* = 1. The data in the "masked by 1" memory in the last LUT8 are duplicated from the complementary address in the "masked by 0" memory in the first LUT8, and thus, the data can be identified by the masked address $MA<4:0>$ when *old_mask* = 1. The same memory duplication and placement occur in the case of the remaining values of SBox and /SBox. As a result, the final output data *MD* can be selected by the masked address $MA<7:0>$ together with the *old_mask* and *new_mask* signals. Both the unmasking and masking operations are executed by the selectors

inside the LUT. The unsafe node of the final selector ($A<7:5>$) in Figure 6 is replaced by a secured masked node $MA<7:5>$. Hence, all dual-rail addresses and data buses (A and $/A$; D and $/D$ in Figure 4) are routed inside the LUT, thus balancing and securing them.

5.1.3 Intra-masking Dual-Rail Memory on LUT Implementation on FPGA Virtex-5

An extension to 256-element masked dual-rail memory, in which the pre-charge, input unmasking, and output masking are included, is shown in Figure 11. Figure 11.a shows the ideal memory extension for a masked dual-rail memory AES SBox. Similar to the model above, the content of SBox is duplicated and stored in inverse order in the first quarter of the memory. The next quarter of the memory stores the contents of /SBox and /SBox in reverse order. The final half of the memory is filled by 0s for pre-charge purposes. The selector with the *old_mask* signal is used to decide

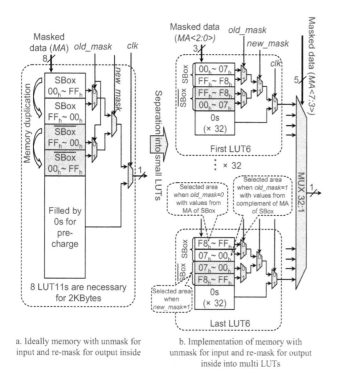

a. Ideally memory with unmask for input and re-mask for output inside

b. Implementation of memory with unmask for input and re-mask for output inside into multi LUTs

Figure 11. Combination of unmask for input and mask for output into 1-bit memory and implementation on LUT6.

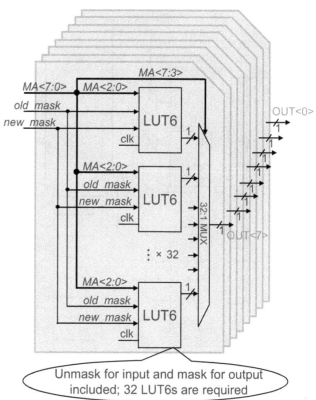

Unmask for input and mask for output included; 32 LUT6s are required

Figure 12. Mounting of the unmask and mask modules into the LUT6 element on Virtex-5.

whether the upper half or lower half of SBox and/SBox is selected. The multiplexer with the *new_mask* input signal is then used to decide if the output comes from SBox or /SBox for output masking purposes. This ideal memory requires an LUT11 (11-input LUT) as the basic element of FPGA. Unfortunately, current FPGA techniques can support only up to LUT6; hence, we must separate the large memory into smaller ones to fit LUT6, as shown in Figure 11.b. The separation method is shown in Figure 10. The LUT area used for SBox is now as small as $8 \times 2 = 16$ bits with only a 3-bit input ($MA<2:0>$) and 1-bit *old_mask*, so the area used for /SBox is small as well. Inside SBox, *old_mask* is used to select the upper part, where the data are the values from the *MA* address of SBox (address is masked by 0), or the lower part, where the data are the values from the complement of the *MA* address of SBox (address is masked by 1). Thus, the final 32:1 selector can use the high bits of the masked address *MA* ($MA<7:3>$) to select the final output. The same method is used to form the /SBox area. The multiplexer with *new_mask* input is used to mask the output of that LUT by selecting either the SBox or /SBox output. Each LUT is filled with the value corresponding to the address shown in Figure 11.b. The first is the responsibility for SBox at the masked address = 00_h–07_h and /SBox at the masked address = FF_h–$F8_h$ in two cases: when the input is masked by 0 and when the input is masked by 1. The next LUT corresponds to the next eight values of SBox and /SBox and so on until the last.

Figure 12 gives a full image for the modified intra-masking dual-rail memory on LUT SBox with 8-bit output implemented on FPGA using LUT6. In general, 256 LUT6s are required for an 8-bit SBox. Each LUT6 can be placed anywhere inside the FPGA and inside a slice during the standard design flow. In order to

reduce the occupied LUT slices of Virtex-5, the LUT6_L, MUX7F, and MUX8F are used together in this implementation. In this case, four LUT6s can be placed in a single slice, and hence, 64 slices are required for an 8-bit SBox.

5.2 Resources and Speed Comparison

Table 2 shows the hardware size of various AES experiments on Virtex-5. The implementations include three advanced tamper-resistant methods, WDDL, MDPL, and threshold, which are available in [10]; two unprotected designs of table-lookup (TBL) and composite field; and our proposed intra-masking DRM on LUT AES. All of the above cores were compiled for a Virtex-5 xc5vlx30-1ff324 device on SASEBO-GII [19] with a standard design flow in ISE13.4. The countermeasure cores given in [10] are supported for ASIC design; hence, an additional interface needed to be included to have them work in the experimental FPGA system given in Table 1.

Table 2. Hardware size comparison

	Implementation method	No. of slice registers	No. of slice LUTs	Freq. MHz	Thrpt. (Mbps)
No counter-measure	TBL	405	891	220.1	2,817
	Composite	397	1,890	138.9	1,778
Counter-measure	WDDL [10]	1,160	7,292	56.3	360
	MDPL [10]	1,205	12,140	44.4	284

	Threshold [10], [12]	1,438	12,627	78.6	503
Proposed	Intra-masking DRM on LUT (LFSR included)	546	5,898	141.2	1,643

The results show that even though our method took up more area than the conventional side-channel unprotected AESs, it required a smaller number of slices than the other protected ones. It was around half the size of the advanced threshold and MDPL [10]. The smallest dual-rail logic implementation method of WDDL also occupied about 24% more hardware than our model.

In terms of speed, the proposed design worked at two-thirds the speed of the no-countermeasure TBL implementation. However, it worked faster than previous designs. In particular, our proposed design achieved a six-fold higher throughput than WDDL and MDPL and a four-fold higher throughput than the threshold in [10] because our design can process one AES round every clock while the references in [10] process each round in two clocks.

5.3 Information Leakage Comparison and Discussion

Figure 13 shows the information leakage comparison between AES devices implemented using our proposed intra-masking DRM on LUT method and the other methods on FPGA. The unprotected AES implementations on FPGA based on single-rail memory table lookup (TBL) and composite field easily leaked their secret information (key) to the side-channel. The CPA attack on implementations on Virtex-5 revealed all 16 SBox keys at round 10 within 1,000 traces for TBL and within 40,000 traces for the composite field. The protected dual-rail logic WDDL implementation on FPGA, which occupies 24% more hardware than our proposed method, also leaked six SBox keys after 942,000 traces. Our proposed intra-masking DRM on LUT implementation achieves the same high security level as MDPL, where no secret key could be extracted even after analysis with 1 million traces. Further experiments were done for the proposed design with 2 million traces. No further keys were specified. The CPA attack results for the threshold in [10], [12] on FPGA are not

available due to some mismatching in modes between the ASIC core and the available FPGA experiment system. However, the threshold implementation should be considered safe given the attack results of [12].

The proposed design uses a simple 8-bit Fibonacci LFSR for random mask generation inside the encryption core. The random number can be guessed if the attacker knows the current value and the time the value is changed. However, such an attack has not yet been claimed for hardware implementation due to the difficulty in measuring the starting time of the LFSR. In addition, the mask generation is independent of the encryption, and hence, it can be replaced by any other random number generation method.

6. SUMMARY AND CONCLUSION

This paper introduces our novel tamper-resistant cryptographic circuit design method called "intra-masking dual-rail memory on LUT." The concept of dual-rail memory is based on an extension of dual-rail logic; however, the straightforward implementation of dual-rail memory was confirmed to leak secret information due to the imbalance in dual-rail buses caused by the placement and routing process; these problems are similar to those encountered by conventional dual-rail logic countermeasures. In order to prevent this leakage, we propose the new method called "intra-masking dual-rail memory on LUT." In this method, dual-rail memory, unmasking, and masking circuits are packed into a single LUT to prevent the imbalance of the dual-rail buses. The FPGA resources used in the implementation of our method were compared with those for other countermeasures such as WDDL, MDPL, and TI, and the experiment results demonstrated that our method consumes the smallest amount of resources. First-order CPA attack, which is a typical powerful side-channel attack, was performed on our method, and the secret key could not be identified even after 1,000,000 traces, while WDDL leaked the first secret key after about 100,000 traces. The "intra-masking dual-rail memory on LUT" method can be applied to not only the AES algorithm but also to DES and future standards. Further, it requires no specific placement and routing on FPGA while maintaining high security under first-order CPA attacks. The current 1-bit mask implementation will be improved to multi-bit mask designs in the future in order to provide protection from advanced high-order side-channel attacks.

In the future, we want to implement and verify the security of the intra-masking dual-rail memory on BRAM.

7. ACKNOWLEDGMENTS
We thank the CREST project for supporting this research.

8. REFERENCES
[1] Brier, E., Clavier, C., and Olivier, F., 2004. Correlation power analysis with a leakage model. *CHES 2004, LNCS* 3156, 16-29.

[2] Bucci, M., Giancane, L., Luzzi, R., and Trifiletti, A., 2006. Three-phase dual-rail pre-charge logic. *Lecture Notes in Computer Science, 2006, Volume 4249, Cryptographic Hardware and Embedded Systems (CHES 2006)*, 232-241.

[3] Canright, D., 2005. A very compact S-Box for AES. *Lecture Notes in Computer Science, 2005, Volume 3659, Cryptographic Hardware and Embedded Systems (CHES 2005)*, 441-455.

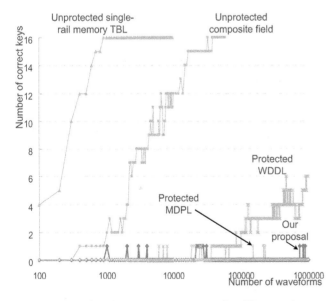

Figure 13. Information leakages under CPA attack

[4] Canright, D., and Batina, L., 2008. A very compact "perfectly masked" S-box for AES. *Proceedings of the 6th International Conference on Applied Cryptography and Network Security* (ACNS'08) 5037, 446-459.

[5] Chaudhuri, S., et al. 2008. Physical design of FPGA interconnect to prevent information leakage. *Lecture Notes in Computer Science, 2008, Volume 4943, Reconfigurable Computing: Architectures, Tools and Applications*, 87-98.

[6] Chen, Z., and Zhou, Y., 2006. Dual-rail random switching logic: a countermeasure to reduce side channel leakage. *Lecture Notes in Computer Science, 2006, Volume 4249, Cryptographic Hardware and Embedded Systems (CHES 2006)*, 81-94.

[7] Cryptographic hardware project in TOHOKU University, available at http://www.aoki.ecei.tohoku.ac.jp/crypto/

[8] Regazzoni, F., Wang, Y., and Standaert, F-X, 2011. FPGA implementations of the AES masked Against power analysis attacks. *Second International Workshop on Constructive Side-Channel Analysis and Secure Design (COSADE 2011)*, 56-66.

[9] Guilley, S. et al. 2008. Place-and-route impact on the security of DPL designs in FPGAs. *IEEE International Workshop on Hardware-Oriented Security and Trust (HOST 2008)*, 26-32.

[10] Information and Physical Security Research Group in YOKOHAMA National University, available at http://ipsr.ynu.ac.jp/circuit/index.html

[11] Kocher, P., Jaffe, J., and Jun, B. 1999. Differential power analysis. *Proceedings of CRYPTO'99, volume 1666 of Lecture Notes in Computer Science*, 388–397.

[12] Mimura, H., and Matsumoto, T., 2011. Security comparison among AES cryptographic circuits with different power analysis countermeasures. *2011 Symposium on Cryptography and Information Security (SCIS 2011)*, in Japanese.

[13] Moradi, A., et al. 2011. Pushing the limits: a very compact and a threshold implementation of AES. *Advances in Cryptology - EUROCRYPT 2011 - 30th Annual International Conference on the Theory and Applications of Cryptographic Techniques* 6632, 69-88.

[14] National Institute of Standards and Technology (U.S.) *Advanced Encryption Standards* (AES), FIPS Publication.

[15] Nikova, S., Rijmen, V., and Schlaffer, M. 2011. Secure hardware implementation of nonlinear functions in the presence of glitches. *Journal of Cryptology* 24, 2, 292-321.

[16] Popp, T., and Mangard, S. 2005. Masked dual-rail pre-charge logic: DPA-resistance without routing constraints. *Proceedings of Cryptographic Hardware and Embedded Systems (CHES), 7th International Workshop*, 172-186.

[17] Popp, T. et al. 2007. Evaluation of the masked logic style MDPL on a prototype chip. *Lecture Notes in Computer Science, 2007, Volume 4727, Cryptographic Hardware and Embedded Systems – CHES 2007*, 81-94.

[18] Saeki, M., Suzuki, D., Shimizu, K., and Satoh, A., 2009. A design methodology for a DPA-resistant cryptographic LSI with RSL techniques. *Lecture Notes in Computer Science, 2009, Volume 5747, Cryptographic Hardware and Embedded Systems (CHES 2009)*, 189-204.

[19] SASEBO project in Research Center for Information Security(RCIS), available at www.rcis.aist.go.jp/special/SASEBO/

[20] Satoh, A., Morioka, S., Takano, K., and Munetoh, S. 2001. A compact Rijndael hardware architecture with S-Box optimization. *Lecture Notes in Computer Science*, 2001, 2248/2001, 239-254.

[21] Sauvage, L. et al. 2009. Successful attack on an FPGA-based WDDL DES cryptoprocessor without place and route constraints. *Design, Automation & Test in Europe Conference & Exhibition (DATE'09)*, 640-645.

[22] Schaumont, P., and Tiri, K., 2007. Masking and dual-rail logic don't add up. *Lecture Notes in Computer Science, 2007, Volume 4727, Cryptographic Hardware and Embedded Systems (CHES 2007)*, 95-106.

[23] Schramm, K., and Paar, C., 2006. Higher order masking of the AES. *Lecture Notes in Computer Science, 2006, Volume 3860, Topics in Cryptology (CT-RSA 2006)*, 208-225.

[24] Sokolov, D., Murphy, J., Bystrov, A., and Yakovlev, A., 2004. Improving the security of dual-rail circuits. *Lecture Notes in Computer Science, Volume 3156, Cryptographic Hardware and Embedded Systems (CHES 2004)*, 255-317.

[25] Suzuki, D., and Saeki, M., 2006. Security evaluation of DPA countermeasures using dual-rail pre-charge logic style. *Lecture Notes in Computer Science (CHES 2006)*, 4249, 255-269.

[26] Suzuki, D., Saeki, M., Ichikawa, T., 2004. *Random Switching Logic: A Countermeasure against DPA Based on Transition Probability*. Cryptology ePrint Archive, Report 2004/346.

[27] Tiri, K. and Verbauwhede, I., 2004. A logic level design methodology for a secure DPA resistant ASIC or FPGA implementation. *Proceedings of the Conference on Design, Automation and Test in Europe, Volume 1 (DATE'04)*, 246-251.

[28] Yu, P., and Schaumont, P., 2007. Secure FPGA circuits using controlled placement and routing. *Hardware/Software Codesign and System Synthesis (CODES+ISSS), 2007 5th IEEE/ACM/IFIP International Conference*, 45-50.

Speedy FPGA-Based Packet Classifiers with Low On-Chip Memory Requirements

Chih-Hsun Chou[1], Fong Pong[2], and Nian-Feng Tzeng[1]

[1]Center for Advanced Computer Studies
University of Louisiana at Lafayette
Lafayette, Louisiana 70504, USA
vesaliusmac@gmail.com
tzeng@cacs.louisiana.edu

[2]Broadcom Corp.
2451 Mission College Boulevard
Santa Clara, California 95054, USA
fpong@broadcom.com

ABSTRACT

This article pursues speedy packet classification with low on-chip memory requirements realized on Xilinx Virtext-6 FPGA. Based on hashing round-down prefixes specified in filter rules (dubbed HaRP), our implemented classifier is demonstrated to exhibit an extremely low on-chip memory requirement (lowering the byte count per rule by a factor of 8.6 in comparison with its most recent counterpart [2]), taking only 50% of Virtex-6 on-chip memory to store every large rule dataset (with some 30K rules) examined. In addition, it achieves a higher throughput than any known FPGA implementation, reaching more than 200 MPPS (millions packet lookups per second) with 8 processing units and 8 memory banks in the HaRP pipeline to support the line rate over 130 Gbps under bi-directional traffic in the worst case with 40-byte packets. By reducing memory probes per lookup, enhanced HaRP can further boost the classification speed to 255 MPPS.

CATEGORIES AMD SUBJECT DESCRIPTORS

B.4.1 [**Input/Output and Data Communications**]: Data Communications Devices – *receivers*.

GENERAL TERMS

Performance, design, experimentation.

KEYWORDS

FPGAs, filter datasets, hash tables, memory efficiency, pipelined design, set-associativity.

1. INTRODUCTION

Packet classification is the technique for classifying the packets into different categories based on a set of pre-defined rules according to multiple fields in the packet header [13, 15]. It is an essential function for traffic management, access control, intrusion prevention, and many other network services. Typical packet header fields involved in a classifier are: source IP (SIP) address, destination IP (DIP) address, source port address range, destination port address range, protocol type, among others. A rule dataset in a core router may contain tens of thousands of rules ordered by priority, with each rule having its involved header fields specified or unspecified (as a wildcard). Upon receiving a packet, the classifier searches over its rule dataset for the matching rule with the highest priority, according to packet header field data. Classification tends to be time-consuming because of multiple fields involved and of large rule datasets commonly found in core routers, which operate at the line rates up to 100Gbps (e.g., Juniper's T160 Core Router [5]). The line rate of 100 Gbps requires classification to perform one lookup of a 40-byte packet in 6.4 *ns* (considering equal bi-directional traffic), under the worst case with the packet length of 40 bytes. As a result, classification can easily become the performance bottleneck of the Internet, calling for fast classifier design.

Hardware-based solutions are favorable for high-speed classification, utilizing ternary content addressable memory (TCAM), field programmable gate array (FPGA), or application specific integrated circuit (ASIC) in support of fast lookups. TCAM solutions are popular in the industry, and algorithms have been introduced to deal with their notorious problems related to range expansion [7] and incremental updates [14]. However, TCAMs are relatively expensive (due to low bit density), slow in operation, and inherently power hungry [1], as will be demonstrated by examples in Section 2.3. On the other hand, an ASIC solution usually involves considerable development time and effort and also suffers from limited flexibility, unable to easily adapt to changes to the classification procedure.

FPGA hardware has gained attention for high-performance classification lately, due to its potential in support of the 100Gbps line rate and its flexibility in accommodating various classification algorithms [2, 11]. For an FPGA-based classifier to achieve the desirable performance level, its employed classification algorithm ought to address the following design issues thoughtfully:

- memory efficiency: preferred for housing the rule dataset to get high performance, on-chip memory is rather limited, thereby favoring a memory-efficient classification algorithm so as to fit the whole rule dataset in on-chip memory,

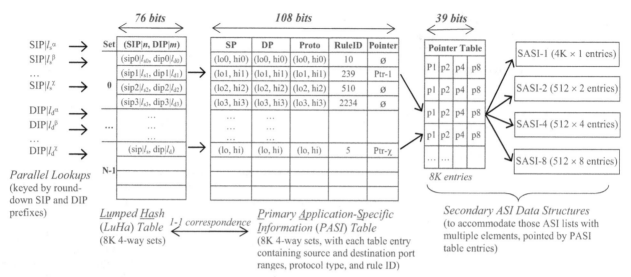

Fig. 1. Overview of the HaRP* data structures implemented for 30K rules.

- parallelism exploitation: memory accesses dominate the lookup process, making it essential to exploit their available parallelism aggressively,
- pipelined design: a high clock rate requires the algorithm to be implemented in pipelining, with low and balanced complexity for pipeline stages.

A previous FPGA-based packet classification algorithm decomposes each lookup into three steps [11], with the first step for longest prefix matching over the SIP and the DIP fields and the next two steps for mapping LPM (longest prefix matching) results to the rule number followed by rule validation. The first step is performed over the compressed rule table held in on-chip memory, whereas the mapping step involves the hash table stored in off-chip memory and the validation step requires the complete rule dataset kept in off-chip memory as well. As a result, each packet classification requires two external memory accesses in the last two steps [11]. As off-chip memory is slower than its on-chip counterpart, with its bandwidth constrained, keeping the rule dataset in off-chip memory limits its achievable classification throughput. A recent packet classification aims to keep the rule dataset entirely in on-chip memory without resorting to any off-chip memory for high performance, made possible by rule grouping to reduce rule duplications for a low memory requirement [2]. It is shown on Xilinx Virtex-5 to achieve a high throughput of over 100 Gbps, without off-chip memory accesses during classification lookups.

Lately, a packet classification algorithm based on hashing round-down SIP and DIP prefixes specified in rules, dubbed HaRP as shown in Fig. 1, has been introduced [10], where superior memory efficiency results from (1) hashing round-down prefixes and (2) collapsing all hash units into one lumped hash (LuHa) table. HaRP takes less than 750KB of memory for holding any of the three large filter datasets examined (each with 30K rules) [10]. Given its low memory requirement, HaRP lends itself particularly suitable for

FPGA implementation, with an entire rule dataset held in FPGA on-chip memory to exhibit high performance classification. Being hash-based, HaRP entails multiple table accesses per packet classification and those accesses can be parallelized effectively, when the LuHa table is composed of multiple memory banks, as will be detailed in later sections. The number of parallel access units involved dictates classification performance, and more memory banks support more access units to attain higher parallelism. Additionally, for the HaRP implementation to get a throughput level required by the top line rate of 100Gbps present in modern core routers, a balanced pipelining design is necessary, with its stages kept simple enough and memory banks properly situated near those parallel access units (without undesirably long routing paths) to get fast timing.

FPGA has become attractive for realizing real-time network processing engines [3, 9, 12], due to its ability to reconfigure and to offer massive parallelism. However, if a large data volume is to be stored in on-chip memory for high performance (like a rule dataset for classification), it is crucial to properly specify those FPGA memory blocks which constitute memory banks in support of parallel data accesses, given that memory blocks are distributed along strips on FPGA chips [16]. Also, LuHa table accesses are indexed by hash results and are likely to experience collision (when multiple hash results fall into the same memory bank). Collisions are unavoidable and will increase the average lookup time. In addition, among all physical elements on an FPGA chip, memory blocks usually have the largest gate delay [11], thus making the routing paths from/to the memory blocks dictate the overall speed (and thereby lookup performance) of any FPGA design with a high on-chip memory requirement.

This article deals with design and implementation of fast classification on the Xilinx Virtext-6 FPGA board based on HaRP and its enhanced variant to lower the number of LuHa table accesses per packet. Three real-life seed filter sets obtained from the public [13], namely, covering Access Control List (ACL), Firewall (FW), and IP Chain (IPC), are employed to evaluate our implemented classifier. Due to its superior memory efficiency, our implemented classifier utilizes only 50% of Virtex-6 on-chip memory to store large rule datasets (each with up to 30K rules). It is demonstrated by evaluation results to exhibit an extremely low on-chip memory requirement (reducing the byte count per rule by a factor of 8.6 in comparing with its most recent counterpart reported in [2]). The implemented HaRP pipeline with 8 processing units and 8 memory banks achieves the highest throughput among known FPGA implementations (reaching more than 200 MPPS, to support the line rate exceeding 130 Gbps under bi-directional traffic in the worst case with 40-byte packets). This is in sharp contrast to earlier implementations where memory-efficiency is often traded for throughput, rendering our implemented classifier to be *four times higher in its efficiency* (defined as the ratio of throughput to byte count per rule) than that of the second best implementation known so far [4]. In addition, enhanced HaRP is considered by reducing memory probes per lookup to further elevate the throughput level, attaining up to 40+% throughput gains but subject to far more overflows in the hash table. With an aid of pseudo set-associativity, enhanced HaRP can exhibit a lookup speed of 255 MPPS (to support the line rate beyond 160 Gbps) while containing hash table overflows.

The rest of this article is organized as follows. Section 2 gives pertinent background, including a brief review of HaRP packet classification introduced earlier [10]. Our FPGA-based design and implementation details are provided

in Section 3, following by resource use and performance results presented in Section 4. Enhanced HaRP for further boosting its lookup performance by lowering the mean number of table probes is stated in Section 5, with its evaluation results included and discussed therein. Section 6 concludes this article.

2. PERTINENT BACKGROUND

This section provides brief reviews of HaRP for packet classification and the Xilinx Virtex-6 device, pertinent to subsequent discussion. TCAM implementation examples on the Xilinx device are also presented.

2.1. Review of HaRP for Packet Classification

A classification rule usually involves multiple fields. This work assumes five classification fields present in each rule: (1) source network IP prefix of length n, denoted as SIP$|n$, (2) destination network IP prefix of length m, represented by DIP$|m$, (3) source port range, SP[*low, high*], (4) destination port range, DP[*low, high*], and (5) protocol type range, Proto[*low, high*]. The first two fields specify a pair of communicating networks, and the next three fields apply application-specific constraints.

A memory-efficient classification method by means of generalized <u>ha</u>shing <u>r</u>ound-down <u>p</u>refixes (denoted by HaRP*) following a two-stage pipeline design has been demonstrated [10] to outperform earlier software-oriented techniques. In the first stage of HaRP*, a single set-associative hash table, referred to as the LuHa (<u>lu</u>mped <u>ha</u>sh) table is used to keep the network-prefix pair (SIP$|l_s$, DIP$|l_d$)

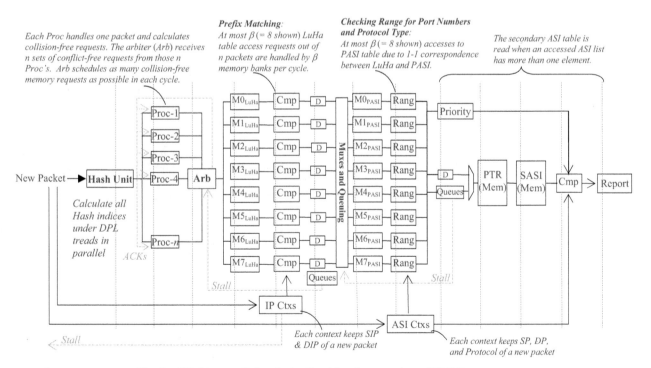

Fig. 2. Highly parallel and pipelined implementation of HaRP.

part of the rules. The other three fields involved in the second stage are stored in corresponding ASI (_Application Specific Information_) data structures.

**LuHa Table Construction.** Consider the set of _designated prefix length_, DPL: $\{l_1, l_2, ..., l_i, ..., l_m\}$, where l_i denotes a prefix length, for the following explanation. As depicted in Fig. 1 and detailed in [10], the LuHa table achieves efficient hash table utilization by permitting _multiple candidate sets_ to accommodate the prefix pair (SIP$|l_s$, DIP$|l_d$) of a given filter rule, and yet maintaining fast search over those possible sets in parallel during the classification process. It is made possible by (1) <u>rounding down</u> Prefix P$|w$ to P$|l_\lambda$, for all DPL elements $l_\lambda \leq w$, $\lambda \in \{1, ..., m\}$, before used to hash the LuHa table for identifying t (= the largest λ) candidate sets, and (2) storing (SIP$|l_s$, DIP$|l_d$) of a filter rule in _one of those multiple_ LuHa candidate sets identified by either SIP$|l_s$ (if not wild carded) or DIP$|l_d$ (if not wild carded), after being rounded down and hashed, as stated in (1).

Since elements (called _treads_) in DPL are determined in advance, the numbers of bits in an IP address of a packet used for hash calculation during classification are clear and their hashed values can be obtained _in parallel_ for concurrent search over the LuHa table. This permits parallel access units to look up the table, if multiple memory banks constitute the table, arriving at a high performance parallel design on the FPGA board.

HaRP* works because it takes advantage of the "_transitive property_" of prefixes – for a prefix P$|w$, P$|t$ is a prefix of P$|w$ for all $t < w$, considerably boosting its pseudo set-associative degree [8, 10]. Under the special case where P$|w$ (with $l_i \leq w < l_{i+1}$) is rounded down to P$|l_b$, for $i \leq b \leq i$, the method is denoted by **HaRP^1**. When the input prefix is further allowed to be rounded down to the next tread l_i (i.e., $i-1 \leq b \leq i$), a scheme called **HaRP^2** results. It means that the input prefix P$|w$ can be stored in hash buckets indexed by either P$|l_i$ or P$|l_{i-1}$. Accordingly, **HaRP*** is defined to allow as many candidate hash buckets (in existence) as possible for holding a given filter rule. Given DPL with 4 treads: {28, 16, 12, 8}, for example, HaRP^1 rounds down the prefix of 011010010001111001× ($w = 18$) to 0110100100011110 (ζ = 16) for hashing, whereas HaRP* rounds down the prefix to 0110100100011110 (ζ = 16), 011010010001 (ζ = 12), and 01101001 (ζ = 8) for hashing. With more candidate sets available potentially, HaRP* makes it possible to choose a small number of treads, which in turns involves fewer hash probes per lookup, thus improving lookup performance [10].

**Construction of ASI Lists.** The second stage of HaRP* comprises an ASI table, keeping the application-specific fields of filter rules. If rules share the same IP prefix pair, their application-specific fields are stored in contiguous ASI entries packed in chunks (of a fixed size) [10]. In essence,

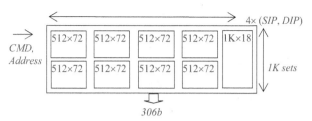

Fig. 3. Constitutive memory bank (module) of LuHa table.

the ASI table is logically a collection of lists (with various lengths), one corresponding to one LuHa table entry, as depicted in Fig. 1.

2.2. Overview of Xilinx Virtex-6 Device

Our hardware HaRP* design targets the Xilinx Virtex-6 device (XC6VLX240T) [16], making efficient use of its on-chip memory and logic cells for high-performance packet classification. A Virtex-6 device contains approximately 240,000 logic cells, forming 37,680 slices, each with <u>four</u> LUTs (lookup tables) and <u>eight</u> flip-flops (FFs). A logic cell possesses combinational logics for realizing such functions as AND, OR, NAND, and addition. Flip flops and the connections to adjacent cells are also implemented by logic cells. The high-speed logic fabric of Virtex-6 permits effective pipelining [16].

Every FPGA device includes reconfigurable on-chip memory blocks (known as _block RAMs_) for implementing anything from random access storage to dual-port architectures, to FIFOs. There are 416 block RAMs in the Virtex-6 (XC6VLX240T) device for use, each containing 36Kb for a total of 14,976 Kb on-chip memory. Those block RAMs are distributed over 8 stripes, so as to have shorter routing delay for designs implemented on it. However, this layout of block RAMs can yield a design with excessively long connections, thereby hurting its performance, if large on-chip memory is required (like packet classification). Particular attention has to be paid for such situations.

2.3. TCAM Implementation Examples

We have employed the Xilinx IP core for implementing the TCAM tables of various sizes to find out hardware resource utilization and associated access timing details on the Xilinx Virtex-6 device (XC6VLX240T). A TCAM table involving 32 entries (or 128 entries), each with a 64-bit input (for an SIP and a DIP under IPv4) plus a 64-bit mask, consumes 3.1% (or 6.6%) of FPGA slices available on the Virtex-6 device. The access time to such a TCAM table with 32 entries (or 128 entries) is 4.80 _ns_ (or 5.94 _ns_). If the TCAM table size rises to 512 (to hold 512 rules), its occupied slice ratio exceeds 24% (higher than that of our whole HaRP pipeline implementation able to hold some 30K classification rules; see the case of φ = 8 and ζ = 8 in Table 2) and its access time extends to 6.5 _ns_ (limiting its clock rate to 154 MHz, in contrast to 300+ MHz for the HaRP pipeline). Those implementation examples confirm that TCAM indeed is expensive and relatively slow, suitable only for small-sized implementation (with no more than, say, 256 entries).

3. FPGA-BASED DESIGN AND IMPLEMENTATION

We have implemented HaRP* on a Vertex-6 XC6VLX240T device, carefully mapping out constitutive tables to keep routing delays checked. The Xilinx Block memory generator [15] was employed to produce appropriate interfaces to on-chip memory for high speed memory accesses.

3.1. Layouts of Data Structures in Memory Blocks

There are four major tables involved, the LuHa table, the PASI table, the SASI table, and the pointer table, with the last three together realizing the ASI functionality of HaRP [10] for efficient on-chip memory utilization and fast accesses to ASI lists of variable length (as detailed later in this subsection). They are all implemented by on-chip Block memory for high-performance classification. Being the largest one, the LuHa table has a total of 8192(sets)×4(ways) entries. Each entry includes 32b SIP and DIP prefixes and two 6-bit prefix length indicators, giving rise to 76 bits in total. The LuHa table shown in Fig. 2 consists of β (= 8 shown) memory modules, with each module holding 1K sets. Such a design permits as many as β collision-free memory accesses to be served by the LuHa table. Each of the memory module is made to output 306 bits per read access, rendering four (SIP|n, DIP|m) pairs resided in one set of the 4-way LuHa table as depicted in Fig. 3.

Because the vast majority of ASI lists contain no more than one element each, the ASI table is realized by three data structures for storage efficiency and fast lookups, as shown in Fig. 1. The primary ASI (PASI) table has a 1-1 correspondence relationship to the LuHa table. Given a LuHa table with k entries, there are k corresponding PASI entries. For each classification, the PASI table is consulted only for those entries matched by prefix pair (SIP|l_s, DIP|l_d) during the first HaRP stage. Each PASI entry is 108-bit wide (see Fig. 1). In our design, each LuHa memory module outputs 4 prefix pairs per cycle for comparison, while PASI supports a more targeted access for one matched LuHa lookup. This discrepancy may require temporary queueing of requests to PASI (when there are multiple matches reported by the same LuHa set, albeit pretty rare to happen). As a result, an economy-wise decision is made to keep the PASI module narrow, instead of widening it to the output of 4×108 bits (which is likely to slow down the pipeline).

Noticeably, each PASI entry contains a pointer to a pointer table. This pointer is active when multiple filter rules share the same (SIP, DIP) prefix pair, resulting in an ASI list that has more than one element. While the PASI table keeps the first element, the additional elements of an ASI entry are kept in the four secondary SASI component tables, which are referred to as the SASI table for simplicity. Each of SASI-1 (or SASI-j, for $j \in \{2, 4, 8\}$) entries contains 1 element (or j elements). A PTR (Pointer Table) entry contains four fields to store the indexes of ASAI-i, for $i \in \{1, 2, 4, 8\}$, as depicted in Figs. 1 and 2.

By this flexible design, memory resources are better utilized to accommodate ASI lists of varying lengths. Together, one PASI entry plus its pointed SASI table entry can hold up to 16 filter rules sharing a given (SIP, DIP) LuHa entry. If the number of filter rules for the same (SIP, DIP) pair exceeds 16, another (SIP, DIP) LuHa entry is created. This can happen, for example, when a large number of filer rules are specified to put constraints on accesses to R (> 16) applications between two communicating sub-networks. During lookups, all (SIP, DIP) LuHa entries are matched such that their corresponding ASI lists are all examined to identify classification rules with the highest priority for use.

Under our design, an ASI lookup works as follow: (1) it fetches the PASI table entry corresponding to the LuHa entry indexed by the (SIP, DIP) pair of an arrival packet, compares the fields of the fetched PASI entry, and then obtains the PTR table pointer, if existing and the compared fields all matched, (2) it fetches the PTR entry via the obtained pointer for indexes to SASI tables, and (3) it gets access to SASI tables via indexes in parallel for all candidate rules, which are then examined to identify the one with the highest priority for use.

3.2. Pipelined Implementation

Fig. 2 illustrates our highly parallel and pipelined implementation under Virtex-6. There are five major functionalities, as stated in sequence below.

Hashing and Tracking States. Upon a new packet arrival, IP addresses, port numbers and the protocol type are stored in an available *context*. If there are n (= 8 shown in Fig. 2) *Proc's* in existence, at most n packets can be processed simultaneously. In the first cycle, all hash indices keyed by SIP and DIP of any arrival packet are calculated by the hash unit under the chosen DPL, as depicted in Fig. 2. For a DPL with m treads, totally $2m$ hashes are performed[†]. The $2m$ hash indices are written into an allocated *Proc*, and those indices in the *Proc* are referred to as a *context*. Given n *Proc's*, our design can handle n concurrent contexts. In other words, a context and its associated *Proc* keep track of the progress of their assigned packet.

Scheduling for LuHa Table Accesses. For n packets each with $2m$ hash table accesses, a two-level scheduler is adopted to avoid a time-critical path (when scheduling $n \times 2m$ accesses totally). In the first level, each *Proc* performs local scheduling: at an odd cycle, memory-collision free requests out of the m hash accesses keyed by an SIP address are chosen; at an even cycle, the same operation is performed for

---

[†] Note that different DPLs with different numbers of treads may be chosen for SIP and DIP prefixes/addresses. Using the same DPL for both SIP and DIP here is meant to keep our discussion simple and focused.

hash accesses keyed by a DIP address. Thus, every *Proc* can forward at most *m* access requests to the second-level *Arbiter* in each cycle. The *Arbiter* selects up to β (being the number of memory banks) collision-free accesses from up to *n×m* access requests forwarded by those *n* independent *Proc's*, and it lunches those selected accesses to the LuHa table. Acknowledgements are returned back to *Proc's* to nullify table access requests being served (see Fig. 2), making room for subsequent packets. As a result, this two-level scheduler permits up to β LuHa table accesses in one cycle, and a larger β tends to yield higher lookup throughputs.

LuHa Table Accesses. As demonstrated in Fig. 2 and explained before, all β LuHa memory banks may be accessed at the same time. Each memory bank outputs 4 prefix pairs, which are compared against the packet inputs. When a match is found, the corresponding PASI entry is then probed. If multiple entries of the same LuHa set report a match, they need to be serialized due to the single port PASI memory and a queue is thus provided for such a purpose, as shown in Fig. 2.

PASI Table Accesses. The β PASI memory banks can accept β reads per cycle. Each memory bank output ASI stored in the entry being accessed. If a match to the port range and the protocol type is found, the rule ID is reported. If the range check fails, searches to the secondary ASI tables continue, provided that the PASI entry contains a valid pointer, which indexes to Pointer Table (as demonstrated in Fig. 1 and marked as "PTR (Mem)" in Fig. 2) where the SASI table entries are specified. This way permits as many as 16 rules and their associated ASI values to be fetched quickly from those SASI tables in parallel, as illustrated in Figs. 1 and 2. Our design preserves superior memory efficiency of HaRP and also achieves fast accesses to a varying number of candidate rules (associated with one LuHa table entry) in a uniform time.

SASI Table Accesses. Accesses to the SASI tables are targeted, fetching only entries specified by the address pointers reported by the pointer table. Range checking on port numbers and the protocol type are performed for all fetched entries in parallel.

4. IMPLEMENTATION EVALUATION
The LuHa table in our implemented classifier is 4-way set-associative. Our evaluation is under the default DPL with 4 treads of {8, 20, 23, 27} for SIP and of {8, 20, 24, 30} for DIP, chosen conveniently, not necessary to yield the best results. From real filter datasets (containing up to 1,550 rules) available in the public domain [13], three synthetic datasets, each with about 30K rules (see Table 1), have been obtained for evaluation, including Access Control List (ACL), Firewall (FW), and IP Chain (IPC), for evaluation under various numbers of *Proc's* and memory banks.

Table 1. On-chip memory usage

Rule dataset	# of rules	Taken on-chip memory (Kbytes)	Usage ratio
ACL	28240	655.2	35.0 %
FW	28473	697.6	37.3 %
IPC	29876	776.8	41.5 %

4.1. Memory Requirement
Details of on-chip memory taken by the four tables in support of HaRP classification under the three datasets are listed in Table 1. Each LuHa entry is 76-bit long, comprising two 32b IP address prefixes and two 6b prefix length indicators. Each PASI entry needs 10 bytes to keep the port ranges and the protocol type, 15-bit rule number (for priority decision), and a pointer to the PTR table. Each PTR table entry contains 4 pointers for indexing SASI component tables, i.e., SASI-1, SASI-2, SASI-4, and SASI-8, whose entries each contain 1, 2, 4, and 8 ASI elements of 95 bits, respectively (see Fig. 1). The memory sizes of the PTR table and the SASI component tables depend on the dataset size.

Our FPGA on-chip memory usage results reveal that each filter rule (under any one of the three datasets examined) on an average takes no more than 26 bytes. They are far favorable in comparison to a recent packet classification design based on SPMT (Set Pruning Multi-Bit Trie) [2], where 10K FW rules utilize 16.88 Mb on-chip memory on a Xilinx Virtex-5 (XC5VFX200T) FPGA device, signifying that each FW rule requires 211 bytes. Our classifier indeed has an extremely low memory, thus enabling a far larger rule dataset to fit in on-chip FPGA memory for speedy classification. More comparative details among various recent FPGA-realized classifiers will be provided in Section 4.3. Clearly, overall on-chip memory required by our FPGA implementation is dictated solely by rule datasets, irrespective of the numbers of Proc's (denoted by φ) and memory banks (denoted by ζ) involved. On the other hand, hardware logic requirement for HaRP implementation is proportional to φ and ζ, which in turn determine lookup throughput outcomes, as stated next.

4.2. Requirement of Hardware Logics
Hardware logics used to realize the HaRP classifier shown in Fig. 2 include LUTs and flip flops (FFs), which are from FPGA slices. The numbers of FFs and LUTs consumed by the implemented classifier under various φ and ζ are depicted in Table 2, where the number of occupied slices (out of 37,680 on the VLX240T FPGA device) is also included. An increase in either φ or ζ leads to higher consumed slice FF and slice LUT counts monotonically, as expected. When (φ, ζ) rises from (4, 4) to (4, 8), the numbers (or percentages) of taken slice FFs and slice LUTs grow respectively to 18,572 and 21,699 (or to 6% and 14%), from 12,414 and 13,977 (or from 4% and 9%), as unveiled in the table. If (φ, ζ) is elevated further to (8, 8), 8% of slice FFs and 20% of slice LUTs will be occupied. Given each slice contains four LUTs and eight FFs, which are not always utilized in full by an implementation, it is useful to know the number of slices involved (either partially or fully) in our classifier implementation. From Table 2, the number

Table 2. Usage of hardware logics

(φ, ζ) (no. of proc's, memory banks)	Consumed hardware breakdowns		
	Slice registers	Slice LUTs	Occupied slices
(1, 1)	4,313 (1 %)	5,124 (3 %)	1,763 (4 %)
(2, 2)	6,995 (2 %)	7,597 (5 %)	2,574 (6 %)
(2, 4)	9,604 (3 %)	10,954 (7 %)	3,548 (9 %)
(4, 4)	12,414 (4 %)	13,799 (9 %)	4,465 (11 %)
(4, 8)	18,572 (6 %)	21,699 (14 %)	6,706 (17 %)
(8, 8)	25,944 (8 %)	31,048 (20 %)	9,374 (24 %)
(8, 16)	41,923 (13 %)	49,048 (32 %)	14,452 (38 %)
(16, 16)	63,856 (21 %)	85,253 (56 %)	24,888 (66 %)

of involved slices is seen to grow from 4,465 (~ 11%) under (φ, ζ) = (4, 4) to 9,374 (~ 24%) under (φ, ζ) = (8, 8). This signifies that one Virtex-6 board can easily accommodate two copies of HaRP classifier with (φ, ζ) = (8, 8). If φ or ζ increases to 16, the percentage of occupied slices rises to 38% or beyond, reflecting that a good portion of available slices on the VLX240T FPGA device is taken. This high usage of hardware logics in fact will make the implemented HaRP classifier run slower due to longer routing paths between its consecutive pipeline stages when mapped to FPGA logics and block RAMs, as detailed next.

On-chip memory on FPGA devices is organized in blocks (i.e., block RAMs/FIFOs), which are scattered over the whole Virtex-6 chip. As hardware logics are to operate on dataset contents stored in on-chip memory, the achievable memory access rate is critical for the pipeline clock rate, dictated by the worst-case routing path delay. An undesirably lengthy delay may result from utilizing FPGA slices unduly distant from those block RAMs employed to hold the filter dataset. To avoid an inefficient HaRP implementation, the design is constrained within one half of the FPGA device to arrive at a compact layout without any excessive path delay. This way makes those occupied FPGA

slices stay near to slices which realize the HaRP functions, rendering a design with the main clock exceeding 300 MHz.

Our implementation puts constraints on those tables basic to HaRP classification, as depicted in Fig. 4(b). For comparison, layout results with and without constraints are demonstrated respectively in Fig. 4(a) and Fig. 4(c). Without a constraint, on-chip block RAMs are seen to be often distant from those taken slices that realize on-chip memory access and processing logics, leading to long route delays for paths from a pipeline stage to the next, as shown in Fig. 4(a), where an exceedingly long path marked by a white line with arrow indicates a memory access route (i.e., from the arbiter stage to the LuHa table access stage, see Fig. 2). If constraints are put to tables realized by blocks of on-chip memory, as illustrated in Fig. 4(b), an improved layout results, with its worst-case path (denoted by the white line with arrow in Fig. 4(c) to indicate the routes from the LuHa Table output to the comparator in the next pipeline stage) shrunk drastically. Hence, the constrained use of memory blocks yields a better layout with more uniform and shorter routes, permitting a faster clock to deliver higher lookup throughput.

4.3. Implementation Results and Discussion

The major performance metric of interest is lookup throughput, which equals the number of classification lookups per second. As mentioned earlier, the line rate of 100 Gbps necessitates one lookup per 6.4 *ns* in the worst case under bi-directional traffic, amounting to some 156 MPPS (million packets per second) for the shortest packets of 40 bytes in length. The throughput results under various φ and ζ (i.e., numbers of Proc's and memory banks) are listed in Table 3 for the three rule datasets considered, with $\varphi \leq \zeta$ to avoid unnecessary access conflicts at memory blocks, thus capitalizing on parallelism fully. They indicate that the desired lookup speed of 156 MPPS can be achieved for φ = 8 and ζ = 8, where the implemented HaRP design

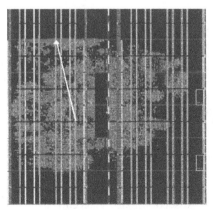

(a) Layout without constraint

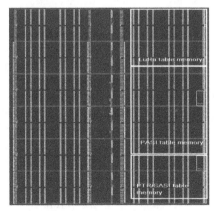

(b) Imposed memory constraints

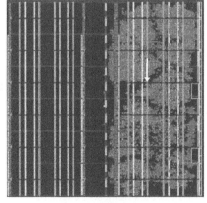

(c) Layout with constraints

Fig. 4. Layout of occupied memory blocks and taken FPGA slices.

(Taken FPGA slices are denoted in blue, while occupied memory blocks and their access logics are represented by red stripes and nearby red dots, respectively. Those stripes in purple and in green denote respectively unused memory blocks and the DSP blocks irrelevant to our implementation. Similarly, objects in other colors are unneeded chip components.)

Table 3. HaRP performance under rule datasets considered

(φ, ζ) (no. of proc's, memory banks)	Lookup Throughput (MPPS)		
	ACL	FW	IPC
(1, 1)	39.7	42.5	40.6
(2, 2)	57.4	57.5	58.1
(2, 4)	69.1	68.3	68.9
(4, 4)	106	105	105
(4, 8)	126	127	128
(8, 8)	204	213	213
(8, 16)	171	178	177

Table 4. Performance comparison of classifiers on FPGA

Approaches	# of ACL rules	Total memory taken (Kbytes)	Memory per rule	Throughput (Gbps)	Efficiency (Gbps/B)
HaRP(8, 8)	28240	655.2	23.2 bytes per rule	130.6	5.63
SPMT [2]	9603	1930	201 bytes per rule	110.7	0.55
Improved HyperCuts [4]	9603	612	63.7 bytes per rule	88.2	1.38
Simplified HyperCuts [6]	10000	286	28.6 bytes per rule	7.22	0.25
2sBFCE [9]	4000	178	44.5 bytes per rule	2.06	0.046
Memory-Based DCFL [3]	128	221	1727 bytes per rule	24.0	0.014

operates in excess of 300 MHz. According to Table 2, the occupied slices then account for only 24% of what are available on the Virtex-6 FPGA device. On the other hand, the rate of occupied slices rises to 38% for φ = 8 and ζ = 16, where the implemented classifier slows down to 220 MHz only, resulting in smaller throughputs to drop as despite its doubled memory banks for fewer access conflicts (see Table 3). This is because those slices employed for realizing HaRP functions then cannot all be situated near their fetched memory blocks. As a result, detrimentally long routes exist in the implementation and thus the main clock rate is dropped (to 220 MHz from more than 300 MHz under φ, ζ ≤ 8), lowering the overall throughput. For φ = 16 and ζ = 16, a far slower clock rate is obtained because the HaRP implementation then occupies 66% of available slices. While not covered in this article, a rectified HaRP pipeline with one extra stage dedicated to route latency reduction can be added under φ = ζ = 16 for improving its throughput (assorted to a deep pipelining design for a higher clock rate).

In general, the throughput figure of an implemented design is dictated by four factors: the main clock rate, numbers of Proc's and memory blocks, and effectiveness of scheduling memory accesses. If the slices are not heavily utilized (say, with the taken rate < 30%), the main clock rate stays almost identically, since the longest route length then remains unchanged. Under that situation (of a low occupied rate), the overall throughput depends only on the remaining three factors. Our two-stage scheduler for memory accesses intends to select as many conflict-free accesses as possible per cycle for maximal throughput, as detailed in Section 3.2. Under the given scheduler, a higher throughput value results from either a larger φ or a larger ζ, provided that φ is no more than 8. With φ = 4, for example, the implemented classifier enjoys a throughput increase of roughly 25% when ζ grows from 4 to 8, as a result of more conflict-free accesses to memory blocks per cycle (see Table 3). Likewise, for ζ = 8, its throughput has a leap exceeding 60%, if φ rises from 4 to 8, directly benefiting from twice memory access requests generated by Proc's per cycle.

In addition to its high throughput, our implemented classifier enjoys a much lower on-chip memory requirement when compared with a counterpart introduced recently, dubbed SPMT (Set Pruning Multi-Bit Trie) [2], as mentioned in Section 4.1. This high throughput plus low memory requirement makes our HaRP design especially

preferred over other known classification approaches implemented using FPGA, as reflected by the "efficiency" measure (defined in [4] as the ratio of throughput to on-board memory per rule, accounting for both time and space factors) listed in Table 4, where HaRP(8, 8) indicates the HaRP classifier with φ = 8 and ζ = 8. While the table includes only the results under the ACL rule dataset, the result trends hold under other rule datasets. Apparently, high efficiency may result from an approach with either a large throughput or a small memory amount per rule. Unlike the SPMT design which trades the memory requirement for an increased throughput in comparison to Improved HyperCuts [4] (see the 3rd and 4th rows of Table 4), the HaRP classifier enjoys both memory requirement reduction (per rule) and a throughput hike when compared with any prior approach, yielding significantly better efficiency. An enhanced HaRP design able to further elevate the lookup throughput is treated next.

5. ENHANCED HARP AND ITS PERFORMANCE

The HaRP classifier considered so far always involves 2×|DPL| memory accesses to the LuHa table for every packet during its lookup. In each cycle, only those non-conflict accesses to different memory banks are chosen to proceed by the developed two-stage scheduling mechanism, with those remaining conflict accesses, if any, scheduled in the subsequent cycle (together with those 2×|DPL| memory accesses of the next packet). The lookup of a packet is done only after its associated memory accesses are all served. While a large φ or ζ can yield a higher throughput (as demonstrated in the last section), an enhanced HaRP design is treated here by *lowering the number of memory accesses per lookup dynamically* (instead of having a fixed 2×|DPL| memory accesses for each packet lookup) under given φ and ζ. This is made possible by taking advantage of the unique property of HaRP in that the filter table lookup process is dictated by the filter rule installation procedure, which can

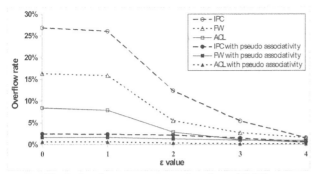

Fig. 5. Overflow rate versus ε under enhanced HaRP.

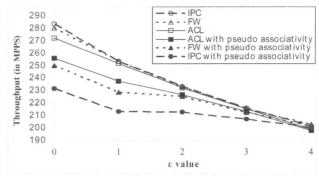

Fig. 6. Lookup throughput versus ε under enhanced HaRP.

follows a control guide to lower the number of candidate sets dynamically for each rule so that the lookup process then involves fewer memory probes accordingly based on the same guide.

In general, the number of candidate sets under enhanced HaRP may range from 1 to 2×|DPL|, determined by the control guide that specifies which treads in DPL are to be "conditioned" before applied for hashing. Two components are involved in the guide: (1) which treads in DPL to be conditioned and (2) how to condition those treads. Let the number of conditioned treads in given DPL be denoted by (|DPL| − ε), for 0 ≤ ε ≤ |DPL|, and the condition simply be "rightmost tread bit being "1."" Naturally, other conditions are possible for throughput improvement as well, e.g., "rightmost tread bit equal to 0" or "rightmost two tread bits being '01' (if two bits are for conditioning)." The following deals with only single-bit conditioning as one example enhancement.

5.1. Example Enhancement Guide and Results

Under enhanced HaRP, an IP prefix (or address) determines its candidate sets for holding the prefix (or for probing the best filter rule to apply) according to the given guide. Apparently, fewer candidate sets result in a higher lookup throughput and also in more LuHa table overflows (during rule installation, since fewer alternative sets are then available for a given prefix) as well. Installing filter rules without excessive overflows thus calls for a larger ε, which in turn contains the throughput as a result. The overflow rates under different ε values will be explored.

The guide may be in different forms, with one considered as follows. Given DPL = {l_1, l_2, …, l_m} with m elements, HaRP rounds down Prefix P of length w (or IP address of length 32) to P|l_t, for all DPL elements $l_\lambda \le w$, $\lambda \in \{1, …, m\}$, before employed to hash the LuHa table for identifying candidate sets, provided that *the rightmost bit of P|l_t equal to 1 for t > ε*. For Prefix P during filter rule installation (or IP address P during filter lookup) under DPL = {8, 20, 23, 27} and ε = 1, as an example, enhanced HaRP considers only those candidate sets (1) indexed by P|20 with its rightmost bit equal to 1, by P|23 with its rightmost bit equal to 1, and by P|27 with its rightmost bit equal to 1 and (2) indexed by P|8, regardless of their rightmost bits (i.e., no

conditioning). This guide for rule installation and lookups lowers the mean number of LuHa table probes (thereby elevating the lookup throughput).

Enhanced HaRP can be accommodated easily in the "Hash Unit" of the implementation pipeline shown in Fig. 2, with the unit deciding whether or not a round-down address should be used for probing a candidate set in the LuHa table. The miss rate versus ε values for the three filter datasets (listed in Table 1) during filter rule installation under enhancement HaRP is shown by the top three curves in Fig. 5, where the LuHa table is 4-way set-associative with 8K sets for a total of 32K entries and the DPL for source IP (or destination IP) prefixes/addresses is {8, 20, 23, 27} (or is {8, 20, 24, 30}). As expected, a larger ε leads to a lower overflow rate for every rule dataset, thereby requiring a smaller TCAM to store those overflow rules (if a small TCAM is assumed to handle the spillovers). Given its |DPL| equal to 4, enhanced HaRP becomes regular HaRP under ε = 4, where all treads are used for indexing candidate sets. The overflow rates of three datasets all drop quickly when ε exceeds 1.

Throughput as a function of ε during lookups under enhanced HaRP with φ = 8 and ζ = 8 is shown in Fig. 6. As seen in the figure, the throughput of every filter dataset is lower for a larger ε, and it reaches the result of regular HaRP with ε = 4 (i.e., being the value listed in the row of (φ, ζ) = (8, 8) in Table 3). Enhanced HaRP with ε = 0 (or 1) under ACL, for example, enjoys some 36% (or 25%) improvement in the lookup throughput, at the expense of a far higher overflow rate (see Fig. 5), and thus a substantially larger spillover TCAM, during LuHa table installation.

5.2. Enhanced HaRP with Pseudo Set-Associativity

A pseudo associative technique [8] can be applied to contain overflows during rule installation, making the 4-way set-associative LuHa table behave like 8-way set-associativity by treating both a set (indexed by a hash value) and its companion (indexed by the 2's complement of the hash value) as candidates to store a filter rule during installation, thus lowering the overflow rate. A companion candidate set is checked only if its associated candidate set is unavailable to hold the rule. During rule lookups, an indexed candidate is probed first before its companion candidate set is examined in the next cycle, should the earlier probe fail to match a rule. When both candidate sets are unavailable for a filter rule (or fail to match any rule), an overflow (or a miss) happens during rule installation (or lookups).

With pseudo set-associativity, enhanced HaRP exhibits far smaller overflow rates under the three filter datasets examined throughout the ε range shown in Fig. 5. The bottom three curves in the figure clearly signify that doubling candidate sets for each filter rule avoids the need of large TCAMs to hold overflow rules for enhanced HaRP. Meanwhile, pseudo set-associativity is seen to let enhanced HaRP outperform its regular HaRP counterpart smartly for a small ε (say, ≤ 1), as demonstrated by the bottom three curves in Fig. 6. Under ACL, for example, enhanced HaRP with pseudo set-associativity still enjoys a throughput gain of some 27% with $\varepsilon = 0$, reaching a lookup rate of 255 MPPS (to support the wire speed over 160 Gbps under bi-directional traffic in the worst case with 40-byte packets). Despite elevating the bandwidth requirement for LuHa table accesses upon classification lookups (due to the possible need of examining companion candidate sets), pseudo set-associativity indeed benefits enhanced HaRP in both overflow reduction and throughput improvement.

6. CONCLUSION

This article has investigated into design and implementation of speedy classification based on HaRP, whose constitutive lookup tables are mapped carefully onto on-chip block RAMs of an Xilinx Vertex-6 FPGA device (XC6VLX240T) to keep routing delays checked. The implemented classifier has an extremely low memory requirement, based on evaluation results obtained using three large datasets generated by real-life seed filter sets available to the public [13]. Its *efficiency* (defined as the ratio of throughput to byte count per rule) *is four times higher* than that of the second best FPGA implementation known so far [4]. An enhanced variant to lower the number of hash table accesses per packet is considered, shown to further elevate the classification throughput with overflows in the hash table contained via pseudo set-associativity.

ACKNOWLEDGEMENT

The authors thank Itthichok Jangjaimon for his assistance in obtaining some FPGA implementation results included in this article.

REFERENCES

[1] F. Baboescu, S. Singh, and G. Varghese, "Packet Classification for Core Routers: Is there an alternative to CAMs," *Proceedings of 22nd IEEE International Conference on Computer Communications (INFOCOM 2003)*, pp. 53–63, Mar./Apr. 2003.

[2] Y.-K. Chang, Y.-S. Lin, and C.-C. Su, "A High-Speed and Memory Efficient Pipeline Architecture for Packet Classification," *Proceedings of 18th IEEE International Symposium on Field-Programmable Custom Computing Machines (FCCM 2010)*, pp. 215-218, May 2010.

[3] G. S. Jedhe, A. Ramamoorthy, and K. Varghese, "A Scalable High Throughput Firewall in FPGA," *Proceedings of IEEE Symposium on Field-Programmable Custom Computing Machines (FCCM 2008)*, Apr. 2008.

[4] W. Jiang and V. K. Prasanna, "Large-Scale Wire-Speed Packet Classification on FPGAs," *Proceedings of 17th ACM/SIGDA Int'l Symposium on Field Programmable Gate Arrays (FPGA '09)*, Feb. 2009.

[5] Juniper Networks, Inc., "DataSheet of 100-Gigabit Ethernet PIC," 2011. URL – http://www.juniper.net/us/en/local/pdf/datasheets/1000346-en.pdf.

[6] A. Kennedy et al., "Low Power Architecture for High Speed Packet Classification," *Proceedings of ACM/IEEE Symposium on Architectures for Networking and Communications Systems (ANCS 2008)*, Nov. 2008.

[7] K. Lakshminarayanan, A. Rangarajan, and S. Venkatachary, "Algorithms for Advanced Packet Classification with Ternary CANs," *Proceedings of ACM Annual Conference of Special Interest Group on Data Communication (SIGCOMM '05)*, pp. 193–204, Aug. 2005.

[8] Y. J. Lee and B.-K. Chung, "Pseudo 3-way Set Associative Cache: A Way of Reducing Miss Ratio with Fast Access Time," *Proceedings of IEEE Canadian Conference on Electrical and Computer Engineering*, May 1999.

[9] A. Nikitakis and I. Papaefstathiou, "A Memory-Efficient FPGA-Based Classification Engine," *Proceedings of IEEE Symposium on Field-Programmable Custom Computing Machines (FCCM 2008)*, Apr. 2008.

[10] F. Pong and N.-F. Tzeng, "HaRP: Rapid Packet Classification via Hashing Round-Down Prefixes," *IEEE Transactions on Parallel and Distributed Systems*, vol. 22, pp. 1105-1119, July 2011.

[11] V. Pus and J. Korenek, "Fast and Scalable Packet Classification Using Perfect Hash Functions," *Proc. of 17th International ACM Symposium on Field Programmable Gate Arrays (FPGA '09)*, pp. 229-235, Feb. 2009.

[12] D. E. Taylor, "Survey and Taxonomy of Packet Classification Techniques," *ACM Computing Surveys*, vol. 37, no. 3, pp. 238–275, Sept. 2005.

[13] D. E. Taylor and J. S. Turner, "ClassBench: A Packet Classification Benchmark," *Proceedings of 24th IEEE International Conference on Computer Communications (INFOCOM 2005)*, Mar. 2005.

[14] G. Wang and N.-F. Tzeng, "TCAM-Based Forwarding engine with Minimum Independent Prefix Set (MIPS) for Fast Updating," *Proceedings of IEEE International Conference on Communications (ICC '06)*, June 2006.

[15] Xilinx Corporation, "IP-SysLogic-BlockMem Generator: Block Memory Generator v3.3," *AR#* 33298, Sept. 2009. URL – http://www.xilinx.com/support/answers/33298.htm.

[16] Xilinx Corporation, "Virtex-6 Family Overview," March 2011. URL – http://www.xilinx.com/product/silicon-devices/fpga/virtex-6/index.htm.

A Real-time Stereo Vision System Using a Tree-structured Dynamic Programming on FPGA

Minxi Jin
Systems and Information Engineering
University of Tsukuba
1-1-1 Ten-ou-dai Tsukuba 305-8573 Japan
jinminxi@darwin.esys.tsukuba.ac.jp

Tsutomu Maruyama
Systems and Information Engineering
University of Tsukuba
1-1-1 Ten-ou-dai Tsukuba 305-8573 Japan
maruyama@darwin.esys.tsukuba.ac.jp

ABSTRACT

Many hardware systems for stereo vision have been proposed. Their processing speed is very fast, but the algorithms used in them are limited in order to achieve the high processing speed by simplifying the sequences of the memory accesses and operations. The error rates by them can not compete with those by software programs. In this paper, we describe an FPGA implementation of a tree-structured dynamic programming algorithm. The computational complexity of this algorithm is higher than those by previous hardware systems, but the processing speed of our system is still fast enough for real-time applications, and its error rate is competitive with software algorithms.

Categories and Subject Descriptors

B.7.1 [**Integrated Circuits**]: Types and Design Styles—*Algorithms implemented in hardware*; I.3.1 [**Computer Graphics**]: Hardware Architecture—*Parallel Processing*

General Terms

Performance

Keywords

FPGA, stereo vision, real-time

1. INTRODUCTION

The aim of stereo vision systems is to reconstruct the 3-D geometry of a scene from images of two separate cameras. When the two cameras are calibrated properly, for each pixel in the images, its matching pixel exists on the same line of the other image. If $L(x+d,y)$ (the pixel at $(x+d,y)$ in the left image) is the most similar to $R(x,y)$ (the pixel at (x,y) in the right image), this means that these two pixels are the same point of an object, and the distance to the object can be calculated from disparity d and the parameters of the two cameras. Many dedicated hardware systems have been developed to realize real-time stereo vision. In most

of those systems, correlation-based algorithms (SSD (Sum of Squared Differences), SAD (Sum of Absolute Differences) or Census transform) and dynamic programming (DP) algorithms have been used to find the matching because of their simplicity (they require only raster scan). However, in these local search algorithms, the disparity of each pixel is decided by only comparing pixels on the corresponding lines in the two images, and the error rate by these algorithms is high. In some hardware systems, algorithms based on belief propagation (BP) were implemented in order to improve the error rate. The quality by those algorithms is much better, because the pixels are compared two-dimensionally, but still can not compete with sophisticated software programs. Furthermore, these algorithms require a lot of memory.

In this paper, we describe an FPGA system for real-time stereo vision. Our goal is to demonstrate how accurate and fast stereo vision systems can be configured on FPGAs.

2. TREE STRUCTURED DP ALGORITHM

We have implemented a simplified version of the tree-structured dynamic programming proposed in [2]. Let $X \times Y$ be the image size, and D the maximum disparity. When the right image is used as the base, D $SADs$ ($SAD(x, y, d)_{\{d=0, D-1\}}$) are calculated for each $R(x,y)$ as follows.

$$SAD(x,y,d) =$$
$$\sum_{dx=-r}^{r} \sum_{dy=-r}^{r} \{ \; ABS(R(x+dx,y+dy), L(x+d+dx,y+dy)) \times$$
$$\delta(R(x,y), R(x+dx,y+dy)) \; \}$$
$$ABS(P,Q) = |P_R - Q_R| + |P_G - Q_G| + |P_B - Q_B|$$
$$\delta(P,Q) = 1 \quad \text{if} \;\; ABS(P,Q) < Threshold_{ABS}$$
$$0 \quad \text{otherwise}$$

Then, the penalties by the smoothness are added to the SADs considering the connection between the pixels. In [2], two kinds of connections of neighbor pixels (horizontal tree and vertical tree) are considered. In our implementation, only the horizontal tree connection is used. In this connection, first, the penalties along the x axis are calculated. In Fig. 1(A), for the pixels labeled '0', their SADs are used as their scores. The SADs of the pixels labeled 'k' are modified according to the scores of the pixels labeled 'k-1' along the arrows. Then, the scores of the pixels labeled 'k' are modified by the scores of the pixels labeled 'k-1' along the y axis, and the final scores of pixel (x,y) are calculated (Fig. 1(B)).

For reducing the computational complexity, $\overrightarrow{m}(x,y,d)$, $\overleftarrow{m}(x,y,d)$ and $m(x,y,d)$ of all pixels are calculated using the following equations first.

$$\overrightarrow{m}(0,y,d) = SAD(0,y,d)$$

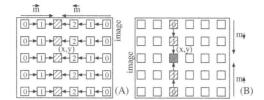

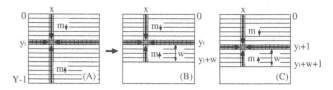

Figure 2: A method to reduce memory banks

Figure 1: Tree-based connection of neighbor pixels

$$\overrightarrow{m}(x,y,d) = SAD(x,y,d) + \min_k \{\overrightarrow{m}(x-1,y,k) + p(d,k)\}$$

$$\overleftarrow{m}(X-1,y,d) = SAD(X-1,y,d)$$

$$\overleftarrow{m}(x,y,d) = SAD(x,y,d) + \min_k \{\overleftarrow{m}(x+1,y,k) + p(d,k)\}$$

$$m(x,y,d) = \min_k \{\overrightarrow{m}(x-1,y,k) + p(k,d)\} +$$
$$SAD(x,y,d) +$$
$$\min_k \{\overleftarrow{m}(x+1,y,k) + p(k,d)\}$$
$$= \overrightarrow{m}(x,y,k) + \overleftarrow{m}(x,y,k) - SAD(x,y,d)$$

In these equations, $p(d,k)$ is used for evaluating the smoothness of the disparities and accumulated along the scan direction (its detail is given afterward). Then, $m(x,y,d)$ show the scores considering the horizontal connection of the pixels. Next, using $m(x,y,d)$ instead of $SAD(x,y,d)$, $m_\downarrow(x,y,d)$ and $m_\uparrow(x,y,d)$ are calculated in the same way as $\overrightarrow{m}(x,y,d)$ and $\overleftarrow{m}(x,y,d)$.

$$m_\downarrow(x,0,d) = m(x,0,d)$$
$$m_\downarrow(x,y,d) = m(x,y,d) + \min_k \{m_\downarrow(x,y-1,k) + p(d,k)\}$$
$$m_\uparrow(x,Y-1,d) = m(x,Y-1,d)$$
$$m_\uparrow(x,y,d) = m(x,y,d) + \min_k \{m_\uparrow(x,y+1,k) + p(d,k)\}$$

Finally, $M(x,y,d)$ is calculated for each pixel in the same way as $m(x,y,d)$, and the k which minimizes $M(x,y,d)$ is chosen as the disparity at (x,y).

$$M(x,y,d) = \min_k \{m_\downarrow(x,y-1,k) + p(k,d)\} + m(x,y,d) +$$
$$\min_k \{m_\uparrow(x,y+1,k) + p(k,d)\}$$
$$= m_\downarrow(x,y,k) + m_\uparrow(x,y,k) - m(x,y,d)$$

$$disparity(x,y) = k \text{ which minimizes } M(x,y,k)$$

$p(d,k)$ should be different according to $|d-k|$. In order to reduce the computational complexity, only four values are used for $p(d,k)$.

$$\overrightarrow{m}(x,y,d) = SAD(x,y,d) +$$
$$\min_k \{\overrightarrow{m}(x-1,y,k) + p(d,k)\} -$$
$$\min_k \{\overrightarrow{m}(x-1,y,k)\}$$
$$= SAD(x,y,d) +$$

$$\min \left\{ \begin{array}{l} \overrightarrow{m}(x-1,y,d) \\ \overrightarrow{m}(x-1,y,d\pm 1) + C_p \\ \min_k \{\overrightarrow{m}(x-1,y,k)\} + f_p(R(x,y),R(x-1,y)) \end{array} \right\} -$$

$$\min_k \{\overrightarrow{m}(x-1,y,k)\}$$

where $f_p(P,Q) = \{C_h \text{ if } |P-Q| < Threshold_n,$
$\qquad\qquad\qquad C_l \text{ otherwise}\}$
$\qquad C_l < C_p < C_h$ are constants

In this equation, $\min_k \{\overrightarrow{m}(x-1,y,k)\}$ is added to keep the data width of $\overrightarrow{m}(x,y,d)$ small.

In order to avoid the errors caused by the occlusions, the same procedures as above are re-executed using the left image as the base, and the same matchings found in both searches are considered as the true disparities. Then, each

non-true disparity is replaced by one of its neighbor true disparities along the x line (the smaller one is chosen).

3. IMPLEMENTATION

This tree structured DP algorithm has two major problems; large memory size and long feedback loop in the computation. By applying the following procedure from $y=0$ to $y=Y-1$, $\overleftarrow{m}(x,y,d)$, $\overrightarrow{m}(x,y,d)$, $m(x,y,d)$ and $m_\downarrow(x,y,d)$ can be efficiently calculated with $D\times 3$ line buffers, which are used for storing $SAD(x,y,d)$, $\overleftarrow{m}(x,y,d)$ and $m_\downarrow(x,y,d)$.

1. Scan the line from $X-1$ to 0, and calculate
   (a) $SAD(x,y,d)$ and
   (b) $\overleftarrow{m}(x,y,d)$ using $SAD(x+1,y,d)$.
   Store $SAD(x,y,d)$ and $\overleftarrow{m}(x,y,d)$ in the line buffers.

2. Scan the same line from 0 to $X-1$, and calculate
   (a) $\overrightarrow{m}(x,y,d)$ using the stored $SAD(x,y,d)$,
   (b) $m(x,y,d)$ using $\overrightarrow{m}(x,y,d)$ and the stored $\overleftarrow{m}(x,y,d)$, and
   (c) $m_\downarrow(x,y,d)$ using $m(x,y,d)$ and the stored $m_\downarrow(x,y-1,d)$ (or $m(x,0,d)$ as $m_\downarrow(x,0,d)$ when $y=0$)
   Store $m_\downarrow(x,y,d)$ instead of $m_\downarrow(x,y-1,d)$.

In this procedure, we can calculate $M(x,y,d)$ by providing $m_\uparrow(x,y,d)$ to each line. The most efficient way for providing them is to calculate $m_\uparrow(x,y,d)$ from $y=Y-1$ to $y=0$ in order, and store them in advance. However, the size of an array to store them is too large to implement an FPGA. In our implementation, for providing $m_\uparrow(x,y_t+1,d)$ to $y=y_t$, we start to calculate $m_\uparrow$ from $y=y_t+w$ instead of $y=Y-1$ as shown in Fig. 2(B). At $y=y_t+w$, $m(x,y_t+w,d)$ is used as $m_\uparrow(x,y_t+w,d)$, and $m_\uparrow(x,y_t+k,d)_{(k=w-1,1)}$ are calculated in order. For providing $m_\uparrow(x,y_t+2,d)$ to $y=y_t+1$, $m_\uparrow$ is calculated from $y=y_t+w+1$ again (Fig. 2(C)). With this approach, we can calculate $M(x,y,d)$ without using the arrays, but we need to calculate $m_\uparrow(x,y,d)$ w times for each pixel. According to our experiments, $w=7$ is enough for achieving low error rates.

Another problem is the long feedback loop in the computation. For calculating $\overrightarrow{m}(x,y,d)$, we need to calculate $\min_k \{\overrightarrow{m}(x-1,y,k)\}$, which requires a selector tree of depth $= \log D$. Furthermore, the unit for calculating $\overrightarrow{m}(x,y,d)$ consists of two stages in order to realize high operational frequency. Therefore, the total feedback delay becomes 8 clock cycles when $32 < D \leq 64$. This delay can be reduced to 2 clock cycles if $\min_k \{\overrightarrow{m}(x-1,y,k)\}$ were calculated in advance, and can be used immediately.

In our approach, the search is divided into two phases. In the first phase, the calculation of 8 lines are interleaved. As shown in Fig. 3(A), 8 lines ($y+7$ to $y+0$) are scanned from right to left (from $X-1$ to 0) first. During this scan, for 8 pixels on each column of the 8 lines, $SAD(x,y+t,d)$ and $\overleftarrow{m}(x,y+t,d)$ ($t=7$ to 0) are calculated in this order.

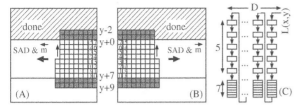

Figure 3: The first phase (finding the minimum)

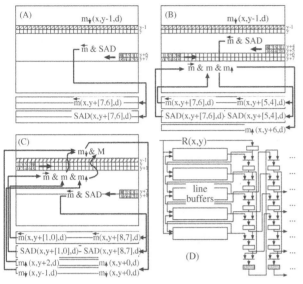

Figure 4: The second phase

With this calculation order, $\overleftarrow{m}(x,y+t,d)$ starts to be calculated 8 clock cycles later than $\overleftarrow{m}(x+1,y+t,d)$. Therefore, $\min_k\{\overleftarrow{m}(x+1,y+t,k)\}$ can be provided to $\overleftarrow{m}(x,y+t,d)$ just in time by the pipelined unit. In this scan direction, only $\min_k\{\overleftarrow{m}(x,y+t,k)\}$ ($t=7$ to 0) is stored in an off-chip memory bank. Then, the search direction is reversed (Fig. 3 (B)), and $\min_k\{\overrightarrow{m}(x,y+t,k)\}$ is calculated and stored. After that, the next 8 lines are processed in the same way until all lines are processed. For calculating $SADs$ in these scans, $R(x,y)$ and $L(x,y)$ of $8+2r$ lines are necessary. They are supplied using register arrays and shallow FIFOs (in Fig. 3(C), the arrays and FIFOs for the left image when scanning from left to right is shown). For reducing the circuit size, $8+2r$ pixels on a column are passed to the next column using a serial link, and it takes $8+2r$ clock cycles. Therefore, $8+2r$ clock cycles are required for processing 8 pixels.

In the second phase, the calculation of 2 lines are interleaved. In this phase, two scans (RtoL and LtoR) are executed in parallel as shown Fig. 4. First, two lines ($y+7$ and $y+6$) are scanned from right to left (RtoL scan) (Fig. 4(A)). During this scan, for two pixels on each column, $SAD(x,y+t,d)$ and $\overleftarrow{m}(x,y+t,d)$ ($t=7$ to 6) are calculated again. Here, $\overleftarrow{m}(x,y+t,d)$ is calculated using $\min_k\{\overleftarrow{m}(x+1,y+t,k)\}$ in the off-chip memory bank. Thus, $\overleftarrow{m}(x,y+t,d)$ of one line can be obtained in every two clock cycles, which means all pipeline stages are full-filled by interleaving the calculations of two lines. $SAD(x,y+t,d)$ and $\overleftarrow{m}(x,y+t,d)$ are stored in lines buffers ($4\times D$ line buffers are used in total). When this scan reaches to $x=0$, the next scan (LtoR) on the same two lines starts from $x=0$ (Fig. 4(B)). In this scan, for $t=7$ to 6, (1) $\overrightarrow{m}(x,y+t,d)$ is calculated using

$SAD(x,y+t,d)$ which was stored in the buffer during RtoL scan, $\overrightarrow{m}(x-1,y+t,d)$ which was calculated in the previous column in this scan, and $\min_k\{\overrightarrow{m}(x-1,y+t,k)\}$ which was stored in the off-chip memory banks in the first phase, (2) $m(x,y+t,d)$ is calculated using $\overrightarrow{m}(x,y+t,d)$ calculated in (1), and $\overleftarrow{m}(x,y+t,d)$ and $SAD(x,y+t,d)$ which were stored in the buffers during RtoL scan, (3) $m_{\uparrow}(x,y+t,d)$ is calculated from $m_{\uparrow}(x,y+t+1,d)$ (when $t=7$, $m(x,y+7,d)$ is used as $m_{\uparrow}(x,y+7,d)$). The calculations of $m_{\uparrow}(x,y+t,d)$ for $t=6$, $4,2,0$ are delayed for 8 clock cycles for obtaining correct $\min_k m_{\uparrow}(x,y+t+1,k)$. $m_{\uparrow}(x,y+t,d)$ for $t=6,4,2,0$ is stored in D line buffers. In parallel with LtoR, RtoL scan for the next two lines ($y+5$ and $y+4$) is executed as shown in Fig. 4(B). The same buffers can be shared by these two scans.

These two scans are repeated, and when the LtoR scan reaches to $y+1$ and $y+0$ (Fig. 4(C)), RtoL scan moves to $y+8$ and $y+7$ for calculating $m_{\uparrow}$ for $M(x,y+1,d)$. In the RtoL scan on $y+1$ and $y+0$, $m_{\downarrow}(x,y+0,d)$ is also calculated using $m(x,y+0,d)$ and $m_{\downarrow}(x,y-1,d)$ which is stored in the buffer. Then, $M(x,y,d)=m_{\downarrow}(x,y,d)+m_{\uparrow}(x,y,d)-m(x,y,d)$ is calculated, and k which minimizes $M(x,y,d)$ is chosen as the disparity at (x,y). Fig. 4(D) shows the line buffers and the register array for the right image (the registers are shared with the first phase shown in Fig. 3(C)).

When we finished the two phases (the disparity of each pixel is stored in an off-chip memory bank in the second phase), the two phases are repeated again by using the left image as the base. In this case, the scan direction of each scan is reversed for calculating the disparity without changing the data paths of the circuit.

4. THE EXPERIMENTAL RESULTS

We have implemented a circuit for $X \leq 512$ and $D=60$ on Xilinx XC4VLX160 on RC2000-4 from Agility. The circuit uses 109.7K LUTs (72%) and 132 block RAMs (46%) (28.8K and 46 when $D=16$). The number of LUTs is almost proportional to D and r, and the number of block RAMs is almost proportional to X and D. XC4VLX160 has 288 block RAMs. Thus, up to $X=1024$ can be processed, though D can not be enlarged according to X.

The circuit runs at 216.233MHz. The computation time by our circuit is almost $(X\times Y\times(2r+1)/8\times2+X\times Y\times(w+1))\times2$ ((the first phase + the second phase)$\times2$). The detection and expansion of the true disparities can be executed in parallel with the second phase when the left image is used as the base. The processing speed of our circuit is 128.0 fps for 320$\times$240 pixel images, and 32.0 fps for 640$\times$480 pixel images when $w=7$.

Table 1 compares the error rate by several algorithms[1]. In Table 1, BP+MLH, DP and SSD+MF can be considered to give the upper limits of BP, DP and correlation-based (SAD and SSD give almost the same results) algorithms which were implemented on previous hardware stereo vision systems. Census[3] shows the error rates based on a recent FPGA system using census transform. Zhang et al.[10] implemented a sophisticated local algorithm on FPGA. Its error rate is comparable with our system. But, it is difficult to maintain its low error rate for large size images without increasing the SAD window size, which requires more hardware resources. The error rate by GPUs (*2 and *3) is also comparable with our system, but the processing speed of our system is faster than them, and D can be larger.

Table 1: Performance comparison (Error rate by each algorithm)

algorithm	Tsukuba			Venus			Teddy			Cones			
	n.o.	all	disc	n.o.	all	disc	n.o.	all	disc	n.o.	all	disc	
ADCensus[5]	1.07	1.48	5.73	0.09	0.25	1.15	4.10	6.22	10.9	2.42	7.25	6.95	*1
AdaptingBP[4]	1.11	1.37	5.79	0.10	0.21	1.44	4.22	7.06	11.8	2.48	7.92	7.32	
SimpleDPTree[2]	1.86	2.56	-	0.42	0.76	-	7.31	12.7	-	4.00	9.74	-	
RealTimeBP[9]	1.49	3.40	7.87	0.77	1.90	9.00	8.72	13.2	17.2	4.61	11.6	12.4	*2
Zhang et al.[10]	3.84	4.34	14.2	1.20	1.68	5.62	7.17	12.6	17.4	5.41	11.0	13.9	
our method (w=7)	1.43	2.51	6.60	2.37	2.97	13.1	8.11	13.6	15.5	8.12	13.8	16.4	
RealTimeGPU[8]	2.05	4.22	10.6	1.92	2.98	20.3	7.23	14.4	17.6	6.41	13.7	16.5	*3
BP+MLH[7]	4.17	6.34	14.6	1.96	3.31	16.8	10.2	18.9	24.0	4.93	15.5	12.3	
DP[6]	4.12	5.04	12.0	10.1	11.0	21.0	14.0	21.6	20.6	10.5	19.1	21.1	*4
SSD+MF[6]	5.23	7.07	24.1	3.74	5.16	11.9	16.5	24.8	32.9	10.6	19.8	26.3	*5
Census[3]	9.79	11.6	20.3	3.59	5.27	36.8	12.5	21.5	30.6	7.34	17.6	21.0	*6

1. n.o.(non-occluded regions) are the errors are only for the non-occluded regions.
2. all (all regions) are the errors in all regions (excluding borders of the image).
3. disc (discontinuity) are the errors only for the regions near depth discontinuities.
BP=Belief Propagation, DP=Dynamic Programming, MLH=Mid-Level Hyothesis,
SSD=Sum of Squared Difference, MF=Min-Filter
  *1: the lowest error rate reported in [1]   *2: 16fps for 320×240 pixel images when D=16
  *3: 43fps for 320×240 pixel images when D=16   *4,5: algorithms widely used on hardware systems
  *6: 230 fps for 640×480 pixel images using XC4VLX200

Figure 5: Tsukuba   **Figure 6: Venus**   **Figure 7: Teddy**   **Figure 8: Cones**

Our processing speed can be improved by storing $m_\downarrow$ of k lines, and calculating $m_\uparrow$ for $[w+k+1]/2$ lines, though it requires $20\times(k-1)$ more block RAMs. The processing speed will become 234.6 and 58.7 fps for 320×240 and 640×480 pixels images when k=4.

Fig. 5, 6, 7 and 8 show the disparity maps by our system.

5. CONCLUSIONS AND FUTURE WORK

In this paper, we have described an FPGA implementation of a tree-structured dynamic programming algorithm. The error rate by our system is much smaller than previous hardware systems, and comparable with software programs (though still behind the best algorithms). The processing speed of our system is fast enough for real-time applications. The circuit size and the number of the block RAMs used in the circuit are kept small by limiting the matching along y axis, and repeatedly calculating them. The operational frequency is also kept fast by dividing the search into two phases. Improving the error rate is our main future work. Latest FPGAs have more block RAMs compared with their LUTs. We need to find a better balance point of the implementation for those FPGAs.

6. REFERENCES

[1] http://vision.middlebury.edu/stereo/eval/.
[2] M. Bleyer and M. Gelautz. Simple but effective tree structures for dynamic programming-based stereo matching. *VISAPP*, 2:415–422, 2008.
[3] S. Jin, J. Cho, X. D. Pham, K. M. Lee, S.-K. Park, M. Kim, and J. W. Jeon. Fpga design and implementation of a real-time stereo vision system. *IEEE Transactions on Circuits and Systems for Video Technology*, 1:15–26, 2010.
[4] A. Klaus, M. Sormann, and K. Karner. Segment-based stereo matching using belief propagation and a self-adapting dissimilarity measure. *ICPR*, 2006.
[5] X. Mei, X. Sun, M. Zhou, S. Jiao, H. Wang, and X. Zhang. On building an accurate stereo matching system on graphics hardware. *GPUCV*, 2011.
[6] D. Scharstein and R. Szeliski. A taxonomy and evaluation of dense two-frame stereo correspondence algorithms. *IJCV*, 2002.
[7] O. Stankiewicz and K. Wegner. Depth map estimation software version 2. *ISO/IEC MPEG meeting M15338*, 2008.
[8] L. Wang, M. Liao, M. Gong, R. Yang, and D. Nister. High-quality real-time stereo using adaptive cost aggregation and dynamic programming. *3DPVT*, 2006.
[9] Q. Yang, L. Wang, R. Yang, S. Wang, M. Liao, and D. Nister. Global stereo matching using hierarchical belief propagation. *BMVC*, 2006.
[10] L. Zhang, K. Zhang, T. S. Chang, G. Lafruit, G. Kuzmanov, and D. Verkest. Real-time high-definition stereo matching on fpga. *FPGA*, 2011.

Incremental Clustering Applied to Radar Deinterleaving: A Parameterized FPGA Implementation

Scott Bailie[*]
MIT Lincoln Laboratory
244 Wood St.
Lexington, MA
bailie@ll.mit.edu

Miriam Leeser
Dept. of Electrical and Computer Eng.
Northeastern University
Boston, MA
mel@coe.neu.edu

ABSTRACT

ICED (Incremental Clustering of Evolving Data) is a novel incremental clustering algorithm designed for data whose characteristics change over time. ICED is an unsupervised clustering technique that assumes no prior knowledge of the incoming data, and supports removing clusters that contain stale data. The user controls the FPGA implementation through a combination of compile time parameters (number of clusters) and run time parameters (distance threshold, fade cycle length). ICED has been applied to a radar application: pulse deinterleaving. ICED is the first implementation of incremental clustering on an FPGA of which we are aware. The implementation runs 39 times faster than an equivalent C implementation on a 3GHz Intel Xeon processor, and is capable of processing radar data in real time.

Categories and Subject Descriptors

H.3.3 [**Information Storage and Retrieval**]: Information Search and Retrieval—*clustering*

Keywords

Clustering, Electronic Warfare, Pulse Deinterleaving, FPGA

1. INTRODUCTION

Clustering is the unsupervised partitioning of similar data samples into groups called clusters. Clustering is a popular technique for data mining and machine learning. Many clustering algorithms require a complete dataset to be present and require multiple passes through the data in order to produce a result; such algorithms are *non-incremental*. In contrast, an incremental clustering algorithm considers each

---

[*]This work is sponsored by the Department of the Air Force under Air Force Contract FA8721-05-C-0002. The opinions, interpretations conclusions and recommendations are those of the author and are not necessarily endorsed by the United States Government.

input data point only once, at which point it assigns it to a cluster. Incremental clustering is better suited to streaming data. With streaming data, the characteristics of the data may change with time, and cluster centers should be updated to remove the affects of old, stale data. FPGAs are well suited to processing streaming data. With an FPGA implementation, processing can be placed close to the sensed data, keeping latency low. Parameters can allow an FPGA solution to adapt to its environment both statically at compile time and dynamically at run time. The input data can be processed and the results, which are often smaller than the inputs, can then be communicated for further processing to another FPGA module or a host computer. We present the implementation of an incremental clustering algorithm, ICED, on a Xilinx Virtex 5 FPGA and apply it to identifying transmitters in a radar environment. Previous research on clustering on an FPGA has focused on non-incremental algorithms [2, 3, 5]. ICED is the first implementation of incremental clustering on an FPGA [1]. Our results show that the FPGA implementation can apply incremental clustering in real time.

2. BACKGROUND

2.1 Radar Pulse Deinterleaving

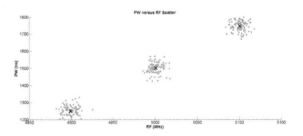

Figure 1: PW/RF Plot For Three Emitters

We process pulsed radar data with the goal of separating pulses from different radar transmitters into different clusters. This technique, radar pulse deinterleaving, is critical for effective electronic warfare (EW) signal processing because it helps to identify enemy radars, allowing intelligent allocation of resources to tasks such as jamming. In modern EW environments, which are both dense and rapidly changing, deinterleaving needs to be completed with very low latency to support timely decision making.

A stream of pulses from a particular radar will have similar characteristics. An EW receiver performs several measurements on each input pulse including time of arrival (TOA), pulse width (PW), pulse repetition interval (PRI), and frequency (RF). These measurements are collected into a Pulse Descriptor Word (PDW). Each received PDW can be treated as a point in an m-dimensional space where m is the number of parameters used. The goal is to detect clusters of points where each cluster corresponds to a unique radar emitter. Fig. 1 illustrates the plot of PDW parameters for $m = 2$ using PW and RF as the characteristics. Each distinct cluster corresponds to an emitter and the cluster centers are the assumed actual RF and PW for each. The clusters in the figure are clearly separated which is expected in pulse deinterleaving. When pulses from multiple emitters coincide in time at a receiver, the resulting PDWs can contain inaccurate measurements which are unrepresentative of the true emitters. These overlapping pulses are expected to be relatively infrequent and therefore should be easily filtered out.

2.2 ICED Algorithm

ICED borrows from many existing clustering algorithms. Choices for ICED were made based on the data that we are processing and the clustering objective, deinterleaving radar pulse data, as well as the knowledge that the implementation would be in FPGA hardware. We use RF and PW data with the goal of separating pulses from different transmitters into different clusters. The ICED algorithm is incremental: each input sample is processed once and no long-term storage of data is used. A secondary goal is to identify and filter out clusters formed due to noise or interference. Since radar transmitters may disappear from the environment over time, ICED allows older clusters to fade away as the operating environment evolves.

The ICED algorithm is initialized by setting all cluster coordinates and weights to zero, except for an initial cluster which is assigned the first input. As inputs are received, they are assigned to the nearest active cluster provided that the distance to the cluster center does not exceed a threshold provided by the user. When a new input is assigned to a cluster, the weight is incremented and the cluster center is updated based on the values of the new sample.

Fading removes clusters that are no longer relevant. Each time a data point is added to an existing cluster the weight is incremented by one, and periodically the weights of all clusters are decremented. The weights of clusters receiving assignments faster than the decrement rate will continue to increase over time, while the weights of clusters receiving assignments at a slower rate will decrease. Once its weight reaches zero, a cluster is deemed stale and deleted. The decrement period is a parameter supplied by the user. In addition to removing outdated clusters, this method can effectively remove clusters created due to spurious data. Such clusters should have relatively low assignment rates and therefore fade quickly.

New Cluster Creation. If an input is received whose distance to the nearest cluster exceeds the threshold, a new cluster must be created. If all cluster hardware is currently allocated, a stale cluster must be identified to be overwritten. We overwrite the cluster with the lightest weight under the assumption that it likely represents noise or outliers as opposed to an actual emitter. This may not always be the case; future work will address more complex methods for choosing a cluster to evict.

Parameters, Inputs and Variables. The ICED algorithm uses two parameters provided by the user: D: distance threshold and L: fade cycle period. The distance threshold is used to determine when a new cluster must be created. The fade cycle period is the length of time after which the weight of every cluster is decreased by one, and affects the rate at which stale clusters are deleted.

3. FPGA IMPLEMENTATION

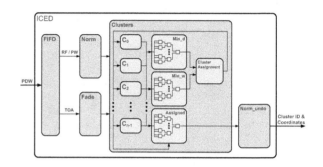

Figure 2: ICED - Top-level block diagram

We have implemented ICED on a Xilinx Virtex-5 FPGA. Fig. 2 shows a block diagram of the top-level modules. As PDWs are received, they are placed in a FIFO. After reading a PDW from the FIFO, parameter normalization and fade calculations are performed. These results are passed to the *Clusters* module which measures the distance to all active clusters and decides to which cluster the input should be assigned. Finally, the coordinates of the assigned cluster are converted from normalized to native units and output along with a cluster ID. The goals of the project, to deinterleave radar pulses in real time and with minimum latency is achieved by exploiting parallelism. Cluster fading and input normalization are implemented in parallel. Distances of each incoming radar pulse to all cluster centers are computed simultaneously and the smallest distance determined. Each cluster computes its updated position assuming the current input will be assigned to every cluster. Once the actual assignment is known, the update can be instantaneous as a result of this speculative execution.

Our design is implemented in a modular fashion so that alternative decisions can easily be implemented. Examples include normalization, distance measurement, and cluster coordinate calculation. While we implement a Manhattan distance metric, it would be easy to choose a different metric. The code is parameterized so that the number of clusters, etc. can be adapted to the particular problem being solved. Host accessible registers provide for real-time configurability of distance thresholds, fade cycle lengths, maximum cluster weights, and parameter ranges for normalization.

A pulse deinterleaver is a single block in a more complicated EW system. The deinterleaver inputs are PDW parameters as measured by the EW receiver's parameter estimation block. We cluster based on RF and PW. The deinterleaver is also provided with TOA for each pulse, which is used in the cluster fade operation. For each input, the pulse deinterleaver outputs the coordinates of the assigned cluster and the cluster ID number. The coordinates of all active clusters are also available.

3.1 Modules in the Design

Normalization and Fade Modules. The PDW parameters are extracted from the output of the FIFO; RF and PW are passed to the normalization module (*Norm*) and TOA is passed to the fade cycle calculation module (*Fade*). Both *Norm* and *Fade* modules require division, which is an expensive operation in an FPGA. The divisors are not powers of two, so a full division must be implemented. The *Fade* module calculates the whole number of fade cycles since the previous input. The result is passed to the *Clusters* module and subtracted from the weight of the cluster centers.

Clusters Module. The core of the implementation is the *Clusters* module (Fig. 2), which is composed of three major components: (1) An array of n *Cluster_center* modules, $C_0 \ldots C_{n-1}$; (2) Three binary trees, *Min_d*, *Min_w*, and *Assigned*; and (3) The *Cluster Assignment* module.

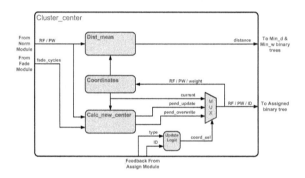

Figure 3: Cluster center block diagram

The *Cluster_center* modules are responsible for measuring the distance to inputs and calculating and updating their center coordinates. n independent *Cluster_center* modules are implemented based on the maximum number of clusters (n) supplied at compile time. The *Cluster_center* module (Fig. 3) receives normalized RF and PW values and the number of fade cycles. RF and PW are sent to the *Dist_meas* module which calculates the distance from the input to the current cluster center. The *Calc_new_center* module decreases the current cluster weight based on the number of fade cycles, and determines the pending cluster coordinates and weight for two cases. The first is if the current input is added to the cluster (*pend_update*), and the second is if the cluster needs to be overwritten (*pend_overwrite*) based on the need to start a new cluster.

Cluster Assignment. The output of the *Dist_meas* module from each cluster and the weight of each cluster are sent to the *Min_d* and *Min_w* binary trees, respectively, to determine the nearest and lightest clusters. Using the nearest and lightest information, the *Cluster_assignment* module decides to which cluster the current input should be assigned, and whether the cluster should be updated, overwritten or unchanged. This decision is fed back to each *Cluster_center* module and used as multiplexer control to select the *current*, *pend_update*, or *pend_overwrite* coordinates and weight. Next, each *Cluster_center* updates (if required) and outputs its coordinates to the *Assigned* binary tree. This tree is responsible for determining the coordinates and ID of the

"winning" cluster. The final step is to output the assigned cluster ID and coordinates in their original native units.

4. EXPERIMENTS AND RESULTS

Our design flow consists of design entry using Verilog, synthesis and implementation using Xilinx ISE, and simulation using Mentor Graphics ModelSim. Test data was generated using MATLAB which allowed full control over test scenarios and produced a more controlled experimental setup with fewer variables than in a real-world setup with measured data. Hardware testing was performed on the Innovative Integration X5-400M XMC module, which contains a Xilinx Virtex-5, running at 200MHz. We used PCI Express to communicate with the host. All results are based on real designs running in hardware. Though our pulse deinterleaver is clustering pre-processed PDWs, a fully integrated system could perform pulse measurement, deinterleaving, and response on the same board.

We use MATLAB to generate input data. The MATLAB model provides a 13-bit RF and 16-bit PW measurement corresponding to resolutions of approximately 4KHz and 16ns, respectively. User scripted scenarios provide TOA, RF, and PW measurements for each simulated detection and the resulting PDWs are stored to a file. This file is transferred to the board over the PCI Express interface. The user defined parameters, distance threshold and fade cycle length, are set from the host.

4.1 Clustering Results

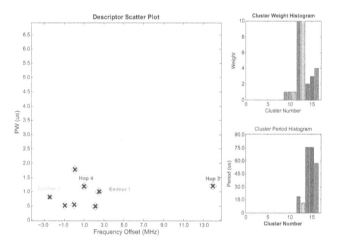

Figure 4: Test Scenario Results

We evaluate cluster quality, hardware resources required, and performance of FPGA hardware compared to a software approach. We investigated a scenario consisting of three emitters, two fixed RF emitters and one frequency agile emitter which hops between four RF values. Frequency agility is a method where the emitter changes its frequency over time in an effort to avoid being tracked. This scenario stresses our algorithm with a large number of false clusters representing overlapping pulses. Additionally, the relatively short dwell time of the frequency hopper helps evaluate the ability of the ICED algorithm to identify the hopper as an emitter rather than an outlier. Fig. 4 shows a snapshot of the clustering soon after the hopper has moved to its fourth frequency. The plot shows the two fixed emitters, Emitter 1 and Emitter 2, as well as the third and fourth locations of the hopper. The other four clusters represent false emitter locations. This case shows good performance for an input data stream containing a significant number of measurements corresponding to overlapping pulses. We expect the ability to filter out some of the false emitters; however, not much can be done when the input is quite noisy. For example, if just as many PDWs are received for a false emitter as a true emitter it is nearly impossible to distinguish them.

4.2 FPGA Resource Consumption

Table 1: FPGA Utilization

	4 clusters	8 clusters	16 clusters
Slice Regs	18834(32%)	26267(45%)	41146(70%)
LUTs	15166(26%)	22087(38%)	37500(64%)
BRAM	21(9%)	29(12%)	44(18%)
DSP48E	120(19%)	200(31%)	360(56%)

In addition to the *ICED* module, our implementation includes a logic framework to support a number of board interfaces. We include these components in our analysis. Table 1 shows the consumption of resources for implementations of 4, 8, and 16 cluster centers along with the percentage of total available resources used. There is a linear relationship between resource consumption and the number of implemented clusters. Extending to a zero cluster implementation gives a good estimate of the board support overhead combined with the ICED logic unaffected by the number of clusters implemented (normalization, inverse normalization, fade, etc). The limiting factor in increasing the number of clusters is the number of slice registers, but the number of LUTs and DSP48Es are not far behind. The theoretical limit would be 25 clusters based on near 100% utilization of LUTs.

4.3 Performance

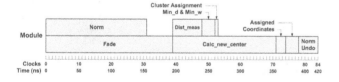

Figure 5: Module Processing Latencies

We measured performance of the 16 cluster implementation by analyzing the overall latency and by comparing processing between the FPGA and software implementations. To keep up with realistically stressing EW cases, the latency must be no greater than 500 ns. The clock frequency used by the *ICED* module is 200 MHz, which would require a latency no greater than 100 clock cycles. Fig. 5 shows a timeline consisting of the major processing steps. Stacked boxes represent operations which are performed in parallel. The 16 cluster implementation requires 84 clocks to complete, which corresponds to 420 ns.

In determining performance, MATLAB was used to generate the input PDWs. For software, we used a C implementation configured with the same parameters as the hardware. The result of the C implementation is an assignment of each

input to a cluster plus the value of all clusters. The software was run on a single core of a 3.00 GHz Intel Xeon 5160. The run time for 80,000 inputs was 13.0s, corresponding to $16.25\mu s$ per input. The FPGA hardware implementation was verified both in simulation using Xilinx ISE and in hardware on an Innovative Integration X5-400M module running at 200MHz. For FPGA hardware, the following steps are performed: (1) The host sends input PDWs to the HW (input FIFO) over the PCI Express bus. (2) The input is read from the FIFO, the fade is performed, the input is assigned and the cluster coordinates are updated. (3) The results (assigned cluster plus coordinates of all clusters) are read back to the host for post-analysis verification. The resulting FPGA speedup over software is 39x. Note that, in a working hardware scenario, we expect PDWs to be received directly on the FPGA from a pulse measurement module and the outputs would be streamed continuously to an EW processing module that could also be implemented in hardware. Therefore, the cost of host data transfers is not considered when determining the speedup of the FPGA hardware over the software implementation. Our FPGA implementation is able to keep up with PDWs at the rate that we expect to be receiving them from the environment.

5. CONCLUSIONS AND FUTURE WORK

We have implemented a novel incremental clustering algorithm that takes advantage of the strengths of FPGAs by clustering data received from sensors. ICED is the first implementation of incremental clustering on an FPGA of which we are aware. The algorithm is incremental to deal with streaming data, and retires clusters that represent stale data through a fading mechanism. We take advantage of the reconfigurability and flexibility of FPGAs in several ways. The number of clusters is configurable at compile time. Parameters including normalization range, fade cycle length, and maximum cluster radius are provided in user defined registers and can change at run time. The implementation is modular so that design decisions can easily be altered in future implementations. We apply ICED to an electronic warfare application: radar pulse deinterleaving. Our results show good performance, with a Xilinx Virtex 5 implementation of ICED running 39 times faster than the equivalent C program on a 3GHz Intel Xeon processor. We plan to apply ICED to the clustering of other applications with streaming, evolving data.

6. REFERENCES

[1] S. Bailie. An FPGA Implementation of Incremental Clustering For Radar Pulse Deinterleaving. Master's thesis, Northeastern University, 2010.

[2] M. Estlick, M. Leeser, et al. Algorithmic Transformations in the Implementation of K-means Clustering on Reconfigurable Hardware. In *Field Programmable Gate Arrays*, pages 103–110, 2001.

[3] W.-J. Hwang, C.-C. Hsu, et al. High speed c-means clustering in reconfigurable hardware. *Microprocessors and Microsystems*, 34:237–246, October 2010.

[4] T. Saegusa and T. Maruyama. An FPGA Implementation OF K-Means Clustering for Color Images Based on KD-Tree. In *Field Programmable Logic and Applications (FPL)*, pages 411–417, 2006.

X-ORCA: FPGA-Based Wireless Localization in the Sub-Millimeter Range

Matthias Hinkfoth
matthias.hinkfoth2@uni-rostock.de

Enrico Heinrich
enrico.heinrich@uni-rostock.de

Sebastian Vorköper
sebastian.vorkoeper@uni-rostock.de

Volker Kühn
volker.kuehn@uni-rostock.de

Ralf Salomon
ralf.salomon@uni-rostock.de

Faculty of Computer Science and Electrical Engineering
University of Rostock
18051, Rostock, Germany

ABSTRACT

Recent research has developed a new, entirely digital architecture, called X-ORCA, that determines the phase shift of two periodic signals with a resolution as good as about 20 ps. This paper incorporates the X-ORCA system into a wireless experimental setup to form a localization system. The practical experiments utilize a 2.484 GHz transmitter and run the X-ORCA core on a Cyclone II FPGA. The results indicate that this simple localization system easily yields a spatial resolution in the sub-millimeter range.

Categories and Subject Descriptors

B.7.1 [**Types and Design Styles**]: Algorithms implemented in hardware

General Terms

Experimentation

1. INTRODUCTION

In the field of FPGA-based time measurement, tapped delay lines (TDLs) [4] constitute a *de facto* standard. A tapped delay line is a chain of simple flip-flops that are connected to each other by means of active delay elements, such as inverters, NANDs, or the like. In this application domain, the duration of the interval is defined by two dedicated signals, called start and stop. In their simplest form, the clock inputs of all flip-flops are directly connected to the stop signal, whereas all the enable signals are indirectly connected to the start signal through the chain of active delay elements. That is, the start signal "ripples" through the chain of active delay elements, and every flip-flop determines which of the two signals, i.e., the stop signal or the delayed start signal, comes first. Depending on

the actual wiring, the chosen FPGA, and some other factors, tapped delay lines achieve a time resolution as small as 50 ps to 1 ns.

Previous research has developed a new time measurement architecture, called BOUNCE, that essentially substitutes the active delay elements with *passive on-chip signal wires*. BOUNCE is thus not limited by the technology-specific signal propagation time of the active delay elements. Rather, the propagation time is a universal, temperature-independent constant $c_w \approx (2/3)c_0$ that is approximately two third of the speed of light $c_0 \approx 3 \cdot 10^8$ m/s in copper, aluminum, and other sorts of metal. On a simple Stratix II FPGA, BOUNCE has achieved a time resolution of about 20 ps [9].

Starting off at some of BOUNCE's core principles, recent research [5] has build a new system, called X-ORCA, that is tailored to the measurement of the *phase shift* of two periodic signals. As Section 2 describes, X-ORCA is a highly parallel architecture that consists of a large number of rather rough phase shift estimators. By averaging over all estimates, X-ORCA finally yields quite a precise resolution. On a simple Cyclone II FPGA that is clocked at 85 MHz, X-ORCA yields a resolution of about 20 ps, if both input signals are transmitted over *wires*.

The relation between the phase shift $\Delta\varphi$ of two periodic signals and the localization of a transmitter is evident and straight forward: Depending on the sender's position, an emitted signal arrives at two or more receivers at different points in time, which can be observed as a phase shift. Conversely, the transmitter's location can be derived from a known phase shift, since both signals originate from a common source.

This paper investigates the performance of the X-ORCA architecture if incorporated into a wireless, radio frequency transmission system. This system is described in Section 3, and utilizes a 2.484 GHz transmitter in order to emit a 5 MHz localization signal. A potential application is the indoor localization of WLAN devices without modifications in infrastructure. Section 4 reports the practical results, which indicate that this localization system yields a spatial resolution in the sub-millimeter range.

The chosen experimental setup is a first laboratory test, which has certainly its shortcomings. These shortcomings, including potential reliefs, are discussed in detail in Section 5. This discussion also lays out some avenues for future research.

2. BACKGROUND: THE X-ORCA SYSTEM

Recent research has developed an architecture, called X-ORCA,

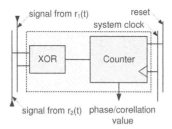

Figure 1: The digital phase shift (correlation) detector consists of a simple XOR gate and a counter. Since the XOR gate controls the counter's enable input, the counter value v_i is proportional to observable phase shift $\Delta\varphi$ and the total number of clock cycles.

that is tailored to the determination of the phase shift $\Delta\varphi$ between two periodic signals $r_1(t)$ and $r_2(t)$ that have frequency f. This section describes this architecture in detail as far as necessary for the understanding of this paper. For further details and wire-based results, the interested reader is referred to the literature [5].

The development of the X-ORCA architecture has drawn significant inspiration from the barn owl auditory system [6], which consists of a large number of neuronal coincidence detectors that are all connected to two distinct axonal structures that connect to the left and right ear, respectively. A key element of the two axonal connectors is that they are anti-parallel within the auditory system. That is, the signals from both ears travel into opposite directions within the auditory cortex. Therefore, all correlation detectors are exposed to different timing patterns, which directly depend on the *location* of the sound source. Conversely, the auditory cortex can derive the location of the sound source from the responses of the neuronal correlation detectors. In essence, the auditory cortex determines the neuron that has the highest activation, and yields its associated location (angle).

Unfortunately, neuronal correlation detectors do not have simple equivalents in digital hardware. However, in case of periodic signals with frequency f, a correlation detector can be approximated by a combination of an XOR gate and a counter, as is illustrated in Fig. 1. As is well known, the output of an XOR gate yields a logical one, if both inputs differ from each other, and a logical zero otherwise. Because of this behavior, the average duration $t = \Delta\varphi/(\pi f)$ of the logical one is proportional to the phase shift $\Delta\varphi$ of the two signals. This duration can be easily estimated, if the XOR gate is connected to the enable input of a counter that in turn is clocked at the sample frequency f_s. After a significantly long sampling period with a total of N_t tics, the final counter value $v = N_t\Delta\varphi/\pi$ is proportional to the phase shift $\Delta\varphi$.

Even though such a correlation detector (or estimator) works well, its accuracy would be clearly limited by the chosen sampling frequency f_s. The same holds for the neuronal correlation detectors of the barn owl auditory system. However, the barn owl overcomes this problem by employing a large number of correlation detectors along the two anti-parallel axonal "wires". These axonal wires induce additional internal time delays onto the sound signals.

Like its role model, X-ORCA also employs a large number of correlation detectors that are all connected to two anti-parallel wires as is illustrated in Fig. 2. The internal wires induce additional time delays δt to the two signals $r_1(t)$ and $r_2(t)$. Therefore, all correlation detectors observe different phase shifts $\Delta\varphi_i$, even though the system is fed with only one global unique phase shift $\Delta\varphi$.

Since the internal wire lengths are rather small, e.g., in the range of 200 μm, the additional time delays $\delta t_i \ll T_s = 1/f_s$ are way

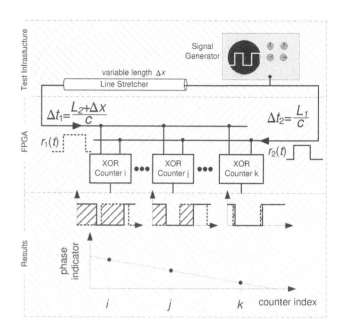

Figure 2: X-ORCA places all phase detectors along two reciprocal (anti-parallel) "delay" wires w_1 and w_2 on which the two signals $r_1(t)$ and $r_2(t)$ travel with approximately two third of the speed of light $c_w \approx \frac{2}{3}c_0$. Because the two wires w_1 and w_2 are *reciprocal*, all phase detectors have different internal delays δt_i.

smaller than the duty cycle of the sampling signal. This leads to two effects: (1) several neighboring phase detectors might have equivalent final counter values v_i, but (2) increasing the *external* phase shift $\Delta\varphi + \delta\varphi$ by a small fraction $\delta\varphi$ causes a few phase detectors to change their counter values, which can be detected by a subsequent processing stage that considers *all* counter values v_i.

The first X-ORCA prototype was implemented on a Cyclone II FPGA [2]. On the top-level view, the X-ORCA prototype consists of 140 phase detectors, a common data bus, a Nios II soft core processor [1], and a system PLL that runs at 85 MHz. The Nios II processor manages all the counters of the phase detectors, and reports the results to a PC. It should be noted that X-ORCA's internal "delay wires" w_1 and w_2 were realized as internal wires connecting the device's logic blocks, without additional active delay elements as previously announced.

The realization of the X-ORCA architecture on an FPGA requires a little more care, since most design tools are not able to generate a working instance right away: They remove all correlation detectors but one, or place them such that the internal wire lengths to the input pads are all identical. Therefore, the designer might mark all XOR gates and counters as relevant, place the detectors manually on the chip, and manually assign the signal paths in order to achieve a high variance of the internal delays δt_i.

During the first evaluation, the transmitter was realized as a simple function generator that emits a sinusoidal signal. In order to emulate different transmitter locations, the function generator was connected to the X-ORCA system via a regular wire as well as a line stretcher [8], as illustrated in Fig. 2. Such a line stretcher can be extended or shortened, and can thus change the signal propagation time accordingly.

In one set of experiments, the (localization) signal had a frequency of $f = 19$ MHz. With 140 counters, the system was able to detect a phase shift of $\Delta\varphi = \pm 0.34°$, which corresponds to a time delay of ± 50 ps. Figure 3 presents a compression of the entire

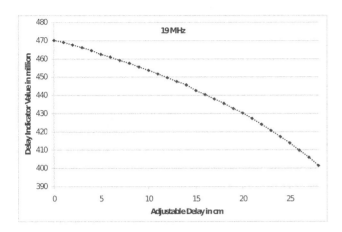

Figure 3: The figure shows the delay value indicator resulting from adjustable delay line lengths when fed with two 19 MHz signals.

set of experiments: Every dot represents the sum $v_{tot} = \sum_i v_i$ of all $n = 140$ counter values v_i; that is, all 140 counter values, which would appear as a line-like graph, are collapsed into one single dot. The graph shows 29 measurements in which the line stretcher was extended in steps of 1 cm. It can be seen that a length difference of $\Delta x = 1$ cm decreases v_{tot} by about 2 million. Even though the variation Δx was done in steps of 1 cm, the significant changes in the counter values *suggest* that the actual resolution might be even better. Further results reported in [5] have shown that the attainable resolution is at least 20 ps.

3. WIRELESS LOCALIZATION

As has been discussed above, X-ORCA is able to measure the phase shift between two signals with an accuracy of about 20 ps. During this short amount of time, an electromagnetic signal travels about 6 mm in air. Because of this fine-grained spatial resolution, this paper evaluates the X-ORCA architecture in a wireless experimental setup. The setup is illustrated in Fig. 4, and consists of a transmitter T_x, a receiving antenna R_1 with a fixed position, and a receiving antenna R_2 with a variable position. Modulation and demodulation are done by Lyrtech quad band RF transceivers combined with Lyrtech VHS-DAC boards [7]. The phase shift $\Delta\varphi$ between both signals $r_1(t)$ and $r_2(t)$ is measured by X-ORCA. Absorber walls helped to reduce multipath propagation caused by the omnidirectional antennas.

The ultimate idea is to directly utilize the high-frequency carrier signal for the localization purpose. However, since the available FPGAs cannot directly process 2.5 GHz signals, the experimental setup is using an $f = 5$ MHz low-frequency (localization) signal, which is modulated onto a $f_c = 2.484$ GHz carrier (channel 14 of the 802.11b standard). After the internal modulation, the antenna emits the following RF signal: $x(t) = \cos[2\pi(f + f_c)t]$. This signal appears at the receiving antennas with slightly different delays where it is demodulated with the same carrier.

A key point of the chosen setup is that the carriers are phase locked at both antennas of the receiver. In terms of communication theory this leads to the following analysis. At the transmitter side, the following complex low-pass signal (single side band) with $f = 5$ MHz is generated:

$$x_{LP}(t) = \cos(2\pi f t) + j\sin(2\pi f t) = e^{j2\pi f t} \quad (1)$$

Afterwards, the localization signal is modulated onto the carrier

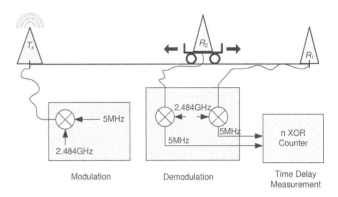

Figure 4: The wireless experimental setup consists of a transmitter T_x, a receiver with a fixed position antenna R_1, and with a variable position antenna R_2. The phase shift $\Delta\varphi$ between both signals $r_1(t)$ and $r_2(t)$ is measured by X-ORCA.

frequency of $f_c = 2.484$ GHz. The resulting RF signal is denoted by

$$x_{BP}(t) = \text{Re}\left\{x_{LP}(t) \cdot e^{j2\pi f_c t}\right\} = \cos[2\pi(f + f_c)t] . \quad (2)$$

At each antenna R_k, the signal

$$y_{BP,k}(t) = \cos[2\pi(f + f_c)(t - t_k)] + n_k(t) \quad (3)$$

is received. Since the two receiving antennas have different distances to the common transmit antenna, different propagation delays t_k occur. Due to the short distance between transmit and receive antennas, the noise $n_k(t)$ (see Eq. (3)) that results from electronic components can be neglected, since it is much smaller than the incoming signal.

The signals in Eq. (3) are demodulated by multiplication with the carrier $e^{-j2\pi f_c t}$ and are subsequently low-pass filtered. The resulting equivalent baseband signal is given by

$$y_{LP,k}(t) = e^{j2\pi f t - (f + f_c)t_k} . \quad (4)$$

Please note that the receiver's antennas are phase locked. As a result, the signals from both antennas are coherently shifted into the base band. Therefore, the phase difference between the two signals is maintained. The phase difference between both receiving antennas amounts to

$$\Delta\varphi = 2\pi(f + f_c)(t_1 - t_2) = 2\pi(f + f_c)\Delta t, \quad (5)$$

which depends on the delay Δt between the received signals and is proportional to the distance between both antennas assuming that all three antennas are located on a straight line. Measuring the phase difference in Eq. (5) delivers the distance between both antennas

$$\Delta s = c_0 \Delta t = c_0 \frac{\Delta\varphi}{2\pi(f + f_c)} \quad (6)$$

with c_0 denoting the speed of light.

4. RESULTS

In the practical tests, the second antenna was moved towards the first antenna in steps of 5 mm. Then, the evaluation was done over 100,000 clock cycles with a sampling frequency $f_s = 59.09$ MHz. Figures 5 and 6 summarize the results of the $n = 200$ counter values v_i. In Figure 6, the counter values v_i are compressed as has already been done in Fig. 3: every dot represents the average v_{avg} of all counters $v_{avg} = (1/n)\sum_i v_i$.

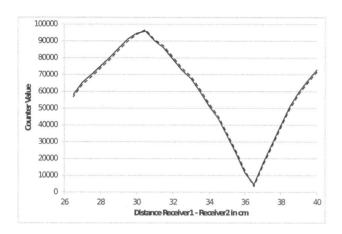

Figure 5: The counter values of two distinct counters, which indicate the phase shift $\Delta\varphi$ and thus the distance between the two receiving antennas as function of their real distance. The distance was changed in steps of 5 mm. The consideration of multiple counters removes the phase ambiguity within one period.

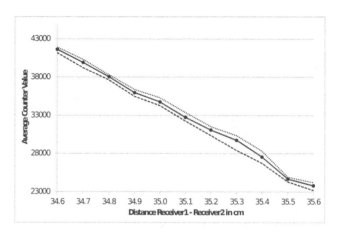

Figure 6: This figure shows a blow up of Fig. 5 between 34.6 and 35.6 cm. The distance was changed in steps of 1 mm. Three graphs are shown, which depict the average, the minimal, and maximal counter values of all ten runs.

This indicates that the localization setup is quite reliable. Figure 6 shows a blow up of Fig. 5 between 34.6 and 35.6 cm. In this blow up, the distance was changed in steps of 1 mm.

Figure 5 also shows that the effective range of the chosen configuration is 12 cm. It furthermore appears that the measurement is ambiguous: the values $v_{avg} = (1/n)\sum_i v_i$ are pairwise identical. This is, however, an artifact of the summation process of this representation. As indicated in Figure 5, the individual counter values are unambiguous, and thus allow for the precise reconstruction of the true distance.

Finally, it should be noted that the obtained resolution and measurement range are due to the properties of the X-ORCA system as well as the chosen quadrature amplitude modulation scheme.

5. DISCUSSION

This paper has briefly reviewed the X-ORCA architecture, which is tailored to the precise determination of the phase shift $\Delta\varphi$ of two periodic signals. This paper has also utilized the X-ORCA architecture in a wireless localization setup. the practical experiments have shown that the entire system can easily achieve a localization resolution of better than 1 mm.

The reported results have been achieved in the very first wireless laboratory experiments. These experiments were merely intended to provide a proof–of–concept. Therefore, the wireless components offer plenty of room for improvement. In order to avoid the consequence, such as fading, amplitude distortion, and echoes, due to multipath propagation, an absorber wall was used. Future research will be dedicated to enhancing the system's robustness in order to be applicable in everyday scenarios. A second issue concerns the localization signal. The first experiments have utilized a regular sinusoidal signal, which gives rise to ambiguities by its very nature. A potential relief consists in integrating pseudo-random codes, as are used, for example, in GPS. In addition, long distance measurements require a calibration method that takes e.g., cable lengths into account.

A second avenue of future research concerns the actual phase shift measurement system X-ORCA. The observable non-linearity in the results curves of both the wired and the wireless experiments reveals the need for further research. For example, its resolution

can be enhanced at least theoretically, if the simple averaging process is exchanged by a more elaborate process, which considers all counter values individually. This particularly concerns the full consideration of the internal time delays δt_i of every single correlation detector. However, this approach might require a more complex calibration process.

Finally, future research will be devoted to a seamless integration into dedicated wireless communication modules. One option for this approach seems to be the utilization of a software-defined radio module, such as the *Universal Software Radio Peripheral 2* [3].

APPENDIX

A. ACKNOWLEDGEMENTS

This work was supported in part by the DFG graduate school 1424.

B. REFERENCES

[1] Altera Corp., San Jose, CA. *Nios II Processor Reference Handbook*, 2007. Altera Document NII5V1-7.2.

[2] Altera Corp., San Jose, CA. *Nios Development Board Cyclone II Edition Reference Manual*, 2007. Altera Document MNLN051805-1.3.

[3] Ettus Research LLC, Mountain View, CA. *USRP2 datasheet*.

[4] C. Favi and E. Charbon. A 17ps time-to-digital converter implemented in 65nm FPGA technology. In P. Chow and P. Y. K. Cheung, editors, *FPGA*, pages 113–120. ACM, 2009.

[5] E. Heinrich, M. Lüder, R. Joost, and R. Salomon. X-ORCA - a biologically inspired low-cost localization system. In *10th International Conference on Adaptive and Natural Computing Algorithms, Part II*, pages 373–382, 2011.

[6] R. Kempter, W. Gerstner, and J. L. van Hemmen. Temporal coding in the sub-millisecond range: Model of barn owl auditory pathway. *Advances in Neural Information Processing Systems*, 8:124–130, 1996.

[7] Lyrtech, Quebec City, Canada. *Quad Dual Band RF Transceiver*.

[8] Microlab, Parsippany, NJ. *Line Stretchers, SR series*, 2008.

[9] R. Salomon and R. Joost. BOUNCE: A new high-resolution time-interval measurement architecture. In *IEEE Embedded Systems Letters (ESL), Vol. 1, No. 2*, pages 56–59, 2009.

Communication Visualization for Bottleneck Detection of High-Level Synthesis Applications

John Curreri, Greg Stitt, Alan George
NSF Center for High-Performance Reconfigurable Computing
Department of Electrical and Computer Engineering
University of Florida
{curreri, gstitt, george}@chrec.org

ABSTRACT
High-level synthesis tools increase FPGA productivity but can decrease performance compared to register-transfer level designs. To help optimize high-level synthesis applications, we introduce a bottleneck detection tool that provides a developer with a visualization of communication bandwidth between all application processes, while identifying potential bottlenecks via color coding. We evaluated the tool using third-party applications to identify and optimize bottlenecks in just several minutes, which achieved speedups ranging from 1.25x to 2.18x compared to the original FPGA execution. Overhead was modest with less than 2% resource overhead and 3% frequency overhead.

Categories and Subject Descriptors
B.5.2 [**Register-Transfer-Level Implementation**]: Design Aids – *Verification*

General Terms
Measurement, Performance, Design, Verification.

Keywords
High-level synthesis; visualization; performance analysis; bottleneck detection.

1. INTRODUCTION
High-level synthesis (HLS) tools [2][8] increase productivity by allowing developers to program field-programmable gate arrays (FPGAs) using higher abstractions than register-transfer-level (RTL) code. However, higher abstractions can result in performance bottlenecks that are challenging for developers to identify due to a lack of visibility into the synthesized circuit structure and runtime behavior.

Although software developers commonly use performance analysis tools to identify bottlenecks, there is a lack of such tools for FPGAs, especially for circuits generated from high-level synthesis. One of the key capabilities missing from existing FPGA performance-analysis approaches is visualization of communication bandwidth between different application processes. Without this information, a developer often must guess at the problem in order to perform optimizations, or has to add extra code to profile the circuit, which would be impractical for large applications. Alternatively, the developer must analyze the synthesized RTL code, which is not practical for many developers.

In this paper, we enable such visualization and bottleneck detection for high-level synthesis. The presented tool provides the developer with a high-level bandwidth visualization of all communication between application processes for both the CPU and FPGA and graphically identifies bottlenecks via color coding. To measure bandwidth, the tool automatically adds hardware counters and software timers to the application code, executes the application, and then uses the measurements to generate the communication visualization of the application. Optimizing the bottlenecks shown by the visualization required only several minutes of developer effort with speedups ranging from 1.25x to 2.18x compared to the original FPGA application with a modest 2% resource overhead and 3% frequency overhead.

2. RELATED RESEARCH
Visualization and analysis tools have been heavily researched [6] for identifying software bottlenecks. To our knowledge, there are few studies on performance analysis of HLS-generated circuits. Performance analysis has been developed for ASICs by Calvez et al. [1] and FPGA circuits by DeVille et al. [4], but neither study targets HLS tools. Related work exists for performance analysis [7] of HDL applications, but that work is not appropritae for HLS applications due to lack of source-code correlation and bandwidth visualizations. Previous work exists for performance analysis of HLS applications and visualization using a modified version of PPW [3]. The techniques in this paper complement this previous timing-based performance analysis by providing instrumentation for bandwidth measurement and communication visualizations.

3. COMMUNICATION VISUALIZATION
This section discusses the visualizations and corresponding bottleneck detection and instrumentation techniques.

3.1 Visualizations
High-level synthesis generates an application-specific graph of processes mapped to different devices (e.g., FPGA, CPU, memory), where edges between processes correspond to communication. The presented visualizations display this graph along with measured bandwidths for each edge.

Although the proposed techniques potentially apply to any high-level synthesis tool with constructs equivalent to parallel

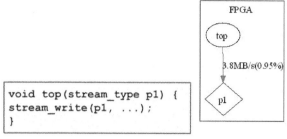

```
void top(stream_type p1) {
stream_write(p1, ...);
}
```

Figure 1: Streaming-communication call visualization

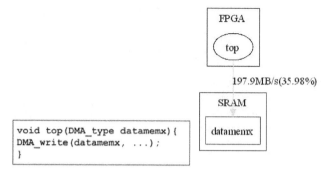

```
void top(DMA_type datamemx){
DMA_write(datamemx, ...);
}
```

Figure 2: Streaming-communication call visualization

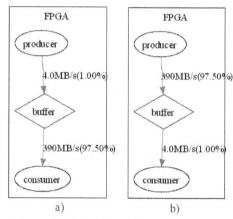

Figure 3: Example of (a) full and (b) empty streaming buffers.

processes, in this paper we target Impulse-C. In Impulse-C, communication channels fall into two categories: streaming and DMA transfers. Streaming transfers use buffers between two communicating processes. Direct memory access (DMA) transfers move data to/from dedicated memory such as SRAM.

The visualization tool initially constructs a directed graph based on the source code of the application. The tool visualizes resources such as CPUs, FPGAs, and memories using boxes. The tool visualizes processes as nodes (i.e., ovals) that are placed into the box of the corresponding resource. For each communication API call in the application code, the tool creates a corresponding edge between nodes. Streaming buffers are shown as diamonds.

Figure 1 shows an example of streaming communication in Impulse-C, along with the corresponding visualization of the process *top* writing to a stream buffer *p1*. Figure 2 shows an example of a DMA-communication call inside a function and the corresponding visualization of an FPGA processs *top* writing to an SRAM using a buffer *datamemx*. Memory used for DMA is shown as a separate box from the CPU or FPGA and the buffer is displayed as a separate box inside of the corresponding memory.

As shown in Figure 3, the tool annotates each edge with the measured bandwidths, as discussed in Section 3.3, in addition to the percentage of maximum possible bandwidth. Note that the percentages correspond only to the time that each process transfers data. For example, a process may execute for 100 cycles without data transfers, which would not affect the measured bandwidth. Therefore, a percentage of 100% does not necessarily suggest a saturated channel, but rather the efficient use of the channel. For example, if multiple processes read from a single memory at different times, both read channels could potentially achieve 100% bandwidth. To simplify bottleneck detection, the edges of the graph corresponding to data transfers are color coded with varying shades of red below 50% bandwidth and shades of green above 50% bandwidth.

3.2 Bottleneck Detection
This section describes analysis techniques to detect communication bottlenecks, which can be used manually or automated to suggest potential optimizations.

Streaming transfer bottlenecks can occur when a streaming buffer is full, in which case input bandwidth is lower than output bandwidth, as shown in Figure 3(a). Although such a bandwidth difference may be counterintuitive due to buffers normally being used to balance bandwidth, as stated in the previous section, the bandwidths only correspond to times that each process transfers data. In this situation, lower input bandwidth occurs because the producer process must block when the buffer is full, which increases the time for transfers and reduces bandwidth. To

optimize this bottleneck, a designer can increase the buffer size, which allows the producer process to execute for longer before the buffer fills. Alternatively, the consumer process can be optimized via pipelining, different stream widths, etc.

Empty stream buffers are another common bottleneck, which cause the consumer process to block, resulting in low output bandwidth, as shown in Figure 3(b). To optimize this bottleneck, the producer bandwidth should be increased using numerous possibilities including changing streaming widths, pipelining, or switching to a different communication method.

An additional bottleneck can be caused by low bandwidths on both sides of the streaming buffer. This bottleneck can be caused by multiple problems, such as streaming bursts of data into streaming buffers that become full and empty at different times. Section 4.2 gives an additional example. To reduce this bottleneck, a combination of both optimization methods can be used.

Memory buffers have common bottlenecks due to simultaneous transfers, in which case processes making DMA calls must block, resulting in decreased bandwidth. To reduce this bottleneck, synchronization can be added between processes to more effectively share bandwidth. Using small DMA transfer sizes can also cause bottlenecks, which can be optimized by sending larger chunks of data to increase bandwidth.

In some situations, DMA transfers to external memory can result in bottlenecks appearing on other edges. For example, for a process writing to memory, the physical bandwidth of the memory may become saturated, in which case streaming buffers upstream will eventually become full.

3.3 Instrumentation and Measurement

To measure bandwidth of each communication channel, the tool instruments the high-level code to measure the amount of data transferred and the total transfer time, which includes idle time as a result of blocking. In software, the tool adds a wrapper around each communication call type (e.g., stream read) that measures the time and records the transfer size.

FPGA processes are instrumented by adding monitoring circuits to each communication call. During execution, the monitoring circuits store measured bandwidths locally in registers, which are extracted by the microprocessor after the application has finished. For Impulse-C, the invocation of a communication API call is specified by a specific state in a state machine. Counters are used to monitor cycles spent in a particular call, from which total transfer time can be determined given the clock frequency.

Some communication calls have a static transfer size while others can change dynamically. For a static transfer size, the fixed size of each transfer can be parsed from the source code and the total data can be determined by the measured number of invocations. For dynamic transfer sizes, a counter is used to sum the total bytes transferred each time the process invokes the API call.

To measure the maximum bandwidth of a particular type of communication, the tool runs benchmarks that perform four types of transfers between processes for both streaming and DMA communication: CPU to CPU, from CPU to FPGA, from FPGA to CPU, and FPGA to FPGA.

4. EXPERIMENTAL RESULTS

Although ideally the proposed techniques would be integrated into a high-level synthesis tool, Impulse-C is proprietary, so we instead added instrumentation to the application code using Perl scripts. A Java GUI front end is used to select files and instrumentation features such as which processes and states should be monitored. The resulting visualization uses Graphviz.

We evaluated the techniques on two different platforms. The first platform was the XtremeData XD1000, which contains a dual-processor motherboard with an Altera Stratix-II EP2S180 FPGA in one of the Opteron sockets. The second platform was one node of the Novo-G supercomputer [5], which uses a GiDEL PROCStar III card with four Stratix-III EP3SE260 FPGAs. Impulse-C 3.3 is used for the XD1000 while Impulse-C 3.6 with an in-house platform support package is used for Novo-G.

4.1 Triple DES

Triple DES is a block cipher used for encryption. The application consists of a modified version of code provided by Impulse-C. We evaluated this application on the XD1000 platform.

Figure 4 shows the resulting visualization, which solely uses streaming communication. By analyzing the visualization, we identified a bottleneck (shown by the two arrows) corresponding to transfers between the CPU and FPGA. This bottleneck was caused by the low FPGA-to-CPU bandwidth relative to the FPGA internal bandwidth. The stream buffer *blocks_decrypted_ic* has a higher input bandwidth than output bandwidth indicating that streaming communication is blocked because the buffer is empty.

To optimize the application, we exploited DMA transfers to increase transfer rates between the CPU and FPGA, which enabled a 2.18x speedup and required about half an hour to

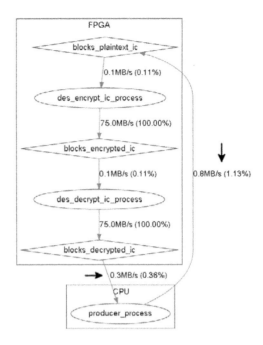

Figure 4: Visualization of streaming Triple DES.

complete (not including synthesis times). Logic overhead for the instrumented code was 2%. Clock frequency overhead was 3%.

4.2 Molecular Dynamics

Molecular Dynamics (MD) simulates interactions between atoms and molecules over discrete time intervals. For our experiments, the simulation computes forces of 16,384 molecules. Serial C MD code was obtained from Oak Ridge National Lab and optimized to run on the XD1000 FPGA using Impulse-C.

By analyzing the visualization in Figure 5, we identified a bottleneck resulting from all of the streaming buffers with the letter *p* becoming full. Also, the streaming buffers with the letter *a* have low bandwidth both on the inputs and outputs due to the streaming buffer becoming full and the *bottom* process blocking during stream reads. The *bottom* process requires data from all *a* streams to be ready simultaneously for its stream read state. If one *a* buffer becomes empty, then the other can become full while the *bottom* process waits for data.

To reduce the bottleneck, we increased the buffer size, which only required several minutes of effort and achieved a speedup of 1.25x. The speedup compared to the serial baseline running on the 2.2 GHz Opteron improved from 6.2x to 7.8x.

4.3 Backprojection

Backprojection is a DSP algorithm for tomographic reconstruction of data via image transformation. We evaluated this application on all four FPGAs of the ProcStar-III board in the Novo-G supercomputer. Although the figure is omitted for brevity, the resulting visualization showed an obvious bottleneck where PCI data transfers are 8-25% of peak speeds due to full streaming buffers.

This bottleneck could potentially be reduced by increasing the buffer size. However, we were unable to perform this optimization due to a lack of full Impulse-C support on Novo-G that limits streaming buffer sizes. Similarly, we could increase

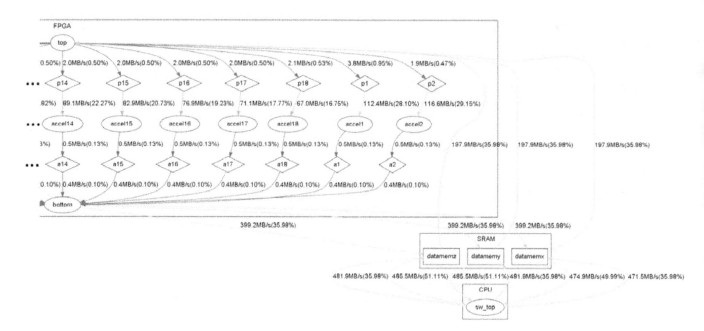

Figure 5: Bandwidth visualization for Molecular Dynamics (half of FPGA cropped to enlarge image).

bandwidths using DMA transfers, but the current Impulse-C support also excludes DMA transfers.

5. CONCLUSIONS

In this paper, we introduced a communication visualization tool for high-level synthesis that allows a developer to quickly locate communication bottlenecks. The application's processes and communication calls are visualized as a directed graph with edges between the CPU, FPGA and buffers. By analyzing the graph for bandwidth distribution and ratios, developers can identify bottlenecks and often quickly apply optimizations. Case studies showed the detection and optimization of bottlenecks, which resulted in FPGA application speedups ranging from 1.25x to 2.18x. In addition, the tool provides source-code correlation, which hides the high-level-synthesis-generated code from the application developer. Future work includes automating analysis and providing suggestions for optimization. The visualization can also be expanded by including performance analysis of computation for each process.

6. ACKNOWLEDGMENTS

This work was supported in part by the I/UCRC Program of the National Science Foundation under Grant No. EEC-0642422. The authors gratefully acknowledge vendor equipment and/or tools provided by Altera, Impulse Accelerated Technologies, Nallatech, and Xilinx.

7. REFERENCES

[1] Calvez, J.P. and Pasquier, O. Performance monitoring and assessment of embedded HW/SW systems. In *Proc. International Conference on Computer Design (ICCD)*, 52-57, 1995.

[2] Canis, A., Choi, J., Aldham, M., Zhang, V., Kammoona, A., Anderson, J. H., Brown, S., and Czajkowski, T.. Legup: high-level synthesis for FPGA-based processor/accelerator systems. In *FPGA '11: Proceedings of the 19th ACM/SIGDA International Symposium on Field Programmable Gate Arrays*, 33–36, 2011.

[3] Curreri, J., Koehler, S., George, A., Holland, B., and Garcia, R. Performance analysis framework for high-level language applications in reconfigurable computing. *ACM Transactions on Reconfigurable Technology and Systems (TRETS)*, 3, 1, (Jan. 2010), 1-23.

[4] DeVille, R., Troxel, I., and George, A.D. Performance monitoring for run-time management of reconfigurable devices. *International Conference on Engineering of Reconfigurable Systems and Algorithms (ERSA)*, 175-181, 2005.

[5] George, A., Lam, H., and Stitt, G.. Novo-g: at the forefront of scalable reconfigurable supercomputing. *Computing in Science Engineering*, 13, 1 (Jan-Feb 2011), 82-86.

[6] Heath, M. and Etheridge, J. Visualizing the performance of parallel programs. *Software, IEEE*, 8, 5, (Sep. 1991), 29-39.

[7] Koehler, S., Curreri, J., and George, A.D. Performance analysis challenges and framework for high-performance reconfigurable computing. *Parallel Computing*, 34, 4-5, (2008), 217-230.

[8] Villarreal, J., Park, A., Najjar, W., and Halstead, R.. Designing modular hardware accelerators in c with roccc 2.0. In *Field-Programmable Custom Computing Machines, Annual IEEE Symposium on*, 127-134, 2010.

CONNECT: Re-Examining Conventional Wisdom for Designing NoCs in the Context of FPGAs

Michael K. Papamichael
Computer Science Department
Carnegie Mellon University
Pittsburgh, PA, USA
<papamix@cs.cmu.edu>

James C. Hoe
Electrical & Computer Engineering Department
Carnegie Mellon University
Pittsburgh, PA, USA
<jhoe@ece.cmu.edu>

ABSTRACT

An FPGA is a peculiar hardware realization substrate in terms of the relative speed and cost of logic vs. wires vs. memory. In this paper, we present a Network-on-Chip (NoC) design study from the mindset of NoC as a synthesizable infrastructural element to support emerging System-on-Chip (SoC) applications on FPGAs. To support our study, we developed CONNECT, an NoC generator that can produce synthesizable RTL designs of FPGA-tuned multi-node NoCs of arbitrary topology. The CONNECT NoC architecture embodies a set of FPGA-motivated design principles that uniquely influence key NoC design decisions, such as topology, link width, router pipeline depth, network buffer sizing, and flow control. We evaluate CONNECT against a high-quality publicly available synthesizable RTL-level NoC design intended for ASICs. Our evaluation shows a significant gain in specializing NoC design decisions to FPGAs' unique mapping and operating characteristics. For example, in the case of a 4x4 mesh configuration evaluated using a set of synthetic traffic patterns, we obtain comparable or better performance than the state-of-the-art NoC while reducing logic resource cost by 58%, or alternatively, achieve 3-4x better performance for approximately the same logic resource usage. Finally, to demonstrate CONNECT's flexibility and extensive design space coverage, we also report synthesis and network performance results for several router configurations and for entire CONNECT networks.

Categories and Subject Descriptors

C.2.1 [**Computer-Communication Networks**]: Network Architecture and Design—*Packet-switching networks*

General Terms

Design, Experimentation, Performance

Keywords

Network-on-Chip, NoC, FPGA, System-on-Chip, SoC

1. INTRODUCTION

The rapidly-growing capacity of Field Programmable Gate Arrays (FPGAs), combined with the steady introduction of hardwired support for a multitude of diverse interfaces and functionalities, has promoted FPGAs to an attractive and capable platform for hosting even extended System-on-Chip (SoC) designs [9]. As the scale of designs targeting FPGAs grows, designers need a systematic and flexible Network-on-Chip (NoC) infrastructure to support communication between the tens and in the future potentially hundreds of interacting modules. In this paper, we present our investigation in synthesizable NoC designs specifically architected and tuned for FPGAs for use with the development of SoCs and other demanding systems applications, such as full-system prototyping [5] and high performance computing [4].

The research literature offers a large body of work on NoCs mapped onto FPGAs for the purpose of NoC simulation studies [25, 31] and for the purpose of prototyping and SoC emulation [24, 30, 15]. In these cases, the actual performance and efficiency of the originally ASIC-oriented NoC designs when mapped to FPGA is not a first-order concern; instead these prior works are motivated to instantiate largely unmodified ASIC-oriented NoC designs to ensure modeling fidelity. There have been only relatively few papers that point out FPGA-specific NoC design issues. We refer to them in our discussions in Sections 3 and 6.

Although the FPGA design flow and the ASIC design flow have much in common in their similar-looking RTL-based design and synthesis environments, they are in fact very different when it comes to making design decisions affecting cost and performance optimizations. A "literal" adaptation of an ASIC-optimized RTL-level NoC design on an FPGA will almost certainly prove to be suboptimal. What may be a compactly optimized router on an ASIC can incur a disproportionately high cost when synthesized for an FPGA because of the FPGA's very different relative cost trade-off between logic, wires and memory. Worse yet, ASIC-motivated optimizaitons will likely not be as effective due to the FPGA's also very different relative speeds in logic, wires and memory. FPGA design optimizations are further complicated by quantization effects because user logic and memory are realized using discretized underlying physical structures of fixed capacity and geometry.

In this work, we take full consideration of FPGAs' special hardware mapping and operating characteristics to identify their own specialized NoC design sweet spot, which we will show is very different from the conventional wisdom stemming from NoC designs on ASICs. Specifically, the considerations that have motivated this work to rethink NoC design for FPGAs are (1) FPGAs' relative abundance of wires compared to logic and memory; (2) the scarcity of on-die storage resources in the form of a large number of modest-sized buffers; (3) the rapidly diminishing return on performance from deep pipelining; and (4) the field reconfigurability that allows for an extreme degree of application-specific fine-tuning.

To support this investigation, we created the CONNECT NoC design generator that can generate synthesizable RTL-level designs of multi-node NoCs based on a simple but flexible fully-parameterized router architecture. The CONNECT NoC architecture embodies a set of FPGA-motivated design principles that uniquely influence key NoC design decisions, such as topology, link width, router pipeline depth, network buffer sizing, and flow control.

In the results section, we compare FPGA synthesis resource usage and network performance results for two instances of CONNECT NoCs to a high-quality state-of-the-art ASIC-oriented NoC design, to demonstrate the effectiveness of the FPGA-specialized tuning and features of the CONNECT NoC architecture. In addition, to highlight the flexibility and adaptability of the CONNECT NoC architecture, we also include synthesis and network performance results for a variety of diverse router and network configurations. Overall, the results of our investigation support that through FPGA specialization, we can gain approximately a factor of two savings in implementation cost without experiencing any significant performance penalty—in many cases, the CONNECT FPGA-tuned router design can actually lead to better performance at a lower implementation cost.

The rest of this paper is organized as follows. Section 2 provides a brief review of key NoC terminology and concepts. Section 3 introduces the motivations behind the CONNECT NoC architecture, and Section 4 presents the architecture of CONNECT-based routers and NoCs. In Section 5 we evaluate a CONNECT-based 4x4 mesh network against an equivalent NoC implemented using publicly available high-quality state-of-the-art RTL and show FPGA synthesis and network performance results for various CONNECT router and NoC configurations. Finally, we examine related work in Section 6, discuss future directions in Section 7 and conclude in Section 8.

2. NOC BACKGROUND

This section offers a brief review of key NoC terminology and concepts relevant to this paper. For a more comprehensive introduction please see [8]. Readers already familiar with NoCs may continue directly to Section 3.

Packets. Packets are the basic logical unit of transmission at the endpoints of a network.

Flits. When traversing a network, packets, especially large ones, are broken into flits (flow control digits), which are the basic unit of resource allocation and flow control within the network. Some NoCs require special additional "header" or "tail" flits to carry control information and to mark the beginning and end of a packet.

Virtual Channels. A channel corresponds to a path between two points in a network. NoCs often employ a technique called virtual channels (VCs) to provide the abstraction of multiple logical channels over a physical underlying channel. Routers implement VCs by having non-interfering flit buffers for different VCs and time-multiplexed sharing of the switches and links. Thus, the number of implemented

VCs has a large impact on the buffer requirements of an NoC. Employing VCs can help in the implementation of protocols that require traffic isolation between different message classes (e.g. to prevent deadlock [7]), but can also increase network performance by reducing the effects of head-of-line blocking [22].

Flow Control. In lossless networks a router can only send a flit to a downstream receiving router if it is known that the downstream router's buffer has space to receive the flit. "Flow control" refers to the protocol for managing and negotiating the available buffer space between routers. Due to physical separation and the speed of router operation, it is not always possible for the sending router to have immediate, up-to-date knowledge of the buffer status at the receiving router. In credit-based flow-control, the sending router tracks credits from its downstream receiving routers. At any moment, the number of accumulated credits indicates the guaranteed available buffer space (equal to or less than what is actually available due to delay in receiving credits) at the downstream router's buffer. Flow control is typically performed on a per-VC basis.

Input-Output Allocation. Allocation refers to the process or algorithm of matching a router's input requests with the available router outputs. Different allocators offer different trade-offs in terms of hardware cost, speed and matching efficiency. Separable allocators [8] form a class of allocators that are popular in NoCs. They perform matching in two independent steps, which sacrifices matching efficiency for speed and low hardware cost.

Performance characterization. The most common way of characterizing an NoC is through load-delay curves, which are obtained by measuring packet delay under varying degrees of load for a set of traffic patterns. A common metric for load is the average number of injected flits per cycle per network input port. Packet delay represents the elapsed time from the cycle the first flit of a packet is injected into the network until the cycle its last flit is delivered. For a given clock frequency, load and delay are often reported in absolute terms, e.g. Gbits/s and ns.

3. TAILORING TO FPGAS

Compared to ASICs, an FPGA is a peculiar hardware realization substrate because it dictates a very different set of design tradeoffs between logic, wires, memory and clock frequency. In this section we focus on specific FPGA characteristics and show how they have influenced fundamental CONNECT design decisions.

3.1 "Free" Wires

FPGAs are expected to be able to handle a wide range of designs with varying degrees of connectivity. Consequently, as previously also noted by other work [26, 20], FPGAs are provisioned, even over-provisioned, with a highly and densely connected wiring substrate. For the average application, this routing resource is likely to be underutilized. In these cases, one could view wires as plentiful or even "free", especially relative to the availability of other resources like configurable logic blocks and on-chip storage (flip-flops and SRAMs). This relative abundance of wires speaks against the conventional wisdom in NoC design where routers are typically viewed as internally densely connected components that are linked to each other through narrow channels that try to multiplex a lot of information through a small set of wires.

Implications. A NoC for FPGAs should attempt to make maximal use of the routing substrate by making the datapaths and channels between routers as wide as possible to consume the largest possible fraction of the available (otherwise unused) wires. Moreover, flow control mechanisms could also be adapted to use a wider interface, which as we show later, can indirectly also reduce router storage requirements. Design decisions such as widening the datapath can even have an indirect effect on issues like packet format. For instance, information that would otherwise be carried in a separate header flit could instead be carried through additional dedicated control wires that run along the data wires. Furthermore, on FPGAs (as well as ASICs actually), the boundaries between one router and another are not sharp—there is not a 10-foot cable separating them like in the old days. We will see later where the CONNECT NoC architecture allows the logic in one router to reach directly into another router for a more efficient implementation of buffer flow control.

3.2 Storage Shortage

Modern FPGAs provide storage in two forms: (1) SRAM macros with tens of kilo-bits of capacity, and (2) small tens-of-bits SRAMs based on logic lookup tables. In this paper, we refer to the former as Block RAMs and the latter as Distributed RAMs, following Xilinx's terminology. The bulk of the available storage capacity (in terms of bits) come in the form of a modest number of Block RAMs. These monolithic memory macros can not be subdivided. This leads to an inefficiency because a full Block RAM must be consumed, even if only a fraction of its capacity is required. Compared to Block RAMs, the Distributed RAMs are very expensive, especially when forming large buffers, since every Distributed RAM consumed is taking away from valuable logic implementation resources. This sets up a situation where NoCs on FPGAs pay a disproportionately high premium for storage because NoCs typically require a large number of buffers whose capacities are each much bigger than Distributed RAMs but much smaller than the Block RAMs. This premium has the consequences of not only limiting the scale of NoCs that can be practically built, but also reducing the resources available to the user logic.

Implications. Given the comparatively high premium for storage, a NoC tuned for FPGA should have a higher threshold for optimizations that increase buffer size in exchange for performance or functionality (e.g., number of VCs), especially when the increase requires consuming Block RAMs that are likely to be in high-demand from the user logic as well. The CONNECT NoC Architecture avoids using Block RAMs entirely and uses Distributed RAMs exclusively for its packet buffers. Furthermore, the choice of buffer sizing and configurations in the CONNECT NoC architecture takes into consideration the specific dimensions and sizes of the Distributed RAM building blocks to make the most efficient use of each consumed LUT.

3.3 Frequency Challenged

A design on an FPGA will operate at a much lower clock frequency than when implemented in ASIC; this was one of the gaps studied in [19]. First of all, Look-Up Tables used to implement arbitrary logic functions are inherently slower compared to fixed-function ASIC standard cells. Secondly, in order to emulate arbitrary logic blocks, FPGAs often need to chain a large number of LUT elements, which in turn requires using long interconnect wires. The time spent travers-ing these wires often ends up being the largest fraction of the critical path in FPGA designs.

Implications. From the perspective of this work, the most important implication from the difference in performance between ASICs and FPGAs actually manifests most strongly in the rapid diminishing return when attempting to deeply pipeline a FPGA design to improve its frequency. In most cases, beyond a small number of stages, it becomes impossible to further subdivide into balanced finer pipeline stages due to the quantization effects of the underlying realization structures and the difficulty in controlling physical details like logic placement, wiring routing, and driver sizing. We will see later that in fact, for FPGA synthesis, the single-stage router used in the CONNECT NoC architecture reaches lower, but still comparable frequency as an ASIC-tuned 3-stage-pipelined router. The FPGA's performance penalty from running at lower frequency is much more efficiently made up by increasing the width of the datapath and links. The shallow pipeline in the CONNECT NoC architecture has the added benefit of reducing network latency as well as greatly reducing the number of precious flip-flops consumed by a router.

3.4 Reconfigurability

The reconfigurable nature of FPGAs sets them apart from ASICs and creates unique opportunities and challenges for implementing an FPGA-oriented NoC. Given the flexibility of FPGAs, an effective NoC design is likely to be called to match up against a diverse range of applications. Fortunately, the NoC itself, making use of the same reconfigurability, can also go to an extreme degree of application-specific customizations that no one would ever consider for a design to be committed to ASICs.

Implications. Instead of a single design instance library IP, the CONNECT NoC Architecture relies on a design generator that can produce NoC instances specifically adapted to match the application or even the specific run-by-run workload. To cover the needs of such a diverse and rapidly changing set of applications, the CONNECT NoC generator is fully parameterized and more importantly topology-agnostic, which means that individual routers can be composed to form arbitrary custom network topologies. Moreover, to minimize changes in the user logic, all CONNECT networks adhere to the same simple standard common interface. From the user's perspective the NoC appears to be a plug-and-play black box device that receives and delivers packets. Rapid prototyping and design space exploration become effortless as any CONNECT network can be seamlessly swapped for another CONNECT network that has the same number of endpoints.

4. CONNECT NOC ARCHITECTURE

CONNECT-based NoCs are meant to be part of larger FPGA-based systems and, as such, must co-exist with the rest of the FPGA-resident components. This means that CONNECT NoCs need to balance between two conflicting goals: (1) provide sufficient network performance to satisfy the communication requirements of the target application; and (2) minimize the use of FPGA resources to maximize the resources available to the rest of the system. CONNECT addresses both goals by making the NoC implementation as efficient as possible, following the principles discussed in the previous section. When compared to ASIC-optimized NoC designs, in many places, the CONNECT NoC architecture goes directly against conventional NoC design wisdom. These differences can be attributed to two fundamental CONNECT router design decisions, which are summarized below:

- **Single pipeline stage.** Instead of the typical three to five stage pipeline found in most contemporary VC-based router designs, CONNECT employs a single stage router pipeline, leading to lower hardware cost, lower latency and opportunities for simpler flow control and more efficient buffer usage, due to the reduced round-trip time between routers.

- **Tightly-Coupled Routers.** Instead of treating the NoC as a collection of decoupled routers connected through narrow links, CONNECT tries to maximize wire usage, by using wider interfaces, leading to tighter coupling between routers. This includes carrying flit control information (that would traditionally be carried in separate header flits) on additional wires that run along the data wires. This decoupling is also the driving idea behind CONNECT's "peek" flow control mechanism, that allows routers to directly peek at the buffer occupancy information of their downstream receiving routers.

4.1 CONNECT Router Architecture

Driven by the special characteristics of FPGAs, we developed a simple router architecture to serve as the basic building block for composing CONNECT networks. Our router design was implemented using Bluespec System Verilog (BSV) [3], which allowed us to maintain a flexible parameterizable design. CONNECT routers are heavily configurable and among other parameters, they support:

- Variable number of input and output ports
- Variable Number of virtual channels (VCs)
- Variable flit width
- Variable flit buffer depth
- Two flow control mechanisms
- Flexible user-specified routing
- Four allocation algorithms

Router Datapath. Figure 1 shows the architectural block diagram of a CONNECT router. Communication with other routers happens through input and output port interfaces, which can vary in number depending on the router configuration. Each input or output port interface consists of two channels; one channel for sending or receiving data and one side channel running in the opposite direction used to handle flow control. Input and output interfaces are either connected to network endpoints or are used to form links with other routers in the network.

CONNECT routers are organized as a single-stage pipeline to minimize hardware costs, minimize latency and simplify flow control. During each clock cycle a router receives and stores new flits from its input ports and forwards previously received flits through its output ports. Upon entering the router, each incoming flit is first processed by the routing logic to be tagged with the proper output port and is then stored in flit buffers that are organized per input and per virtual channel. To determine which flits will be scheduled to depart from the router, arbitration logic considers flit buffer occupancy and credit availability to decide which flits will traverse the switch and be forwarded through the switch to the output ports. In addition to scheduling flits, the arbitration logic is also responsible for respecting VC

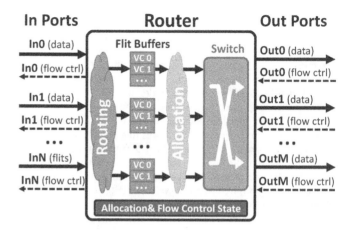

Figure 1: CONNECT Router Architecture

and port priorities, as well as preventing flits from different multi-flit packets from being interleaved on the same virtual channel (when virtual links are enabled).

Each individual router component is optimized to make the most efficient use of FPGA resources. Below we discuss implementation details about some specific router components of interest.

Packet Routing. Routing in CONNECT routers is handled by look-up tables that hold output ports for each possible destination in the network. Look-up based routing provides flexibility to construct arbitrary networks with custom routing. Even though routing look-up tables can grow to require a large number of entries for large networks, CONNECT implements them in an efficient manner by exploiting the geometry of FPGA Distributed RAMs. Each Distributed RAM is typically a single-bit wide memory element with 16 to 64 entries, depending on the specific FPGA family. Since routing tables tend to have many entries (one entry per network node), each being 2-3 bits wide (wide enough to encode a router output port), they map very efficiently to Distributed RAM and in almost all cases they occupy less than 10 LUTs. In many regular topologies, such as mesh or torus, these look-up tables could be easily replaced by topology-specific routing functions implemented in logic.

Flit Buffers. Flit Buffers in CONNECT are organized per input. CONNECT efficiently implements flit buffers using Distributed RAM by implementing multiple logical FIFOs, one per VC, in each single-read single-write Distributed RAM. Each Distributed RAM is split into fixed regions, and each VC-specific FIFO is implemented as a circular buffer within one of these regions. The head and tail pointers required to provide a logical FIFO abstraction occupy minimal area and are stored in discrete registers. This careful buffer space management, allows CONNECT to efficiently scale to large numbers of VCs.

Buffer Allocation. CONNECT supports four variations of separable input-output allocation algorithms [8]. The allocation module consists of two submodules; one that handles input arbitration and one for output arbitration. During each clock cycle the two input and output allocation submodules are triggered in sequence and the results of one submodule are fed to the other in order to produce a valid matching of eligible inputs with available outputs.

4.2 Highlights and Discussion

Below, we focus on some of the most interesting features of the CONNECT NoC Architecture.

Topology-agnostic. A major benefit of allowing any number of input or output ports and being flexible with respect to the routing algorithm is the ability to support arbitrary topologies. As long as flit widths match, CONNECT routers can be hooked to each other and form custom topologies that can better serve the needs of the application at hand. Similarly, all CONNECT networks that connect the same number of endpoints are interchangeable, which can greatly accelerate design space exploration.

Virtual Channels. In order to meet the diverse communication requirements of various applications, CONNECT has support for multiple VCs[1], which, as explained earlier, are implemented in a very FPGA-efficient manner. Multiple VCs are fundamental for ensuring deadlock freedom, implementing protocols that require traffic isolation between different message classes (e.g., memory requests and responses) and can also be used to increase network performance by reducing the effects of head-of-line blocking [22].

Virtual Links. In order to ease the implementation of receive endpoints in NoCs that use multi-flit packets and employ multiple VCs, CONNECT offers a feature called "Virtual Links". When enabled, this feature guarantees contiguous transmission and delivery of multi-flit packets. In other words, this guarantees that once a packet starts being delivered it will finish before any other packet is delivered. Enabling virtual links can cause a slight increase in hardware cost, but, in return, can significantly reduce the reassembly buffering and logic requirements at the receive endpoints.

Peek Flow Control. In addition to offering traditional credit-based flow control, CONNECT also supports another flow control mechanism, which we call "peek" flow control. Although effective, credit-based flow control can be inefficient in the context of CONNECT, as credit-based flow control is designed to: (1) tolerate long round-trip delays, caused by multi-cycle link latencies and deep router pipelines and (2) minimize the number of required wires between neighboring routers by multiplexing flow-control information pertaining to different VCs over the same set of wires.

In peek flow control, instead of having routers exchange credits, routers effectively expose the occupancy information of all of their buffers to its upstream sending routers. This way, instead of maintaining credits and using them as a proxy to determine how much buffer space is available at the downstream receiving routers, sending routers can directly observe the buffer availability. The peek flow control scheme reduces storage requirements by eliminating the multiple credit counters that are normally maintained for each output and VC pair.

CONNECT currently employs a single-bit version of peek flow control, which is very similar to stop-and-go queuing [14] or simple XON/XOFF flow control schemes [13]. The flow control information exposed by each router corresponds to a single bit per buffer that indicates if the specific buffer is full or not. If round-trip communication delay is high, such a simplistic scheme can severely under-utilize the available

buffer space. This is not an issue for CONNECT routers which only introduce a single cycle of delay.

5. EVALUATION AND RESULTS

To demonstrate the effectiveness of CONNECT's FPGA-centric design choices, we first compare FPGA synthesis results and network performance of two instances of a CONNECT-based NoC against a high-quality state-of-the-art ASIC-oriented NoC [29], both before and after modifying their ASIC-style RTL for efficient FPGA synthesis while maintaining bit and cycle accuracy to their original RTL. To further evaluate the CONNECT NoC architecture and highlight its flexibility and extensive design space coverage, we examine multiple router configurations, as well as entire CONNECT networks and report FPGA synthesis results and network performance results. Synthesis results include FPGA resource usage and clock frequency estimates for a moderately sized Xilinx Virtex-6 LX240T FPGA (part xc6vlx240t, speed grade -1) and a large Xilinx Virtex-6 LX760 FPGA (part xc6vlx760, speed grade -2). To assess network performance, we drive the NoCs with various traffic patterns and show the resulting load-delay curves.

5.1 Methodology

FPGA synthesis results are obtained using Xilinx XST 13.1i. Network performance results are collected through cycle-accurate RTL-level simulations. Each load-delay curve is generated through multiple simulations that sweep a range of different network loads. For each simulation, a traffic generator feeds traffic traces through the NoC endpoints and collects statistics as the packets are drained from the network. For each experiment the simulator is initially warmed up for 100,000 cycles, after which delay measurements are collected for 1,000,000 cycles. The duration of warmup and measurement periods were empirically set to be long enough to ensure that the reported metrics had stabilized.

5.2 Comparing to ASIC State-Of-The-Art

To put the FPGA hardware cost and network performance of CONNECT into perspective, we compare it against publicly available RTL of a high-quality state-of-the-art VC-based router [29], which we will refer to as SOTA. This router is written in highly-parameterized Verilog and is modeled after the VC-based router described in [8]. It employs a 3-stage pipeline and supports many advanced features, that are not present or not applicable in CONNECT, such as a larger collection of allocators or adaptive routing. The router supports single or multi-dimensional mesh and torus topologies, as well as the flattened butterfly topology [18].

In the presentation below, we first compare FPGA synthesis results for different configurations of a single router in isolation. We then compare a 4x4 mesh network built using CONNECT routers against a similarly configured 4x4 mesh composed of SOTA routers. Our comparison includes FPGA synthesis results, as well as network performance results under synthetic traffic patterns.

Router comparison. Since the original SOTA router RTL is ASIC-oriented and was thus not optimized for FPGA synthesis, to make the comparison more fair, we modified the SOTA router RTL by applying RTL coding discipline suitable for FPGA synthesis. In particular, our changes only affect storage elements; we ensured that all register files properly mapped to Distributed RAM, instead of discrete registers or Block RAM.

---

[1]In addition to user-exposed VCs (a.k.a. message classes), NoCs often also employ a number of internal VCs within each router to improve network performance. Such VCs are typically only visible and allocated within the network and are not exposed to the network clients. To reduce hardware cost, CONNECT exposes all VC handling to the network clients. As a result, applications seeking to use additional VCs for performance improvements need to manually handle VC allocation at the NoC endpoints.

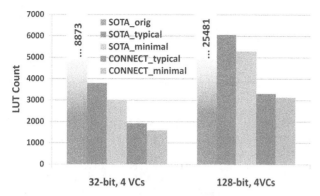

Figure 2: FPGA cost of SOTA and CONNECT 5-port router variants.

The bar graph in Figure 2 shows the difference in FPGA hardware cost in terms of the number of LUTs for a 32-bit 4-VC and a 128-bit 4-VC CONNECT and SOTA router. Both routers have 5 input and output ports, which corresponds to the typical configuration used in mesh or torus topologies. In all cases we configure the CONNECT and SOTA routers to be as similar as possible, including buffer depths, link widths, allocator type and number of VCs. For each design we examine two router variations: 1) typical, which uses SOTA's default parameter settings (for any remaining parameters) and 2) minimal, which picks those parameter settings that absolutely minimize hardware cost. Even though the minimal configurations do not necessarily constitute realistic design points, they provide a sense of the absolute lower-bound in terms of SOTA's hardware cost. We also include the results for the original RTL (SOTA_orig), before applying the above-mentioned FPGA coding style changes.

In the case of CONNECT, the only change between typical and minimal is the allocator choice; in minimal we use a variation of a separable output-first allocator that minimizes LUT count at the cost of lower router performance. In the case of the SOTA router, for the minimal configuration, we performed a sweep of all router parameters and picked the combination that yielded the lowest FPGA hardware cost.

First of all, it is interesting to note the vast reduction in the SOTA hardware cost from just applying proper RTL coding discipline, which ranges from 57% to 76%. After correcting for FPGA-coding style, the SOTA routers are still almost twice as costly compared to the equivalent CONNECT routers in terms of LUT usage. For typical configurations, CONNECT routers use between 40% to 50% fewer LUTs, for typical and minimal configurations respectively. It is worth mentioning that a 128-bit wide CONNECT router uses approximately the same amount or even fewer FPGA resources than its 32-bit SOTA counterpart. As we will see later, these "saved" FPGA resources can be used to build a more aggressive CONNECT NoC that uses approximately the same FPGA resources as a SOTA NoC, but can offer 3-4x higher network performance.

Mesh Network Comparison. To compare the two designs at the network level we use CONNECT and SOTA routers to build three 4x4 mesh networks with 4 VCs, 8-entry flit buffers and seperable allocators. Table 1 shows synthesis results for the resulting networks targeting Xilinx Virtex-6 LX240T and LX760 FPGAs. When configured with the same 32-bit flit width (SOTA and CONNECT_32), the SOTA network is more than twice as costly in terms of LUT usage, but can achieve approximately 50% higher

clock frequency. The potential performance loss due to the maximum clock frequency difference can be easily regained by adapting other CONNECT NoC parameters, such as flit width. To demonstrate this, we also include results for a 128-bit wide version of the CONNECT mesh NoC (CONNECT_128), that uses aproximately the same FPGA resources as its SOTA 32-bit counterpart, but offers three to four times higher network performance.

We should point out that the endpoints in a SOTA network are required to precompute routing information for each packet injected into the network. This incurs a small additional hardware cost that is, however, excluded from our reported results, since it affects the network endpoints and not the network itself.

4x4 Mesh w/ 4VCs	Xilinx LX240T		Xilinx LX760	
	%LUTs	MHz	%LUTs	MHz
SOTA (32-bit)	36%	158	12%	181
CONNECT_32 (32-bit)	15%	101	5%	113
CONNECT_128 (128-bit)	36%	98	12%	113

Table 1: Synthesis Results for CONNECT and SOTA Mesh Network.

To compare the example CONNECT and SOTA NoCs in terms of network performance, we examine the load-delay behavior of the networks under uniform random traffic, where the destination for each packet is randomly selected, and an instance of the unbalanced traffic pattern, where a fraction of the generated packets determined by the *Unbalance Factor* are local and are sent to neighboring nodes. In our experiments we set the *Unbalance Factor* to 90%, which represents a system where nodes communicate heavily with their neighbors and occasionally also send packets to other randomly chosen nodes in the system. We size packets to half the flit buffer depth, which corresponds to 4 flits, and pick the VC randomly for each injected packet.

Since NoCs are typically used within larger systems hosted on an FPGA, their clock frequency is oftentimes dictated by other components and constraints in the system. This is especially true in FPGA environments, where the clock frequency gains of running each component at its maximum frequency are likely to be outweighed by the added synchronization latency increase and hardware cost. To properly capture this potential frequency disparity, we report network performance results for both 1) assuming the studied NoCs are all running at a common clock frequency of 100MHz, possibly dictated by some other system component, and 2) assuming each NoC is running in isolation and can be precisely clocked at its maximum frequency, which provides an upper bound for performance.

All packets in the SOTA network require an additional header flit that carries control information, which brings the total number of flits per packet to five; one header flit and four data flits. CONNECT does not require this extra header flit; instead it carries flit control information "on the side" using wider links. Since the header overhead can change depending on the specific packet size, we also report the SOTA_raw curve, which eliminates SOTA's header overhead and captures raw flit throughput, providing an upper bound for the fully amortized performance of SOTA.

Figures 3 and 4 present load-delay curves for the CONNECT and SOTA networks all running at the same frequency of 100MHz under the two traffic patterns introduced above. Interestingly, when operating at the same frequency, even CONNECT_32, which shares the same 32 bit flit width

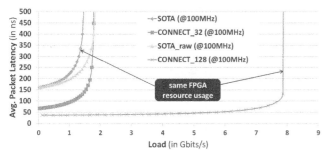

Figure 3: Load-Delay Curves for SOTA & CONNECT @ 100MHz with Unif. Random Traffic.

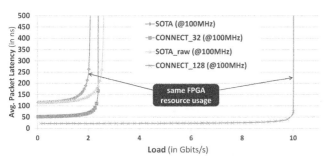

Figure 4: Load-Delay Curves for SOTA & CONNECT @ 100MHz with Unbalanced 90% Traffic.

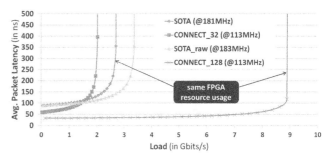

Figure 5: Load-Delay Curves for SOTA & CONNECT @ Max. Freq. with Unif. Random Traffic.

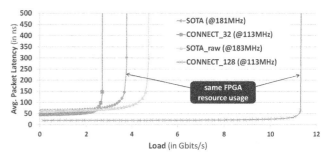

Figure 6: Load-Delay Curves for SOTA & CONNECT @ Max. Freq. with Unbalanced 90% Traffic.

with SOTA and occupies about half the FPGA resources, yields better network performance, both in terms of latency and saturation throughput. This is due to the additional header flit overhead on the SOTA network. When compared to SOTA_raw, which excludes the header overhead, CONNECT's performance is comparable to SOTA.

Figures 5 and 6 show the equivalent results when each network is running at its maximum clock frequency. In this case CONNECT_32 still offers significantly lower latency (59% lower) for the majority of operating loads, because of its reduced pipeline stages. At higher loads, as the load approaches the saturation point, SOTA outperforms CONNECT_32, due to its higher clock frequency.

However, notice that in all cases CONNECT_128, which occupies about the same FPGA resources as SOTA, can easily outperform all other networks by a wide margin across all traffic patterns and regardless of frequency adjustments; it consistently offers three to four times higher saturation throughput and more than three times lower latency.

Overall, for comparable configurations, CONNECT can offer similar network performance to SOTA with consistently lower latency at approximately half the FPGA resource usage. Alternatively, for the same FPGA resource budget, CONNECT can offer much higher performance than SOTA – three to four times higher saturation throughput and more than three times lower latency. In all cases the unbalanced traffic pattern, which consists of mostly local traffic, increases the saturation throughput across all networks, which is expected for the mesh topology that performs better under increased neighbor-to-neighbor traffic.

5.3 CONNECT Router Synthesis Results

As mentioned earlier, in addition to the mesh topology studied above, CONNECT supports a variety of different router and network configurations to better suit the diverse communication needs of emerging SoCs. To get a better feel for the cost and performance of different CONNECT-

based routers, Table 2 shows FPGA resource usage and clock frequency synthesis results for a range of different CONNECT router configurations targeting a Xilinx Virtex-6 LX760 FPGA. All reported results are for a single router to be used in a 64-node network. As expected, increasing the number of router ports, VCs, flit width or buffer depth all contribute to higher LUT counts and negatively impact clock frequency.

The number of router ports has the largest impact in hardware cost, followed by the number of VCs. Both of these parameters influence the buffering requirements, as well as the allocation and flow control logic. Changes in flit width and buffer depth only affect buffering requirements and as such have a lower relative impact. It is interesting to note that buffer depth affects LUT count in a more unpredictable manner due to quantization effects of Distributed RAMs; intuitively, wider memory arrays scale smoothly in terms of LUT cost, while taller memory arrays scale in a more abrupt step-wise manner.

5.4 CONNECT Network Synthesis Results

In this section, to demonstrate the flexibility and extensive design space coverage of CONNECT, we examine a few different examples of CONNECT-based networks in terms of hardware cost and network performance. Table 3 lists the selected network configurations, which range from a low-cost low-performance ring network (Ring16) all the way to a high-performance fully-connected network (HighRadix16), as well as an indirect multistage network (FatTree16). The number next to each network name indicates the number of supported network endpoints. The HighRadix16 network corresponds to a network with eight fully connected routers, where each router is shared by two network endpoints, i.e. with a concentration factor of two.

Table 4 shows synthesis results for these eight sample network configurations targeting a moderately sized Xilinx Virtex-6 LX240T FPGA, as well as a larger Xilinx Virtex-6

Flit Width		32 bits								128 bits							
Num. VCs		2 VCs				4 VCs				2 VCs				4 VCs			
Buf. Depth		4	8	16	32	4	8	16	32	4	8	16	32	4	8	16	32
2 In/Out	LUTs	242	292	340	485	373	427	564	952	562	612	659	936	693	739	1020	1861
Ports	MHz	306	284	247	218	260	240	217	195	306	283	245	221	260	232	219	193
4 In/Out	LUTs	688	813	893	1236	938	1137	1454	2408	1424	1550	1629	2230	1672	1872	2460	4310
Ports	MHz	180	183	154	143	169	167	147	139	180	183	154	143	166	165	147	139
6 In/Out	LUTs	1893	2005	2161	2812	2000	2351	2861	4399	3705	3839	4018	4987	4055	4442	5364	8439
Ports	MHz	150	142	130	126	122	123	115	109	149	140	127	122	122	124	117	110
8 In/Out	LUTs	3012	3171	3439	4149	3767	3953	4849	6565	5368	5544	5780	7035	6134	6322	7753	11280
Ports	MHz	117	114	103	101	107	102	103	95	117	114	102	101	107	102	103	94

Table 2: Synthesis results for various CONNECT router configurations.

Network	Routers	Ports/Router	VCs	Width
Ring64	64	2	4	128
DoubleRing16	16	3	4	32
DoubleRing32	32	3	2	32
FatTree16	20	4	2	32
Mesh16	16	5	4	32
Torus16	16	5	2	64
HighRadix8	8	8	2	32
HighRadix16	8	9	2	32

Table 3: Sample Network Configurations.

LX760 FPGA. For each network we report the LUT usage as a percentage of the total amount of LUTs on the respective FPGA, as well as synthesis clock frequency.

The synthesis results indicate that all networks easily fit within both FPGAs, with plenty of room to spare for placing many other pieces of user logic. In fact, when targeting the LX760 FPGA, all networks occupy less than 10% of the available LUTs. Finally, it is also worth mentioning that CONNECT networks do not occupy even a single Block RAM, which leaves a great amount of on-chip storage available to other FPGA-resident components.

	Xilinx LX240T		Xilinx LX760	
Network	%LUTs	MHz	%LUTs	MHz
Ring64	30%	175	9%	200
DoubleRing16	9%	139	3%	158
DoubleRing32	11%	146	4%	169
FatTree16	12%	117	4%	143
Mesh16	15%	101	5%	113
Torus16	25%	91	8%	100
HighRadix8	20%	73	5%	76
HighRadix16	28%	67	9%	75

Table 4: Synthesis results for sample networks.

5.5 CONNECT Network Performance

In this section, we focus on a subset of four networks (DoubleRing16, Mesh16, FatTree16 and HighRadix16), that all support 16 network clients, and as such would be interchangable when used as the interconnect within an FPGA-based system. To study network performance we use the same two traffic patterns described earlier, uniform random and unbalanced (with an UnbalanceFactor of 90%), which can be thought of as corresponding to two different classes of FPGA applications, each with different degrees of local communication. Once again, we size packets to half the flit buffer depth, which corresponds to 4 flits, and pick the VC randomly for each injected packet.

Figure 7 shows the load-delay curves for the four selected networks under uniform random traffic. Given the low bi-

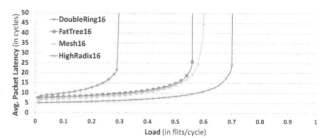

Figure 7: Load-Delay Curves for CONNECT Networks with Uniform Random Traffic.

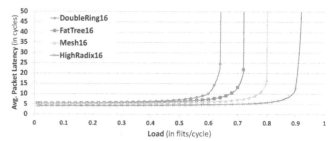

Figure 8: Load-Delay Curves for CONNECT Networks with Unbalanced 90% Traffic.

section bandwidth of the double ring topology, the DoubleRing16 network is the first to saturate at a load of approximately 30%. The Mesh16 and FatTree16 networks can sustain much higher loads before they saturate at roughly 55% load. This can be both attributed to the higher connectivity and bisection bandwidth of the mesh and fat tree topology, as well as the higher number of VCs in the case of the Mesh16 network (4 instead of 2). Finally, as expected, the HighRadix16 network achieves the highest performance, offering lower latency across all loads and saturating at a load of 70%. This should come as no surprise, given that the HighRadix16 network is fully-connected (maintains single-hop point-to-point links between all routers in the network), which means that the only source for loss of performance is output contention [8].

Figure 8 shows the equivalent load-delay curves under the unbalanced traffic pattern, which favors mostly neighbor-to-neighbor communication. As expected, the increased locality allows all networks to perform better, with the DoubleRing16 network experiencing the largest relative performance gains. In fact, under unbalanced traffic the DoubleRing16 network outperfoms the more FPGA resource intensive Mesh16 and FatTree16 networks.

Even though these results are mainly presented to demonstrate the flexibility of CONNECT and, as such, are not exhaustive or might omit other implementation details, such as frequency-related constraints, they do show that NoC performance can be highly dependent on network topology and configuration, but more importantly on the specific traffic patterns and requirements of an application. This observation is especially important in the context of FPGAs, where NoC topology and configuration can be easily adapted to suite the requirements of the given application.

6. RELATED WORK

Although there has been extensive previous work that combines FPGAs and NoCs, a large part of this work either examines the use of FPGAs for efficient NoC modeling [25, 31] or simply presents a larger FPGA-based design that also happens to include an application-specific ad-hoc NoC [30]. There is only a limited amount of previous studies that focus on FPGA-tailored NoC architectures to support SoC emulation or other FPGA applications.

In the context of FPGA-oriented NoC architectures, No-Cem [11, 12] presents a very simple router block that can be used to compose larger networks on FPGAs. Compared to CONNECT it lacks more advanced features, such as support for virtual channels or selectable allocation and flow control schemes. More importantly, for equivalent networks, compared to CONNECT, it appears to incur a much higher cost after a rough normalization for the differeneces in the FP-GAs used. (A more exact quantitiatve comparision is hard because the NoCem synthesis results were obtained on the much older Virtex-2 FPGAs.)

PNoC [15] is an interesting proposal for building lightweight networks to support FPGA-based applications. Even though PNoC can also yield low-cost FPGA-friendly networks, the fundamental difference compared to CON-NECT is that it can only be used to create circuit-switched networks, instead of packet-based. Circuit-switched networks can be useful for the classes of FPGA applications that have structured non-conflicting traffic patterns and are willing to tolerate the additional setup and tear-down delay and connection management associated with circuit-switched networks.

In the context of FPGA-related NoC studies, previous work has developed analytical models for predicting NoC performance on FPGAs [21], as well as examined the effect of various NoC parameters, such as topology and number of nodes, on the performance of an FPGA-resident multiprocessor system [20]. Morever, previous work has also studied the trade-offs between FPGA implementations of packet-switched and time-multiplexed networks [17].

Previous work has also looked at leveraging or modifying the FPGA configuration circuitry to build efficient FPGA-based NoCs. Metawire [28] overlays a communication network on top of the configuration network in a Virtex-4 FPGA. However, such an approach yields lower performance and is inevitably tied to the specific FPGA architecture and vendor. Francis et al. [10] propose replacing the statically configured FPGA wiring with time-division multiplexed wiring that can enable the implementation of efficient low-overhead NoCs for future FPGAs.

Finally, there is also a large body of commercial interconnect approaches, such as Spidergon STNoC [6], ARM's AMBA [2] or even FPGA-specific approaches, such as the CoreConnect [16] PLB and OPB buses, commonly found in Xilinx FPGAs, or Altera's Qsys [1]. CONNECT offers a more lightweight, fine-grain and flexible FPGA-tailored solution for building soft NoCs, that can synergistically coexist with the above approaches to cover the diverse communication needs of emerging SOCs.

7. FUTURE DIRECTIONS

As FPGAs continue to gain more traction as computing and SoC platforms, the role of NoCs will inevitably become more central in future FPGA-based systems. This can happen both in the form of efficient flexible architectures tailored for soft-logic implementations, such as the one presented in this paper, but can potentially also trigger the transition to future FPGA devices with fixed hardened NoCs.

Our immediate plan is to release a current version of CON-NECT in the form of a web-based flexible RTL NoC generator.i Moreover, we are interested in experimenting with a 2-stage pipeline router design that will yield improved clock frequency while still keeping FPGA resource usage at a minimum. Another interesting future direction is to study how we can apply the FPGA-oriented design guidelines and disciplines used in CONNECT to other common FPGA components in order to improve their efficiency.

As a longer term goal, we are interested in examining the form of future FPGA devices and their underlying switching fabric. Other researches have suggested that future FPGAs will consist of islands of reconfigurable logic connected through a dedicated high-performance NoC [27], which raises a few fundamental interesting questions: What will this NoC look like and how will its architecture and implementation be affected by the use of silicon interposers [23] in modern FPGAs? Which parts does it make sense to implement in hard-logic and which parts should be left to be implemented in soft-logic?

8. CONCLUSION

In this paper, we presented CONNECT, a flexible and efficient approach for building NoCs for FPGA-based systems. CONNECT embodies a set of design guidelines and disciplines that try to make the most efficient use of the FPGA substrate and in many cases go against ASIC-driven conventional wisdom in NoC design. We compare a high-quality state-of-the-art NoC design against our design both in terms of FPGA cost, as well as network performance for a similarly configured 4x4 mesh NoC. Across a wide range of configuration parameters, we find that CONNECT consistently offers lower latencies and can achieve comparable network performance at one-half the FPGA resource cost; or alternatively, three to four times higher network performance at approximately the same FPGA resource cost. Moreover, to demonstrate the flexibility and extensive design space coverage of CONNECT we report synthesis and network performance results for a wide range of router configurations and a variety of diverse CONNECT-based networks.

9. ACKNOWLEDGMENTS

Funding for this work was provided by NSF CCF-0811702 and NSF CCF-1012851. We thank the anonymous reviewers and Carl Ebeling for their useful comments that helped shape this paper. We thank the members of the Computer Architecture Lab at Carnegie Mellon (CALCM) and Daniel Becker from the Stanford CVA group for their comments and feedback. We thank Xilinx for their FPGA and tool donations and Bluespec for their tool donations and support.

Finally, we also thank Derek Chiou for organizing the 2011 MEMOCODE Contest that sparked the inspiration for this work.

10. REFERENCES

[1] Altera. Qsys System Integration Tool. http://www.altera.com/products/software/quartus-ii/subscription-edition/qsys/qts-qsys.html.

[2] ARM. AMBA Open Specifications. http://www.arm.com/products/solutions/AMBAHomePage.html.

[3] Bluespec, Inc. Bluespec System Verilog. http://www.bluespec.com/products/bsc.htm.

[4] E. S. Chung, J. C. Hoe, and K. Mai. CoRAM: An In-Fabric Memory Abstraction for FPGA-based Computing. In *Nineteenth ACM/SIGDA International Symposium on Field-Programmable Gate Arrays (FPGA)*, 2011.

[5] E. S. Chung, M. K. Papamichael, E. Nurvitadhi, J. C. Hoe, and K. Mai. ProtoFlex: Towards Scalable, FullSystem Multiprocessor Simulations Using FPGAs. *ACM Transactions on Reconfigurable Technology and Systems*, 2009.

[6] M. Coppola, R. Locatelli, G. Maruccia, L. Pieralisi, and A. Scandurra. Spidergon: A Novel On-Chip Communication Network. In *International Symposium on System-on-Chip*, 2004.

[7] W. Dally and C. Seitz. Deadlock-Free Message Routing in Multiprocessor Interconnection Networks. *IEEE Transactions on Computers*, 1987.

[8] W. Dally and B. Towles. *Principles and Practices of Interconnection Networks*. Morgan Kaufmann, 2004.

[9] P. Del valle, D. Atienza, I. Magan, J. Flores, E. Perez, J. Mendias, L. Benini, and G. Micheli. A Complete Multi-Processor System-on-Chip FPGA-Based Emulation Framework. In *IFIP International Conference on Very Large Scale Integration*, 2006.

[10] R. Francis, S. Moore, and R. Mullins. A Network of Time-Division Multiplexed Wiring for FPGAs. In *Second ACM/IEEE International Symposium on Networks-on-Chip (NOCS)*, 2008.

[11] G. Schelle and D. Grunwald. Onchip Interconnect Exploration for Multicore Processors Utilizing FPGAs. In *2nd Workshop on Architecture Research using FPGA Platforms (WARFP)*, 2006.

[12] G. Schelle and D. Grunwald. Exploring FPGA Network on Chip Implementations Across Various Application and Network Loads. In *International Conference on Field Programmable Logic and Applications (FPL)*, 2008.

[13] M. Gerla and L. Kleinrock. Flow Control: A Comparative Survey. *IEEE Transactions on Communications*, 1980.

[14] S. J. Golestani. A stop-and-go queueing framework for congestion management. In *Proceedings of the ACM symposium on Communications architectures & protocols*, SIGCOMM, 1990.

[15] C. Hilton and B. Nelson. PNoC: A Flexible Circuit-Switched NoC for FPGA-based Systems. *IEE Proceedings Computers and Digital Techniques*, 2006.

[16] IBM. The Coreconnect Bus Architecture. https://www-01.ibm.com/chips/techlib/techlib.nsf/products/CoreConnect_Bus_Architecture, 1999.

[17] N. Kapre, N. Mehta, M. deLorimier, R. Rubin, H. Barnor, M. Wilson, M. Wrighton, and A. DeHon. Packet Switched vs. Time Multiplexed FPGA Overlay Networks. In *14th Annual IEEE Symposium on Field-Programmable Custom Computing Machines*, 2006.

[18] J. Kim, J. Balfour, and W. Dally. Flattened Butterfly Topology for On-Chip Networks. In *40th Annual IEEE/ACM International Symposium on Microarchitecture (MICRO)*, 2007.

[19] I. Kuon and J. Rose. Measuring the Gap Between FPGAs and ASICs. *IEEE Transactions on Computer-Aided Design of Integrated Circuits and Systems*, 2007.

[20] J. Lee and L. Shannon. The Effect of Node Size, Heterogeneity, and Network Size on FPGA based NoCs. In *International Conference on Field-Programmable Technology (FPT)*, 2009.

[21] J. Lee and L. Shannon. Predicting the Performance of Application-Specific NoCs Implemented on FPGAs. In *Nineteenth ACM/SIGDA International Symposium on Field-Programmable Gate Arrays (FPGA)*, 2011.

[22] M. Karo; M. Hluchyj; S. Morgan. Input Versus Output Queuing on a Space-Division Packet Switch. In *IEEE Transactions on Communications*, 1987.

[23] M. Matsuo, N. Hayasaka, K. Okumura, E. Hosomi, and C. Takubo. Silicon interposer technology for high-density package. In *50th Electronic Components and Technology Conference*, 2000.

[24] M. Moadeli, A. Shahrabi, W. Vanderbauwhede, and P. Maji. An analytical performance model for the Spidergon NoC with virtual channels. *Journal of Systems Architecture*, 2010.

[25] M. K. Papamichael, J. C. Hoe, and O. Mutlu. FIST: A Fast, Lightweight, FPGA-Friendly Packet Latency Estimator for NoC Modeling in Full-System Simulations. In *Fifth IEEE/ACM International Symposium on Networks on Chip (NoCS)*, 2011.

[26] M. Saldana, L. Shannon, J. S. Yue, S. Bian, J. Craig, and P. Chow. Routability of Network Topologies in FPGAs. *IEEE Transactions on Very Large Scale Integration (VLSI) Systems*, 2007.

[27] H. Schmit. Extra-dimensional Island-Style FPGAs. In *International Conference on Field Programmable Logic and Applications (FPL)*, 2003.

[28] Shelburne, M.; Patterson, C.; Athanas, P.; Jones, M.; Martin, B.; Fong, R. MetaWire: Using FPGA Configuration Circuitry to Emulate a Network-on-Chip. In *International Conference on Field Programmable Logic and Applications (FPL)*, 2008.

[29] Stanford Concurrent VLSI Architecture Group. Open Source Network-on-Chip Router RTL. https://nocs.stanford.edu/cgi-bin/trac.cgi/wiki/Resources/Router.

[30] L. Thuan. NoC Prototyping on FPGAs: A Case Study Using an Image Processing Benchmark. In *International Conference on Electro/Information Technology*, 2009.

[31] D. Wang, N. Jerger, and J. Steffan. DART: A programmable architecture for NoC simulation on FPGAs. In *Fifth IEEE/ACM International Symposium on Networks on Chip (NoCS)*, 2011.

A Performance and Energy Comparison of FPGAs, GPUs, and Multicores for Sliding-Window Applications

Jeremy Fowers, Greg Brown, Patrick Cooke, Greg Stitt
University of Florida
Department of Electrical and Computer Engineering
Gainesville, FL 32611
{jfowers, rickpick, pcooke, gstitt}@ufl.edu

ABSTRACT

With the emergence of accelerator devices such as multicores, graphics-processing units (GPUs), and field-programmable gate arrays (FPGAs), application designers are confronted with the problem of searching a huge design space that has been shown to have widely varying performance and energy metrics for different accelerators, different application domains, and different use cases. To address this problem, numerous studies have evaluated specific applications across different accelerators. In this paper, we analyze an important domain of applications, referred to as sliding-window applications, when executing on FPGAs, GPUs, and multicores. For each device, we present optimization strategies and analyze use cases where each device is most effective. The results show that FPGAs can achieve speedup of up to 11x and 57x compared to GPUs and multicores, respectively, while also using orders of magnitude less energy.

Categories and Subject Descriptors

C.3 [**Special-purpose and Application-based Systems**]: Real-time and embedded systems, C.4 [**Performance of Systems**]: Design studies.

Keywords

FPGA, GPU, multicore, sliding window, speedup, parallelism.

1. INTRODUCTION

Over the past decade, computing architectures have started on a clear trend towards increased parallelism and heterogeneity, with most mainstream microprocessors now including multiple cores, and system architectures commonly integrating accelerators such as graphics-processing units (GPUs) [2][4][21] and field-programmable gate arrays (FPGAs) [3][6][10][28] over PCIe and even on the same chip [19][27]. Numerous studies have shown that such architectures can accelerate applications by orders of magnitude compared to sequential software [2][3][4][5][26].

However, the multitude of accelerator options has significantly increased application design complexity due to the need for extensive design-space exploration to choose an appropriate

device. Although GPUs have become a common accelerator due to widespread availability, low cost, and a simplified programming model compared to FPGAs, numerous device characterization [5][26] and application studies [2][3][4][20] have shown that metrics for different devices can vary significantly for different applications. Therefore, design-space exploration of different devices for different applications is critical to prevent designers from choosing inappropriate devices.

One challenge that makes such exploration difficult is that there is rarely a globally optimal device for a particular application. Instead, applications generally have a set of Pareto-optimal implementations that tradeoff numerous metrics such as performance, power, energy, cost, size, reconfigurability, application-design complexity, fault tolerance, etc. Furthermore, such exploration is complicated by numerous use cases. For example, an embedded system performing convolution may involve much smaller input sizes than convolution in high-performance computing, which would likely have different optimal or Pareto-optimal implementations.

In this paper, we perform an extensive analysis of sliding-window applications to determine the most effective devices for different use cases by considering performance and energy, different input sizes, different precisions, and different interconnects (e.g. PCIe, same chip). Sliding-window applications are a subdomain of digital signal processing that involve sliding a smaller signal (i.e., a window) across all positions in a larger signal (e.g., image), while generally performing a computationally intensive function at each window position. We evaluate sliding-window applications due to their frequent usage in digital signal processing, which is common on multicores, FPGAs, and GPUs.

The results show that an Altera Stratix III E260 FPGA is generally the fastest device for sliding-window applications compared to an NVIDIA GeForce 295 GTX GPU and quad-core Xeon W3520, with speedups of up to 11x and 57x, respectively. For an Information Theoretic Learning [24] based application, the FPGA was the only device capable of real-time usage. Furthermore, the FPGA used orders of magnitude less energy than other devices in many situations, providing the only realistic embedded system implementation for high-definition video.

The main contributions of the paper are summarized as follows:

- Highly optimized circuit architectures for FPGA implementations of sliding-window applications
- Optimization strategies for sliding-window applications on GPUs and multicores
- Analysis of performance and energy for different use cases such as different input sizes, precisions, and

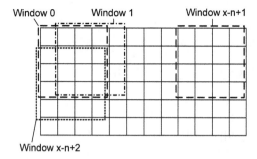

Window 0   Window 1   Window x-n+1

Window x-n+2

Figure 1: Input access patterns of sliding-window applications, where a "window" slides over all possible positions in the image.

```
Input: image of size x×y, kernel of size n×m
for (row=0; row < x-n; row++) {
  for (col=0; col < y-m; col++) {
    // get n*m pixels (i.e., windows
    // starting from current row and col)
    window=image[row..row+n-1][col..col+m-1];
    output[row][col]=f(window,kernel);
  }
}
```

Figure 2: Pseudo-code for typical sliding window applications, assuming fully immersed windows, where the window function f() varies depending on the application.

interconnect types, including estimations for emerging single-chip CPU/GPU devices.

- To our knowledge, we present the first sliding-window implementations demonstrated to achieve real-time usage with high-definition video, even while supporting larger window sizes than previous work.

The remainder of the paper is organized as follows. Section 2 discusses previous work. Section 3 describes the sliding-window applications that we evaluate. Section 4 describes the custom circuit architectures for our FPGA implementations, in addition to optimization strategies for the GPU and multicore implementations. Section 5 presents experimental results.

2. PREVIOUS WORK

Numerous studies have evaluated application performance for FPGAs [4][10] and GPUs [20][25]. Much work has focused specifically on image and video processing [4][15][23][25]. For example, Sinha et al. [25] evaluated the Kanade-Lucas-Tomasi (KLT) Feature Tracker algorithm on a GPU, which tracks specified features in a given image. Porter et al. [23] implemented several stereo matching algorithms on an FPGA, including sum of absolute differences (SAD). We also evaluate SAD, but using different use cases on different devices. [23] measured relative cost to perform real-time computations using a custom technology-independent cost function.

Previous work has also compared performances of FPGAs, GPUs, and CPUs. Baker et al. [3] evaluated a matched filter algorithm on a Cell processor, FPGA, and GPU, concluding that the Cell provided the best performance and energy efficiency, but the GPU exhibited the best performance per dollar. Pauwels et al. [21] compared two complex vision-based algorithms requiring real-time throughput. The multi-stage algorithm involved a Gabor filter, stereo disparity estimates, local image features, and optical flow. That study found that although FPGAs were faster for certain single-stage algorithms, the GPU exhibited better performance when executing the entire multi-stage algorithm.

Several studies have considered different use cases of some of the same applications as this paper. The authors of [6] implemented 2D convolution and color correction on a GPU and FPGA to determine if GPUs can replace FPGAs in video processing. The authors optimized both implementations, and measured throughput using kernel sizes up to 11x11 for the 2D convolution. They concluded that the FPGA had better performance at higher kernel sizes than the GPU. Our study differs by evaluating multiple sliding-window applications, including 2D convolution, while also considering different precisions, larger image and

kernel sizes that represent current use cases, and newer devices including multicore microprocessors.

Yu et al. [28] introduced an analytical approach to determine potential FPGA performance of sliding-window applications, while also creating on-chip buffers to exploit data reuse. In this paper, we evaluate custom FPGA circuits with similar buffering techniques for various common and emerging applications, while also comparing to GPUs and multicores.

In [2], Asano et al. studied image-processing techniques on a multicore CPU, GPU, and FPGA. The implementations included a 2D filter algorithm, SAD stereo vision disparity, and k-means clustering. The 2D filter's performance was measured up to a 15x15 kernel size, 241 SAD operations, and 48x4 distances in k-means clustering. In contrast to the SAD implementation in this paper, that previous study implemented a stereo-vision specific SAD algorithm. In that study, the FPGA had better performance for SAD and k-means, but was outperformed by the GPU for the 2D filter. The CPU outperformed the GPU in both SAD and k-means. Our study extends this previous work by providing a more in depth analysis of sliding-window applications. We present a generalized circuit architecture for sliding-window applications over a wider range of image and kernel sizes that apply to current and emerging use cases of sliding-window applications [12]. Additionally, we provide superior performance at significantly higher image and kernel sizes, and are the first to our knowledge to deliver real-time sliding-window processing of high-definition video on a single GPU or FPGA. We also evaluate a new application based on Information Theoretic Learning [24], which is an emerging area highly amenable to FPGA implementation.

3. SLIDING-WINDOW APPLICATIONS

For all applications, the input is a 2D *image* with dimensions $x \times y$. Although sliding-windows applications also apply to 1D examples and signals other than images, this input is representative of many applications. Each application also takes as input a 2D *kernel* of size $n \times m$, whose purpose varies depending on the application (e.g., an image to search for, a set of constants, etc.). Each application slides a *window* of the same size as the kernel across all possible positions in the image, as shown in Figure 1, where the data associated with each window are the underlying image pixels. For each window, the application performs some application-specific *window function*, shown as *f()* in Figure 2, based on the current window and the kernel. Although the number of outputs is application specific, sliding-window applications often generate one output per window, as shown in the pseudo-code. In some cases, the exact ranges of sliding windows are application specific. In this paper, we consider use cases where the kernel is fully immersed in the image, meaning that the window never exceeds the image boundaries. We chose this use case because windows that extend past image boundaries generally require a padded image. Such padding commonly requires software pre-processing that is not relevant to the device comparison, which we therefore excluded.

Note that for the remainder of the paper we use these terms:
- n and m: kernel dimensions
- x and y: image dimensions

Sliding-window applications tend to be highly memory intensive due to the need to gather each window. For example, a 40×40 window consists of 1,600 pixels. For a 1000×1000 image, there are $(1000-40+1)^2$ = 923,521 windows. Therefore, the total amount of pixels an application must read from memory is approximately 1,477,633,600. For 16-bit images, these reads correspond to approximately 3 terabytes of data, much of which must be accessed from memory non-sequentially.

Similarly, sliding-window applications are often computationally intensive due to complex window functions. For example, 2D convolution multiplies each pixel of a window with a constant in the kernel and then accumulates the results. For a 40×40 window, each window requires 1,600 multiplications and 1,599 additions. For a 1000×1000 image, there are 923,521 different windows, thus requiring approximately 3 teraoperations.

Many sliding-window applications have a similar behavior as shown in Figure 2. Therefore, in the following sections, we simply define the window function $f()$ for each application, along with characteristics of the image, kernel, and output.

3.1 Sum of Absolute Differences (SAD)

Sum of absolute differences (SAD) is used in content-based image retrieval [8][29] and other image-processing applications as a measure of similarity between two images. For example, a security system may search a video stream for other images (i.e., kernels) from a database of criminals. For each kernel, the output with the lowest SAD value represents the closest match.

As the name suggests, the window function for SAD calculates the absolute difference between each window pixel and kernel pixel, and then accumulates the differences for the entire window. Therefore, each output is a measure of similarity between the corresponding window and the kernel image, where lower values represent a closer match. The output upon completion is a two-dimensional data structure of dimensions $(x-n+1)\times(y-m+1)$, which corresponds to the total number of windows. Although specific applications would post-process the output, we instead simply generate the output due to the large variation in SAD applications.

For all evaluations, we consider 16-bit grayscale images for both the image and the kernel image, while exploring numerous combinations of image and kernel sizes.

3.2 2D Convolution

2D convolution is a common operation in digital signal processing and scientific computing used in computing systems ranging from small embedded systems to high-performance embedded computing systems (e.g., satellites) to supercomputers.

For 2D convolution, the window function multiplies each image pixel with a constant in the kernel. The method for generating each output pixel is shown by the following formula:

$$output[a][b] = \sum_{i=0}^{n-1} \sum_{j=0}^{m-1} image[a+i][b+j] \times kernel[n-i][m-j]$$

This formula multiplies each image pixel with a constant at the same location in a "flipped" version of the kernel. For a 3×3 window, 2D convolution multiplies pixel (0,0) of the window with the constant at (2,2) in the kernel, followed by the multiplication of pixel (0,1) with constant (2,1), etc.

After multiplying the window with the flipped kernel, 2D convolution accumulates the products, which generates a single output. The entire output for fully immersed windows is an image of size $(x-n+1)\times(y-m+1)$. For 2D convolution, we consider 16-bit grayscale images and kernels consisting of various precisions, including 32-bit floating point and 16-bit fixed point.

One optimization commonly implemented for 2D convolution consists of performing convolution twice with one-dimensional kernels, which is possible when the two-dimensional kernel is separable. In this paper, we evaluate non-separable kernels, due to their larger computational requirements.

Similarly, applications often perform large 2D convolutions using FFT convolution due to a lower time complexity. Although we do evaluate 2D FFT convolution for the GPU and multicore, we did not evaluate 2D FFT convolution on the FPGA analysis due to the lack of a 2D FFT core. Therefore, reported FPGA speedups represent a pessimistic lower bound. Additionally, we found that our 2D FFT multicore implementation, coded with the FFTW 3.2.2 2D FFT function [9], did not perform significantly better than the OpenCL sliding-window 2D convolution for the kernel sizes we tested. While we would expect significant speedup for larger kernels, we omitted the FFT CPU results from this paper because they are not applicable to our use cases.

3.3 Correntropy

Correntropy [14] is a measure of similarity based on Information Theoretic Learning (ITL) [24]. Correntropy can be used for many purposes [12][14], but we evaluate an application similar to Section 3.1 that searches an image to find the closest match of another image. Correntropy is defined as:

$$k(image_{i,j} - kernel_{a,b}) = \frac{1}{\sigma\sqrt{2\pi}} exp(-\frac{(image_{i,j} - kernel_{a,b})^2}{2\sigma^2})$$

For the application in this paper, function $k()$ is a Gaussian. Based on these equations, the correntropy application performs the following computation for each window. First, correntropy finds the difference between each pixel in the window ($image_{i,j}$) and each corresponding pixel in the kernel image ($kernel_{a,b}$). However, instead of summing these differences, each difference is used as a parameter to the Gaussian function. For an exact match (i.e., a difference of zero), the Gaussian function will return 1, which corresponds to a perfect measure of similarity. For larger differences, the similarity measure will drop increasingly fast depending on the exact characteristics of the Gaussian curve. After computing the similarity for each pixel of the window, the application sums the similarities for all pixels to create a single value that represents the similarity for the entire window. Although different applications would process the similarity values in different ways, the application in this paper outputs the the two largest values and their corresponding window locations.

4. DEVICE IMPLEMENTATIONS

In this section, we present implementation strategies for the three sliding-window applications from the previous section. Section 4.1 describes custom circuit architectures for the FPGA. Section

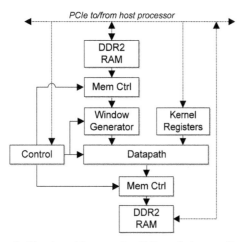

Figure 3: Circuit architecture for sliding window applications.

4.2 describes GPU implementations and optimizations. Section 4.3 is similar for OpenCL on multicore processors.

For the FPGA analysis, we target a GiDEL ProcStar III board with a 65 nm Altera Stratix III E260 FPGA. The board has four FPGAs, in addition to 3 external memories per FPGA, although we only use one FPGA to keep device comparisons fair. The board is connected over PCIe x8 to a 2.26 GHz quad-core Xeon E5520 CPU with 6 GB of RAM.

For the GPU analysis, we target a 55 nm EVGA GeForce GTX 295 PCIe x16 board with Compute Capability 1.3. This board also has multiple devices, but we limit analysis to a single device for fair comparisons. All implementations use CUDA Version 3.2. All GPU examples were tested using a Red Hat Enterprise 5 Linux 64-bit server with 12 GB of RAM and a 45 nm 2.67 GHz Intel Xeon 4-core W3520 with 8 threads via Hyper-Threading

The OpenCL multicore implementations use the same system as the GPU, but with Windows 7 Enterprise 64-bit instead of Linux in order to use the latest OpenCL Intel SDK Version 1.1.

Although evaluating an older 65 nm FPGA results in a slightly unfair comparison, the FPGA is still generally the most effective device for most use cases, as shown in Section 5. Also, although the slower processor used with the ProcStar III potentially makes the FPGA results pessimistic, the CPU was responsible for less than 1% of execution time.

For the SAD and correntropy applications, we limit window sizes to 45×45 due to shared memory limitations on the GPU. Interestingly, the FPGA implementations have resource restrictions that support only slightly larger windows. For 2D convolution, we restrict window sizes to 25×25 due to limited FPGA multipliers, as discussed later. The FPGA circuits can support arbitrary kernel sizes with trivial extensions, but we limit the analysis to sizes that the FPGA can execute in parallel.

4.1 FPGA Circuit Architecture

Because much of the sliding-window functionality is shared across multiple applications, all custom circuits used for the FPGA evaluation use the architecture shown in Figure 3. This architecture consists of a controller and pipelined datapath that takes as input a window and the kernel for the application.

To keep the pipelined datapath from stalling, the circuit must provide a new window every cycle, which requires very high

bandwidth. For example, a 40×40 window of 16-bit data requires 3,200 bytes per cycle, or 320 GB/s for a 100 MHz clock, which cannot be provided by external memory. However, the *window generator* buffers the overlap between consecutive windows inside of the FPGA, significantly reducing bandwidth requirements. Although previous studies have introduced various window generators [28][30], we use a buffer similar to [7] that aims to maximize performance at the cost of extra area.

The window generator buffers n-1 complete rows of the image using on-chip RAMs that act as specialized FIFOs. Like all FIFOs, these specialized FIFOs pop from the front and push to the back. However, pop operations do not actually delete the data and simply move the front pointer to the next element.

To use the window generator, the controller initially starts a sequential read from an external DDR2 memory that stores the image. The window generator stores arriving pixels in a FIFO corresponding to the current row. When the current FIFO is full (i.e., the entire row is buffered), the window generator starts storing pixels in the next FIFO. After the nth FIFO has received pixels, the window generator begins to create windows. Specifically, whenever there are pixels in all n FIFOs, the window generator pops a pixel from each FIFO into an $n×m$ set of shift registers used to store the current window. After the window generator pops m pixels from each FIFO, the shift registers contain a valid window. The window generator continues to pop pixels, producing a new window each cycle, until all the FIFOs are empty, which corresponds to the end of one row of windows. At this point, the window generator adjusts internal pointers to move each FIFO up one row, while moving the first FIFO to the back and discarding the contents because the remaining windows will not require data from the first row. Because the other FIFOs already contain the buffered data for a row of the image, the window generator resets the front pointer for each FIFO to the first pixel, effectively making the FIFOs full again without having to reread data from memory. After resetting the front pointers, the window generator repeats this process for the remaining windows.

To buffer n rows of the image, the window generator requires $n*y$ words of memory. For example, for a 1920×1080 image with 40×40 windows, the window generator requires 1920*40 memory words. For 16-bit grayscale images, these words require only 154 kilobytes of on-chip memory, which is a small amount for current FPGAs. Although other window-generation techniques use less area [28], those techniques trade off performance. Because on-chip memory was not a bottleneck for the evaluated applications, we used the described approach to maximize performance.

For each window, the datapath performs the application-specific window function using the window and the kernel, which the circuit also stores in $n×m$ registers. After some latency, the datapath produces an output each cycle, which a memory controller writes to a second external DDR2 memory.

The circuit connects to a PCIe bus that allows the host microprocessor to read and write data into the external memories, the kernel registers, and the controller. For all applications, the host software transfers the image into the input DDR2 memory and then initializes the kernel. Next, the software enables the controller to start the computation and then polls the controller until the datapath has produced all outputs. Finally, the software reads back all outputs from the second DDR2 memory.

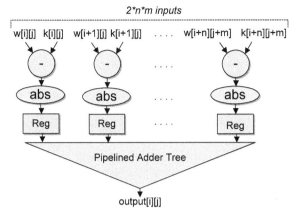

Figure 4: Datapath for sum of absolute differences (SAD).

4.1.1 SAD

For the SAD application, we created the datapath shown in Figure 4. The datapath takes $2*n*m$ inputs, where half of the inputs are the window pixels (shown as $w[]$), and the other half are kernel pixels ($k[]$). The datapath initially subtracts every corresponding pair of window and kernel pixels, and then takes the absolute value, which is stored in a register. The datapath then passes all $n*m$ absolute differences to a pipelined adder tree that contains registers at each level. The datapath then outputs the result from the adder tree. For all SAD evaluations, we use an image and kernel consisting of 16-bit grayscale images of varying sizes. The adder tree generates carry bits at each level of the tree to ensure that overflow cannot occur. Therefore, for 16-bit inputs, each output is $16+\log_2(n*m)$ bits. We could have potentially reduced area requirements by using 16-bit adders in the adder tree, but preventing overflow is important for many use cases and also provides a lower bound on FPGA performance.

The total number of parallel operations for the SAD datapath is $n*m$ subtractions, $n*m$ absolute values, and $n*m-1$ additions. For a 40×40 window, the datapath executes 1,600 subtractions, 1,600 absolute values, and 1,599 additions every cycle after the initial pipeline latency of $1+\log_2(n*m)$.

This SAD circuit, for a 1920×1080 image and 45×45 kernel, complete with all IP for the GiDEL board, uses 137,260 LUTs (67%), 156,377 registers (76%), 2,256,464 block memory bits (15%), and zero DSP blocks on the Stratix III E260. Resource utilization increases linearly with kernel size and the limiting resource is logic elements. The circuit was operated at frequencies between 100 and 115 MHz depending on the kernel size.

4.1.2 2D Convolution

The datapath for 2D convolution is similar to Figure 4, with the difference that the subtraction and absolute value operations are replaced by a multiplication. In addition, convolution flips the order of the kernel in the inputs as described in Section 3.2. The pipelined adder tree is identical to the description in the previous section. For the convolution examples, we evaluate 16-bit grayscale images, and kernels consisting of both 32-bit floating-point values and 16-bit fixed-point values. To reduce resource usage from fixed-to-float conversions, we pre-process the 16-bit image in software and transfer 32-bit float pixels to the FPGA.

The total number of parallel operations for this datapath is $n*m$ multiplications and $n*m-1$ additions every cycle. For the floating-point kernels, the output is also floating point. For the fixed-point

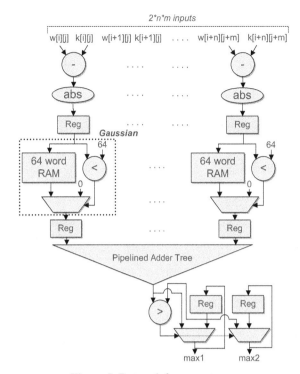

Figure 5: Datapath for correntropy.

kernels, the output is $16+\log_2(n*m)$ bits due to the adder tree accounting for overflow.

When using a 1920×1080 image and a 16-bit fixed-point 25×25 kernel, this circuit with all GiDEL IP uses 33,547 LUTs (17%), 57,122 registers (28%), 1,601,104 block memory bits (11%), and 738 DSP blocks (96%) on the Stratix III E260. For this circuit, the limiting resource is multipliers in the DSP blocks. The circuit was operated at frequencies between 104 and 115 MHz depending on the kernel size.

The floating-point version uses significantly more DSP blocks to achieve the same kernel size. The resources available allow up to a 13×13 kernel, and the circuit uses 129,024 LUTS (63%), 126,821 registers (62%), 1,633,872 block memory bits (11%), and 676 DSP blocks (88%). The limiting resource is DSP blocks and the circuit used frequencies between 103 and 114 MHz.

4.1.3 Correntropy

Figure 5 illustrates the correntropy datapath. The initial stages of the pipeline calculate the absolute difference of each pair of window and kernel pixels, which is identical to SAD. The datapath then connects the absolute difference to a lookup table that implements a Gaussian curve. The datapath uses the absolute difference to take advantage of the symmetry of the Gaussian curve. By ignoring negative values, we reduce the size of the lookup table by 50%. Choosing the exact size of the lookup tables is highly application dependent, but we chose a size of 64 words based on the curves required by a correntropy-based optical flow application [12]. In addition to the lookup table, the datapath uses a comparator and mux that treats points on the Gaussian for differences of greater than 64 as zero. The output of the Gaussian lookup is an 8-bit fixed-point value between 0 and 1 that represents the similarity between each pair of pixels. We chose the 8-bit precision based on the requirements of [12] and point out that other applications may require different precisions. These similarity values are then summed using the same pipelined adder

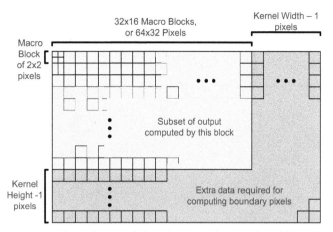

Figure 6. Organization of shared memory for each thread block.

tree as the previous examples. Finally, the datapath monitors the output of the adder tree and saves the two largest similarity values for all possible windows, along with the corresponding window positions (not shown). Each output is $8+\log2(n*m)$ bits.

The correntropy datapath performs $n*m$ subtracts, $n*m$ absolute values, $n*m$ Gaussian lookups, and $n*m$-1 additions every cycle.

For a 1920×1080 image and 45×45 kernel, the correntropy circuit, with all GiDEL IP, uses 141,633 LUTs (69%), 143,137 registers (70%), 2,256,464 block memory bits (15%), and zero DSP blocks on the Stratix III E260. Resource utilization increases linearly with kernel size, and the limiting factor is logic elements. The circuit was operated at frequencies between 101 and 111 MHz depending on the kernel size.

4.2 GPU

A graphics processing unit (GPU) is a highly parallel architecture which can run thousands of threads. An overview of GPU functionality is described in [20]. We use the CUDA framework [17] to implement applications for GPUs.

A complete discussion of the CUDA code is omitted for brevity. Instead, we focus on optimizations for the CUDA memory hierarchy, which previous work has shown can significantly improve performance [20]. The CUDA memory hierarchy consists of local memory, shared memory, texture memory, constant memory, and global memory. Ideally, threads should use local or shared memory, which have the lowest latency, but their limited size and their restriction of only sharing data within a thread block requires applications to also use the other memories. By contrast, global memory has the largest size due to the use of external memory, and can be accessed by all thread blocks, but also has the highest latency. Texture memory is a cached version of global memory, which is more suitable for 2D locality.

The presented GPU implementations use a specialized memory organization that maximizes the usage of shared memory for the numerous repeated accesses in sliding-window applications. This organization stores the entire kernel in shared memory, as well as a subset of the image needed by the corresponding thread block, while storing the entire image in texture memory rather than global memory due to a lower penalty for uncoalesced reads and improved access times from cache hits.

The basic functionality of the GPU implementations is described as follows, which is based on [15] and illustrated in Figure 6. Each thread block initially loads a subset of the image from

texture memory into shared memory, then generates the corresponding subset of the output pixels. Individual threads generate small groups of output pixels that we refer to as macroblocks, which are stored to global memory. Shared memory stores all image pixels used by a thread block, in addition to the kernel, which requires $(a+n-1)*(b+m-1)+n*m$ words, where $a×b$ are the dimensions of the output pixels generated by the thread block. For this paper, we determined output groups of 64×32 performed well on the targeted device. Therefore, each thread block uses $(64+n-1)*(32+m-1)+n*m$ words of shared memory.

Another consideration for the GPU is macroblock size. Smaller macroblocks increase threads per block, but increased thread count may also increase shared memory bank conflicts. We empirically determined that 2×2 macroblocks performed well on the targeted device for sliding-window applications, which differed from previous work [15] that used 8×8 macroblocks. Based on this macroblock size, each thread block consists of 32×16=512 threads, which is the maximum amount.

One limitation of this memory organization is that shared memory limits the maximum window size. For example, for the evaluated GPU, the shared memory supports window sizes up to 45x45. For larger windows, implementations must use other memories, which would likely significantly reduce performance.

The SAD application's threads compute the sum of absolute differences between the kernel and the four windows in the corresponding macroblock.

The 2D convolution threads are similar, but with each thread performing multiply accumulates as described in Section 3.2. For the GPU, we also evaluate a frequency-domain implementation of 2D convolution using a 2D FFT as described in [22], which used the CUFFT library [18]. The frequency-domain implementation computes the 2D FFT of the kernel and of the image, then performs a point-wise multiplication of the frequency-domain signals. The implementation then performs an inverse FFT on the resulting products to produce the output. Pre and post-processing is required to account for small kernel sizes and to extract the output desired. Note that we refer to the original time-domain implementation as sliding-window convolution and the frequency-domain version as FFT convolution.

The correntropy implementation for the GPU extends the SAD implementation by adding the intermediate step of taking the Gaussian of each absolute difference before accumulating. We optimized performance by storing the Gaussian function in a lookup table in shared memory, using the same size lookup table as the FPGA implementation. Absolute differences cannot be predicted here, so shared bank conflicts likely cannot be prevented, but we still expect shared memory to provide the best performance due to low latency.

One challenge with the GPU implementation of correntropy is locating maximum similarity values. On the FPGA this functionality only required two registers, a comparator, and muxes. However, for the GPU implementation, there is no way to communicate between thread blocks other than global memory, and there is no way to synchronize or guarantee the order in which thread blocks execute. Therefore, finding the maximum value from multiple thread blocks must take place after the sliding-window outputs have been produced. We implement this maximum function as a reduction problem, where each thread compares a subset of the outputs simultaneously. Each thread temporarily stores the results, which the implementation then uses for a smaller reduction. This reduction process continues until it computes the two maximum values. We implemented this reduction by altering a highly optimized reduction adder from NVIDIA [11].

4.3 Multicore

We used the OpenCL parallel programming standard [16] for the multicore implementations. Similar to CUDA's thread organization, OpenCL organizes threads into a 1, 2, or 3 dimensional grid called an NDRange. This NDRange is divided into work-groups, which are further divided into work-items. The work-items are the threads that run on a device, and each work-item has access to three types of memory (listed from greatest to smallest latency): global, local, and private. Global memory is available to all threads. Each work-group has local memory that is shared among threads in the group. Private memory stores individual thread data.

Like the CUDA implementations, leveraging the NDRange and memory hierarchy effectively are vital for optimizing OpenCL applications. To ensure good performance, we followed all guidelines specified in [13]. Since caching OpenCL memory objects on a CPU is managed automatically by hardware, managing the memory hierarchy is limited to coalescing memory accesses. As a result, we focused our optimizations on minimizing communication between threads. Each implementation uses the same following structure. The NDRange is a 2D grid with the same dimensions as the output. As recommended in [13], we stored the image, kernel, and output as buffers in global memory. Each work-item computes the result of one window. Unlike the GPU implementations, the OpenCL compiler automatically groups the work-items into work-groups as well as unrolls loops and vectorizes operations when applicable [13].

The implementation of each work-item was a straightforward specification of the window function for each application. The correntropy implementation used a global lookup table buffer for the Gaussian calculations. This lookup table was the same size as the one used for the GPU and FPGA. Like the GPU correntropy implementation, locating the maximum similarity values required a two-phase reduction where each work-item locates the maximum values for a section of the output.

5. EXPERIMENTAL RESULTS

The experiments section is organized as follows. We first define the experimental setup (Section 5.1). We then evaluate application performance individually for each device in terms of frames per second (Section 5.2), while also providing a speedup analysis. Next, we estimate speedup for emerging single-chip CPU/GPU systems and standalone FPGAs (Section 5.3). We then discuss the energy efficiency and performance in embedded systems (Section 5.4) of all implementations.

5.1 Experimental Setup

Details of the targeted systems are given in Section 4. We also estimated performance for emerging single-chip systems that integrate CPUs and GPUs, in addition to standalone FPGAs not requiring PCIe and a host processor. For simplicity, we refer to these GPU and FPGA systems as *single-chip systems*. We obtained upper-bound performance estimates for these systems by removing PCIe transfer times.

In addition to the implementations in Section 4, we also evaluate a sequential C++ implementation on the same microprocessor as

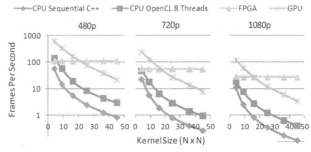

Figure 7. Performance of the SAD implementations measured in frames per second (1/execution time). Each chart corresponds to the results for all kernel sizes across one image size. The y-axis uses a log 10 scale for clarity.

the multicore examples, which we use as a baseline for speedup comparisons. These baseline implementations were compiled using g++ 4.1.2 with -O3 optimizations.

All implementations were evaluated for image sizes of 640×480 (480p), 1280×720 (720p), and 1920×1080 (1080p), which are common video resolutions. The SAD and correntropy implementations were evaluated at kernel sizes of 4×4, 9×9, 16×16, 25×25, 36×36, and 45×45. 2D convolution was evaluated at kernel sizes of 4×4, 9×9, 16×16, 25×25. Section 4 explains the maximum kernel sizes for each example.

5.2 Application Case Studies

In this section, we evaluate the performance of each application on each device in terms of frames per second (FPS). The frame rate is derived by inverting the execution time for one frame.

5.2.1 Sum of Absolute Differences

Figure 7 shows the frame rates for each implementation of SAD. All of the implementations were able to achieve real-time frame rates of 30 FPS or greater at small image and kernel sizes. However, the CPU rapidly decreased in performance. The GPU supported real-time usage when either the image or kernel size was small, but fell below 30 FPS for kernels larger than 25×25 in 720p and 1080p images. The FPGA was the only device able to maintain real-time performance over all of the input sizes tested.

The frame rates of the FPGA were constant across all kernel sizes for the same image size because the circuit computed one window each cycle regardless of kernel size. For larger kernel sizes that the circuit cannot compute in parallel, FPGA performance would decrease linearly as kernel size increased. FPGA frame rate decreased linearly with larger images, due to larger PCIe transfers

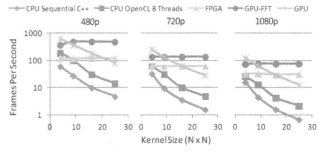

Figure 8. Performance of the 2D Convolution implementations measured in frames per second (1/execution time). Each chart corresponds to the results for all kernel sizes across one image size. The y-axis uses a log 10 scale for clarity.

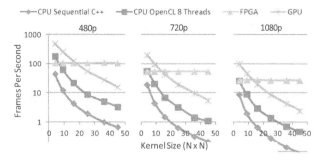

Figure 9. Performance of the correntropy implementations measured in frames per second (1/execution time). Each chart corresponds to the results for all kernel sizes across one image size. The y-axis uses a log 10 scale for clarity.

and the increased number of windows.

CPU and GPU frame rates decreased linearly with kernel size (width×height) and image size (width×height). The kernel and image sizes in Figure 7 each increase quadratically, causing the CPU and GPU graphs to decrease quadratically for each image size and $O(n^4)$ overall. This trend occurs because these implementations calculate every subtraction and addition using a limited pool of parallel resources that quickly becomes saturated as kernel size increases. The GPU always delivers a faster frame rate than the CPU running OpenCL, which in turn is always faster than the CPU sequential C++ baseline.

5.2.2 2D Convolution

The frame rates for each implementation of 2D Convolution, using 16-bit fixed-point kernels, are given in Figure 8. The trends for 2D convolution were similar to SAD, except on a smaller scale due to the more limited set of kernel sizes. As mentioned in Section 4.1.2, the FPGA supports a maximum window of 25×25 due to a shortage of multipliers.

The GPU-FFT and FPGA implementations were able to maintain frame rates over 30 across all input sizes tested. The two CPU implementations were only able to provide real-time performance for 4×4 and 9×9 kernels and had low frame rates overall for 1080p images. The GPU sliding-window (i.e., time domain) implementation provided the highest frame rates for 4×4 kernels and was able to deliver 30 FPS for all kernel and image size combinations except the maximum of 25×25 and 1080p.

It should be noted that the GPU-FFT implementation performs independently of kernel size when the kernel fits within the FFT size (i.e. the 1080p version could operate with a 128×128 kernel in the same amount of time as the 25×25 kernel).

2D convolution using floating-point kernels was also evaluated. The sequential C++ baseline took an average of 2x longer to use floating point. The OpenCL and GPU implementations performed within 5% of their execution times for 16-bit fixed-point kernels. The FPGA used an average of 20% more time for the same kernel sizes, due entirely to the additional cost of moving a 32-bit image over the PCIe bus as described in Section 4.1.2.

5.2.3 Correntropy

The FPS of each implementation of correntropy is given in Figure 9. The trends for correntropy and SAD were extremely similar.

The FPGA delivered real-time performance across all feature sizes at 480p and 720p, and 27 FPS for all kernel sizes at 1080p. The GPU provided more than 30 FPS for 25×25 and lower at 480p, 16×16 and lower for 720p, and 9×9 and lower for 1080p.

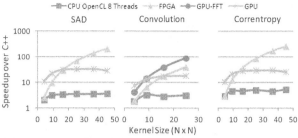

Figure 10. Speedup of all implementations over the sequential C++ baseline for SAD, 2D Convolution, and correntropy at all kernel sizes tested on 720p images. A log 10 scale is used on the y-axis for improved clarity.

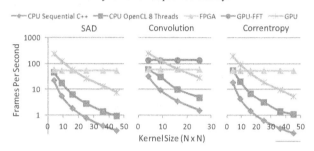

Figure 11. Performance of all implementations in frames per second for SAD, 2D Convolution, and correntropy at all kernel sizes tested on 720p images. A log 10 scale is used on the y-axis for improved clarity.

The CPU implementations only provided real-time frame rates at the smallest kernel sizes in 480p and 720p.

5.2.4 Discussion

Performance, as indicated by speedup over the CPU sequential C++ implementation for each application, is shown for 720p images in Figure 10. The trends for 480p, 720p, and 1080p were similar enough that it is only necessary to display one image size.

The data shows that, for each application, the sliding-window GPU implementation (i.e., time domain) was faster than the FPGA for 4×4 and 9×9 kernel sizes and roughly equivalent in the 16×16 case. The FPGA gained significantly over the GPU at 36×36 and larger kernels, reaching a maximum speedup over the baseline of 240x, 45x, and 298x for SAD, 2D convolution, and correntropy, respectively. While the GPU had nearly constant speedup, the FPGA increased its speedup linearly with kernel size due to its kernel-size independent performance.

CPU OpenCL implementations provided steady speedup over the baseline, with a maximum of 3.9x, 3.7x, and 5.3x for SAD, 2D convolution, and correntropy, respectively. This consistency was supplied by performing similar operations spread out over the 4 CPU cores. OpenCL was marginally faster than the FPGA at 4x4 kernels and significantly slower at all other sizes.

The GPU-FFT implementation for 2D Convolution was faster than the FPGA for all kernel sizes tested, with an average of 3x better performance than the FPGA. As mentioned previously, a FFT implementation on the FPGA may reduce this speedup.

The data in Figure 11 shows that performance can vary by application for each device, despite each application sharing the same basic structure and memory access pattern. The sequential C++ CPU took an average of 1.7x longer to execute SAD than 2D convolution because of the extra steps for calculating the absolute

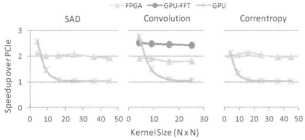

Figure 12. Speedup of single-chip implementations over their PCIe equivalents for SAD, 2D Convolution, and correntropy at all kernel sizes tested on 720p images.

value. Correntropy took significantly longer for the sequential C++ than either SAD or 2D Convolution because of the additional step of accessing the Gaussian lookup table. Additionally, tracking the maximum value required extra comparisons. The same trends apply to the CPU OpenCL implementation.

The FPGA implementations took nearly the same amount of time to execute regardless of the sliding-window function because the pipelined architecture amortizes any extra steps as latency without affecting throughput. The GPU implementation for SAD executed slightly faster than the correntropy implementation because the Gaussian lookups and the comparisons for establishing the maximum output became an expensive reduction operation, as mentioned in Section 4.2.

5.3 Single-Chip Systems
Figure 12 presents speedup of emerging single-chip CPU/GPU devices in addition to standalone FPGAs over traditional PCIe accelerators, which we collectively refer to as single-chip systems. The graphs in Figure 12 are limited to 720p images only because the trends for all image sizes were similar.

The results show significant improvements compared to accelerator boards due to the elimination of PCIe transfer times, which accounted for as much as 65% of execution time for the GPU and 64% for the FPGA.

The single-chip, sliding-window GPU implementations experienced the greatest speedup at low kernel sizes, which resulted from low computation compared to data set size. The speedup decreased quickly as data transferred over the PCIe x16 bus was amortized against quadratically larger computations.

The standalone FPGA implementations had a nearly constant speedup averaging 2x over PCIe versions. Speedup was constant because the execution time did not change as kernel size changed, leading to no amortization of data-transfer times. The GPU-FFT convolution implementation followed the same trend.

It should be noted that as of this writing, the available CPU/GPU chips on the market, such as the AMD Fusion APU and NVIDIA Tegra 2, do not come close to the performance of the NVIDIA GeForce GTX 295 used in these experiments. Still, the severity of the PCIe bottleneck points to huge potential for devices that hardware can accelerate without bus transfers.

5.4 Energy Comparisons
To evaluate energy consumption, we calculate the energy for a given implementation by multiplying the execution time of each implementation by the reported worst-case power consumption of the corresponding device. Although such an analysis may be pessimistic, sliding-window applications are likely to reach these

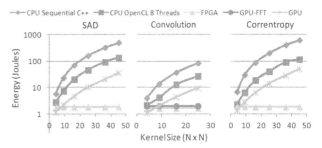

Figure 13. Energy consumed to process one frame for SAD, 2D Convolution, and correntropy at all kernel sizes tested on 720p images. A log 10 scale is used on the y-axis.

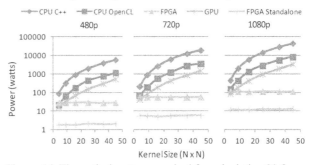

Figure 14. Theoretical wattage required for calculating 30 frames per second for correntropy. A log 10 scale is used on the y-axis.

worst-case power levels due to their memory-intensive and computation-intensive behavior. The CPU implementations have a worst case power of 130 watts, the GPU implementations use 274.5 watts (130 watt CPU + 144.5 watt GPU), and the FPGA uses 100 (80 watt CPU + 20 watt FPGA) [1]. When used as a standalone device the FPGA consumes 20 watts.

The data in Figure 13 shows the stratification between the energy efficiency of each device. The FPGA was clearly the most energy-efficient device, with one and two orders of magnitude lower energy than the sliding window GPU and CPU, respectively, at the 45×45 kernel size. The GPU-FFT implementation was able to obtain comparable energy efficiency to the FPGA for convolution because of its better performance.

Next, we evaluate the amenability of each device for real-time embedded systems usage by determining the theoretical power required to provide 30 frames per second. *Note that many of the devices were not capable of providing such performance, causing the resulting power to far exceed the worst-case power of the device.* We calculate this data by multiplying the energy for 1 frame by 30 frames per second (FPS). Figure 14 presents the power analysis for correntropy, which was selected for its applications in resource-limited embedded systems [12]. The results show that an embedded system using correntropy for target tracking under a realistic power budget can only be achieved using an FPGA, as the other devices required orders of magnitude more power for larger kernels sizes. In addition, the wattage for non-FPGA systems was optimistic because those implementations were not capable of providing 30 FPS without parallelizing across multiple devices, for example 2 GPUs in an SLI configuration.

The FPGA was able to produce 30 FPS correntropy results for 2, 5.5, and 12 watts for 480p, 720p, and 1080p, respectively. The CPU and GPU required several orders of magnitude higher power, using a *theoretical* 8 kW and 3 kW, respectively, for the 1080p 45×45 case.

A system consisting of a standalone FPGA is practical for this application because the correntropy architecture described in Section 4.1.3 is capable of receiving data directly from a camera with the same Stratix III E260 used in these experiments. By contrast, the single-chip GPU estimation is excluded from this comparison because current state-of-the-art embedded GPU solutions, such as the NVIDIA Tegra 2, do not come close to the GeForce GTX 295 in performance and do not support CUDA as of this writing. We plan such analysis as future work.

6. CONCLUSIONS

In this paper, we compared performance and energy of sliding-window applications when implemented on FPGAs, GPUs, and multicore devices, under a variety of different use cases. For most cases, the FPGA provided significantly faster performance, except for small inputs sizes, with speedups up to 11x and 57x compared to GPUs and multicores, respectively. GPUs provided the best performance when the basic sliding-window functionality could be replaced by frequency-domain algorithms. FPGAs provided the best energy efficiency in almost all situations, and were in some cases orders of magnitude better than other devices. For large input sizes, FPGAs were the only device capable of realistic embedded system usage. The consistency of the results across the 3 applications studied suggests that the trends described in this paper can be applied to other sliding-window applications, with only minor differences caused by the operation applied to the sliding window. To our knowledge, we also demonstrated the first real-time sliding-window implementations to operate on high definition video with kernels up to 45×45.

7. ACKNOWLEDGEMENTS

This work was supported by the National Science Foundation grant CNS-0914474.

8. REFERENCES

[1] Altera, Inc. 2011 Stratix III Early Power Estimator. http://www.altera.com/support/devices/estimator/st3-estimator/st3-power-estimator.html.

[2] Asano, S., Maruyama, T., and Yamaguchi, Y. 2009. Performance comparison of FPGA, GPU and CPU in image processing. In *Proc. of Int. Conf. on Field Prog, Logic and App.* FPL '09. 126-131.

[3] Baker, Z.K., Gokhale, M.B., and Tripp, J.L. 2007. Matched filter computation on FPGA, Cell and GPU. In *Proc. of the IEEE Symp. on Field-Prog. Custom Computing Machines.* FCCM'07. 207-218.

[4] Chase, J., Nelson, B., Bodily, J., Zhaoyi W., and Dah-Jye, L. 2008. Real-time optical flow calculations on FPGA and GPU architectures: a comparison study. In *Proc. of the Int. Symp. on Field-Prog. Custom Computing Machines.* FCCM '08. 173-182.

[5] Che, S., Li, J., Sheaffer, J.W., Skadron, K., and Lach, J. 2008. Accelerating compute-intensive applications with GPUs and FPGAs. In *Proc. of the Symp. on Application Specific Processors.* SASP'08. 101-107.

[6] Cope, B., Cheung, P.Y.K., Luk, W., and Witt, S. 2005. Have GPUs made FPGAs redundant in the field of video processing? In *Proc. of the IEEE Int. Conf. on Field-Prog. Technology.* 111-118.

[7] Dong, Y., Dou, Y., and Zhou, J. 2007. Optimized generation of memory structure in compiling window operations onto reconfigurable hardware," in *Proc. of the Int. Symp. on Applied Reconfigurable Computing,* ARC '07. 110–121.

[8] Friemel, B.H., Bohs, L.N., and Trahey, G.E. 1995. Relative performance of two-dimensional speckle-tracking techniques: normalized correlation, non-normalized correlation and sum-absolute-difference. In *Proc. of the IEEE Ultrasonics Symp..* 2, 1481-1484.

[9] Frigo, M., and Johnson, S. 2009. FFTW Library. http://fftw.org

[10] Guo, Z., Najjar, W., Vahid, F., and Vissers, K. 2004. A quantitative analysis of the speedup factors of FPGAs over processors. In *Proc. of the ACM/SIGDA Int. Symp. on Field Prog. gate arrays.* FPGA '04. 162-170.

[11] Harris, M. 2007. "Optimizing Parallel Reduction in CUDA," NVIDIA Developer Technology.

[12] Hunt, L. 2009. Fault-aware machine vision in small unmanned systems. In *Proc. of the Florida Conf. on Recent Advances in Robotics.* FCRAR'09.

[13] Intel. 2010. *Writing Optimal OpenCL Code with Intel OpenCL SDK: Performance Guide.* http://software.intel.com/file/37171/.

[14] Liu, W., Pokharel, P., and Principe, J. 2007. Correntropy: Properties and applications in non-Gaussian signal processing. *IEEE Tranactions on. Signal Process*ing, 55, 11 (Nov. 2007), 5286–5298.

[15] Mehta, S., Misra, A., Singhal, A., Kumar, P., and Mittal, A. 2010. A high-performance parallel implementation of sum of absolute differences algorithm for motion estimation using CUDA. *HiPC Conf. 2010.*

[16] Munshi, A. *The OpenCL Specification.* http://www.khronos.org/registry/cl/specs/opencl-1.0.29.pdf.

[17] NVIDIA. 2001. CUDA. http://developer.nvidia.com/object/cuda.html.

[18] NVIDIA. 2011. CUDA CUFFT Library. http://developer.nvidia.com/cuda-toolkit-40.

[19] NVIDIA. 2011. NVIDIA Tegra 2. http://www.nvidia.com/object/tegra-2.html.

[20] Owens, J.D., Houston, M., Luebke, D., Green, S., Stone, J.E., and Phillips, J.C. 2008. GPU computing. *Proc. of the IEEE.* 96, 5, 879-899.

[21] Pauwels, K., Tomasi, M., Diaz Alonso, J., Ros, E., and Van Hulle, M. 2011. A comparison of FPGA and GPU for real-time phase-based optical flow, stereo, and local image features. *IEEE Transactions on Computers.* 99.

[22] Podlozhnyuk, V. 2007. *FFT-based 2D convolution.* White Paper. NVIDIA Corporation.

[23] Porter, R.B. and Bergmann, N.W. A generic implementation framework for FPGA based stereo matching. In *Proc. of the IEEE Speech and Image Technologies for Computing and Telecommunications,* TENCON '97. 461-464.

[24] Principe, J., Fisher III, J., Xu, D. 2000. Information theoretic learning. In S. Haykin (Ed.), *Unsupervised adaptive filtering.* New York, NY: Wiley.

[25] Sinha, S., Frahm, J.M., and Pollefeys M. 2006. *GPU-based Video Feature Tracking and Matching.* Technical Report TR06-012, University of North Carolina at Chapel Hill.

[26] Underwood, K.D. and Hemmert, K.S. 2004. Closing the gap: CPU and FPGA trends in sustainable floating-point BLAS performance. In *Proc. of the IEEE Symp. on Field-Prog. Custom Computing Machines,* FCCM'04. 219-228.

[27] Xilinx. 2010. *Virtex-4 Family Overview v3.1.* (Aug 30, 2010). http://www.xilinx.com/support/documentation/data_sheets/ds112.pdf

[28] Yu, H. and Leeser, M. 2006. Automatic sliding window operation optimization for FPGA-based computing boards. In *Proc. of the IEEE Symp. on Field-Prog. Custom Computing Machines.* FCCM '06. 76-88.

[29] Zhang, J., He, Y., Yang S., and Zhong, Y. 2003. Performance and complexity joint optimization for H.264 video coding. In *Proc. of the Int. Symp. on Circuits and Systems.* ISCAS '03. 2, 888-891.

[30] Zhi G., Betul B., and Walid N. 2004. Input data reuse in compiling window operations onto reconfigurable hardware. In *Proc. of the ACM SIGPLAN/SIGBED Conf. on Languages, compilers, and tools for embedded systems.* LCTES '04. 249-256.

A Mixed Precision Monte Carlo Methodology for Reconfigurable Accelerator Systems

Gary C.T. Chow
Dept. of Computing
Imperial College London
SW7 2AZ, United Kingdom
cchow@doc.ic.ac.uk

Anson H.T. Tse
Dept. of Computing
Imperial College London
SW7 2AZ, United Kingdom
htt08@doc.ic.ac.uk

Qiwei Jin
Dept. of Computing
Imperial College London
SW7 2AZ, United Kingdom
qj04@doc.ic.ac.uk

Wayne Luk
Dept. of Computing
Imperial College London
SW7 2AZ, United Kingdom
wl@doc.ic.ac.uk

Philip H.W. Leong
School of Electrical and
Information Engineering
University of Sydney
Sydney, Australia
philip.leong@sydney.edu.au

David B. Thomas
Dept. of Electrical and
Electronic Engineering
Imperial College London
SW7 2AZ, United Kingdom
d.thomas1@imperial.ac.uk

ABSTRACT

This paper introduces a novel mixed precision methodology applicable to any Monte Carlo (MC) simulation. It involves the use of data-paths with reduced precision, and the resulting errors are corrected by auxiliary sampling. An analytical model is developed for a reconfigurable accelerator system with a field-programmable gate array (FPGA) and a general purpose processor (GPP). Optimisation based on mixed integer geometric programming is employed for determining the optimal reduced precision and optimal resource allocation among the MC data-paths and correction data-paths. Experiments show that the proposed mixed precision methodology requires up to 11 % additional evaluations while less than 4 % of all the evaluations are computed in the reference precision; the resulting designs are up to 7.1 times faster and 3.1 times more energy efficient than baseline double precision FPGA designs, and up to 163 times faster and 170 times more energy efficient than quad-core software designs optimised with the Intel compiler and Math Kernel Library. Our methodology also produces designs for pricing Asian options which are 4.6 times faster and 5.5 times more energy efficient than NVIDIA Tesla C2070 GPU implementations.

Categories and Subject Descriptors

C.0 [**Computer System Organization**]: System architecture; G.1.0 [**Numerical analysis**]: Multiple precision arithmetic

General Terms

Design, algorithms, performance

Keywords

Monte Carlo, mixed precision

1. INTRODUCTION

Monte Carlo (MC) simulations are a class of algorithms based on randomisation which are extensively used in many high performance computing applications in science, engineering and finance. High performance computing is often needed to solve these problems since they are computationally expensive. MC simulations are well suited to Field Programmable Gate Arrays (FPGAs), due to the parallel nature of MC algorithms and the availability of cost-effective random number generators for FPGAs. It has been shown that FPGA-based Monte Carlo applications can offer 1-2 orders of speedup over their software counterparts running on high-end CPUs [12, 21].

The ability to support customizable data-paths of different precisions is an important advantage of reconfigurable hardware. Reduced-precision data-paths usually have higher clock frequencies, consume fewer resources and offer a higher degree of parallelism for a given amount of resources compared with full precision data-paths. Although the use of reduced precision can lead to higher performance, it also affects the accuracy of the results. Most FPGA Monte Carlo designs exploit this trade-off and use data-paths that are sufficiently accurate to produce outputs within the required error tolerance [11, 17, 21]. However, when very accurate outputs are required, high precision data-paths with lower performance are unavoidable. This makes FPGAs less attractive in MC applications with high accuracy requirements.

This paper introduces a novel mixed precision methodology for accurate Monte Carlo simulations. The key difference between the proposed methodology and previous FPGA Monte Carlo designs lies in the way finite precision errors are handled. Instead of keeping the output error within a certain tolerance, the FPGA data-path is initially constructed with an aggressively reduced precision. This produces a result with finite precision error exceeding the error tolerance. An auxiliary sampling process using both a high precision reference and the reduced precision is then used to correct the error. The output accuracy of the proposed technique is not limited by the precision of the data-paths. The proposed methodology can also exploit the synergy between different processors in a reconfigurable accelerator system. Reference precision computations required in the auxiliary sampling can be carried out by a general purpose processor (GPP) in a host PC, while reduced precision computations target customized data-paths on the FPGA. This allows different processors to work in precisions for which they are specialised, leading to higher overall performance.

The major contributions of this paper are:

- an error analysis that separates finite precision error and sampling error for reduced precision Monte Carlo simulations, and a novel mixed precision methodology to correct finite precision errors through auxiliary sampling (Section 2 and Section 3).

- techniques for partitioning workloads of different precisions for auxiliary sampling to a reconfigurable accelerator system consisting of FPGA(s) and GPP(s) (Section 4).

- an optimisation method based on an analytical model for the execution time of a Monte Carlo simulation on a reconfigurable accelerator system, and Mixed Integer Geometric Programming to find optimal precision for the FPGA's data-paths and optimal resource allocation (Section 5).

- evaluation of the proposed methodology using four case studies, with performance gains of 2.9 to 7.1 times speedup over FPGA only designs using double precision arithmetic. The mixed precision designs are also 44 to 163 times faster and 41 to 170 times more energy efficient compared with software design on a quad-core GPP (Section 6 and 7).

2. BACKGROUND

This section provides an error analysis for Monte Carlo simulations. The total error ϵ_{total} of a Monte Carlo simulation can be divided into two components. Sampling error ϵ_S is the error due to having a finite number of samplings and finite precision error ϵ_{fin} is due to non-exact arithmetic. It is assumed that when a sufficiently accurate precision, such as IEEE-754 double precision, is used, the finite precision error is negligible. We call this value the **reference precision**. We also review how finite precision error is handled in related work. Let us begin with sampling error. Monte Carlo methods are used to simulate random processes and estimate the distribution of the results. Consider a sequence of mutually independent, identically distributed random variables, X_i from a MC simulation. If, $S_N = \sum_{i=1}^{N} X_i$, and the expected value, I, exists, the Weak Law of Large Numbers states that if $p(x)$ is the probability of x, for $\epsilon > 0$, the approximation approaches the mean for large N [8],

$$\lim_{N \to \infty} p\left(|\frac{S_N}{N} - I| > \epsilon \right) = 0 \quad (1)$$

Moreover, if the variance σ^2 exists, the Central Limit Theorem states that for every fixed a,

$$\lim_{N \to \infty} p\left(\frac{S_N - NI}{\sigma \sqrt{N}} < a \right) = \frac{1}{\sqrt{2\pi}} \int_{-\infty}^{a} e^{-z^2/2} dz \quad (2)$$

that is, the distribution of the standard error is normal.

In practice, we must deal with finite N. If the sampling function f represents a mathematical expression defining the

quantity being sampled, $\vec{x_i}$ is the input vector of length s from a uniform distribution [1] $[0,1)^s$, N is the number of sample points and $\langle f_H \rangle_N$ is the sampled mean value of the quantity, the conventional MC sampling process[2] can form an approximation to I,

$$I \approx \langle f_H \rangle_N = \frac{1}{N} \sum_{i=1}^{N} f_H(\vec{x_i}) \quad (3)$$

Thus a sampling error $\epsilon_S(\langle f_H \rangle_N) = I - \langle f_H \rangle_N$ with approximately normal distribution is introduced:

$$\epsilon_S(\langle f_H \rangle_N) \sim \mathcal{N}(0, \sigma_{f_H}^2/N) \quad (4)$$

Equation 4 shows that the bound of the sampling error can be constructed as a confidence interval. Given the same confidence level, the interval is proportional to the standard deviation of the sampling function, σ_{f_H}, and inversely proportional to the square root of the number of sample points, N. Hence quadrupling the number of sample points halves the confidence interval of the sampling error $\epsilon_S(\langle f_H \rangle_N)$. We assume there is no precision error associated with the sampling error. In FPGA designs, the sampling function f is usually evaluated using a low reduced precision, f_L, compared to the high reference precision, f_H. The reduced precision design is smaller and faster, at the expense of higher error. However, reduced precision increases the error. We call the difference between a reference precision computation and a reduced precision computation, $f_H(x) - f_L(x)$, the finite precision error.

Methods for dealing with finite precision error in FPGA-based MC simulations can be classified into two categories. In the first category, only standard precisions such as the IEEE single/double precision are used in sampling data-paths [10, 12]. Users are responsible for determining whether the finite precision error is acceptable, because the FPGA MC engines will follow the result of software exactly.

In the second category, error bounds of the finite precision error are constructed and the precision of the sampling data-path is adjusted such that the error bounds are smaller than the error tolerance. In [11], the maximum relative error of the sampling data-path is used to construct the error bound. The maximum relative error can be characterised using analytical methods such as interval [14] or affine arithmetic [7]. However, these approaches do not take into account that finite precision errors from different sample points might have different signs and would cancel out each other. Hence there is usually an over-estimation of finite precision error in Monte Carlo simulation.

In [17], test runs with a pre-defined number of sample points are used to empirically estimate the maximum percentage error due to finite precision effect empirically. The finite precision error of MC simulations using the same datapath and the same number of sample point are then assumed to share the same error bound. This assumption is not al-

---

[1] Some MC simulations require non-uniformly distributed $\vec{x}$ values, for example in many option pricing simulations normally distributed $\vec{x_i}$ are required.

[2] Throughout the paper, we use the subscript H and L to denote quantities evaluated with the reference precision arithmetic and the reduced precision arithmetic respectively. We use $\langle X \rangle$ to denote the sampled mean value of a random variable X and $\langle X \rangle_N$ to denote the sampled mean value of X calculated by N samples.

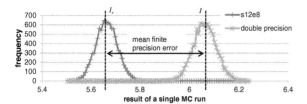

Figure 1: Distribution of 10k runs of a reduced precision and a double precision Monte Carlo.

ways valid and thus the empirical error bound can only be used as a reference rather than a rigorous bound.

In [18], a design is proposed with both high precision and reduced precision data-paths for computing cumulative distribution functions (CDFs). The two CDFs are compared using a Kolmogorov-Smirnov test, the distance score of which is then used to control the precision of the reduced precision data-path adaptively such that finite precision error is within the range of error tolerance.

In [4] a mixed-precision approach for comparison is presented. It is different from this work since it does not involve MC simulation.

The benefits for reduced precision designs are well-known. For instance, it has been shown [5] that appropriate wordlength optimisation can improve the area of adaptive filters and polynomial evaluation circuits by up to 80%, power reduction of up to 98%, and speed of up to 36% over common alternative design strategies. This research shows that impressive gains can also be obtained by exploiting reduced precision for complex designs supporting Monte Carlo simulation.

3. MIXED PRECISION METHODOLOGY

Our novel mixed precision methodology is motivated by two ideas. First, we can correct the finite precision error when both its magnitude and sign are known. Second, in Monte Carlo simulations, we are only interested in the finite precision error in the final result but not the finite precision errors of individual sample points.

When a reduced precision data-path is used in a Monte Carlo simulation, the reduced precision expected value I_r is approximated by the following equation, where N_L is the number of sample points:

$$I_r \approx \langle f_L \rangle_{N_L} = \frac{1}{N_L} \sum_{i=1}^{N_L} f_L(\vec{x_i}) \qquad (5)$$

Due to the effect of finite precision error, the reduced precision sample mean $\langle f_L \rangle_N$ cannot be used to approximate the expected value I directly as I might not equal to I_r. We define the difference of the two expected means as the mean finite precision error, $\mu_{\epsilon_{fin}}$, where

$$\mu_{\epsilon_{fin}} = I - I_r \qquad (6)$$

Figure 1 shows the distributions of Monte Carlo simulations using a reduced precision (s12e8)[3] data-path and a double precision data-path of for pricing Asian options. In each MC simulation, $N = 32{,}768$ sample points are used and each of the reduced and double precision MC simulation is

---

[3]In this paper, we use the notation $sAeB$ to denote a floating point representation, where A is the number of significand bits and B is the number of exponent bits.

repeated for 10,000 times with different random seeds. As shown in the figure, the magnitude of the mean finite precision error $\mu_{\epsilon_{fin}}$ between the expected value of I and I_r is significant. When reduced precision data-paths are used for the Monte Carlo simulation without the correction by the auxiliary sampling, the true value of the simulation will lie within $\pm\mu_{\epsilon_{fin}}$ of the sampled value. Moreover, this uncertainty cannot be reduced by increasing the number of sample points and is a fundamental limit of conventional reduced precision MC simulations.

To find both the magnitude and the signs of the mean finite precision error $\mu_{\epsilon_{fin}}$, we define an auxiliary sampling function $f_a(\vec{x})$:

$$f_a(\vec{x}) = f_H(\vec{x}) - f_L(\vec{x}) = \epsilon_{fin}(\vec{x}) \qquad (7)$$

where ϵ_{fin} is the finite precision error for each $\vec{x}$. With a sufficient large sample size N_a, we can approximate the mean finite precision error $\mu_{\epsilon_{fin}}$:

$$\mu_{\epsilon_{fin}} \approx \langle f_a \rangle_{N_a} = \frac{1}{N_a} \sum_{i=1}^{N_a} f_a(\vec{x_i}) \qquad (8)$$

The sampling error of this auxiliary sampling $\epsilon_S(\langle f_a \rangle_{N_a}) = \mu_{\epsilon_{fin}} - \langle f_a \rangle_{N_a}$ is approximately normal distributed:

$$\epsilon_S(\langle f_a \rangle_{N_a}) \sim \mathcal{N}(0, \sigma_{f_a}^2/N_a) \qquad (9)$$

Finally, we can approximate the true mean I by two sets of sampling:

$$I_{mixed} = \langle f_L \rangle_{N_L} + \langle f_a \rangle_{N_a} \qquad (10)$$

$$E(I_{mixed}) = E(\langle f_L \rangle_{N_L}) + E(\langle f_a \rangle_{N_a})$$
$$= I_r + (I - I_r) = I \qquad (11)$$

As shown in Equation 11, the expected value of the auxiliary sampling is $I - I_r$. Hence the expected mean of the mixed precision approximation I_{mixed} is exactly the same as the expected mean I computed in the reference precision. Equation 10 can thus be viewed as the reduced precision sample mean plus a correction for the mean finite precision error.

Since two samplings are used in the proposed mixed precision methodology, there are two sampling errors in the result and they can be found using Equation 13 and 14. As both sampling errors are approximately normally distributed, their sum is also approximately normally distributed and has a variance equal to the sum of their individual variances as shown in Equation 15 if uncorrelated random numbers are used. By using the proposed mixed precision methodology, we effectively replace the finite precision error of reduced precision data-paths by the sampling error of the auxiliary sampling. A confidence interval can also be constructed using the combined variance.

$$\epsilon_S(I_{mixed}) = \epsilon_S(\langle f_L \rangle_{N_L}) + \epsilon_S(\langle f_a \rangle_{N_a}) \qquad (12)$$

$$\epsilon_S(\langle f_L \rangle_{N_L}) \sim \mathcal{N}(0, \sigma_{f_L}^2/N_L) \qquad (13)$$

$$\epsilon_S(\langle f_a \rangle_{N_a}) \sim \mathcal{N}(0, \sigma_{f_a}^2/N_a) \qquad (14)$$

$$\epsilon_S(I_{mixed}) \sim \mathcal{N}(0, \sigma_{f_L}^2/N_L + \sigma_{f_a}^2/N_a) \qquad (15)$$

Although the proposed mixed precision methodology is analysed mathematically, we also show its desired effect through experiments. Using Equation 15, we find that a mixed precision MC run using a precision of s12e8 with $N_a = 1078$ and

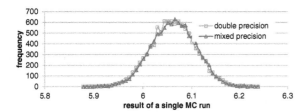

Figure 2: Distribution of 10k runs of a mixed precision and a double precision Mont Carlo.

$N_L = 33{,}773$ should yield the same error as a double precision sampling with $N = 32{,}768$. We repeat both the mixed precision and the double precision MC 10,000 times using different random seeds, and their distributions are shown in Fig. 2. Note that both distributions have roughly the same variance and mean. The result agrees with our mathematical model and no finite precision error exists between the double precision Monte Carlo and our mixed precision Monte Carlo runs.

The proposed mixed precision methodology provides several advantages over previous FPGA designs.

1. The final result is adjusted with an approximated mean finite precision error $\mu_{e_{fin}}$. This is a novel approach which enables us to obtain a more accurate result from the reduced precision result instead of passively finding the error bound.

2. Since there are only sampling errors in the output, we can achieve very accurate result by increasing the number of sample points N_L and N_a. The output accuracy is no longer limited by the reduced precision.

3. The methodology is applicable to any Monte Carlo simulation because no accuracy analysis is required for the relative error and the methodology is totally independent of the function f.

Although the proposed mixed precision methodology enables us to aggressively exploit reduced precision data-paths while maintaining the accuracy of the final result using auxiliary sampling, each auxiliary sampling still requires a costly evaluation of the sampling function f at the reference precision.

The effectiveness of the proposed technique depends heavily on how resources are allocated among the reduced precision hardware and auxiliary sampling hardware. To find the optimal resource allocation, we should consider a number of factors such as the cost of evaluating f_L and f_H, the area available on the FPGA, the bandwidth between the FPGA and GPP, and the reduced precision values being used.

In the next section, we propose different schemes for partitioning workloads. An analytical model is developed in Section 5 based on the partitioning schemes which enables us to find the optimal resource allocation and optimal reduced precision using mixed integer geometric programming.

4. WORKLOAD PARTITIONING

General-purpose processors (GPPs) are optimised for standard precisions such as IEEE-754 single/double precision. GPPs can also employ reduced precision via multiple precision software libraries such as MPFR [9]. Multiple standard precision instructions are required to complete a reduced precision computation even if the reduced precision format

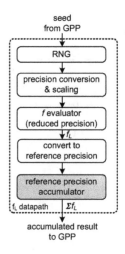

Figure 3: Reduced precision sampling data-path.

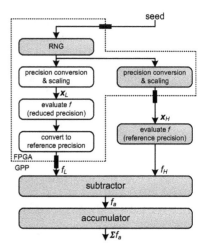

Figure 4: Workload partitioning of the auxiliary sampling. States in reference precision are shaded.

has a smaller wordlength. Hence, it is usually not cost effective to use GPPs for reduced precision computations. On the other hand, FPGA data-paths are customizable. Lower precision are usually preferred over higher precision ones because they usually have higher clock frequency, consume less resources and allow higher degrees of parallelism given the same amount of resources. It is thus better to perform reduced precision computations on the FPGA and leave reference precision computations to the GPP.

Since the sampling of $\langle f_L \rangle_{N_L}$ involves only reduced precision evaluations of f, we assume it is achieved by using reduced precision sampling data-paths on FPGA as shown in figure 3. A seed is fed into the random number generator from the GPP. The random numbers are converted into the reduced precision format and scaled to the sampling domain. Although only a small fraction of bits generated by the RNG are used in reduced precision sampling, we keep the bit-width of the RNG the same as that for reference precision sampling. The scaled random number is then evaluated by the reduced precision sampling function evaluator. The accumulation is performed in reference precision to avoid loss of accuracy due to insufficient dynamic range in the accumulator. Finally, the accumulated result is sent back to the GPP. Multiple reduced precision sampling data-paths can be used with different seeds, and the averaging of the final results is done in the GPP.

Figure 4 shows the workload partitioning of the auxiliary sampling. It consists of 4 main stages: (1) random number generation, (2) evaluation of the sampling function f in reference and reduced precision, (3) computing the difference e between f_L and f_H in reference precision, and (4) accumulation of the difference. Auxiliary sampling is the process used to estimate the average finite precision error ($\mu_{\epsilon_{fin}}$) between the reduced and the reference precision data-paths under the same set of random inputs. We implement the random number generator using the FPGA and sent results back to the GPP. This method utilises highly efficient RNG generation on FPGAs which are an order of magnitudes faster than GPP based RNGs [16]. The trade-off for this partitioning method is increased bandwidth. For each sample point of the auxiliary sampling, we need to transfer s reference precision random numbers and one reference precision evaluation result from the FPGA to the GPP where s is the dimension of the sampling function.

Problem parameters	
σ_{tol}	output error tolerance, in terms of standard deviation of the output
s	dimension of the sampling function
σ_{f_L}	standard deviation of reduced precision sampling
σ_{f_a}	standard deviation of auxiliary sampling
L	the number of significand bits being used in the reduced precision data-paths
Resource allocation parameters	
$p_L \in \mathbb{Z}$	number of reduced precision sampling data-paths
$p_{aux} \in \mathbb{R}$	**effective** number of auxiliary sampling data-paths
FPGA parameters (for each FPGA)	
A_{total}	total available area
A_{com}	cost of communication infrastructure
R_S	slack ratio
$freq$	clock frequency
c	number of clock cycles to compute a sample point
A_{red}	cost of a reduced precision sampling data-paths as shown in figure 3
A_{aux}	cost of auxiliary sampling data-path
GPP parameters	
T_{aux}	time required to compute a sample point
System parameters	
N_{core}	number of cores in each GPP
N_{gpp}	number of GPPs in the system
N_{fpga}	number of FPGAs in the system
BW_{gpp}	bandwidth between the GPP and I/O the hub (in terms of number of reference precision data / sec)
BW_{fpga}	bandwidth between each FPGA and I/O the hub
Output	
t	time required for the system to get output with specific error tolerance

Table 1: Parameters in our analytical model.

5. MIXED PRECISION OPTIMISATION

In this section, we develop analytical models for determining the required execution time of the proposed mixed precision method on a reconfigurable accelerator system. Figure 5 shows the system architecture for the reconfigurable system in our analytical model. The GPP is connected to an I/O hub (i.e. North Bridge) through a high bandwidth

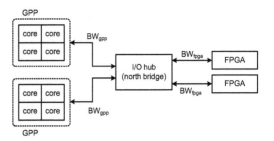

Figure 5: System architecture of the reconfigurable accelerator system in our analytical model.

communication channel such as the Intel QPI or the AMD HyperTransport link. The FPGAs are connected to the I/O hub through another bus, usually PCI express. Thus communication between the GPP and the FPGA has to pass through the two kinds of communication link.

Table 1 shows the parameters in our analytical model. It should be noted that all FPGA cost related parameters such as A_{total} and A_{red} should be applied to every kind of FPGA resource that is involved. For example, there will be 4 different A_{total} parameters for FPGA's look up table (LUT), registers, embedded DSP blocks and block memory respectively. Some other assumptions are made in the model. First, we assume a fixed amount of FPGA resources is used for the communication infrastructure between the FPGA and the I/O hub. Second, we assume the entire FPGA is running at a single clock frequency. Finally, we assume that a certain percentage of the FPGA's resource (the slack ratio) is left intentionally unused to avoid over-congestion in placement and routing.

Since the aggregated throughput of the auxiliary sampling on GPP does not always match the throughput of an auxiliary sampling data-path on the FPGA, we assume the effective number of auxiliary sampling data-paths can take fractional values. For example, $p_{aux} = 0.75$ means there is one auxiliary sampling data-path on the FPGA but only 75% of its outputs are computed by the GPP. The remaining 25% of the outputs are discarded.

Let TH_{red} and TH_{aux} be the aggregated throughput of reduced precision sampling and auxiliary sampling of the entire system. Using Equation 15, the required execution time for the system to produce an output with error equal to σ_{tol} can be found by Equation 16:

$$\sigma_{tol}^2 = \frac{\sigma_{f_L}^2}{t \times TH_{red}} + \frac{\sigma_{f_a}^2}{t \times TH_{aux}} \implies$$
$$t = \frac{\sigma_{f_L}^2}{\sigma_{tol}^2 \times TH_{red}} + \frac{\sigma_{f_a}^2}{\sigma_{tol}^2 \times TH_{aux}} \quad (16)$$

The aggregated throughput of the reduced precision sampling and the auxiliary sampling of all FPGAs can be modelled as:

$$TH_{red} = N_{fpga} \times p_L \times freq/c$$
$$TH_{aux} = N_{fpga} \times p_{aux} \times freq/c \quad (17)$$

The execution time for the mixed precision methodology is:

$$t(p_L, p_{aux}) = \frac{c}{\sigma_{tol}^2 \times N_{fpga} \times freq} \times \left(\frac{\sigma_{f_L}^2}{p_L} + \frac{\sigma_{f_a}^2}{p_{aux}} \right) \quad (18)$$

The following constraint should be applied to ensure the architecture described by the resource allocation parameters can fit within the FPGA. We round p_{aux} to the next larger integer. The constraint (19) is transformed into two new constraints (20-21) using a new integer variable p_{aux_i} to avoid the ceiling function:

$$p_L \times A_{red} + \lceil p_{aux} \rceil \times A_{aux} \leq A_{total} \times (1 - R_S) - A_{com} \quad (19)$$

$$p_L \times A_{red} + p_{aux\_i} \times A_{aux} \leq A_{total} \times (1 - R_S) - A_{com} \quad (20)$$

$$p_{aux\_i}^{-1} \times p_{aux} \leq 1 \quad (21)$$

The number of auxiliary samplings that each GPP can perform is N_{core}/T_{aux} and the aggregated throughput of all GPPs is $N_{gpp} \times N_{cores}/T_{aux}$. Hence the effective number of auxiliary sampling data-paths on each FPGA is constrained by the following equation:

$$N_{fpga} \times p_{aux} \times freq/c \leq N_{gpp} \times N_{core}/T_{aux} \quad (22)$$

One evaluated value of f and s random numbers must be sent every cycle to the GPP to complete the subtraction and accumulation for each auxiliary sampling, hence the bandwidth constraints are:

$$p_{aux} \times freq/c \times (s+1) \leq BW_{fpga} \quad (23)$$

$$N_{core}/T_{aux} \times (s+1) \leq BW_{gpp} \quad (24)$$

The optimal resource allocation among the reduced precision sampling and the auxiliary sampling can be found by applying the following optimisation:

$$\min_{p_L \in \mathbb{Z}, p_{aux} \in \mathbb{R}, p_{aux\_i} \in \mathbb{Z}} t(p_L, p_{aux}, p_{aux\_i})$$

$$s.t. \text{ constraints (20)-(24) are satisfied}$$

Since the objective function $t(p_L, p_{aux})$ and all the constraints are posynomial, the optimisation can be solved using mixed integer geometric programming (MIGP) [2]. The globally optimal p_L, p_{aux} values and the optimal precision can be found using enumeration from Algorithm 1, where L_{min} and L_{max} are the minimum and maximum choice of reduced precision in the system respectively.

Algorithm 1 Enumeration process for optimal reduced precision and optimal resource allocation.

1: $t_{global} \leftarrow$ huge_value
2: **for** $L = L_{min} \rightarrow L_{max}$ **do**
3:     apply MIGP on $t(p_L, p_{aux})$ for the minimum execution time t_{min} in precision L
4:     **if** $t_{min} < t_{global}$ **then**
5:        $t_{global} = t_{min}$
6:        $p_{aux}(global) = p_{aux}$, $p_L(global) = p_L$
7:     **end if**
8: **end for**

---

6. CASE STUDIES

6.1 Asian option pricing

The first case study for our mixed precision methodology is an arithmetic mean Asian call option pricing problem. An Asian call option is characterised by S_0, the current price of the underlying asset; K, the strike price; T, time to maturity and $steps$, the number of observation points to maturity. The arithmetic mean of the asset's current price and the prices at all the observed points computed. At maturity, if the mean is larger than the strike price, the option pays the owner the mean less the strike price. Otherwise, if the strike

price is larger, the payoff of the option is zero. Unlike the geometric mean Asian option pricing problem, there is no closed form analytical solution for this problem and Monte Carlo simulation is a common way for pricing this option.

Algorithm 2 shows the sampling function for pricing Asian options using the Black-Scholes model. Since the intermediate variables $drift$ and $vsqrtdt$ are the same for every sample point, they are pre-computed to reduce computation workload. We do not compute the actual arithmetic mean in each sample point. Instead, the strike price is multiplied by $(steps+1)$ and it is compared with the sum of the prices. This optimisation removes the division operation from the sampling function as it is expensive to implement. We use a fully pipelined design and it takes on average $(steps+1)$ clock cycles for the FPGA to complete a single sampling of the Asian option problem.

Algorithm 2 Sampling function for the Asian call option pricing problem

---

Input: S_0 = current price of the underlying asset, K = strike price of the option, v = volatility of the underlying asset, r = interest rate, $steps$ = number of time step, δt = time period between two time steps
$W \sim \mathcal{N}(0,1)$ Gaussian random number
Output: p_steps = payoff of the option multiplied by $(steps+1)$

1: $drift \leftarrow (r - v^2/2)\delta t$, $vsqrtdt \leftarrow v\sqrt{\delta t}$
2: $S_i \leftarrow S_0$, $S_{sum} \leftarrow S_0$
3: **for** $i = 1 \rightarrow steps$ **do**
4:     $S_i \leftarrow S_{i-1} \times exp(drift + vsqrtdt \times W)$
5:     $S_{sum} \leftarrow S_{sum} + S_i$
6: **end for**
7: $p\_steps \leftarrow max(0, S_{sum} - K \times (steps+1))$

6.2 The GARCH volatility model

Our second case study is for pricing of a fixed strike lookback call option under the GARCH model. This option pays the owner $max(S_{ceil} - K)$ at maturity, where K is the strike price and S_{ceil} is the maximum day closing price of the underlying asset within the lifetime of the option.

In the original Black-Scholes model, the volatility of an asset is assumed to be constant. However, this assumption may not be realistic. A solution is to employ a stochastic volatility model such as the generalised autoregressive conditional heteroskedasticity (GARCH) model proposed by Bollerslev [1]. We use the common GARCH(1,1) model where the volatility of the asset v_i at time step i can be modelled. Let α and β be pre-calibrated model constants, v_0 be the volatility at the start time and λ be a random number following a $\mathcal{N}(0,1)$ distribution.

$$v_i^2 = v_0 + \alpha v_{i-1}^2 + \beta v_{i-1}^2 \lambda^2$$
$$= v_0 + v_{i-1}^2 (\alpha + \beta \lambda^2) \quad (25)$$

The implementation of lookback option pricing is similar to Asian option pricing, except that $drift$ and $vsqrtdt$ are updated every time step according to Equation 25. An additional random number source is also required.

6.3 Collateralized Mortgage Obligation

Our third case study concerns pricing Collateralized Mortgage Obligation (CMO) [15]. A CMO is a security which generates cashflow to the owner from interest and prepayments from a pool of mortgages. The actual payoff of CMO depends on the classes of the CMO, commonly referred to as tranches, and a set of pre-specified rules [6]. We adopt the algorithm in [15] in our FPGA design. An arctan function is required in each random walk of the interest rate. This function is often not efficiently implemented in GPP maths libraries On the other hand, lookup table based function evaluation is efficient on FPGA, especially when the required precision is low. Algorithm 3 shows the sampling function of the CMO pricing problem. The variables can be found in Table 2.

u_k	discount factor for month k
m_k	cash flow for month k
i_k	interest rate for month k
w_k	fraction of remaining mortgages prepaying in month k
r_k	fraction of remaining mortgages at month k
c_k	(remaining annuity at month k) / c
c	monthly payment

Table 2: Variables in the sampling function of CMO pricing problem.

Algorithm 3 Sampling function for the CMO pricing problem

Input: c = monthly payment, K_0, K_1, K_2, K_3, K_4 = constants of the model, I_0 = initial interest rate, M = length of the mortgages, σ = standard deviation of interest rate
Pre-computed constants $c_k = \sum_{j=0}^{M-k} (1 + I_0)^{-j}$, remaining annuity at month k) / c
$W \sim \mathcal{N}(0, 1)$ Gaussian random number
Output:
PV = present value of the security
1: $sum \leftarrow 0$, $i_0 \leftarrow K_0 \times I_0$
2: $r_1 \leftarrow 1$, $u_1 \leftarrow (1 + I_0)^{-1}$
3: **for** $k = 1 \rightarrow M$ **do**
4:    $i_k \leftarrow K_0 \times exp(\sigma W) \times i_{k-1}$
5:    $w_k \leftarrow K_1 + K_2 \times arctan(K_3 \times i_k + K_4)$
6:    **if** $k \geq 2$ **then**
7:       $r_k \leftarrow r_{k-1} \times (1 - w_k)$
8:       $u_k \leftarrow u_{k-1} \times (1 + i_{k-1})^{-1}$
9:    **end if**
10:   $m_k \leftarrow c \times r_k ((1 - w_k) + w_k \times c_k)$
11:   $sum \leftarrow sum + u_k \times m_k$
12: **end for**
13: $PV \leftarrow sum$

6.4 Numerical integration

Our last case study is multi-dimensional integral evaluation using the Monte Carlo integration method. Multi-dimensional integrals arise in many areas such as engineering, biology, chemistry and physics modellings and they are not always solvable with analytical methods. Equation 26 shows a multi-dimensional integration where a_i and b_i are the lower and upper bounds of the integration domain of the ith dimension. To evaluate the integral using Monte Carlo simulation, random input vectors are generated within the integration domain and the average value of the integration function f is sampled. The approximated value for the integral can then be found by multiplying the average with the hypercube V of the integration domain as shown in Equation 28. The Monte Carlo integration method is preferable over quadrature based integration methods for high dimensional integrals, because MC integrations always converge with a rate of $\mathcal{O}(1/\sqrt{N})$ and the complexity of MC integration does not increase exponentially as quadrature based numerical integration methods. We refer to [3] for an extensive introduction of Monte Carlo integration method.

$$I = \int_{a_1}^{b_1} dx_1 \int_{a_2}^{b_2} dx_2 \cdots \int_{a_n}^{b_n} dx_n f(x_1, x_2, \cdots x_n) \quad (26)$$

$$I \sim V \times \langle f \rangle \quad (27)$$

$$V = \prod_{i=1}^{n} (b_i - a_i) \quad (28)$$

We use Genz's "Discontinuous" multi-dimensional integral in this case study (29). This is a common test integral used in evaluation of different numerical integration methods. In our tests we use $n = 8$ as the dimension and an integration domain $[0, 1]^8$. Fully parallelised designs are used in our FPGA implementations and the data-paths can compute a single sample point per clock cycle, with constants c_i and w_i:

$$f_{dis} = \begin{cases} 0 & \text{if } x_0 > w_0 \text{ or } x_1 > w_1 \\ exp(\sum_{i=1}^{n}(c_i \times x_i)) & \text{otherwise} \end{cases}$$
$$(29)$$

7. EVALUATION

7.1 Reconfigurable accelerator system

We use the MaxWorkstation reconfigurable accelerator system from Maxeler Technologies for our evaluation. It has a MAX3424A card with a Xilinx Virtex-6 SX475T FPGA. The card is connected to an Intel i7-870 CPU through a PCI express link with a measured bandwidth of 2 GB/s. The Intel GPP has 4 physical cores.

	current	I	II	III
N_{gpp}	1	8	1	1
N_{fpga}	1	1	4	8
$BW_{gpp}(GB/s)$	2	2	4×2	8×2
$BW_{fpga}(GB/s)$	2	8×2	2	2

Table 3: Parameters of the current system and other hypothetical systems.

An important advantage of having an analytical model for our mixed precision methodology is that system designers can predict the performance of a hypothetical system based on parameters of the current system and the analytical model. Table 3 shows the parameters for our current system and three hypothetical systems. The hypothetical systems are constructed in such a way that the aggregated computational power of the FPGAs or the GPPs are 4 or 8 times higher than the current system, and the bandwidth is scaled proportionally.

The Intel Compiler (ICC) and the Intel Math Kernel Library are used in our software implementations. We use the SFMT random number generator and the Box-Muller transformation in the Intel Vector Statistical Library (VSL) for the random number generation. Every effort has been made to ensure the software implementations are optimised, and

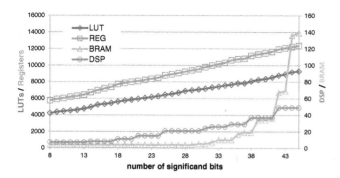

Figure 6: Cost of reduced precision sampling data-paths of the Asian option problem.

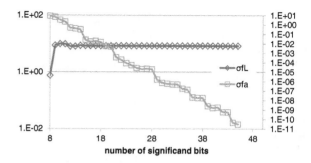

Figure 7: The standard deviations of the reduced precision sampling and the auxiliary sampling verses different precisions.

the comparisons are fair and accurate. For the FPGA implementations, we use the MaxCompiler as our development system. This adopts a streaming programming model similar to [13] and supports customisable data formats so that floating point can be exploited with different precisions. All the FPGA results reported in this paper are post place and route results.

The error tolerance σ_{tol} of the three financial case studies is set to 2.5e-3 such that 99.99% (4σ) of the time the error is less than a cent, given that the pricing is in dollars. For the numerical integration case study, the tolerance is set to 2.5e-4 since most scientific applications require high accuracy.

7.2 Applying optimisation

There are a few steps to apply Algorithm 1 in Section 5 in order to use the proposed mixed precision methodology optimally on a reconfigurable accelerator system. The first step is to find the system parameters such as N_{core}, N_{gpp}, N_{fpga}, BW_{gpp} and BW_{fpga}. These parameters can usually be found in the specification of the reconfigurable accelerator system.

The second step is to collect application specific FPGA and GPP parameters. In the MaxCompiler system, we describe the precision of the entire sampling data-path using a global variable and scripts are used to automatically generate data-paths with varying number of significand bits. Figure 6 shows the place and routed result of reduced precision data-paths of the Asian option problem. It is clearly shown that all the resource requirement increase with precision. Moreover, due to the function approximator in the exponential function, the block memory usage increases exponentially with the precision. The figure shows the A_{red} parameters of different precisions used in our model. Other FPGA parameters such as the A_{aux} parameters can be found using a similar method. The cost of communication infrastructure A_{com} is assumed to be constant. We also estimate the GPP parameter T_{aux} by writing a software benchmark program, which implements the data-flow in Figure 4c for certain iterations, and the average time required for an iteration is used as T_{aux}.

The next step is the estimation of the standard deviations for reduced precision sampling, σ_{f_L}, and for auxiliary sampling, σ_{f_a}. An FPGA bit-stream with auxiliary sampling data-paths of different precisions is loaded and the results of the sampling function evaluations in different reduced precisions are sent back to the host PC. Using these results and the reference precision sampling function evaluations result

from the GPP, the two standard deviations can be estimated using the two-pass algorithm on the host PC [20]. Figure 7 shows how the two standard deviations change with different precisions in the Asian option problem, using the parameters ($S_0 = K = 100, T = 1, v = 0.2, r = 0.05, steps = 360$).

It is interesting to note that the standard deviation of the auxiliary sampling σ_{f_a} decreases exponentially with increasing precision. However the standard deviation of the reduced precision sampling σ_{f_L} is low when the precision is low and increases rapidly to reach a constant maximum value with further increases in precision. The same pattern is also observed in the standard deviations of other case studies. A possible explanation is that when the reduced precision is low, different values are compressed to the same numerical representation and hence the standard deviation is reduced. The standard deviation grows with reduced precision because there are more possible representations, and will finally converge to a value where the value is the same as the standard deviation of the reference precision sampling (i.e. σ_{f_H}). The observed exponential reduction of σ_{f_a} could be explained by the fact that finite precision error decreases exponentially with the number of significand bits in floating point formats.

Using parameters collected in the previous steps, we can apply geometric programming to find the optimal precision and resource allocation. A major assumption of this flow is that the two standard deviations do not change with input parameters (e.g. strike price of an option). If this assumption does not hold, we can profile the common σ_{f_L} and σ_{f_a} combinations and generate an optimal bit-stream for each of these combinations. When the input parameters change, we profile the two standard deviations again, run the geometric programming solver and load the bit-stream closest to the optimal configuration.

It is important to note that the choice of error tolerance σ_{tol} affects the execution time of our mixed precision methodology as shown in Equation 16. However, the optimal reduced precision and resource allocation do not change with the error tolerance. Hence there is no need to rerun the geometric programming.

7.3 Performance: parallelism versus precision

Figure 8 shows the execution time and the degree of parallelism of the Asian option pricing problem for different reduced precision in the current system as evaluated by our analytical model. The optimal reduced precision in this benchmark is s12e8. The performance curve and the optimal point can be explained by considering Figures 6 and 7.

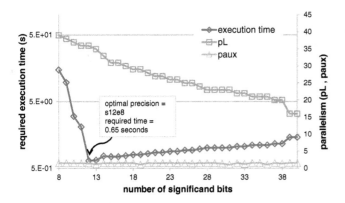

Figure 8: Results of Asian option pricing versus different number of significand bits.

If the reduced precision is lower than the optimal one, the auxiliary sampling error σ_{f_a} is high, and more computations must be included. This will take a longer time even though parallelism is increased due to smaller data-paths. If the reduced precision is higher than the optimal one, the decrease of the auxiliary sampling error is marginal and cannot offset the disadvantage of reduced parallelism.

	current	I	II	III
execution time (s)	0.65	0.65	0.19	0.09
optimal precision	s12e8	s12e8	s15e8	s16e8
p_L/p_{aux}	23.9	13	73.9	147.9

Table 4: Execution time, optimal reduced precision and the p_L/p_{aux} ratio of the same Asian option pricing under different system parameters.

We also investigate the relationship between the optimal reduced precision and system parameters. Table 4 shows that when the aggregated FPGA computational power is increased (II and III), the optimal reduced precision will increase because the system can perform more reduced precision sampling (i.e. higher p_L/p_{aux} ratio), and thus afford a higher σ_{f_L}. Investing in higher aggregated GPP computational power (I) seems to have a marginal effect in this benchmark as the sampling error in reduced precision already dominates the sampling error in the auxiliary sampling.

7.4 Comparison: GPP/FPGA double precision

Table 5 shows comparisons of the 4 MC case studies running on a GPP only system with double precision arithmetic, an FPGA only system with double precision arithmetic, and a reconfigurable accelerator system with our proposed mixed precision methodology. All designs are run for a specific time so that the 3 systems have the same accuracy. As shown in the table, the mixed precision methodology requires 5-11 % additional sample function evaluations but only 1-4 % of total evaluations are computed in reference precision. This clearly shows the trade-offs between number of computations and the contribution of each computation in increasing the accuracy of the final result. Using the mixed precision methodology, we achieve 2.9 to 7.1 times speedup over the double precision FPGA designs and 44 to 163 times speed up over the **quad-core** GPP designs.

We also compare the energy efficiency of the three settings. The average power consumption is measured using

a remote power measuring socket from Oslon® electronics with an measuring interval of 1 second. As shown in Table 5, although the mixed precision designs using both the FPGA and the GPPs have the highest power consumption compared with the GPP only or the FPGA only settings, they consume the least total energy to achieve the required accuracy because the execution times are significantly reduced thanks to our technique for workload partitioning. Our mixed precision methodology achieves 1.4 to 3.1 times energy saving compared with the FPGA only designs with double precision, and 41 to 170 times energy saving compared with GPP only designs, while meeting the same output accuracy requirement.

7.5 Comparison: GPU

We also compare our mixed precision methodology on a reconfigurable accelerator system with a graphics processing unit (GPU). Table 6 compares the execution time and power consumption of our mixed precision methodology with an NVIDIA Tesla C2070 GPU for pricing Asian options. The GPU has 448 cores running at 1.15 GHz. Both the Tesla 2070 GPU and the Virtex-6 FPGA are fabricated using 40nm technology. We use the same GPU Asian option pricing design as described in our previous work [19]. Since random number generation and pricing calculation take place on the GPU, communication is not a bottleneck as only the final accumulated result in double-precision need to be sent from the GPU to the GPP. Using our mixed precision methodology, an Virtex-6 ST475X FPGA and an i7-870 GPP are able to out-perform the GPU by 4.6 times. We also achieved 5.5 times energy saving compared with the GPU.

	GPP only	GPU	FPGA only	FPGA + GPP
precision	double	double	double	mixed
execution time (s)	29	3	4.7	0.65
power (W)	183	236	85	192
energy (kJ)	5.3	0.71	0.4	0.13
normalised speedup	1x	9.7x	6.2x	44.6x
normalised energy	40.7x	**5.5x**	3.1x	1x

Table 6: Comparison with GPP and GPU.

8. CONCLUSION

This paper proposes a novel mixed precision methodology for Monte Carlo simulation in reconfigurable accelerator systems. The technique is applicable to any Monte Carlo application and exploits the synergy between FPGA and GPP to produce results of the desired accuracy. An analytical model and optimisation method is developed for locating the optimal precision and optimal resource allocation. Experimental results on four realistic case studies show that auxiliary sampling would only require 5 % to 11 % additional evaluations, and less than 4 % of total evaluations are computed in the reference precision (Table 5). We demonstrate that reconfigurable accelerator system using our methodology can be up to 4.6 times faster than state-of-the-art GPU, 7.1 times faster than a baseline FPGA design using double precision, and 163 times faster than optimised software running on a quad-core GPP. It can also be up to 5.5 times more energy efficient than a GPU and 170 times more energy efficient than a quad-core software implementation.

	Asian option			GARCH			CMO			Numerical integration		
	SW	FP	Mixed	SW	FP	Mixed	SW	FP	Mixed	SW	FP	Mixed
clock freq. (GHz)	2.93	0.175	0.175^1	2.93	0.175	0.175^1	2.93	0.175	0.175^1	2.93	0.175	0.16^1
num. of cores[2]	4	5	36/1.5	4	5	24/0.9	4	5	20/0.65	4	5	16/0.18
num. of f_L evaluations (M)	0	0	12	0	0	321	0	0	7.2	0	0	2320
num. of f_H evaluations (M)	11.3	11.3	0.47	317	317	11.6	6.75	6.75	0.23	2230	2230	26.8
num. of total evaluations (M)	11.3	11.3	12.5	317	317	333	6.75	6.75	7.43	2230	2230	2347
additional evaluation (%)	-	-	10.6	-	-	4.8	-	-	10	-	-	5.2
evaluations in reference precision (%)	100	100	3.8	100	100	3.5	100	100	3.1	100	100	1.1
execution time (sec.)	29	4.7	0.66	1560	131	26.6	117	2.8	0.72	95.8	2.6	0.9
normalised speedup	1x	6.2x	44x	1x	12x	59x	1x	42x	**163x**	1x	37x	106x
mixed precision gain	-	1x	**7.1x**	-	1x	4.9x	-	1x	3.9x	-	1x	2.9x
power consumption (W) [3]	183	85	192	179	90	181	175	94	171	184	90	189
energy consumption (kJ)[4]	5.3	0.4	0.13	280	11.8	4.8	20.4	0.26	0.12	17.6	0.23	0.17
normalised energy	41x	**3.1x**	1x	58x	2.5x	1x	**170x**	2.2x	1x	104x	1.4x	1x

[1] Only the FPGA clock frequencies are reported and the 4 GPP cores are all running at 2.93 GHz.

[2] For the mixed precision design, all the 4 GPP cores are used and the number of reduced precision sampling and auxiliary sampling data-paths (pL/p_{aux}) are shown.

[3] The idle power consumption of the system is $80W$.

[4] Energy consumption = power consumption × execution time.

[5] The optimal precision of the 4 mixed precision designs is **s12e8**.

Table 5: Comparison of MC simulations using GPP only system (SW), double precision FPGA only system (FP) and mixed precision methodology using both GPP and FPGA (Mixed).

Future work includes applying the proposed methodology to other sampling methods such as the Quasi-Monte Carlo methods. Other directions of further research involve extending the proposed methodology to cover heterogeneous systems consisted of GPPs, FPGAs and GPUs, and automating the steps of the methodology.

Acknowledgments

The research leading to these results has received funding from the Croucher Foundation, Maxeler Technologies, Xilinx, the UK EPSRC, and the European Union Seventh Framework Programme under grant agreements number 257906 and 248976.

9. REFERENCES

[1] T. Bollerslev. Generalized autoregressive conditional heteroskedasticity. *Journal of Econometrics*, 31(3):307–327, 1986.

[2] S. Boyd, S.-J. Kim, L. Vandenberghe, and A. Hassibi. A tutorial on geometric programming. *Optimization and Engineering*, 8:67–127, 2007.

[3] R. E. Caflisch. Monte Carlo and quasi-Monte Carlo methods. *Acta Numerica*, 7:1–49, 1998.

[4] G. Chow, K. Kwok, W. Luk, and P. Leong. Mixed precision processing in reconfigurable systems. In *Proc. FCCM*, pages 17 –24, 2011.

[5] G. A. Constantinides. Word-length optimization for differentiable nonlinear systems. *ACM Trans. Des. Autom. Electron. Syst.*, 11:26–43, 2006.

[6] F. J. Fabozzi. *The Handbook of Mortgage-Backed Securities*. McGraw-Hill, 2005.

[7] C. Fang, T. Chen, and R. Rutenbar. Floating-point error analysis based on affine arithmetic. In *Proc. IEEE International Conference on Acoustics, Speech, and Signal Processing*, volume 2, pages II–561–4, 2003.

[8] G. S. Fishman. *Monte Carlo Concepts, Algorithms, and Applications*. Springer, 1995.

[9] L. Fousse, G. Hanrot, V. Lefèvre, P. Pélissier, and P. Zimmermann. MPFR: A multiple-precision binary floating-point library with correct rounding. *ACM Trans. Math. Softw.*, 33, 2007.

[10] M. Gokhale, J. Frigo, C. Ahrens, and R. Minnich. Monte Carlo radiative heat transfer simulation on a reconfigurable computer. In *Proc. FPL*, pages 95–104, 2004.

[11] A. Gothandaraman, G. D. Peterson, G. Warren, R. J. Hinde, and R. J. Harrison. FPGA acceleration of a quantum Monte Carlo application. *Parallel Computing*, 34(4-5):278 – 291, 2008.

[12] A. Kaganov, A. Lakhany, and P. Chow. FPGA acceleration of multifactor CDO pricing. *ACM Trans. Reconfigurable Technol. Syst.*, 4:20:1–20:17, 2011.

[13] O. Mencer. ASC: a stream compiler for computing with FPGAs. *IEEE Trans. CAD*, 25(9):1603 –1617, 2006.

[14] R. E. Moore. *Interval arithmetic and automatic error analysis in digital computing*. PhD thesis, Stanford University, 1963.

[15] S. H. Paskov. New methodologies for valuing derivatives. In M. Dempster and S. Pliska, editors, *Mathematics of derivative securities*, number 15 in Publications of the Newton Institute. Cambridge Univ. Press, 1997.

[16] D. B. Thomas, L. Howes, and W. Luk. A comparison of CPUs, GPUs, FPGAs, and massively parallel processor arrays for random number generation. In *Proc. FPGA*, pages 63–72, 2009.

[17] X. Tian and K. Benkrid. Design and implementation of a high performance financial Monte-Carlo simulation engine on an FPGA supercomputer. In *Proc. ICFPT*, pages 81–88, 2008.

[18] X. Tian and C.-S. Bouganis. A run-time adaptive FPGA architecture for Monte Carlo simulations. In *Proc. FPL*, 2011.

[19] A. H. Tse, D. B. Thomas, K. H. Tsoi, and W. Luk. Efficient reconfigurable design for pricing Asian options. *SIGARCH Comput. Archit. News*, 38:14–20, 2011.

[20] E. W. Weisstein. Sample variance computation. http://mathworld.wolfram.com/ SampleVarianceComputation.html, accessed Sept. 2011.

[21] G. Zhang et al. Reconfigurable acceleration for Monte Carlo based financial simulation. In *Proc. ICFPT*, pages 215 –222, 2005.

Saturating the Transceiver Bandwidth: Switch Fabric Design on FPGAs

Zefu Dai   Jianwen Zhu
Department of Electrical and Computer Engineering
University of Toronto, Toronto, ON, Canada, M5S 3G4
{zdai,jzhu}@eecg.utoronto.ca

ABSTRACT

Driven by the demand of communication systems, field programmable gate array (FPGA) devices have significantly enhanced their aggregate transceiver bandwidth, reaching terabits per second for the upcoming generation. This paper asks the question whether a single-chip switch fabric can be built that saturates the available transceiver bandwidth.

In answering this question, we propose a new switch fabric organization, called Grouped Crosspoint Queued switch, that brings significant memory efficiency over the state-of-the-art organizations. This makes it possible to build high bandwidth, high radix switches directly on FPGA that rivals ASIC performance. The proposal was validated at small scale by a 16x16 160Gps switch on the available Virtex-6 device, and simulated at a larger scale of fat-tree switching network with 5Tbps capacity.

Categories and Subject Descriptors

C.1.2 [**Multiple Data Stream Architectures**]: Interconnection architectures

General Terms

Design

Keywords

Switch Fabric, Input Queued, Output Queued, Crosspoint Queued, Transceiver

1. INTRODUCTION

Recent evolution of field programmable gate array (FPGA) devices has seen a tremendous growth of transceiver speed, boasting 28 Gbps per link and Terabits per second aggregate bandwidth per device. Understandably, this is largely driven by the communication sector, reportedly FPGA's largest customer.

While the two largest FPGA vendors are drumming up the competition on delivering the highest IO bandwidth, the question remains on whether it can be fully utilized on key applications. For example, although there are announcements and reports on 100G-400G line cards, little was reported whether, and how, high radix (port count) switch

FPGA'12, February 22–24, 2012, Monterey, California, USA.
Copyright 2012 ACM 978-1-4503-1155-7/12/02 ...$10.00.

fabric can be built with FPGAs that can saturate their available IO bandwidth.

The choice of switch fabric architecture is heavily influenced by the ratio of link speed and memory access speed [11]. This is because the minimum packet size is fixed and does not scale with the link speed. With higher link speed, packets arrive and leave faster. As a result, memories that implement packet buffers have to be accessed at a higher speed. To cope with the memory speed challenge, the switch fabric architecture used by application-specific integrated circuit (ASIC) based chips has evolved from the Output Queued (OQ) to Combined Input and Output Queued (CIOQ) and then to the Combined Input and Crosspoint Queued (CICQ) architecture. This is exemplified by the three generations of the IBM Prizma switch [10].

The literature on FPGA implementation of the switch fabric seems rather sparse. Early demonstrations from Actel and Xilinx [4, 24] follow the straightforward crosspoint crossbar architecture. The latest whitepaper from Altera speaks about the available device bandwidth [5], hinting the possibility of high-radix switches, but did not give design details or achieved performance of the switch itself. The NetF-PGA provides an excellent platform for building low-radix switches and routers [16], but as a platform, do not discriminate specific switch architectures. The most comprehensive designs available using Virtex FPGA [25, 23] employ the CICQ architecture. But is good choice for ASICs automatically a good choice for FPGAs?

We argue that the CICQ is a bad choice for the FPGA implementation of high radix switch fabric. This is primarily due to its requirement of N^2 crosspoint buffers, where N is the port count. The SRAM resources on FPGAs will simply run out, for the large number of N permitted by modern FPGA devices. Given the known complexity of scheduling logic, the Input Queued (IQ) and CIOQ switch architectures are not favorable either.

In light of this, it becomes necessary to thoroughly investigate the switch fabric design, if the FPGA industry intends to adds it to the list of ASIC replacements, which seems to be attractive given its ubiquitous usage in carrier network, internet, data center network, and high-performance computing.

In this paper, we show that indeed it is possible to construct a high radix switch fabric using FPGAs that saturates their transceiver bandwidth: for Xilinx's Virtex-7, this promises terabits per second single-chip switching performance. In addition, in contrast to the common belief that FPGAs may suffer significant performance disadvantage, we show that the performance can rival its ASIC counterpart. More specifically, we make the following contributions:

- We propose a new switch fabric organization, called Combined Input and Grouped Crosspoint Queued architecture (GCQ), and demonstrate that it can, given the same memory resources, outperform the CICQ architecture, the state-of-the-art switch fabric architecture;

- We show that how FPGA hard resources, instead of its logic fabric, can be best utilized to "impedance match" the GCQ architecture and give ASIC-like performance.

The rest of the paper is organized as follows. In Section 2, we review important milestones in packet switch architecture. In Section 3, we discuss the proposed switch fabric architecture in detail. In Section 4, we describe FPGA implementation issues. In Section 5, we provide evaluation result.

2. BACKGROUND

A switch fabric performs two major tasks: 1) provide datapath connections between inputs and outputs; and 2) resolve congestion. Usually, a crossbar is used to provide datapath connections between different inputs and outputs. And packet buffers are used to store data temporarily at times of congestion, i.e., when data from multiple inputs destinate to the same output simultaneously. The importance of high performance switch fabric has led to the publication of numerous switch architectures. They can be categorized into 3 basic types based on their buffer organizations:

Output Queued switch which buffers data at output ports;

Input Queued switch which buffers data at input ports;

Crosspoint Queued switch which buffers data at the crossbar;

2.1 Basic Definitions

For the following sections, we consider a switch fabric with N input and N output ports, each running at a link speed of R, e.g. 10 Gb/s. We assume packets can be segmented into fixed-length cells of size C, which are referred to as *flits*. The arrival time between flits at any input (C/R) is called a *time slot*. The *internal speedup* S of a switch is defined as the ability to remove up to S flits from any input buffer and store up to S flits to any output/crosspoint buffer at any time slot.

Table 1: Roadmap for optical fiber and SERDES speed.

	Process	nm	45	30	22
Link Char.	Link Speed	Gb/s	80	160	320
	Max Link Length	m		10	
	In Flight Data	bytes	1107	2214	4428
Optical Params	Data Wavelengths		8	16	32
	Optical Data Rate	Gb/s		10	
Electrical Params	SERDES Speed	Gb/s	10	20	32
	SERDES Channels		8	8	10

The link speed has increased rapidly over the past years from OC-3 (155Mb/s) to OC-768 (40Gb/s). However, when going beyond 10Gb/s, higher link speeds in the optical fiber are achieved by ganging multiple channels of 10Gb/s together. And the processing of such link speeds is usually done by dividing a single link to multiple parallel switches, each processing part of the data at a lower data rate, i.e. 10Gb/s. Table 1 shows the technology roadmap for optical fiber and SERDES [6]. The Dense Wavelength-Division Multiplexing technology vastly increases the number of channels available in a single fiber. But the speed of a single channel stays the same. As a consequence, low radix, high line rate (fat) switches can be implemented using parallel low-radix, low line rate (thin) switches. Therefore, in this paper, we focus on the design of more challenging high-radix switches with thin ports, rather than low radix switches with fat ports.

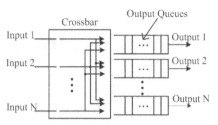

Figure 1: The Output Queued switch.

2.2 The OQ Switch

The OQ switch is the reference switch which can achieve the best possible switching performance. It has high requirements on memory access time, but low demand on packet scheduling logic. Example OQ switches include the early generation of IBM's PRS switch [9] published in 1995 and Fulcrum's FocalPoint FM4000 switch [7] published in 2009. Both switches employ the Centralized Shared Memory (CSM) architecture (which belongs to the OQ switch architecture) to achieve high throughput and low latency, as well as efficient support for multicast. The CSM architecture minimizes the amount of memory needed for congestion buffering as the memory is shared across all ports. However, the number of ports it can support is limited by the memory access time. The early generation of the IBM PRS switch supports 16 ports. Although Fulcrum's FocalPoint FM4000 switch supports 24 ports, it has exceeded the limit of memory access time. And there is a jitter of up to 70 ns when multiple output ports need to access the same memory location in parallel.

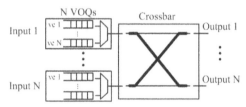

Figure 2: The Input Queued switch.

2.3 The IQ Switch

The IQ switch becomes attractive as the memory technology fails to satisfy the increasing demand on memory speed. It is advantageous that buffers in IQ switches only need to run at $2R$, compared to $2NR$ in OQ switches. However, due to the Head-of-Line (HOL) blocking problem, the throughput of the IQ switch could be limited to approximately 58% [14]. Although the Virtual Output Queue (VOQ) technique [22] can solve this blocking problem, complex scheduling algorithm is required to achieve high performance. In fact, ideal scheduling algorithms [19] are too complex to implement and practical algorithms usually take multiple stages. For example, the *Tiny Tera* IQ switch designed by McKeown et. al. at 1996 [18] uses a 3-step iSLP [17] scheduling algorithm to achieve high throughput for unicast traffic. In addition to the complex scheduling problem, the IQ switch does not support multicast well. Since all buffers are running at $2R$, a broadcast packet will take N time slots for each of its flits to pass through the switch. Otherwise, a dedicate logic has to be built for multicast traffic as was done in the *Tiny Tera* switch.

2.4 The CIOQ Switch

To improve the performance of the IQ switch, the Combined Input and Output Queued (CIOQ) switch has been studied and became one of the most popular switch architecture. The idea is: with an internal speedup of S, it is

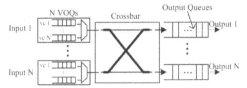

Figure 3: **The Combined Input and Output Queued switch.**

possible for the CIOQ switch to emulate an OQ switch with a memory speed of $(S + 1)R$. Previous work has shown that with $S = 2$ and a complex centralized scheduling algorithm, the CIOQ can emulate an OQ switch [12]. However, practical implementations usually use larger internal speedup to simplify the scheduling algorithm. For example, the second generation of IBM's Prizma switch [20] is a CIOQ switch with a speedup of N. Using high internal speedup, the switch requires only a simple localized scheduler at each input port to achieve high throughput. However, despite its popularity in both academia and industry, the CIOQ switch requires either a complex centralized scheduling logic or high memory access speed.

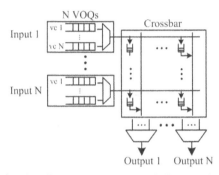

Figure 4: **The Combined Input and Crosspoint Queued switch.**

2.5 The CICQ Switch

The Combined Input and Crosspoint Queued (CICQ) switch was proposed to address both the complex scheduling and high memory access speed problems. The CICQ switch deploys a packet buffer in each crosspoint of its switching crossbar, so that flits from input queues are delivered to the crosspoint buffers first instead of going to output ports directly. As a result, the input and output scheduling are decoupled and no centralized scheduler is needed. Also, each crosspoint buffer only needs to run at $2R$ as it is shared by a single input and a single output port. Therefore, the CICQ switch can scale to support high port speed. Moreover, it can achieve high throughput and low latency with reasonable size crosspoint buffers [13]. These appealing features have increased the industry's interest in the CICQ switch. For example, FORTH implemented a 32×32 single chip CICQ switch [21] in 2004. And IBM's third generation Prizma switch also adopted the CICQ architecture [3] to build a 4 Tb/s single stage switch. However, the major problem of the CICQ switch lies in its requirement of N^2 crosspoint buffers, which makes it less scalable in terms of port number. Also, by distributing packet buffers across N^2 crosspoint memories, it becomes difficult to balance the workload of different buffers, resulting in low memory space efficiency in packet buffers.

2.6 The Hierarchical Crossbar Switch

In an attempt to address the port number scalability problem of the CICQ architecture and build high radix switch, Kim et. al. proposed the Hierarchical Crossbar (HC) archi-

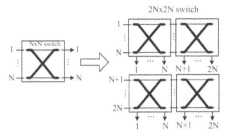

Figure 5: **Single stage port extension.**

tecture in 2005 [15]. The HC aims to reduce memory cost of the CICQ switch by partitioning the buffered crossbar into sub-switches, and implementing each sub-switch using a CIOQ architecture. It was estimated by the authors that by partitioning a 64×64 crossbar into $64\ 8 \times 8$ sub-switches, there is a 40% saving in memory. The HC switch can be viewed as a single stage port extension version of the CIOQ switch, as illustrated in Figure 5. Although memory saving can be achieved, complex scheduling logic is still needed in the sub-switches. In this paper, we follow a similar way of thinking as Kim's work; however, we target FPGA platform, which dictates us to address the problem of switch fabric design from a different angle. In summary, we list the resource requirements of different switch architectures including our proposed switch organization in Table 2. Details of the proposed switch organization will be explained in following sections.

Table 2: **Requirement of different switch architectures.**

	OQ	IQ	CIOQ	CICQ	Proposed
Input Buffer	0	N	N	N	N
Crosspoint Buffer	0	0	0	N^2	$(N/S)^2 \cdot P$
Output Buffer	N	0	N	0	0
Internal Speedup	N	1	S	1	S
Scheduling	D*	C*	C	D	D

*D = Distributed; *C = Centralized

3. MAIN IDEAS

In this section, we describe the main idea of our proposal.

3.1 Memory Is the Switch

Switching fabric often involves crossbars, well known to be wire dominated since wide multiplexers often need to be implemented. Ironically, while FPGAs are arguably made of multiplexers, they are not efficient to implement multiplexers, not to mention the long delay associated with long wires necessary to bring signals from geographical far locations to the central switch. So the conventional wisdom seems to have been that FPGAs would be inefficient, compared to the ASIC switch.

While the FPGA fabric is admittedly slow, FPGAs also integrate large amount of hardcore resources like SRAMs, which can run, in principle, at the same level of speed and power consumption as ASIC chips.

Memories have been widely used as buffers in all switches. The CICQ architecture, the most favourable today for the ASIC switches today, leverages the bandwidth of on-chip SRAMs by allocating a separate SRAM buffer at each crosspoint. Unfortunately, this does not always lead to the best utilization of SRAM bandwidth.

EXAMPLE 1. *The dual-port SRAMs on Xilinx's Virtex-6 FPGA have an access time of approximately 2.5ns. Assuming a line rate of 10Gb/s and a minimum packet size of 40B, the fastest possible packet arrival speed is 12.5ns. Therefore, one SRAM can accommodate multiple ports by way of time multiplexing. Furthermore, if multiple dual-port SRAMs can run in parallel, up to 12 input and output ports can be serviced simultaneously.*

This example illustrates that by clever organization of SRAMs both in time (by time multiplexing) and space (but using parallel memories), one can achieve a decent speedup, and as a result, one buffer can serve multiple port accesses at the same time. This leads us to the first idea that **memory can be shared**: let S be the memory speedup, a shared buffer can replace $S \times S$ crosspoint buffers.

Further examination reveals that it is not necessary to implement $S \times S$ logical crosspoint buffers. In fact, one only needs to implement S logical output buffers. This insight leads to the second idea that **memory is the switch**: while FPGA logic fabric might be slow, the FPGA SRAMs, which run at the ASIC speed, can serve as a small-scale switch: In fact, the hardwired decoding logic and sense-amplifier boosted data bus can serve the same purpose of wide multiplexers! Now the shared buffer serves the dual purposes of buffering and switching, and can be considered as a $S \times S$ OQ *subswitch*. We can therefore organize the high radix switch fabric as a two dimensional switch, in the same spirit of [15]. In the mean time, the total number of buffers has reduced to $(N/S)^2$, a S^2 reduction from the CICQ architecture.

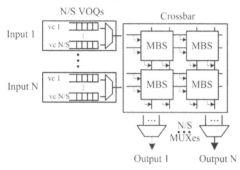

Figure 6: The organization of the proposed switch.

The third idea, which is not new, but comes as added benefit of using shared buffers, is that **memory can be borrowed**: we do not have to reserve fixed buffer size for each logical queue in the shared buffer. By using dynamic memory allocation, a "busy" output queue could occupy more memory spaces than less busy ones, and as a result, can accommodate bursty traffic or congestion better. The input queue buffers then can afford to be smaller due to less congestion burden.

We call each shared buffer a *Memory Based Switch (MBS)*, as it uses memory to implement both switching and buffering, the two major tasks of a switch.

As dipicted in Figure 6, since each shared buffer functions as a small switch, the input queues only need to maintain N/S VOQs. And the size of both input and output schedulers is reduced from $N - to - 1$ to $N/S - to - 1$, as part of the switching is done by the MBSs.

In summary, by leveraging the speedup of the SRAMs on FPGAs, we can reap the following benefits:

1. reduced number of packet buffers.

2. reduced input-queue length.

3. reduced number of VOQs in the input queues.

4. reduced complexity for both the input and output scheduling.

We name the resulting switch fabric architecture, which employs an array of memory-based subswitches, the Combined Input and Grouped Crosspoint Queued (GCQ) switch.

3.2 The Shared Buffer Design

To allow access time sharing, the memories that implement the shared buffers in the crossbar should run S times faster than the flit arrival rate. And the S input and output ports are serviced in a time-multiplexing manner, as shown in Figure 7. Now because the shared buffers in the crossbar are running at a different clock frequency, clock domain crossing logic is needed both before and after the crossbar.

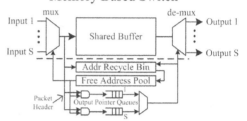

Figure 7: The structure of the shared buffer.

To achieve memory space sharing, a dynamic memory management scheme is implemented using a Free Address Pool, Output Pointer Queues and an Address Recycle Bin. Each incoming data is provided an address out of the Free Address Pool. The data is then written into the shared buffer using allocated address. At the same time, this address is pushed into destination Output Pointer Queues. The output de-multiplexer uses pointers from the Output Pointer Queues to access the shared buffer and sends data to corresponding output ports. Using the Output Pointer Queue structure, it is trivial to support multicast. The address of each multicast data is pushed into multiple Output Pointer Queues simultaneously, while only one copy of the multicast data is stored in the shared memory. The Address Recycle Bin keeps track of all references of each data in the shared buffer. When all references of a data are sent to the output, the address of that data is recycled to the Free Address Pool.

3.3 Scalability

Using the shared buffer design, the GCQ switch can, in theory, reduce the number of buffers in the crossbar by a factor of S^2. However, there are memory overhead associated with the shared buffer design. For example, it may require multiple SRAMs to provide an aggregate bandwidth of $2SR$; the Free Address Pool, Address Recycle Bin and Output Pointer Queues costs extra SRAMs to implement. Assume the implementation of each shared buffer requires P times more SRAMs compared to a single crosspoint buffer in the CICQ swith, the actual memory requirement becomes $(N/S)^2 \cdot P$. For example, if $S = 4$ and $P = 4$, there will be a 75% saving in the memory resource.

Assuming a single SRAM can provide a maximum bandwidth of B, each crosspoint buffer in the CICQ switch needs to achieve a total bandwidth of $2R$, thus requiring $2R/B$ SRAMs. Similarly, each shared buffer in the GCQ switch with a speedup of S costs $2SR/B$ SRAMS to implement. If $2R/B \geqslant 1$, each shared buffer needs S times more SRAMs compared to a single crosspoint buffer. Therefore, $P \geqslant S$. Otherwise if $2R/B \leqslant 1$, each crosspoint buffer in the CICQ switch still need at least 1 SRAM to implement, and the overhead of each Shared Buffer becomes $2RS/B$. Then the total memory saving will be $(1 - 2R/(SB))$.

EXAMPLE 2. *The 18Kb dual-port BRAMS in Xilinx's Virtex-6 have a maximum frequency of 525 MHz, and a data width of 36-bit, providing a total bandwidth of 37.8Gb/s. For a 10Gb/s line rate, the shared buffer of the GCQ switch can be implemented with an SRAM overhead of $0.53S$. For $S = 8$, the memory saving is 93%.*

4. HARDWARE IMPLEMENTATION

This section describes the implementation details of the proposed switch. We first discuss optimizations that are made in order to achieve a feasible hardware implementation. Later, we detail the hardware costs for different implementations of the proposed design. We target a 10 Gb/s link speed as it is and will likely continue to be widely used. Throughout this section, we use Xilinx's FPGA devices.

4.1 Clock and Memory Optimizations

Before hardware is implemented, the design parameter S should be determined according to the target memory technology. The 18Kb Block SRAM (BRAM) available on Xilinx FPGAs has two physical ports and a data width of 36 bits when operates in the Simple Dual Port mode. The BRAMs have an access time of around 2.5 ns on Virtex-6 (-1 speed grade) FPGA devices and 4 ns on Spartan-6 (-3 speed grade) devices. Assuming a minimum packet size of 40 bytes, the dual-port BRAMs can service up to 12 switch ports on Virtex-6 devices and 8 ports on Spartan-6 devices. However, it is impractical for FPGA designs to run at clock frequencies as high as 400 MHz. To make a feasible hardware design, we chose to have $S = 4$ and a data width of 32-byte, which requires 8 BRAMs running at 160 MHz, amounting to a total shared buffer size of 18 KB. With a 32-byte data width, the flit arrival rate is 40 MHz.

As the port logic and shared buffers in the crossbar are running at different clock frequencies, clock domain crossing logic is needed both before and after the crossbar. The input queues can be implemented as asynchronous FIFOs directly to reduce the resource overhead. Due to the large buffer size (18 KB) available in the shared buffers, the input queue buffers only need small amount of memory space to account for pipeline latencies. In order to save BRAMs, we used distributed RAMs to implement all the input queues.

4.2 Dynamic Memory Management Optimizations

To achieve a frequency of 160 MHz, the dynamic memory management logic for each shared buffer in the crossbar has to be simple enough. As shown in Figure 7, besides the data SRAMs, the dynamic memory management logic consists of the Output Pointer Queues, Free Address Pool and Address Recycle Bin. All data SRAMs operate in Simple Dual Port mode, and require straight forward read and write control logic. In current design, we assume First-Come-First-Serve service model, so that the Output Pointer Queues can be implemented as standard FIFOs. The Free Address Pool costs a single BRAM to implement multiple free address queues, and each input port is serviced in a time multiplexing manner.

The Address Recycle Bin however is not as straight forward especially when multicast support is required. For multicast packets, only one copy of the data is stored in the shared buffer, but could be accessed multiple times by different output ports. Since different outputs are not synchronous with each other, these multiple accesses could happen in different time slots. As a result, a counter is needed for each data item to maintain reference status and do proper address recycling.

A naive solution uses a bit vector to represent the destinations of each memory location in the shared buffers, with each bit corresponding to a specific output port. For every data written to shared buffers, its associated destination vector is updated with '1's, whereas each departing flit clears a corresponding bit. For each departing flit, the Address Recycle Bin checks if its destination vector becomes zero, and determines if the address should be recycled into the Free Address Pool. But the destination vectors are not easy

to implement because they are not small and have to be accessed in parallel.

EXAMPLE 3. *In a 4×4 shared buffer, each destination vector consists of 4 bits. The 18 KB shared buffer can store up to 576 flits of 32-byte data and requires 576 destination vectors, which is 2304 bits in total. In each time slot, a maximum of 4 flits will leave the shared buffer. If these 4 flits all read from the same memory location, the corresponding destination vector has to be updated 4 times continuously.*

Implementing the destination vectors using LUTs will increase the area of core logic greatly and make it difficult to meet timing. Although BRAM can be used to implement large bit vector arrays, it has read and write latency of at least 1 cycle, therefore do not support continuous updating. This problem can be solved by leveraging the byte-enable feature of the BRAM. Instead of representing each destination port with a single bit, one byte is used to denote a single destination port. Then the bit vector becomes a byte vector. The continuous updates can be done with byte-enable writes. In order for destination vectors to be checked after each update, the BRAM has to be configured to work in the "WRITE-FIRST" mode, so that each write to the BRAM will result in the updated vector appearing in the output port after certain latency. As a result, the Address Recycle Bin can be implemented with a True-Dual-Port BRAM plus a check logic in the output.

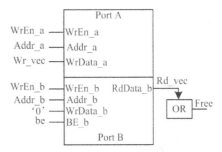

Figure 8: The Address Recycle Bin.

Figure 8 shows the implementation of destination vectors using a True-Dual-Port BRAM. For each incoming data written into the shared buffer, its corresponding destination vector is updated through Port A of the BRAM. Each departing flit will update a single byte in its destination vector using Port B with valid byte-enable signals. The updated vector will then appear in the output of Port B after a read latency. Figure 9 depicts the timing diagram of a continuous update operation. A stream of 4 updates is performed to the same address in the BRAM using different byte-enable signals. The updated data is streamed out after a latency of 1 cycle. The check logic at the output port examines the output data and generates a 'Free' signal when it finds that the output data is equal to zero.

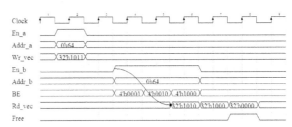

Figure 9: Timing diagram of a WRITE-FIRST SRAM.

Although feasible, the above method wastes a lot of memory space because one byte is used to represent the information of a single bit. For Virtex-6 FPGAs, this problem can be solved by exploiting the BRAM feature of independent read and write port width. For example, we can configure

Port B of the BRAM in Figure 8 to have a write width of 1 bit and a read width of 4 bits. As a result, the byte-enable signals are no longer needed and no memory area is wasted.

By examining the implementation details of all components, we show that the dynamic memory management logic of the shared buffer can be implemented with SRAMs and built-in FIFOs.

4.3 Wire Optimizations

For large switch designs, the internal wire routing is a big challenge. Considering a 16×16 GCQ switch with $S = 4$, and a data width of 32 bytes, each input port requires a 256-bit bus, which needs to be broadcast to 4 shared buffers in the same row of the crossbar. The entire switch requires a 4096-bit broadcasting bus. Given that the transceiver I/Os are typically distributed across different I/O banks, those broadcasting buses may need to travel a long distance and consume many routing resources. To alleviate this problem, we leverage the signal serialization method. As illustrated in Figure 10, the 4 input data buses are serialized into a single 256-bit bus using a 4-to-1 multiplexer, and then connected to 4 shared buffers. The serialized data bus has to run 4 times faster than the input data buses, at 160 MHz.

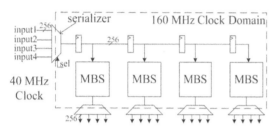

Figure 10: Wire optimization.

Signal serialization can reduce the width of the broadcasting bus, but it does not help with the long distance wiring problem. To cut the long wires down into shorter wires, registers have to be inserted. And the broadcasting bus becomes a multi-stage pipeline. The shared buffers connected to different stages of the pipeline therefore need to adjust their data arrival latency accordingly.

4.4 Hardware Cost

We chose 2 different FPGA devices to evaluate the hardware cost of the GCQ switch: Xilinx Virtex-6 XC6VLX240T-1 and Spartan-6 XC6SLX150T-3. The Virtex-6-240T device has a transceiver bandwidth of 158.4 Gb/s and 832 18Kb BRAMs; the Spartan-6-150T device has a transceiver bandwidth of 25.6 Gb/s and 268 18Kb BRAMs. Implementation and analysis was performed using version 13.1 of the Xilinx ISE Design Suite.

After applying the above described optimizations, the tool successfully routed a 16×16 switch for the Virtex-6-240T device and a 9×9 switch, with $S = 3$, for the Spartan-6-150T device. For our current designs, we assume the logic that transfers data into and out of the FGPA device through transceiver I/Os already exists. And we focus on the on-chip circuit implementation of the switch design. The resource utilization is listed in Table 3.

The 16×16 switch implemented on the Virtex-6 uses 224 18Kb BRAMs in total. As a comparison, a 16×16 CICQ switch has 256 crosspoint buffers. For each crosspoint buffer to run at 160 MHz, it requires at least 2 18Kb BRAMs to provide the data width. This results in a total of 512 BRAMs. Therefore, the proposed switch architecture is able to save 288 BRAMs for $S = 4$, which is 56% of the original requirement.

In the Spartan-6 case, the 9×9 switch costs 95 18Kb BRAMs in total. Compared to a 9×9 CICQ switch, which

Table 3: Resource Utilization.

	Virtex6-240T	Spartan6-150T
N	16	9
S	4	3
Data Width	256 bits	256 bits
Core Frequency	160 MHz	120 MHz
Latency	250 ns	250 ns
Registers	36945(12%)	27028(14%)
LUTs	49537(32%)	37285(40%)
BRAMs	224(27%)	95(36%)
BRAM Saving	288(56%)	67(41%)

requires at least 162 18Kb BRAMs to run at 120MHz, the proposed architecture achieves a saving of 67 BRAMs (41%) for $S = 3$. In Spartan-6, each 18Kb BRAM in Spartan-6 FPGAs can be used as two independent 9Kb BRAMs, enabling more efficient implementation of the control logic in the core switch. For example, the Output Pointer Queues and Free Address Pool favours smaller BRAMs with narrower data width, because they only deal with small pointer size. Therefore, the dynamic memory management logic can be packed into fewer number of 18Kb BRAMs.

Both designs use many LUTs and registers to implement the clock domain crossing logic, and the broadcasting data buses. In total, the designs cost about 32% of total LUT resource of the target Virtex-6 device and 40% of the Spartan-6 device. The cost of the clock domain crossing logic can be further reduced by employing the source synchronous clock domain crossing technique in future design. The transceiver bandwidth is successfully saturated in both devices. The efficient utilization of the hardware resource provides large room for other applications, so that they can easily integrate with the switch design.

Both implementations achieve a port-to-port latency of 250 ns, most of which is spent on the clock domain crossing logic. But, we did not consider the latency of data going into and out of the FPGA through the transceiver I/Os, which normally has a total latency of around 50ns. Compared to the ASIC design of Fulcrum's FocalPoint FM4000 switch published in 2009 [7], which has a latency of 300 ns, the proposed design performs closely.

Since the proposed design is very symmetric in physical layout and only simple control logic is involved, it is possible that the speed and quality of the Place & Route can be greatly enhanced with the help of manually BRAM placement. And a higher internal speedup can be expected. Furthermore, the dynamic memory management logic of the shared buffer in the crossbar is necessitated by the support multicast traffic. If only unicast is required, the memory overhead associated with the dynamic memory management can be eliminated. And the proposed design will achieve better resource saving as discussed in Section 3.3.

5. PERFORMANCE EVALUATION

Having validated the implementation feasibility with a 16x16 switch with 160Gps switching performance, which saturates the bandwidth of the device available to us, we now turn to evaluate the the strength and weakness of the general GCQ switch architecture relative to those in the literatures, in particular, the CICQ architecture.

5.1 Evaluation Setup

To evaluate switch architecture performance, we implemented a cycle-accurate C model of the proposed switch architecture, and integrated in to Booksim, a comprehensive interconnection network simulator [1], supplied as a companion for a classic textbook on the subject [8]. The simulator provides standard traffic generators, different network topologies and performance measurement facilities, therefore enabling the fair comparison of different architectures.

We compare proposed GCQ switch architecture with IQ,

OQ, and CICQ switch architectures. For GCQ, we provide two variants: GCQ-S4, for internal speed up $S = 4$, and GCQ-S8, for $S = 8$. Key simulation configuration settings are summarized in Table 4. Here because for a flit to travel from the input queues to the crossbar it is necessary to cross clock domains, we set aside two cycles of delay. Same is true for credit packets (for flow control). The switch radix is set as 16, same as the one demonstrated in the previous section. We also simulated small packet with size of 1 flit, and size of 16 flits. The traffic is chosen as uniform traffic using an independent and identical Bernoulli process. Finally, the simulator is warmed up with traffic load before real measurement is taken.

Table 4: Simulation settings.

Flit delay	2
Credit delay	2
N	16
Flit size	32 bytes
Packet size	1 flit, 16 flit
Traffic	uniform
Network Topology	fat tree
Routing	Nearest common ancestor

For fair comparison, we assume all switch architecture have the same total buffer space.

We study both the single switch (16x16) performance, as well as network performance. The network topology is chosen as a 3-level fat-tree, which contains 192 switches and 512 ports, resulting in a total of 5 Tb/s switching capacity.

5.2 Throughput Test

We first evaluate if the proposed switch is able to achieve ideal throughput. In this experiment, we fix the buffer space in the the crossbar, and *sweep* the different input queue sizes. We set a single-flit buffer for each crosspoint for the CICQ switch (CICQ-1flit). We set the equivalent (for equal total buffer space) of 16-flit and 64-flit shared buffer for GCQ-S4 and GCQ-S8. Test results are shown in Figure 11 and 12 for different packet sizes. The horizontal axis captures the depth of each input queue and the vertical axis captures the throughput. We are interested in the "knees" of the curve for different architectures.

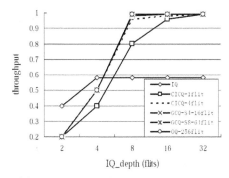

Figure 11: Throughput test with single-flit packets.

5.2.1 Small Packet

Figure 11 shows the test results using traffic consists of single-flit packets. Due to the flit traversing latency, when the input-queue-depth is smaller than 4, all switches have low throughput. The IQ switch has slightly higher throughput than others at small input-queue-depths because it only has a single stage of packet buffering inside the switch, while others buffer each packet twice inside the switches.

When input-queue-depth increases to 32, all switches except the IQ switch reach a saturation throughput close to 100%.

The CICQ switch has a much lower throughput than others at an input-queue-depth of 8. It is because the flit transferring delay between input queues and the crossbar will cause the input queue with small depth to overflow.

To see the small cross point buffer does cause problems, we also show performance of CICQ switch with 4-flit crosspoint buffers (CICQ-4flit): throughput does enhance.

The IQ switch has a saturation throughput fixed at about 58% due to the HOL blocking phenomenon.

5.2.2 Large Packet

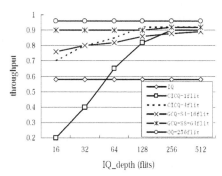

Figure 12: Throughput test with 16-flit packets.

Figure 12 shows the test results using traffic consists of 16-flit packets. With longer packets, there is higher demand on congestion buffers, as it needs much larger packet buffers to temporarily store multiple packets. The CICQ switch with 1-flit crosspoint buffer performs worst compared to other switches when the input-queue-depth is 16, because there is not enough packet buffer to resolve congestion.

The GCQ-S8 and OQ switches achieve very high throughput even with small input queue sizes because their crossbar buffers or output-queue buffers are deep enough to resolve congestion. When input-queue-depth increases to 128-flit and higher, the changes in saturation throughput of all switches become very small, indicating that the performance is now limited by the buffers in the crossbar or output queues. with an input-queue-depth of 128, the total packet buffer size of different switches is 2304 flits.

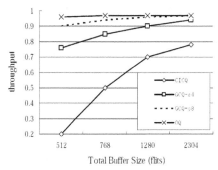

Figure 13: Throughput test with fixed IQ-depth.

5.3 Memory Efficiency

The above taught us that GCQ architecture is more "tolerant" on the input buffer space. Put it on another way, for the same input buffer space, it requires *less* packet buffers in the crossbar in order to achieve full throughput.

In this experiment, we fix the input-queue-depth as 16 flits. We then *sweep* the different cross-point buffer size and compare the throughput. For example, for a total buffer size of 2304 flits, we dedicate 256 flits of space the input queues, which left us with 2048-flit total space for others. For the CICQ switch, that means a maximum of 8-flit in each crosspoint buffer, while the GCQ-S4 and GCQ-S8 can have a maximal buffer size of 128 and 512 flits respectively.

Using traffic of 16-flit packets, the throughput of different switches with different total buffer sizes is shown in Figure 13. The OQ switch arrives at a stable result with a total buffer size of 768 flits. The GCQ-S8 achieves the same throughput of the OQ switch at a total buffer size of 2304 flits. The GCQ-S4 performs slightly worse than the OQ switch with a total buffer size of 2304 flits. The CICQ switch performs much worse due to its N^2 crosspoint buffer requirement. Because the total buffer size will increase fast with a small increase in each of its crosspoint buffer, there is not enough buffer depth to resolve congestion when the input-queue-depth is also small.

It is interesting to note that to achieve 80% throughput, the CICQ requires 2304 flit space, whereas the GCQ-S4 requires only 512 flit space.

5.4 Latency Test

We are also interested in latency result of different switch architectures under different *traffic loads*. We chose to compare the CICQ switch with 4-flit crosspoint buffer, the OQ switch with 1024 flits of total output-queue buffers as well as the GCQ switch with $S = 4$ and different sizes of shared buffers from 8 flits to 64 flits. The input-queue-depth of different switches are set to 16 flits. The IQ switch with 80 flits in each input queue is also tested. Test results are shown in Figure 14 and 15. The horizontal axis captures the injected traffic load and the vertical axis captures the average packet latency.

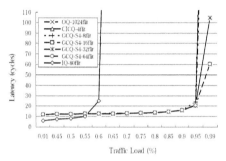

Figure 14: Latency test witch single-flit packets.

Figure 14 shows the test results using traffic consists of single-flit packets. The average latency of different switches is quite stable for a traffic load lower than 95%. For traffic loads higher than 95%, the GCQ-S4-64flit switch achieves the same performance as the OQ switch. The GCQ-S4-8flit switch performs closely to the CICQ switch. However, the total buffer size of the tested CICQ switch is more than 3 times larger than that of the GCQ switch with an 8-flit size of shared buffers.

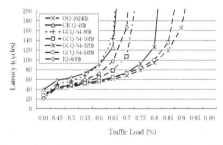

Figure 15: Latency test witch 16-flit packets.

Figure 15 shows the test results using traffic consists of 16-flit packets. Again, the GCQ-S4-8flit switch performs closely to the CICQ switch. And the GCQ-S4-64-flit switch

performs closely to the OQ switch. These tests demonstrates that, with a small internal speedup of 4, the GCQ switch is able to approach the ideal OQ switch performance. And the GCQ switch is much more memory efficient in terms of performance than the CICQ switch.

Looking from another angle, with the same total buffer space, the "knees" of the GCQ curve is much more later than that of the CICQ curve. In other words, the GCQ will sustain low latency at much higher traffic load than the CICQ.

5.5 Network Performance

For scalability test, we constructed a 3-level fat-tree using different switches as the base element. As is shown in Figure 16, the fat-tree has 192 switches and 512 switch ports. Assume a 10 Gb/s port speed, this fat-tree provides a total capacity of 5 Tb/s. The Nearest Common Ancestor with random output selection is used as the routing algorithm in the fat-tree. A packet will go through a maximum of 5 hops to reach any destination port in the tree. Same as previous tests, we use uniform traffic to test the performance of the fat-tree, and results are shown in Figure 17 and 18.

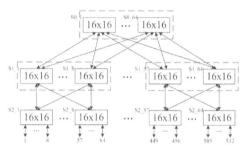

Figure 16: The Fat-tree switch network.

Figure 17 plots the test results using traffic consists of single-flit packets. The fat-trees constructed from different switch elements exhibit similar performance features as their base switches, but have higher packet latency. The GCQ-S4-32flit fat-tree achieves almost the same performance as the GCQ-S4-64flit as well as OQ fat-trees, which indicates the GCQ switch with a shared buffer size of 32 flits already has enough packet buffer for fat-tree topology in this test. The fat-tree constructed from the CICQ switch performs closely to the GCQ-S4-16flit fat-tree when the traffic load is high. The IQ fat-tree saturates at a traffic load of about 60%, which is similar to the single switch performance.

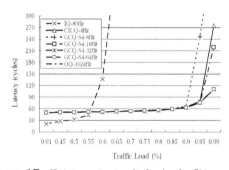

Figure 17: Fat-tree test witch single-flit packets.

Figure 17 shows the test results using traffic consists of 16-flit packets. The GCQ-S4-8flit, GCQ-S4-16flit and CICQ fat-trees quickly saturate at a traffic load of around 60%, because there are not enough buffer depth.The IQ fat-tree benefits from the long packet traffic and saturates at around 65% traffic load. The GCQ-S4-64flit fat-tree performs closely to the OQ fat-tree with a saturation throughput of approximately 85%. These results demonstrate that, the GCQ switch with a shared buffer size of 64 flits performs closely to the ideal OQ switch with the same total packet buffer size, and is less sensitive to changes of burst length of the traffic.

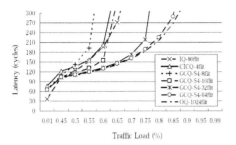

Figure 18: Fat-tree test witch 16-flit packets.

6. CONCLUSION

In this paper, we describe a new switch fabric organization and argue its merits against other organizations in general, the combined input and crosspoint queued (CICQ) switch in particular. The proposal was demonstrated by an implementation of a single-chip 16x16 switch fabric, reaching 160Gps switching capacity. This is the largest permissible for the Virtex6-240T device available to us on a ML605 development board, running at a modest frequency of 40Mhz for port processing logic, and 160Mhz for the BRAMs.

We now return back to our question raised in the beginning of the article: *is it possible to implement a single-chip switch on FPGA that saturates its transceiver bandwidth?*

Consider the largest, the latest (not yet shipped, but advertised) Xilinx Virtex-7 XC7VH870T device, which has a maximum transceiver bandwidth of about 1.4 Tbps (with 72 13.1Gps links and 16 28.05Gps links), and the maximum frequency of 600 MHz for 2820 18-Kb, 36-bit wide dual-port BRAMs [2]. For ease of calculation, consider the implementation of a high radix, 100x100 switch with 10Gps link speed, running 400Mhz for its BRAMs.

For CICQ, each cross point buffer requires 1 BRAM, therefore a total of at least 10,000 BRAMs, not including input buffers. This is far from being possible.

In contrast, with the proposed GCQ architecture with a speed up of $S = 8$, each Shared Buffer requires 6 BRAMs, therefore a total of 1014 BRAMs, about half of what is available on the device.

7. ACKNOWLEDEGEMENTS

The authors like to thank Canwen Xiao for his critiques and help in the network simulation. The authors also like to thank the support of National Sciences and Engineering Research Council of Canada, as well as China Scholarship Council for the first author.

8. REFERENCES

[1] Booksim interconnection network simulator. http://nocs.stanford.edu/cgi-bin/trac.cgi/wiki/resources/booksim.

[2] Xilinx virtex-7 data sheet. http://www.xilinx.com/support/documentation/data_sheets/ds183_Virtex_7_Data_Sheet.pdf.

[3] F. Abel, C. Minkenberg, R. P. Luijten, M. Gusat, and I. Iliadis. A four-terabit packet switch supporting long round-trip times. 2002.

[4] Actel, Inc. Designing high-speed ATM switch fabrics by using Actel FPGAs. http://www.actel.com/documents/hispeedatm_an.pdf, 1996.

[5] Altera, Inc. Integrating 100-GbE switching solutions on 28-nms FPGAs. http://www.altera.com/literature/wp/wp-01127-stxv-100gbe-switching.pdf, 2010.

[6] N. Binkert, A. Davis, N. P. Jouppi, M. McLaren, N. Muralimanohar, R. Schreiber, and J. H. Ahn. The role of optics in future high radix switch design. In *Proceedings of the 38th annual international symposium on Computer architecture*, ISCA '11, pages 437–448, New York, NY, USA, 2011. ACM.

[7] U. Cummings, D. Daly, R. Collins, and V. Agarwal. Fulcrum's FocalPoint FM4000: A scalable, low-latency 10 gige switch for high-performance data centers. In *17th IEEE Symposium on High Performance Interconnects*, 2009.

[8] W. J. Dally and B. Towles. *Principles and Practices of Interconnection Networks*. Morgan Kaufman, 2004.

[9] W. E. Denzel, A. P. J. Engbersen, and I. Iliadis. A flexible shared-buffer switch for ATM at Gb/s rates. *Computer Networks & IDSN Systems*, 27:611–624, January 1995.

[10] A. P. J. Engbersen. Prizma switch technology. *IBM Journal of Research and Development*, March, 2003.

[11] S. Iyer. Load balancing and parallelism for the internet. *PhD. Thesis, Standford University*, July, 2008.

[12] S. Iyer and N. McKeown. Using constraint sets to achieve delay bounds in CIOQ switches. *IEEE Communications Letters*, 7(6), June, 2003.

[13] Y. Kanizo, D. Hay, and I. Keslassy. The crosspoint-queued switch. In *IEEE International Conference on Computer Communications*, Rio de Janeiro, Brazil, 2009.

[14] M. J. Karol, M. G. Hluchyj, and S. P. Morgan. Input vs. output queuing on a space-division packet switch. *IEEE Transactions on Communication*, 35(12):1347–1356, 1987.

[15] J. Kim, W. J. Dally, B. Towles, and A. K. Gupta. Microarchitecture of a high-radix router. In *32nd Annual International Symposium on Computer Architecture*, New York, NY, USA, 2005.

[16] A. W. Lockwood, N. McKeown, G. Watson, G. Gibb, P. Hartke, J. Naous, R. Raghuraman, and J. Luo. NetFPGA - an open platform for gigabit-rate network switching and routing. In *IEEE Microelectronic Systems Education*, San Diego, CA, USA, June 2007.

[17] N. McKeown. Scheduling algorithms for input-queued cell switches. *PhD. Thesis, University of California at Berkeley*, 1995.

[18] N. McKeown, M. Lzzard, A. Mekkittikul, W. Ellersick, and M. Horowitz. The tiny tera: A small high-bandwidth packet switch core. In *Proceedings of Hot Interconnects IV*.

[19] N. McKeown, A. Mekkittikul, V. Anantharam, and J. Walrand. Achieving 100% throughput in an input queue switch. In *IEEE International Conference on Computer Communications*, San Francisco, CA, USA, 1996.

[20] C. Minkenberg and T. Engbersen. A combined input and output queued packet switched system based on PRIZMA switch-on-a-chip technology. *IEEE Communication Magazine*, 38:70–77, 2000.

[21] D. Simos. Design of a 32x32 variable-packet-size buffered crossbar switch chip. *MSc. Thesis, University of Crete*, July, 2004.

[22] Y. Tamir and G. Frazier. High performance multi-queue buffers for VLSI communication switches. In *15th Annual International Symposium on Computer Architecture*, HI, USA, June 1988.

[23] Xilinx, Inc. High-speed buffered crossbar switch design using Virtex-EM devices. http://japan.xilinx.com/support/documentation/application_notes/xapp240.pdf, 2000.

[24] Xilinx, Inc. Building crosspoint switches with CoolRunner-II CPLDs. http://www.xilinx.com/support/documentation/application_notes/xapp380.pdf, 2002.

[25] K. Yoshigoe, K. Christensen, and A. Jacob. The RR/RR CICQ switch: Hardware design for 10-Gbps link speed. In *IEEE International Performance, Computing, and Communications Conference*, 2003.

The VTR Project: Architecture and CAD for FPGAs from Verilog to Routing

Jonathan Rose[1], Jason Luu[1], Chi Wai Yu[4], Opal Densmore[1], Jeffrey Goeders[3],
Andrew Somerville[2], Kenneth B. Kent[2], Peter Jamieson[5] and Jason Anderson[1]

[1]Dept. Electrical and Computer Engineering, University of Toronto
[2]Dept. Computer Science, University of New Brunswick
[3]Dept. Electrical and Computer Engineering, University of British Columbia
[4]Dept. Electrical Engineering, City University of Hong Kong
[5]Dept. Electrical and Computer Engineering, Miami University

ABSTRACT

To facilitate the development of future FPGA architectures and CAD tools – both embedded programmable fabrics and pure-play FPGAs – there is a need for a large scale, publicly available software suite that can synthesize circuits into easily-described hypothetical FPGA architectures. These circuits should be captured at the HDL level, or higher, and pass through logical and physical synthesis. Such a tool must provide detailed modelling of area, performance and energy to enable architecture exploration. As software flows themselves evolve to permit design capture at ever higher levels of abstraction, this downstream full-implementation flow will always be required. This paper describes the current status and new release of an ongoing effort to create such a flow - the 'Verilog to Routing' (VTR) project, which is a broad collaboration of researchers. There are three core tools: ODIN II [10] for Verilog Elaboration and front-end hard-block synthesis, ABC [16] for logic synthesis, and VPR [13] for physical synthesis and analysis. ODIN II now has a simulation capability to help verify that its output is correct, as well as specialized synthesis at the elaboration step for multipliers and memories. ABC is used to optimize the 'soft' logic of the FPGA. The VPR-based packing, placement and routing is now fully timing-driven (the previous release was not) and includes new capability to target complex logic blocks. In addition we have added a set of four large benchmark circuits to a suite of previously-released Verilog HDL circuits. Finally, we illustrate the use of the new flow by using it to help architect a floating-point unit in an FPGA, and contrast it with a prior, much longer effort that was required to do the same thing.

Categories and Subject Descriptors

B.5.2 [**Design Aids**]: Automatic Synthesis, Optimization

General Terms

Algorithms, Design, Architecture, Measurement, Performance

1. INTRODUCTION

The exploration of new programmable architectures, and the development of innovative algorithms required to synthesize circuits into FPGAs requires a robust software flow that permits experimentation. In order to model modern and future architectures, such a flow is necessarily quite complex, and largely beyond the capacity of any single academic enterprise to create, evolve and maintain. By contrast, the related commercial flows are supported by hundreds of full-time engineers. Equally important, to serve the same needs, is a set of relevant large-scale circuit benchmarks that can be used to test architectures and algorithms. This paper describes the status of a global collaboration attempting to provide such a framework - including several innovations within the three main parts of the tool flow, new work to create robust benchmarks, and an illustration of the flow's capability to explore a new kind of hard logic block.

One of the goals of this project is to provide a managed repository where enhancements of the flow by others can be more easily integrated into the suite of tools. Without such an effort, innovations are often orphaned, preventing our field from progressing by building on each other's work. These goals are challenging, because making a flow that is robust requires more effort than a typical academic project and publication requires.

In this paper we first describe several advancements in the core set of tools - Section 2 describe new features of the ODIN II tool [10] which takes in the Verilog description of the circuit and elaborates it into a BLIF netlist. Section 3 describes the use of the ABC tool [16], which performs logic optimization and technology mapping on the soft-logic portion of the BLIF. Section 4 describe advances in the VPR tool [13] which takes the synthesized BLIF and performs physical synthesis and timing analysis. Section 5 gives the basic flow's result for the set of previous and some new Verilog benchmark circuits that are a part of the related release. Section 6 provides a case study of the use of the flow to replicate previous work (done on a branch of the original VPR 4.3[3] flow) to model floating-point logic blocks. Section 7 gives the details of the full release of software, architecture files and benchmark circuits. Section 8 outlines the extensive set of additional features we see as necessary to continue this work, while Section 9 concludes.

2. ODIN II: ELABORATION

The Odin II Verilog elaboration front end [10] has four key roles in the VTR framework:

1. To interpret and convert some of the Verilog syntax into a logical netlist targeting the 'soft logic' on the FPGA.

2. To synthesize other constructs directly into 'hard logic' blocks on the FPGA, making specific use of the logical properties of those blocks to ensure that the logical netlist is physically realizable.

3. To be responsive to the architecture description of the FPGA. This is provided in the architecture description file which contains an extensive description of physical properties of the FPGA, together with a small amount of the logical properties. This includes the routing architecture of the FPGA [3], the internal structure of the logic blocks [13], the global pattern of logic blocks, and the I/O structure. Examples of a portion of an architecture file are given in Section 6.

4. To provide a framework for the verification of the correctness of the software flow.

This section reports on several new capabilities in these roles, including synthesis for memories, multipliers and other hard logic, a macro pre-processor, and a new verification infrastructure.

2.1 Compilation

It is essential for the elaboration step to be aware of the higher-level functionality of hard blocks in existing and hypothetical FPGAs. The Odin II Verilog compiler has been improved to enable more sophisticated mapping of hard multipliers and memories on FPGAs, which we describe in detail in this section, in addition to other 'generic' hard blocks that the architect can model.

2.1.1 Multipliers

Multiplication is a very common operation in digital circuits, and so modern FPGAs have included hard multipliers [2][1] for area-efficiency and higher performance. Odin II detects the multiplication operator (*) and synthesizes it directly into a hard multiplier block on the FPGA, if one exists in the architecture description file. Multiplication can appear in Verilog either as an explicit instantiation, or implicitly as a multiplier operation, as shown in Figure 1. Currently, Odin II can only synthesize unsigned multipliers.

The key issue with the elaboration and synthesis of multiplication is the transformation between the logical specification of the operation and its implementation using the available physical hard blocks [19]. The input circuit can contain any size of multiplier operation, whereas the FPGA architecture will typically have a physical hard block that is fixed in size. For example, a design could contain a logical 128-bit x 128-bit multiply that produces a 256-bit result, while the physical FPGA may contain only 8-bit x 8-bit multiplication hard blocks (as specified in the architecture description file) that provide a 16-bit result. This requires

```
multiply my_mult (a, b, out1);  // explicit

always @(c,d)
begin
    out2 <= c * d;  // implicit
end
```

Figure 1: Explicit & Implicit Multiply in Verilog

a 'splitting' of the large logical multiplier into many smaller multiplications that make use of the hard multipliers, and the synthesis of some soft logic, between the hard multipliers, to create the correct arithmetic function. This splitting operation can best be described through an example: Consider a logical multiplication operation that splits perfectly in half (into two multipliers of exactly half the size) the result is four multiplication operations (each of 50% the size) and three addition operations, as illustrated in Figure 2. The additions must be implemented in generated soft logic. Figure 2 shows the long multiplication form of A x B. If the resulting multiplication operations remain too large to implement in a hard multiplier, this splitting process is repeated recursively until the multiplication operation is small enough to fit.

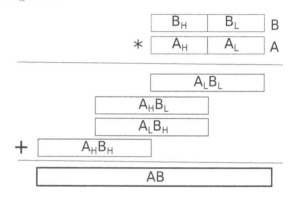

Figure 2: Long multiplication of variables A and B.

It is important to note that, in working with hard and soft multipliers, it has become clear that it is very inefficient to use larger physical hard multipliers to implement small logical multipliers - the resulting circuit is faster and smaller when small logical multipliers are built in soft logic. For this reason, Odin II has a parameter that sets the size of the logical multiplier that is too small to be implemented in a hard physical multiplier. This parameter is also applied to the remaining multiplier after the recursive splitting process described above. This parameter should be a function of the size of the smallest hard multiplier available on the FPGA; however we do not yet automate that setting.

2.1.2 Memories

Memory is a key component of digital circuits, and embedded memories are commonly found in many commercial FPGA architectures [1][2]. Odin II identifies explicit instantiations of memory (implicit memory coded as Verilog arrays is not yet supported) in the input Verilog circuit. Explicit memories, both single and dual port, can be specified by the designer as illustrated in Figure 4. These logical memories are synthesized into the hard blocks that exist in the target FPGA architecture, as specified in the architecture description file. Figure 3 illustrates part of the definition of a hard

```
<model name="single_port_ram">
  <input_ports>
  <port name="we"/>      <!-- control -->
  <port name="addr"/>  <!-- address lines -->
  <port name="data"/>  <!-- data lines -->
  <port name="clk" is_clock="1"/>
  </input_ports>
  <output_ports>
  <port name="out"/>     <!-- output -->
  </output_ports>
</model>
```

Figure 3: Excerpt of FPGA Architecture Description File Showing Model of Single Port Memory

```
single_port_ram my_mem(we, data, addr, clk, out);

dual_port_ram my_mem2(we1, we2, addr1, addr2, data1,
    data2, clk1, out1, out2);
```

Figure 4: Example Verilog Instantiations of Single and Dual Port Memories

block single-port RAM in the FPGA architecture description file, which is used by ODIN II to perform the synthesis. The keywords **single_port_ram** and **dual_port_ram** along with the ports for each shown in Figure 4 have specific meanings that are necessary to perform the memory synthesis within Odin II. A later physical layout section of the description file gives the size of the memory ports, which is also required.

Similar to multipliers, memories will appear in application circuits in many different logical sizes, and some will have to be split to fit into the physical size of memories on the FPGA. This may also require the generation of additional soft logic. The depth of a memory can be split in half by utilizing one of the address lines to select one of two smaller memories to access. The width of a memory can be split by utilizing two memories where the entries are concatenated together to achieve the required data width. Furthermore, most FPGA memories have the ability to trade depth for width, making the full memory synthesis problem somewhat intricate [9]. Odin II can be set to split the logical memories in two ways:

1. It can split a logical memory down into the size of the smallest physical memory, or

2. It can split all memories into the smallest data size possible (ie. 1-bit data). This method relies on downstream tools to pack memories together into the physical hard block size; we describe that more fully in Section 4. This can potentially render better results as multiple logical memories can share a single physical hard block.

2.1.3 Generic Hard Blocks

Multipliers and memories are common entities that are found in many digital circuits. A key goal of the VTR project is to enable exploration of other dedicated hard blocks, in order to explore their effectiveness. Odin II supports the detection and synthesis of such generic hard blocks. To use this feature, the design must explicitly specify the precise usage of the generic hard block (similar to explicit specification of memories in Figure 4) where the logical size is identical to the physical size. The specification of a generic hard block in the FPGA architecture is identical to that of memories and multipliers. A model must be described that

specifies the name of the block along with the input and output port names. Subsequently in the physical layout of the FPGA the size of the ports must be provided. No splitting of generic hard blocks can be provided as this is dependent on the block's functionality. It is an open question as to whether there is a method of specification of a hard block that can concisely include how it can be split.

2.2 Pre-processor

An addition to Odin II in this release is a Verilog pre-processor that can support greater Verilog language coverage. The pre-processor performs an additional parsing of the Verilog circuit description prior to Verilog compilation. The pre-processing step provides support of Verilog compiler directives (i.e. **define**, **ifdef**, **else**, **endif** and **include**). This support has substantially increased the flexibility in writing Verilog benchmarks for the VTR project and has increased the likelihood that an existing circuit description can pass through the VTR flow without requiring modification.

2.3 Verification

As we have developed the VTR flow, it has become important to ensure that the output of Odin II, and ultimately the downstream tools, is correct. It is particularly difficult to verify that the output of Odin II is correct when the user defines new hard blocks with new functionality. This requires an ability to separately specify the logical functionality of those blocks to any kind of verification tool. To address this issue, a logic simulator was developed [17] inside ODIN II. The simulator exercises either the Verilog input (after elaboration into ODIN II's internal data structures) or a BLIF netlist file, as the specification of the circuit. It can use either an optional set of input test vectors that stimulate the circuit, or the simulator will generate a specified number of random vectors.

The logic functionality multiplication and memory hard blocks is built in to the Odin II simulation, as they are the most common blocks. When the FPGA architect wants to create custom hard blocks with unique functionality, the architect must also provide the simulator with a C-language description of the logical functionality. At simulation run-time, these blocks are simulated by loading these compiled codes as run-time libraries.

The output of the simulator is the computed result of the input vectors for the circuit, from either the Verilog or BLIF versions of the circuit. Once complete, any output vectors can be compared for equality to achieve a level of confidence that a circuit is synthesized correctly. In addition, the output of the post-synthesis simulation can be compared against the pre-synthesis simulation of commercial simulators, such as Modelsim from Mentor. In the benchmarks presented in Section 5, we have employed this latter method of verification.

3. ABC: LOGIC SYNTHESIS

We use the ABC logic synthesis framework [16] for technology independent logic synthesis and technology mapping to LUTs and flip-flops. ABC has evolved over the years to produce higher-quality results with a number of innovations. It has also evolved to handle both hard structures on FPGAs, as well as predefined soft structures, as discussed below. ABC represents a logic circuit using a network of two-input `AND` gates and inverters: an `AND`-inverter graph (AIG). We use ABC's `resyn2` script for technology independent optimization, which iteratively calls ABC commands that optimize the AIG to reduce the number of nodes and balance the lengths of its paths, thereby minimizing the maximum number of `AND` gates on any combinational path.

We also employ the WireMap [11] technqiue for technology mapping the AIG into K-input LUTs. WireMap produces depth-optimal mappings for a given AIG, while attempting to minimize the number of *used* inputs in the resulting LUTs. This reduction in LUT inputs benefits both routability and power [11], as well as facilitating a more efficient packing of these smaller LUTs into dual-output fracturable LUT architectures [13].

While soft logic is represented as an AIG in ABC, hard blocks, such as memories and multipliers, are received from the Odin II front-end modelled as *black boxes* in ABC. More recent versions of ABC permit the modelling of timing paths through black boxes which allows the synthesis and mapping steps to optimize the surrounding soft logic while taking that timing into account. We do not yet take advantage of this feature of ABC, but hope to do so in the near future. To do so will require experimentation and measurement of its effectiveness. We will also work similarly to move to modelling hard blocks as *white boxes*, wherein their internal logic functionality is exposed. The logic circuits in the transitive fanin and fanout of the white box can then be further optimized, for example, by leveraging don't-cares arising from the white box logic functionality.

4. VPR: PHYSICAL SYNTHESIS

This release of VTR includes an important new release of the VPR tool that is based on the VPR 6.0 beta release described in [13]. That version introduced new constructs that allow the description and packing of far more complex logic blocks. The architecture description file can now describe an essentially arbitrary interconnection of primitives inside the block, together with unlimited layers of hierarchy and multiple modes of operation within each piece of the hierarchy. That version, however, suffered from a lack of timing-driven physical synthesis - none of the packing, placement or routing phases was timing driven.

In the present release, timing-driven functionality has now been implemented. To do this, the timing analyzer in VPR, and most particularly the timing graph generator inside the timing analyzer, needed to generate timing graphs that reflect the arbitrary graph of connectivity that is now possible inside the complex logic blocks. It must also correctly model the different modes of operation, with different timing numbers for each mode. This has now been done; since the VPR placement and routing algorithms only needed proper timing analysis to work, this enhancement was all that was required to make both of those steps timing-driven. There are two interesting issues that arose in this work that will be described in the two subsequent sections: first, we needed to create a timing-driven packing algorithm that could deal with the arbitrarily complex logic blocks. Second, we had to come up with a clean method for handling the complexity of specifying the timing inside hard and soft logic blocks

4.1 Timing-Driven Packing

We begin by describing the area-driven packing algorithm in [13] and then describe how it was enhanced. The area-driven algorithm first selects a complex logic block, and populates it with primitives from the netlist, one at a time, until the complex block is full. This is repeated until the entire netlist is packed into complex blocks. This algorithm resolves the mode and hierarchy requirements of a complex logic block by using a depth-first traversal of the hierarchy and mode tree to find the appropriate location and mode settings for each netlist block. For each such selection, the algorithm invokes a router to ensure that the specific choice is feasible - in the general case it is necessary to perform routing within the block because there is no guarantee that any arbitrary network of switches will allow the candidate packed primitives to connect. The selection of which netlist primitive to pack next is based on an attraction function that was entirely area driven.

To make this algorithm timing-driven, the attraction function was modified based on a new timing model. The timing model used is a simplified version of the model used by T-VPack [15]. All blocks are modelled to have a logical depth of one. Nets are modelled as having zero delay. The timing analysis engine calculates the normalized criticality of each netlist primitive by dividing by the longest-path logical depth. Tie breakers are employed to determine which block should have higher criticality in the event that two blocks share the same depth criticality. These tie breakers are the same as those found in [15]; they are based on counting the number of critical/near critical paths that pass through a particular block.

The attraction function was modified to account for timing as follows:

$$Attr = \alpha \cdot criticality(B) + (1 - \alpha) \cdot area\_attraction(B) \quad (1)$$

Where B is the candidate netlist block B and $area\_attraction$ is the attraction function used in [13]. The parameter $alpha$ defaults to 0.75 to place a heavier emphasis on criticality.

As with earlier works on packing, the criticality of blocks serves as a good proxy for packing critical edges into the cluster. Due to the strong weighting in the attraction function in favour of criticality, there is a bias for packed netlist blocks to have equal or higher criticality than candidates. If a packed block is critical and the candidate block it is connected to is critical, then the edge to the candidate from the cluster is critical too. Thus, block criticalities alone achieves the goal of absorbing critical edges most of the time while avoiding the larger development effort needed to implement edge criticalities.

We performed an experiment to compare the impact of timing-driven and area-only packing in VPR 6.0 using the legacy T-VPack/VPR 5.0 flow as the baseline. The largest 20 MCNC circuits were mapped onto a transistor optimized homogeneous FPGA from the iFar repository [12] with clusters of ten 6-LUTs. The timing-driven VPR 6.0 flow gave, on average, the same critical path delay as T-VPack/VPR 5.0. The area-driven packer followed by timing-driven placement and routing, on the other hand, produced circuits that were 5% slower on average.

4.2 Timing Specification

The goal of the VPR 6.0 release was to enable the architectural exploration of very complex logic blocks. With the addition of a fully-timing driven flow, it was necessary to ensure that proper timing modelling and analysis was performed at the various stages of the flow. It was necessary to rethink how the timing of the primitives are described in the architecture description file, so that a wide variety of new kinds of blocks can be correctly modelled. Timing paths through a primitive can be either purely combinational, or have registered inputs, and/or have registered outputs, and/or have an internal pipeline, or some combination of these. It turns out that for certain primitives, it is not necessary to completely specify all its timing paths. For example, Figure 5 shows the timing for a primitive with an internal pipeline. Notice that, for the purposes of physical synthesis, the primitive is indivisible so it is not necessary to specify the timing in this level of detail.

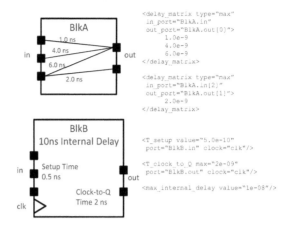

```
<delay_matrix type="max"
   in_port="BlkA.in"
   out_port="BlkA.out[0]">
      1.0e-9
      4.0e-9
      6.0e-9
</delay_matrix>

<delay_matrix type="max"
   in_port="BlkA.in[2]"
   out_port="BlkA.out[1]">
      2.0e-9
</delay_matrix>

<T_setup value="5.0e-10"
   port="BlkB.in" clock="clk"/>

<T_clock_to_Q max="2e-09"
   port="BlkB.out" clock="clk"/>

<max_internal_delay value="1e-08"/>
```

Figure 6: Examples of the different timing scenarios that can be expressed for a primitive.

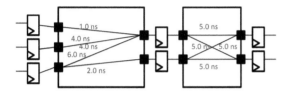

Figure 5: Example of an internal pipeline within a primitive.

Rather than have the architecture file specify the full timing graph itself, we decided that there was sufficient flexibility being able to specify delays in these four ways, dealing only with the input pins and output pins of the primitive:

1. Fully combinational paths - a different combinational delay can be specified from every input pin to every output pin in a timing matrix. If all of the delays to a specific output pin are the same, then this can be specified more concisely.

2. Input pins that feed flip-flops can be specified as having a set-up time to reflect the actual set-up time and any extra combinational delay in the path from the pin to the D input of the flip-flop.

3. Output pins that are fed by the output of flip-flops can be specified as having a clock-to-Q delay, which can also include any extra combinational delay in the path from the Q output to the pin.

4. In the case that there are internal pipeline stages that are not visible with the above specifications, the architect can also specify the minimum clock period for the primitive. By having only one possible specification here, we limit the flexibility of the timing analysis. However, if more accuracy is needed, then a more detailed set of primitives can be used, essentially creating the timing graph as part of the complex block.

Figure 6 provides two examples to illustrate these scenarios. The timing for *BlkA* illustrates a primitive with fully combinational paths. The timing for *BlkB* illustrates the timing information for primitive with an internal pipeline. This primitive has a setup time at the input pins, a clock-to-Q delay at the output pins, and a longest combinational path delay of 10 ns.

5. BENCHMARKS AND FLOW RUN

An important part of this software release is the concomitant release of benchmarks circuits that can be processed through the circuit flow. We have come to realize that the benchmarks associated with a major software effort such as this are both as important as the software itself, and are themselves a form of software. As the language coverage of the ODIN II tool improves, it is clear to us that specific versions of benchmarks must be associated with specific releases of the tool flow. In this section we describe the benchmarks and give the results of the running of the new flow on each.

5.1 The Circuits

Table 1 list the set of 19 benchmarks provided with this release, including four new circuits and several modifications to previously-released circuits. The table gives the circuit name, the number of primary inputs and outputs, the number of 6-input lookup tables in the combinational soft logic, the number of flip-flops, the number of 36x36 multiplier equivalents in each circuit, the number of logical memories, the maximum data width of all the logical memories in the circuit and the number of address bits in the deepest logical memory.

Each benchmark is coded in the Verilog HDL, and in many cases been recoded to meet the language coverage restrictions of ODIN II. There are three significant new additions to the benchmark suite from [13]: The MCML circuit, which is an application that uses Monte Carlo simulation of photons that could be used as part of a Photo-Dynamic Therapy-based cancer treatment plan [14]. The second circuit has provided two separate benchmarks - LU8 and LU32. This is a scalable linear system solver that makes use of the LU Decomposition Method [20]. The third circuit, bgm, is a Monte Carlo simulation for a financial application that uses the BGM interest rate model to price derivatives [8]. The other benchmarks come from a variety of sources: Opencores (or1200, sha), various university research projects (blob_merge, raygentop, boundtop, diffeq1 and diffeq2 ch_instrinsics and stereovisionX) and an FPGA consultant (mkDelayWorker32B, mkSMAdapter5B, and mkPktMerge).

The circuits range in size from 170 to 99,700 6-LUTs. Three of the new circuits added to the benchmark set are of significant size, although we note that even so, we are not keeping up with the size of the modern, largest FPGAs, which contain roughly to 500,000 6-LUTs.

Circuit	# In	# Out	# 6-LUTs	# FFs	# Mult	# Mem	Max width	Max Addr bits
bgm	257	32	30089	5362	11	0	0	0
blob_merge	36	100	6016	735	0	0	0	0
boundtop	275	192	2921	1671	0	1	32	14
ch_intrinsics	99	130	413	233	0	1	8	14
diffeq1	162	96	434	193	5	0	0	0
diffeq2	66	96	277	96	5	0	0	0
LU8PEEng	114	102	21954	6630	8	9	256	14
LU32PEEng	114	102	75530	20898	32	9	1024	14
mcml	36	33	99700	53736	30	10	36	16
mkDelayWorker32B	511	553	5580	2491	0	9	313	14
mkPktMerge	311	156	226	36	0	3	153	14
mkSMAdapter4B	195	205	1977	983	0	3	61	14
or1200	385	394	2963	691	1	2	32	14
raygentop	239	305	2134	1423	18	1	21	14
sha	38	36	2212	911	0	0	0	0
stereovision0	157	197	11462	13405	0	0	0	0
stereovision1	133	145	10366	11789	152	0	0	0
stereovision2	149	182	29849	18416	564	0	0	0
stereovision3	10	30	174	102	0	0	0	0

Table 1: Benchmarks and Data

Circuit	Pack Time (s)	Place Time (s)	MinW Route Time (s)	Route Time (s)	Min W (Tracks)	Crit Path Delay (ns)
bgm	2314	1768	12265	77	168	29.3
blob_merge	624	107	475	7	100	14.2
boundtop	522	37	17	2	72	7.57
ch_intrinsics	52	4	2	0	48	3.72
diffeq1	25	5	7	2	62	19.1
diffeq2	17	3	7	1	52	17.7
LU8PEEng	2526	1322	2564	116	136	149
LU32PEEng	9395	9983	109313	926	204	149
mcml	12674	5901	14173	168	144	109
mkDelayWorker32B	1115	154	1368	36	110	7.43
mkPktMerge	11	8	53	6	50	3.48
mkSMAdapter4B	481	25	44	2	80	7.80
or1200	373	55	92	4	90	24.0
raygentop	431	29	16	2	74	6.46
sha	516	21	20	2	64	15.6
stereovision0	1538	172	93	6	78	4.54
stereovision1	2500	209	539	18	120	5.89
stereovision2	3160	923	157888	276	172	16.9
stereovision3	20	1	0	0	30	3.51

Table 2: Data from Basic Flow Run

One key contribution of this work is the more careful verification of the elaboration stage (through Odin II) of the Verilog HDL code of these benchmarks. For 14 of the 19 circuits, there is an exact simulation match (of randomly generated vectors) between the output of the ODIN II simulator, and Modelsim simulation of the same code and vectors. For the other 5 circuits, (raygentop, boundtop, bgm, mkSMAdapter4B, and mkDelayWorker32B) the variations were minor.

5.2 Running the Flow

These Verilog circuits were run through the VTR flow targeting a hypothetical 40 nm FPGA architecture, which contains soft logic clusters of 10 fracturable LUTs. In this architecture, each fracturable LUT can operate as either a single 6-input LUT or two 5-input LUTs that share all five inputs, similar to the Virtex 6 FPGA [2]. The delays for this cluster were scaled from a 45 nm 6-LUT FPGA found in the iFar repository [12]. The routing architecture consists of segments of only length 4 wires, with Fc(In) set to 0.15 and Fc(Out) = 0.1 [3]. Its delay model was taken from the same iFar model used for the soft logic. The memory block in this architecture is similar to the Altera Stratix IV M144K memory block [1]. It contains 144K bits, and can act either as a single-port or dual port RAM. In single-port mode, the largest data width is 72 bits, and the smallest width is 9 bits; the maximum depth is 16K words. In dual-port mode the maximum width is 36 bits, and the maximum depth is 16K words. The memory speed was based on the speed of the Stratix IV M144K block. Each multiplier in the architecture can operate as one 36x36 or two independent 18x18 multipliers, which in turn can operate as two independent 9x9 multipliers. The multiplier delays were set to be the same as the Stratix IV DSP block.

The circuits are run through the flow in the following way: first through Odin II and ABC to create the pre-packing netlist. Odin II is set to target the specific physical memory and multipliers described above through a related description in the architecture file. Then the minimum channel width (the number of tracks per channel, as is often measured) is determined by running VPR's packing, placement and routing in non-timing-driven mode. (Here, as usual, the router is run repeatedly to find the smallest number of tracks per channel, W, which will succeed in routing.) Finally, the VPR flow is again invoked, using timing-driven routing with the channel width set to 1.3 times W. The latter measurement is used to determine the final critical-path delay of the circuit. The results of this flow are shown in Table 2. The first column lists the circuits in the benchmark. In the VTR flow, the VPR stage dominates the runtime so the next four columns that follow are the packing, placement, minimum channel width routing, and final routing runtimes for VPR in seconds. The last two columns show the key circuit statistics - minimum channel width and critical path delay.

One aspect of these results stand out - the runtime for packing, compared to placement and routing, is very large. This is caused by the part of the packing algorithm that invokes a router to determine if a specific primitive can be connected correctly within the logic block. This feature allows the packer to handle any arbitrary internal routing structure within the logic block, which we feel is important. However, we plan to reduce this runtime when specific flexibile structures, such as crossbars, are present. This is left as future work.

To illustrate the new timing-driven nature of the VPR portion of the flow, we measure the effect of each stage's timing-driven algorithm for packing, placement and routing. To do so, each circuit was run through VPR, holding its channel width at 1.3 * min W given in Table 2 (as is fairly common to create a low-stress routing [3]), but turning the timing-driven setting for each of placement and routing on and off. The results are shown in Table 3. The first column lists the circuit name followed by the critical path delays for each run normalized to the default, fully timing-driven run. The stages of the flow that have timing turned off are labelled after the NT prefix. For example, the column labelling NT PackPlace is a flow with non-timing-driven packing and placement, but with timing-driven routing.

The last row shows the geometric mean of these ratios. We see that turning off timing-driven placement results in the least impact on critical path delay with only a 3% increase on average. We also see that the stages are not independent: turning off timing for all stages results in a 22% increase to critical path delay on average but multiplying the individual

Circuit	Full Timing	NT Route	NT Place	NT Pack	NT PackPlace	NT
bgm	1.00	1.06	1.01	1.14	1.24	1.30
blob_merge	1.00	1.02	1.08	1.14	1.18	1.20
boundtop	1.00	1.04	1.05	1.15	1.34	1.44
ch_intrinsics	1.00	1.02	0.96	1.02	1.09	1.17
diffeq1	1.00	1.06	1.00	1.04	1.04	1.07
diffeq2	1.00	1.05	1.01	1.03	1.00	1.06
LU8PEEng	1.00	1.02	1.01	1.08	1.10	1.13
LU32PEEng	1.00	1.03	1.05	1.09	1.14	1.17
mcml	1.00	1.01	1.02	1.13	1.20	1.23
mkDelayWorker32B	1.00	1.04	1.06	1.16	1.28	1.36
mkPktMerge	1.00	1.00	1.04	1.06	1.02	1.02
mkSMAdapter4B	1.00	1.07	1.02	1.03	1.03	1.12
or1200	1.00	1.01	1.02	1.17	1.20	1.22
raygentop	1.00	1.06	1.00	1.28	1.30	1.30
sha	1.00	1.07	1.03	1.16	1.18	1.22
stereovision0	1.00	1.06	1.03	1.15	1.25	1.33
stereovision1	1.00	1.00	1.06	1.32	1.42	1.43
stereovision2	1.00	1.02	1.15	1.05	1.26	1.29
geomean	1.00	1.04	1.03	1.12	1.17	1.22

Table 3: Impact of timing-driven algorithms in different stages of CAD flow

delay increases from different stages results in a lower gain of 20%.

6. EXAMPLE: FLOATING-POINT BLOCKS

In this section we illustrate the power of the VTR framework showing how a hard block modelled in a previous research project (at great effort) can be modelled in the VTR framework with far less effort. The goal of the previous project [4] was to improve the computational efficiency of floating-point-heavy applications on FPGAs. It explored the architecture of a floating-point hard block, and showed that, for certain floating-point intensive applications, an FPGA employing the new block consumed a factor of 25 times less area and the speed of the resulting circuit increased by four times. In the following we describe the floating-point block, and show how it can be captured in the new complex block architecture description language. We then run the VTR flow and compare the results with the previous work, and comment on the relative effort required.

6.1 Architecture of Floating-Point Block

Figure 7 illustrates the generic architecture of the floating-point block that was explored in [4] that we will model in the VTR flow. The block consists of three basic elements: floating-point multipliers (FMs), floating-point adders (FAs) and wordblocks (WBs). The wordblocks are used for fixed-point arithmetic and logical operations. The FAs, FMs and WBs are connected in series using bus-based routing, which is provided by the multiplexers shown in the figure.

For the purpose of architecture exploration, several parameters are used to explore the architecture of the block, including bus width (N), number of input buses (M), number of output buses (R), number of feedback paths (F), number of blocks (D) and number of FA and FM (P). The work in [6] determined good choices for these parameters.

6.2 Architecture Description in VTR

The description of the block shown in Figure 7 was rendered in the new complex block architecture description language described in [13]. As described in Section 2, the first part of the architecture file gives the atomic primitive constructs that must be instantiated in the Verilog input code,

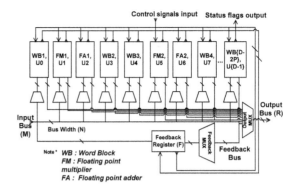

Figure 7: Architecture of the floating-point block

```
1  <model name="fpu_mul">
2  <input_ports>
3      <port name ="clk" is_clock="1"/>
4      <port name="opa"/>
5      <port name="opb"/>
6  </input_ports>
7  <output_ports>
8      <port name="out"/>
9      <port name="control"/>
10 </output_ports>
11 </model>

13 <model name="fpu_add">
14 <input_ports>
15     <port name ="clk" is_clock="1"/>
16     <port name="opa"/>
17     <port name="opb"/>
18 </input_ports>
19 <output_ports>
20     <port name="out"/>
21     <port name="control"/>
22 </output_ports>
23 </model>
```

Figure 8: Architecture Description of Floating Point Block ("fpu_mul" is the primitive of FM and "fpu_add" is the primitive of FA)

and will appear in the netlist that are sent to the packing stage, and is shown in Figure 8. Here the model "fpu_mul" describes the primitive of the floating point multiplier, FM, and "fpu_add" describes the primitive of the adder, FA. In both primitives, the port *clk* provides the clock signal to the registers in the block, and ports *opa* and *opb* are two floating-point inputs of the primitives. The port *out* is the floating-point addition or multiplication output. The port *control* emits status flags that indicate such things as the result is not a number value and overflow.

The FPU block contains the primitives illustrated in Figure 7, and the synthesis flow produces a netlist containing those primitives. The packing step packs those primitives into the physical block. For this to work, there must be a clear description of the contents and interconnection between the primitives in the architecture description file. Figure 9 gives an excerpt of the architecture description file that corresponds to the FAs, FMs and WBs and some of their interconnect. The physical block name of the floating-point block is "block_FPU" and is given in line 1. The "FPU" on line 12 is one of the configuration modes of this block, as different modes can be set for the overall "block_FPU". The bus width for all the floating point quantities is 32 bits ($N=32$) as shown by the **num_pins** construct on lines 2 to 9. There are four input busses ($M=4$ from in1 to in4) and three output busses ($R=3$ from out1 to out3). There are a total of eight primitives ($D=8$), including two floating-point adders and two floating-point multipliers ($P=2$). The description of the first FM (FM1) is from line 25 to 36 and the description of first FA (FA1) is from line 37 to 48. Within those descriptions are the netlist designations of the primitives that must appear in the pre-packing BLIF netlist - for example ".subckt fpu_mul" is required to indicate an FM primitive. The <T_setup> tag (on line 31) provides the setup time of the input port of the primitive block and <T_clock_to_Q> tag (on line 33) is the output port delay time. The maximum delay of the block is specified by <max_internal_delay value> tag (on line 35), as described in Section 4.2. The four wordblocks are simply registers as described from line 18 to 24, which contain 32 registers in each wordblock. As such, these registers appear in the pre-packing netlist as the standard BLIF primitive ".latch." There are three feedback registers named "feedback_reg1", "feedback_reg2" and "feedback_reg3". The description of "feedback_reg1" is from line 51 to 57.

After providing the definitions of the atomic primitive constructs, the interconnection of FAs, FMs and WBs are described in the interconnect section beginning on line 60. Line 62 defines the direct connection of internal control signals. Line 69 describes a single bit of the input multiplexer of "WB1", which allows selection of the input from the four block inputs and the outputs of feedback registers.

6.3 Experiment

In this section we describe an experiment that replicates a prior research effort that explored the architecture of the floating point block [6, 7]. The goal of that experiment was to measure the area and delay impact of the presence (vs. absence) of the floating point block. We employ the same set of eight circuits, shown in Table 4, that were used in the prior research. The table gives the number of FA and FM instances used in each circuit, and the general nature of the circuit. Note that all of these circuits can be completely expressed with just these primitives, so in the case of a FPU-based FPGA, there is almost no need of soft logic blocks to implement these circuits.

Circuit	# of FA	# of FM	Nature
bfly	4	4	DSP kernel
dscg	2	4	DSP kernel
fir	3	4	DSP kernel
mm3	2	3	Linear Algebra kernel
ode	3	2	Linear Algebra kernel
bgm	9	11	Finance application
syn2	5	4	Synthetic circuit
syn7	25	25	Synthetic circuit

Table 4: Floating-point benchmark circuits

In the case that the floating point block is not present, the floating point operations are implemented in the soft logic of the FPGA; in this experiment and the prior research, the soft-logic implementation does not make use of hard integer multipliers. For both FPGAs (with and without the floating point unit), the soft logic CLB consists of four 4-input LUTs, similar to the Virtex-II. The routing architecture used for both FPGAs was Fc(In) = 0.15, Fc(Out) = 0.25, and all length four wire segments. The number of tracks per channel, W, was set to be 72.

We used a 130nm CMOS process technology for estimating the area and timing of both the soft logic CLB and the floating-point block. The area and delay of the soft logic CLB are based on the iFAR architecture file FPGA Repos-

```
1  <pb_type name="block_FPU" height="8">
2  <input name="in1" num_pins="32"/>
3  <input name="in2" num_pins="32"/>
4  <input name="in3" num_pins="32"/>
5  <input name="in4" num_pins="32"/>
6  <output name="out1" num_pins="32"/>
7  <output name="out2" num_pins="32"/>
8  <output name="out3" num_pins="32"/>
9  <output name="control" num_pins="32"/>
10 <clock name="clk" num_pins="1"/>

12 <mode name="FPU">
13 <pb_type name="FPU_slice" num_pb="1">
14 <input name="in1" num_pins="32"/>
15 <input name="in2" num_pins="32"/>
16 .
17 .
18 <pb_type name="WB1" blif_model=".latch" num_pb="32"
      class="flipflop">
19 <input name="D" num_pins="1" port_class="D"/>
20 <output name="Q" num_pins="1" port_class="Q"/>
21 <clock name="clk" num_pins="1" port_class="clock"/>
22 <T_setup value="3.88e-10" port="WB1.D" clock="clk"/>
23 <T_clock_to_Q max="1.557e-10" port="WB1.Q"
      clock="clk"/>
24 </pb_type>
25 <pb_type name="FM1" blif_model=".subckt fpu_mul"
      num_pb="1">
26 <clock name="clk" num_pins="1"/>
27 <input name="opa" num_pins="32"/>
28 <input name="opb" num_pins="32"/>
29 <output name="out" num_pins="32"/>
30 <output name="control" num_pins="8"/>
31 <T_setup value="3.88e-10" port="FM1.opa" clock="clk"/>
32 <T_setup value="3.88e-10" port="FM1.opb" clock="clk"/>
33 <T_clock_to_Q max="1.557e-10" port="FM1.out"
      clock="clk"/>
34 <T_clock_to_Q max="1.557e-10" port="FM1.control"
      clock="clk"/>
35 <max_internal_delay value="2.99e-9"/>
36 </pb_type>
37 <pb_type name="FA1" blif_model=".subckt fpu_add"
      num_pb="1">
38 <clock name="clk" num_pins="1"/>
39 <input name="opa" num_pins="32"/>
40 <input name="opb" num_pins="32"/>
41 <output name="out" num_pins="32"/>
42 <output name="control" num_pins="8"/>
43 <T_setup value="3.88e-10" port="FA1.opa" clock="clk"/>
44 <T_setup value="3.88e-10" port="FA1.opb" clock="clk"/>
45 <T_clock_to_Q max="1.557e-10" port="FA1.out"
      clock="clk"/>
46 <T_clock_to_Q max="1.557e-10" port="FA1.control"
      clock="clk"/>
47 <max_internal_delay value="2.99e-9"/>
48 </pb_type>
49 .
50 .
51 <pb_type name="feedback_reg1" blif_model=".latch"
      num_pb="32" class="flipflop">
52 <input name="D" num_pins="1" port_class="D"/>
53 <output name="Q" num_pins="1" port_class="Q"/>
54 <clock name="clk" num_pins="1" port_class="clock"/>
55 <T_setup value="3.88e-10" port="feedback_reg1.D"
      clock="clk"/>
56 <T_clock_to_Q max="1.557e-10" port="feedback_reg1.Q"
      clock="clk"/>
57 </pb_type>
58 .
59 .
60 <interconnect>
61 <!--Connection sequence:WB1-->FM1-->FA1-->
      WB2-->WB3-->FM2-->FA2-->WB4-->
62 <direct name="direct1" input="FM1.control[7:0]"
63    output="FPU_slice.control[7:0]"> </direct>
64 <direct name="direct2" input="FM2.control[7:0]"
      output="FPU_slice.control[15:8]"></direct>
65
66
67 <!--###### WB1 ########-->
68 <!-- Input Mux WB1 in1 -->
69 <mux name="WB1_in1_mux1" input="FPU_slice.in1[0:0]
      FPU_slice.in2[0:0] FPU_slice.in3[0:0] FPU_slice.in4[0:0]
      feedback_reg1[0:0].Q feedback_reg2[0:0].Q
      feedback_reg3[0:0].Q" output="WB1[0:0].D"/>
70 .
71 .
```

Figure 9: Code for Architecture Description

itory [12]. The area included the routing resources at a channel width equal to 72 tracks, and the area of the CLB tile, including programmable routing was determined to be 5679 um^2.

The area and delay of the floating-point block (with parameters N=32, M=4, R=3, F=3, D=8 and P=2) was estimated, as in [7] by synthesizing the floating-point block into standard cells from UMC in their 130nm CMOS process, using the Synopsys Design Compiler. The area of the FPU block, including programmable routing was determined to be 498,847 um^2. Using these two values, we calculate that one FPU tile requires the same area as 88 CLB tiles. In the discussion that follows, area is expressed in equivalent CLB area units.

The total area consumed by a circuit is the sum of the area taken by the floating-point block (in equivalent CLBs) plus the number of soft logic CLBs used. The maximum delay of the floating-point block is 2.99ns which is specified in <max_internal_delay> in Figure 9.

All eight of the circuits described in Table 4 were implemented in two hypothetical FPGAs: one with a hard floating point hard block, and one without - the latter containing only soft logic CLBs as described above. We compared the speed and area consumed in both cases, and for the purposes of this paper, compared those results with the measurements in [4].

The speed comparison is shown in Table 5. The average speedup of the FPGAs with a hard floating-point block is 12.6 times. This number differs significantly from the average speed ratio measured in the previous work which was only 4 times [4]. In that work, the soft logic FPGA was a Xilinx Virtex-II device which employs fixed carry chains in the adders and the adders contained in the multipliers. Those dedicated carry chains are significantly faster than carry logic as implemented in the CLBs in this experiment - directly in the LUTs, and using the regular intra-CLB routing to connect. This speed difference shows up particularly in the soft multiplier implementation; the speed of the soft multipliers in most of the circuits alone was 35ns, accounting for the difference. This result clearly shows that our future work must include the modelling of high-speed carry logic to support these kinds of experiments.

Circuit	Soft-Only Critical Path (ns)	Hard-Logic Critical Path (ns)	Ratio
bfly	36.1	2.99	12.1
bgm	35.7	2.99	11.9
dscg	36.6	2.99	12.3
fir	36.0	2.99	12.0
mm3	35.3	2.99	11.8
ode	34.5	2.99	11.6
syn2	37.6	2.99	12.6
syn7	50.3	2.99	16.9
Geomean			12.6

Table 5: Delay Comparison of Hard v. Soft Logic

Table 6 shows the area comparison between the pure soft logic and FPU hard block FPGA, where area is measured in equivalent CLBs. On average, the the circuits implemented in the FPGA with the hard FPU block is 18 times smaller than the FPGA with only soft logic. This result is in the same ballpark as the result (25x) in [4].

6.4 Comparison of Effort

The prior research that explored the floating point block studied the optimization of its internal routing and logic

Circuit	Soft-Only Area (CLBs)	Hard-Logic Area (Equiv CLBs)	Ratio
bfly	6405	264	24.3
bgm	16908	792	21.3
dscg	6371	440	14.5
fir	6215	352	17.7
mm3	4556	264	17.3
ode	3609	480	7.5
syn2	6553	264	24.8
syn7	39240	1584	24.8
Geomean			17.9

Table 6: Area Comparison of Hard v. Soft Logic

architecture [6, 7] using a customized version of the VPR flow that was based on VPR 4.2, called VPH [5]. The development time of the VPH tool was roughly one year; the modelling using the new VTR flow took approximately 2 man-weeks of time, a significant reduction in effort. Overall, this experiment shows that VTR framework provides a platform to evaluate new complex blocks such as the floating-point block, and that it is a more efficient way to enable this kind of experiment. It is also useful to note that the prior flow used a commercial synthesis tool from Synplicity as the front end, which can only target existing FPGA architectures. This prevents the exploration of FPGA architecture parameters - for example changing the size of the LUT in the FPGA. The new VTR flow permits the changing of many more parameters in the FPGA architecture from synthesis through placement and routing.

7. RELEASE

The release of this software and benchmarks can be found at the following location:

http://www.eecg.utoronto.ca/vtr/

It contains:

1. The source code for the specific versions of ODIN II, ABC, and VPR that are being released, which are compatible with the benchmarks being released.

2. A few sample architecture files including various memory architectures with different combinations of size and flexibility, a suite of different fracturable LUT architectures, and a few heterogeneous architectures with realistic timing numbers.

3. The 19 benchmark circuits, which are compatible with the release of the software.

4. Example scripts for running experiments as well as regression tests for the software. These tests come with golden results and a range of error bands.

5. A web page with documentation on how to run the various versions of the flow on the released benchmarks.

6. An issue-tracking site that users can report software issues on the flow.

8. FUTURE WORK

There is a great deal of future work to be done on VTR system to include all of the innovations already done, and some new things in the future. These include:

1. Multi-clock timing analysis and optimization.

2. Cross-block Carry-Chains. VPR, with its new more complex logic blocks, can model carry chains (which provided augmented arithmetic speed and density) within a single complex logic block. However, to model the standard practice of building inter-block carry chains, the placement algorithm has to be capable of aligning (typically vertically) the blocks with connecting carry logic. In addition, the front-end synthesis flow must correctly capture and emit carry chains in the correct circumstances.

3. Clock tree architecture. There should be a separately described set of clock tree architectures that can be explored, and used for more realistic modelling of clocks.

4. Verilog language coverage. parameters. One of the most laborious conversion issues for benchmark circuits is the lack of parameters in ODIN II's coverage of the Verilog language.

5. Libraries. To help acquire more circuits, we need to have Verilog libraries of standard cores, such as dividers, square root units, and floating point arithmetic.

6. Bus-based routing. To help connect FPGAs to data-oriented blocks, it would be good to integrate the routing of multi-wire busses into the router.

7. Power/Energy modelling. The back-end flow, VPR, needs to model the energy consumption of all architectures, similar to [18]. This will necessitate a way to properly model the new, more complex logic blocks.

8. Transistor-level modelling. The most accurate way to model many aspects of the FPGA architecture is to have a transistor-level model of the logic and the routing. To be sensible, these models must have proper electrical design, including sizing.

9. White and Black box modelling in ABC. To enhance the quality of logic synthesis, ODIN II will need to transmit information, contained in the architecture file, to ABC, making use of its white and black box synthesis capabilities.

9. CONCLUSIONS

This paper has described new features and benchmarks of the Verilog-To-Routing (VTR) flow, a publicly available synthesis flow that permits exploration of hypothetical FPGA architectures and new CAD algorithms. The release is now fully timing-driven, and comes with a set of larger benchmarks. We have shown how it can be used to model new FPGA logic structures far more easily than previous tools. This is a ongoing, world-wide collaboration which has much more work to do to make the tool suite more viable.

10. REFERENCES

[1] Stratix IV Device Family Overview. http://www.altera.com/literature/hb/stratix-iv/stx4_siv51001.pdf, 2009.

[2] Xilinx Virtex-6 Family Overview. http://www.xilinx.com/support/documentation/data_sheets/ds150.pdf, 2009.

[3] V. Betz, J. Rose, and A. Marquardt. *Architecture and CAD for Deep-Submicron FPGAs*. Kluwer Academic Publishers, Norwell, Massachusetts, 1999.

[4] C.H. Ho, C.W. Yu, P.H.W. Leong, W. Luk and S.J.E. Wilton. Floating-Point FPGA: Architecture and Modeling. *IEEE Trans. on VLSI Systems*, 17(2):1709–1718, Dec 2009.

[5] C.W. Yu. A Tool for Exploring Hybrid FPGAs. In *Proc. International Conference on Field Programmable Logic and Applications (FPL), PhD Forum*, pages 509–510, 2007.

[6] C.W. Yu, A.M. Smith, W. Luk, P.H.W. Leong, S.J.E. Wilton. Optimizing Floating Point Units in Hybrid FPGAs. *IEEE Trans. on VLSI Systems*, to appear.

[7] C.W. Yu, W. Luk, S.J.E. Wilton, P.H.W. Leong. Routing Optimization for Hybrid FPGAs. In *Proc. International Conference on Field Programmable Technology (FPT)*, pages 419–422, 2009.

[8] G.L. Zhang and P.H.W. Leong and C.H. Ho and K.H. Tsoi and C.C.C. Cheung, D. Lee, R.C.C. Cheung and W. Luk. Reconfigurable Acceleration for Monte Carlo Based Financial Simulation. In *Proc. International Conference on Field Programmable Technology (FPT)*, pages 215–222, 2005.

[9] W. K. C. Ho and S. J. E. Wilton. Logical-to-physical memory mapping for fpgas with dual-port embedded arrays. In *Proceedings of the 9th International Workshop on Field-Programmable Logic and Applications*, pages 111–123, London, UK, 1999. Springer-Verlag.

[10] P. Jamieson, K. Kent, F. Gharibian, and L. Shannon. Odin II-An Open-Source Verilog HDL Synthesis Tool for CAD Research. In *IEEE Annual Int'l Symp. on Field-Programmable Custom Computing Machines*, pages 149–156. IEEE, 2010.

[11] S. Jang, B. Chan, K. Chung, and A. Mishchenko. WireMap: FPGA technology mapping for improved routability and enhanced LUT merging. *ACM Trans. on Reconfigurable Technology and Systems*, 2(2):1–24, 2009.

[12] I. Kuon and J. Rose. Automated transistor sizing for fpga architecture exploration. In *Proceedings of the 45th annual Design Automation Conference*, DAC '08, pages 792–795, New York, NY, USA, 2008. ACM.

[13] J. Luu, J. Anderson, and J. Rose. Architecture description and packing for logic blocks with hierarchy, modes and complex interconnect. In *Proceedings of the 19th ACM/SIGDA international symposium on Field programmable gate arrays*, FPGA '11, pages 227–236, New York, NY, USA, 2011. ACM.

[14] J. Luu, K. Redmond, W. Lo, P. Chow, L. Lilge, and J. Rose. Fpga-based monte carlo computation of light absorption for photodynamic cancer therapy. *Field-Programmable Custom Computing Machines, Annual IEEE Symp. on*, 0:157–164, 2009.

[15] A. Marquardt, V. Betz, and J. Rose. Using Cluster-Based Logic Blocks and Timing-Driven Packing to Improve FPGA Speed and Density. *ACM Int'l Symp. on FPGAs*, pages 37–46, 1999.

[16] A. Mishchenko et al. ABC: A System for Sequential Synthesis and Verification. http://www.eecs.berkeley.edu/alanmi/abc, 2009.

[17] P. O'Brien, A. Furrow, B. Libby, and K. Kent. A simple tractable approach to design tool verification through simulation and statistics. In *to appear in IEEE Conference on Field Programmable Technologies*, FPT '11. IEEE, 2011.

[18] K. K. W. Poon, A. Yan, and S. J. E. Wilton. A flexible power model for fpgas. In *Proceedings of the Reconfigurable Computing Is Going Mainstream, 12th International Conference on Field-Programmable Logic and Applications*, FPL '02, pages 312–321, London, UK, UK, 2002. Springer-Verlag.

[19] S. Srinath and K. Compton. Automatic generation of high-performance multipliers for fpgas with asymmetric multiplier blocks. In *Proceedings of the 18th annual ACM/SIGDA international symposium on Field programmable gate arrays*, FPGA '10, pages 51–58, New York, NY, USA, 2010. ACM.

[20] W. Zhang, V. Betz, and J. Rose. Portable and scalable fpga-based acceleration of a direct linear system solver. In *International Conference on Field-Programmable Technology, FPT 2008.*, pages 17 –24, dec. 2008.

Compiling High Throughput Network Processors

Maysam Lavasani
Electrical and Computer
Engineering
The University of Texas at
Austin
maysam@mail.utexas.edu

Larry Dennison
Lightwolf Technologies,
Walpole, MA
larry@lightwolftech.com

Derek Chiou
Electrical and Computer
Engineering
The University of Texas at
Austin
derek@ece.utexas.edu

ABSTRACT

Gorilla is a methodology for generating FPGA-based solutions especially well suited for data parallel applications with fine grain irregularity. Irregularity simultaneously destroys performance and increases power consumption on many data parallel processors such as General Purpose Graphical Processor Units (GPGPUs). Gorilla achieves high performance and low power through the use of FPGA-tailored parallelization techniques and application-specific hardwired accelerators, processing engines, and communication mechanisms. Automatic compilation from a stylized C language and templates that define the hardware structure coupled with the intrinsic flexibility of FPGAs provide high performance, low power, and programmability.

Gorilla's capabilities are demonstrated through the generation of a family of core-router network processors processing up to 100Gbps (200MPPS for 64B packets) supporting any mix of IPv4, IPv6, and Multi-Protocol Label Switching (MPLS) packets on a single FPGA with off-chip IP lookup tables. A 40Gbps version of that network processor was run with an embedded test rig on a Xilinx Virtex-6 FPGA, verifying for performance and correctness. Its measured power consumption is comparable to full custom, commercial network processors. In addition, it is demonstrated how Gorilla can be used to generate merged virtual routers, saving FPGA resources.

Categories and Subject Descriptors: C.5.m [Computer System Implementation]:Miscellaneous

General Terms: Performance, Design, Experimentation

Keywords: FPGA, network processor, templates

1. INTRODUCTION

Specialized hardware can be significantly higher in performance and significantly lower in power than programmable processors [13]. Designing and implementing high quality specialized hardware requires an expert. Unfortunately, few application domain experts are experts in high performance hardware design and vice-versa.

Gorilla helps to bridge the gap between hardware experts and domain experts, providing each with the ability to focus on what they do best and automatically combining the efforts to generate highly efficient hardware. Gorilla is designed to maximize utilization of the most critical hardware resources in the system. In many systems, the most critical resources are the external pins of the chip. Critical resources are encapsulated in *accelerators* that perform core domain-specific functions and contain appropriate critical state. Domain experts define accelerator functionality and algorithms in a stylized C (Gorilla C). Domain experts also write application code that contains explicit calls to accelerators in Gorilla C.

Gorilla uses parameterized *templates*, which are generally defined by hardware experts, to encapsulate efficient hardware structures. Gorilla's compiler accepts a template along with template parameters that include arbitrary functions, state machines, and constants and generates synthesizable Verilog that defines customized hardware. A *canonical architecture* is a template designed to implement a specific class of applications.

The Gorilla compiler is used to combine application and accelerator code with the appropriate templates to generate specialized hardware that implements the application. Gorilla generates Verilog that is intended for FPGAs but could be used to generate ASICs. However, when Gorilla is used to design FPGAs, the dual benefits of high efficiency specialized hardware that competitive with application-specific processors and full programmability through the FPGA fabric are achieved. In fact, Gorilla on FPGAs is more general than application-specific processors.

A family of network processors that achieve similar performance and power to application-specific processors while running on an FPGA at 100MHz is generated to demonstrate the Gorilla methodology. One instance, targeting a single Xilinx Virtex-7 VHX870T, achieves 100Gbps (200 Million-Packets-Per-Second (MPPS)) in a single FPGA while processing any mix of IPv4, IPv6, and multi-labeled MPLS packets. To our best knowledge this is the highest throughput for a FPGA-based network processor generated from a high level specification. A 40Gbps instance was run on a single Xilinx Virtex-6 XC6VLX240T FPGA on an ML605 prototyping board.

The network processing template is far more general than simply network processing. By replacing accelerators, it is especially well suited for streaming applications and data parallel applications whose performance and power scale poorly due to fine grain processing irregularity caused by data dependencies and irregular access to shared resources. The potential for reuse by factoring functionality from the template in the Gorilla style is significant.

The contributions of this paper are as follows:

- The Gorilla methodology that includes a programming model consisting of stylized sequential C calling accelerators written by a domain expert, the concept of highly parameterized templates written and assembled by a hardware expert, and a tool chain that automatically combines the two to generate hardware comparable in quality to hand-written hardware.

- A library of infrastructure templates including hierarchical arbitration, load balancing, reordering queue, and rate adaptation templates. Their parametric nature enables easy exploration of the design space to find the best configuration for a specific application.

- A case study of a network processor family, capable of processing any combination of IPv4, IPv6, and MPLS packets at up to 100Gbps at 100MHz, that has been synthesized, placed-and-routed, and verified for correctness in available FPGAs. An IPv4-only version running at 40Gbps (100MPPS) was run on a Xilinx ML605 prototyping board, measured for performance and power, and verified for correctness. All were generated using the Gorilla methodology.

2. PACKET PROCESSING

Because we use packet processing as our example, we describe it here. Internet routers process incoming packets to determine which output port they should be forwarded to. How a packet is processed is defined by the first part of the packet, known as the *header*. Packets can be of different types, each type requiring different processing steps. A single type of packet could have a variable number of *labels* that change the number of total processing steps. Packets can also be *encapsulated* in other packets, potentially requiring the router to look beyond the first header to complete the processing.

Figure 1 shows a simplified chart showing the processing steps of a router supporting IPv4/IPv6/MPLS. The grey rectangles are steps that require global and/or shared state accesses. The entry point in processing of a packet in this figure is *Dispatch* step. It checks the packet layer 2 protocol and jumps to the appropriate state for processing the protocol header (In figure 1 we are showing only one layer 2 protocol which is Ethernet.) *Ethernet* step detects the layer 3 protocol and also checks the integrity of Ethernet header. Processing IPv4 (one of the possible layer 3 protocols) consists of extracting the packet fields, checking the integrity of the fields, classifying the packet by looking up the source and destination addresses, and finally updating the packet header fields. In many router applications, the source and destination IP addresses together determine the output port, though our simplified code does not reflect that. The destination port for the packet is determined using lookup process and packet is forwarded to that port. Processing IPv6 is similar to IPv4 except that IPv6 has longer addresses. A MPLS header may contain multiple labels. The next action is determined using the result of the MPLS label lookup, whether

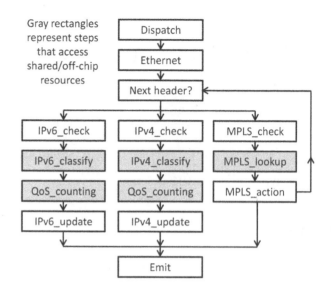

Figure 1: IPv4/IPv6/MPLS Packet Processing

it is processing another MPLS label, deleting the current label, attaching a new label, or processing another encapsulated protocol. The exit point for processing a packet in Figure 1 is the *Emit* step.

3. GORILLA METHODOLOGY

In this section we describe the Gorilla programming model, its tool chain, and the concept of canonical architectures.

3.1 Programming Model

Gorilla domain code consists of a sequence of steps, where each step is C code that (i) calls accelerators that implement specific functionality and handle accesses to expensive and/or shared resources, (ii) performs computation, and (iii) determines the next step. Accelerator calls within a single step must be independent of each other, a condition that is checked by the Gorilla compiler. Gorilla's infrastructure guarantees that each step is fully complete before the next step starts, giving the illusion of a sequential programming model to the domain expert. The results of all previous steps as well as accelerator calls are available to the current step. The same program of steps (but not necessary the same path through the program) is executed on each input data, such as a packet in a network processing application. Figure 2 shows example domain code. The accelerators themselves can be written as a domain program consisting of steps that can call other downstream accelerators. An example of such case is given in Section 4.

3.2 Canonical architecture

A canonical architecture is a highly parameterized template designed to efficiently implement a specific application class. Parameters enable the easy trading off of performance, resource sizing, and selecting options such as scheduling algorithms. The template contains code to implement any of the valid parameter values. The Gorilla tool chain combines a canonical architecture with domain code and parameter values provided by the domain expert to generate an efficient, specialized hardware implementation.

```
IPv4_check() {
  status = IPv4_header_integrity_check(Header);
  if (status == CHKSUM_OK)
    Next_step = IPv4_lookup;
  else
    Next_step = Exception;}

IPv4_lookup() {
  Da_class = lookupx.search(Header.IPv4_dstaddr);
  Sa_class = lookupy.search(Header.IPv4_srcaddr);
  if (Da_class == NOT_FOUND)
    Next_step = Exception;
  else if(Sa_class == NOT_FOUND)
    Next_step = Exception;
  else
    Next_step = IPv4_modify;}

IPv4_modify() {
  if((IP_update_fields(Header) == ZERO_TTL))
    Next_step = Exception;
  else {
    Dport = Da_class.dport;
    Next_step = Emit;
  }}
```

Figure 2: Simplified IPv4 steps

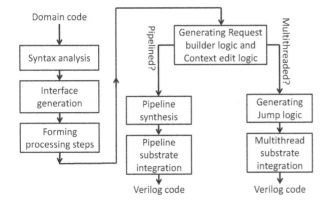

Figure 3: Compilation process for composite templates and user programmed templates

Figure 4 shows a canonical architecture discussed in Section 4.1. Though it was no simpler to write than a single implementation, it generates a wide range of possible implementations of substrates for packet processing applications.

3.3 Gorilla Tool Chain

The Gorilla tool chain consists of domain code compiler and scripts (Figure 3) and was developed using the ANTLR [1] tool and Perl. The input to the tool chain is the (i) domain code for the templates, (ii) all templates in the design and (iii) other parameter values. The compiler generates a specialized hardware implementation using these inputs.

The domain code compiler ensures that there are no dependencies or structural hazards when accessing accelerators in each step. The compiler splits a step to resolve such hazards when necessary. The compiler splits the domain code into three parts: (i) Request Builder logic, (ii) Context Edit logic, and (iii) Jump logic. Request Builder logic computes arguments for the accelerator requests and issues the requests. Context Edit logic updates thread contexts as well

as global variables, using accelerator replies and state from previous steps. Jump logic determines the next step based on the results of the computations in the current step.

The code is currently compiled to Verilog that is then either inserted into a multithreaded template or is used to generate a pipeline. Multithreaded hardware supports multiple "threads", where each thread is processing a specific unit of work, such as a single packet in a network processor. Threads are switched in hardware when progress on the running thread is blocked due to long or variable latency operations or contention for resources. Multithreading enables applications to transparently trade off accelerator usage (e.g., one data element requires 10 accelerator calls to process and another requires none.) As the number of bidders (requesters) for accelerators grows, contention for shared resources increases as well, increasing variability, making multithreading more and more important. The multithreading template automatically handles all multithreading scheduling and storage issues and receives/sends the data through scratch pad memories.

As an alternative, if (i) there are only forward jumps in the domain code and (ii) accelerator delays are constant, the compiler can automatically generate a pipelined implementation from the domain code by assembling the Request Builder, and Context Edit of all processing steps into a straight pipeline. Pipelining avoids the overhead of thread scheduling and thread context memories, delivering high throughput when the maximum latency is fixed and reasonably small. Pipelining is less efficient when latencies are long, there are irregular transitions in control flow, or there are data dependencies between pipeline stages.

4. EXAMPLE: NETWORK PROCESSOR

Fine grain irregularities are common in packet processing due to heavy control flow dependency on data and contention for shared resources, making it an ideal application to demonstrate Gorilla's capabilities. We implemented a packet processing canonical architecture and a router domain code that supports IPv4, IPv6, and MPLS in Gorilla C. The IPv4, IPv6, and MPLS applications have 10, 12, and 13 steps respectively. Most of the steps fully utilize scratch pad memory (128bits wide) bandwidth. There are currently between one to six accelerator calls, depending on the packet protocol and router configuration, for each packet.

The Verilog code that was generated by the Gorilla tool chain is passed through the standard Xilinx tools for simulation and implementation on an FPGA. Two man-years were spent on building the tools and infrastructure templates. Once completed, however, the IPv6 protocol required only two man-weeks and the MPLS protocol four man-weeks.

4.1 NP canonical architecture

The canonical architecture for our Network Processor (NP) is shown in Figure 4. Every component is a highly parameterized template. When packets enter the NP, a programmable pre-processor adds necessary meta data to packet headers and splits the packets into a header and a body, passing the header through the processing pipeline and storing the body in a separate buffer. A header is assigned, in a load balanced fashion, to an arbitrary thread in an arbitrary engine. Engines which are the main packet processing elements contain hardware compiled from the domain code, creating an entirely hardware implementation of the domain

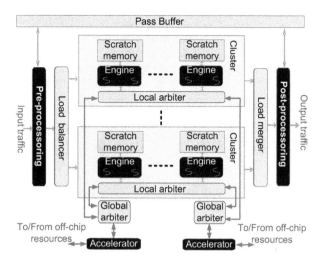

Figure 4: NP canonical architecture: Black modules are domain-specific templates. The sizing of resources as well as resource management policies are settable using parameters in the respective templates.

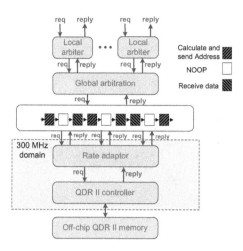

Figure 5: Simplified IPv4 trie lookup architecture including arbitration structures

code. Domain code makes calls to the hardware accelerators that are also compiled by the Gorilla compiler. Most hardware accelerators are pipelined, rather than multithreaded, since the accelerators are given dedicated resources with deterministic latencies. Once processing is complete, the header goes through a programmable post-processor to strip the meta data, get recombined with its body, and forwarded in arrival order.

Header processing domain code is compiled with a multithreaded engine template by the Gorilla compiler. It reads pre-written packet header information from scratch memories that the packet pre-processing domain code compiled with the preprocessor template has written. Special actions are taken by compiler as well as engine address translation to map the packet protocol fields into block RAM memory addresses. When the engine encounters a long latency operation, such as an accelerator call, the engine uses its multithreading support to switch to a ready thread whose accelerator calls have all returned.

Accelerators are essentially the same as in any network processor including IP lookup accelerator or flow counting accelerator. Accelerators that have dedicated off-chip QDR memory with deterministic delay are compiled to pipelined implementations. For example a trie-based lookup consists of different steps to walk through the trie levels. Each step can be written in Gorilla C to process the current node and chase the pointer to the next level node. Retrieving node information from the memory is done as an accelerator call which in this case is a memory (downstream accelerator) for our lookup accelerator.

4.2 Trie Lookup Accelerator

Many techniques have been proposed for high throughput IPv4 lookup engines [8, 12, 15, 40]. To support a large number of prefixes, Gorilla provides a lookup accelerator that stores the forwarding table in off-chip SRAM QDRII [26] that supports a new double word read operation and a new double word write operation every memory clock cycle.

We implemented a common trie lookup algorithm [12] as domain code. The algorithm divides an IPv4 address into three chunks of 20b, 4b, and 8b. At each stage, the QDR address is calculated using the current address chunk and the value returned from the QDR in the previous stage. If the entry is a leaf entry, no further read requests for that particular lookup are made. Since the pipelined lookup accelerator is clocked at 100MHz and the QDR is clocked at 300MHz, three requests from three different stages of the lookup pipeline are sent to QDR memory each 100MHz clock cycle. We use rate adaption logic between the trie lookup pipeline and QDR for this purpose. A simplified version of the generated pipelined architecture for our IPv4 trie lookup algorithm is shown in Figure 5.

Figure 6 shows the capacity requirements, in 18 bit words per entry, for different RIPE [30] routing tables. We use a single QDR chip for each QDR channel and assume a four million word QDR part.

The IPv6 lookup accelerator is similar to the IPv4 lookup accelerator except the number of trie levels is 6 instead of 3. Therefore, each IPv6 lookup accelerator has twice delay as IPv4 lookup and requires two QDR memory channels to provide full bandwidth.

The NP prototype supports multiple MPLS labels per packet. The prototype's MPLS lookup is a two level architecture with the first level acting as a cache for the second level. The first level of MPLS lookup is done using a multi-bin indexed table indexed with a hash function. If the lookup hits at the first level, it takes eight clock cycles. If the lookup misses, it is handled using exactly the same architecture as IPv4 lookup. Clusters direct their missed lookup requests to a global lookup unit which use off-chip QDR memory to lookup the MPLS labels. The MPLS lookup accelerator uses a direct mapped table with one million entries stored in QDR memories.

Our prototype's IPv4 and IPv6 lookup accelerators assume no locality and always access the QDR SRAMs for all memory reads, to ensure that full performance will always be available, regardless of the traffic pattern. Such immunity to performance "divots" is critical in high-end applications, such as core routers; otherwise, they would be highly suscep-

	Prefixes	Trie Config	Level1 entries	Level2 entries	Level3 entries	Total
rrc00, Ripe NCC Amsterdam	344,029	20,4,8	1,048,576	1,053,584	679,680	2,442,000
rrc01, Linx London	338,947	20,4,8	1,048,576	1,020,416	534,016	2,336,000
rrc02, Sfinx Paris	274,115	20,4,8	1,048,576	865,104	210,924	2,019,142
rrc16, Miami	344,029	20,4,8	1,048,576	1,051,792	352,512	2,276,624

Figure 6: Lookup unit storage requirement in 18bit words

MPPS	Accelerator operation	QDRII channels	QDRII+ channels	Required MPLS traffic - QDRII	Required MPLS traffic - QDRII+
100	IPv4 LU	1	1	0	0
100	IPv6 LU	2	1	0	0
100	IPv4 LU-FC	3	2	0	0
100	IPv6 LU-FC	5*	3	27%	0
200	IPv4 LU	2	1	0	0
200	IPv6 LU	4	2	0	0
200	IPv4 LU-FC	6*	3	42%	0
200	IPv6 LU-FC	10*	5*	69%	27%

Figure 7: QDR random transaction rate budgeting(LU: destination lookup, LU-FC: destination lookup, source lookup and flow counting), assuming maximum of four 300MHz QDR channels per FPGA configurations with * cannot fit in current FPGAs

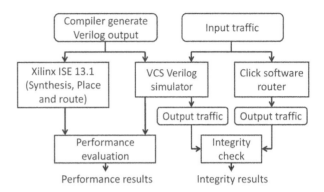

Figure 8: Simulation-Based Verification Process

tible to performance attacks. The control processor accesses QDR memories through the network processor to update IPv4, IPv6, and MPLS tables. Although our current lookup architecture can handle the required lookup throughput for our application, using compressed lookup structures and/or caching can reduce the number of off-chip transactions and, consequently, save power.

4.3 Flow Counting

Network processors in routers keep track of millions of flows for security, management, and QoS purposes [34]. In order to demonstrate Gorilla performance and programmability for flow counting, we implemented a counter scheme in which a 216 bit partitionable counter can be manipulated for each packet. That 216 bit counter can be split into three independent 72 bits counters (36 bits QDR data width × Burst length of 2.)

A pipelined flow counting architecture, which is very similar to the pipelined lookup architecture, is used for counter updates. The only difference is that writes are also required. Therefore a counter update operation includes three reads and three writes each from different stage of the pipeline. Read operations and write operations are each performed by a dedicated data channel provided by QDR SRAMs. The prototype counts the number of packets for each flow ID, which is generated by concatenating the class IDs associated with source and destination lookups. The flow ID is used as the index to the counter array. Because we support 3 72b of counter per packet, up to two other 72b counters for each packet are supported by the prototype.

4.4 FPGA Pins

Figure 7 shows the required number of QDR channels for each of the Gorilla configurations we explored. Although

there are enough pins to go beyond four channels, we did not achieve post place and route timing closure for more than four channels running at 300MHz. Thus, we assume that a maximum of four 300MHz QDR channels are feasible on a single FPGA, making the entries in the table that require more than four channels infeasible using current memory technology and FPGAs. Therefore, 200MPPS packet processing with single level flow counting is memory bound. Hierarchical flow counting can address the problem, but has not yet been implemented in our system.

The QDR consortium recently announced the availability of QDRII+ memories with Random Transaction Rate(RTR) of 600M operations/sec [27] in the near future. This technology could double off-chip bandwidth and, therefore, further improve performance.

Higher throughput can be achieved providing a portion of the traffic has lower memory throughput demands. For example, an MPLS packet with only cluster level lookup does not use QDR channels, allowing other packets to use the extra bandwidth. Figure 7 shows the minimum portion of the MPLS traffic that does not require off-chip memory accesses for delivering the desired performance with both QDRII and QDRII+ standards.

4.5 NP Evaluation

We targeted three different FPGAs designed for networking applications and synthesized various configurations of Gorilla on those three FPGAs using the Xilinx ISE 13.1 tools. The Xilinx Virtex-5 TX240T FPGA (used on the NetFPGA-10G board [22]) is used to implement the IPv4-only NP. We targeted the Xilinx Virtex-6 HX380T for our 50Gbps multi-protocol (IPv4, IPv6, and MPLS) NP and the Xilinx Virtex-7 VHX870T for our 100Gbps multi-protocol NP. In all cases, the core (engines, clusters, load balancing, merging, pre-processing and post-processing) run at 100MHz, while external QDR SRAMs run at 300MHz.

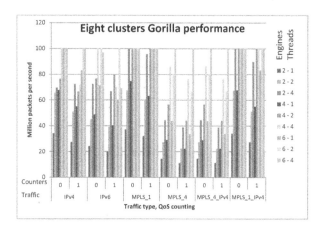

Figure 9: Performance results for 50Gbps sourced traffic with post place and route timing closure

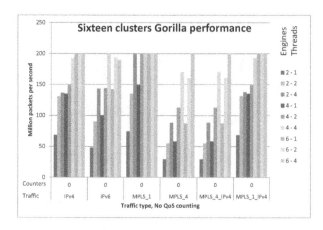

Figure 10: Performance results for 100Gbps sourced traffic with post place and route timing closure

Figure 8 shows the simulation component of our verification process for our NP family. The Click software router [17] was used as the reference system to verify the correctness of NP. In all reported performance results, the functionality of NP is checked against Click software router and post place and route timing closure is met. Synopsys VCS was used to simulate the NP for performance and functionality testing and debugging. Performance results using 100MPPS (50Gbps) and 200MPPS (100Gbps for 64 bytes packets) input traffic rates are shown in Figure 9 and Figure 10 respectively. Packets are dropped whenever performance does not reach the input traffic rate. We use clusters-engines-threads notation to represent a specific configuration. For example in a 16-8-4 configuration there are 16 clusters, 8 engines, and 4 threads.

The largest configuration which fit in Virtex-6 HX380T achieves 50Gbps (100MPPS) supporting all three protocols simultaneously and the largest configuration which fit in Virtex-7 VHX870T achieves 200MPPS supporting all three protocols simultaneously without flow counting.

IPv4 Routing: We generated the IPv4 traffic by extracting only minimum size packets from the CAIDA [3] traffic dump files. In addition to minimum size packet work loads, we tested our IPv4 router using several of equinex-chicago anonymized traffic traces from CAIDA. The major steps in the processing of IPv4 is processing layer two protocol (e.g. Ethernet), extracting and validating the IPv4 header, Looking up source and destination addresses, and modify the packet. Also for IPv4 and IPv6 packets, flow counting adds an extra Gorilla step. There are other steps to handle control packets as well as packets with integrity problem in their headers.

Figure 10 shows that 200MPPS IPv4 routing performance without flow counters can be achieved using three different 16-6-* configurations (16-4-4 configuration has some packet loss.) With flow counting, two of the configurations (16-6-2, and 16-6-4) deliver 200MPPS.

IPv6 Routing Because we did not have IPv6 traces, we IPv4 packets in our minimum-sized IPv4 traffic traces to IPv6 packets. We used a six-level trie for the lookup accelerator with random entries in the route table. Although the IPv6 program is quite similar to the IPv4 program, it reads 128bit addresses instead of 32bit addresses, putting more

pressure on scratch memory. Therefore the header fields may need to be read in multiple steps. Also the trie lookup operation for IPv6 takes twice as long as IPv4 lookup.

MPLS Switching We wrote three Gorilla processing steps to extract the tag, tag lookup, and tag manipulation. These steps read the MPLS header, lookup the MPLS tag, and do the label swap operation respectively. If MPLS headers are stacked, the program jump back from the tag manipulation step to read the next tag.

We generated a variety of MPLS packets with different numbers of stacked labels ranging from one to four. As is shown in Figure 10, most of the Gorilla configurations can handle MPLS_1 traffic, that only contains MPLS packets with one label, without packet loss. MPLS_4, that contains MPLS packets with four labels, needs four tag lookups and four iterations of MPLS steps. As a result only the 16-6-4 configuration can deliver the desired performance when flow counting is off. When flow counting is turned on, the prototype delivers 195.7 MPPS for MPLS4 traffic.

Mixed Traffic In addition to homogeneous IPv4, IPv6, and MPLS test traffic, we generated mixed traffic combining different packet types to study the effect of protocol versatility on the Gorilla generated NP. For example, MPLS-1-IPv4 contains one labeled MPLS packets interleaved with IPv4 packets. When IP traffic is mixed with MPLS traffic, the performance degrades (Figure 10) due to the static inter-cluster packet scheduling algorithm we used in current prototype. We expect that a slightly more dynamic scheduler will fix this problem.

4.5.1 FPGA Resource Utilization

Figure 11 shows FPGA utilization as well as packet processing throughput using a Virtex-5 TX240T with different numbers of engines and threads. We report FPGA resource utilization for two different systems, an *Embedded test system* as well as a *NetFPGA integrated system*.

- *Embedded test system* is an NP without a framer and external memory controllers. Internal packet generators, internal statistical collectors, and internal lookup memories are used instead.

- *NetFPGA integrated system* is an NP targeting the NetFPGA-10G that includes 4*10Gbps Ethernet MAC

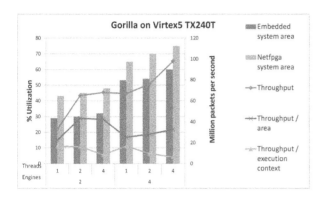

Figure 11: FPGA resource utilization and performance of Gorilla on Virtex-5 TX240T

Engines	Threads	Virtex-6 LUT utilization%	Virtex-7 LUT utilization%
Four	One	27	28
Four	Two	35	34
Four	Four	50	48
Six	One	42	39
Six	Two	53	51
Six	Four	75	73

Figure 12: FPGA utilization of Gorilla on Virtex-6 VHX380T, Clusters=8 and Virtex-7 855T, Clusters=16

controllers connected to FPGA Multi-Gigabit I/Os. This system includes necessary QDR-II controllers for lookup and flow counter accelerators. The integration of the NP with all necessary NetFPGA-10G IP demonstrates the fact that a single FPGA IPv4 router including MAC controllers is feasible.

Figure 11 shows that an 8-4-4 configuration (which delivers 100MPPS) fits in a NetFPGA integrated system with Virtex-5 TX240T. In addition to raw throughput, the figure shows throughput per execution context as well as normalized throughput per area. As is expected, for a particular number of engines, increasing number of threads improves the throughput per area while increasing the number of engines reduces the throughput per area.

Figure 12 shows the FPGA resource utilization of the embedded test system on both Virtex-6 VHX380T, and Virtex-7 855T. We explored an extensive amount of the design space to meet the post place and route timing closures for these two FPGAs. This exploration was only possible due to the parametric nature of the templates in the system.

4.5.2 Scheduling, Arbitration, and Reordering

The Gorilla resource management templates enables easy trading off between area and performance. Figure 13 shows the performance of Gorilla running IPv4 packet routing using two different lookup delays (32 and 128 cycles), and two different thread schedulers. When the number of threads is increased, performance saturates earlier with low latency accelerators than with high latency accelerators. A *round robin* thread scheduler switches to the next thread whenever a thread accesses an accelerator, regardless of whether the next thread is ready or not. A *ready-to-execute* thread

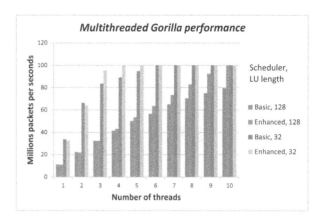

Figure 13: Multithreaded Gorilla throughput scaling for IPv4 with 100MPPS source rate (clusters=8, engines=2)

scheduler switches to a thread which is not stalled every cycle. Although the ready-to-execute thread scheduler performs better for large number of threads, it requires a reorder queue because processing of packets might finish out of order. Consequently, ready-to-execute thread scheduling imposes a considerable area overhead on the design (10%-15% for large configurations.) When accelerator latencies are small, round robin scheduling saves area. However for large latency accelerators we need ready-to-execute scheduling for performance scaling.

Another parametric template in the Gorilla infrastructure is the arbitration structure. Arbitration for accesses to global accelerators is done in a hierarchical fashion. Either fair, yet complex arbiters or low overhead unfair arbiters are possible. Due to the large contention on global arbiters comparing to local arbiters, in most cases strong fairness is only required for global arbiters. This works perfectly for our design because of two reasons. Firstly, the local schedulers, like the thread scheduler and the engine scheduler, are self throttling, ensuring freedom from starvation even with an unfair scheduler. Secondly, there are many more local arbiters than global arbiters. Consequently there are many small, lightweight arbiters and few large arbiters in the design. Using fair local arbiters in a large configuration imposes an 8% area overhead on the whole design.

4.6 Board Implementation

Since we did not have access to high performance networking test gear, our real implementation included an embedded test platform on the FPGA to generate random packets and verify that they were processed correctly. Note that we met timing closure with exactly the same network processor integrated with real Gigabit transceivers and memory controllers on equivalent FPGA with real I/Os, indicating we will run correctly with real interfaces and memory.

We implemented a 16-3-2 configuration of IPv4-only Gorilla with embedded peripherals on a Xilinx Virtex 6 ML605 board [37]. The implementation includes an embedded packet generator unit, an embedded packet collector, and a statistic report generator. The lookup engines (Section 4.2) are using a BRAM-based module which exactly emulates the QDR timing. The working board validates the functionality

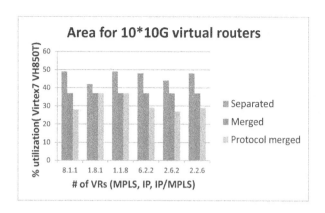

Figure 14: Protocol-Based Virtual Router Consolidation in Gorilla

of NP on real hardware and that the delivered performance is consistent with the Verilog simulation results. The board handles 40Gbps (100MPPS) traffic (injecting a new packet every clock cycle) using 55% of LUT resources.

Using the Xilinx *system monitor* tool, the measured current and voltage driving the FPGA core logic (VCCint and the corresponding driving current) indicated power is less than 4 watts. Although it does not include I/O, the power is stunningly low. The core logic power of Gorilla is comparable or better than the power consumption reported for state of the art network processors [23, 28]. The I/Os should be similar, implying that our total solution will be at least power competitive.

4.7 Performance Analysis

The network processor achieves its high performance from a combination of reasons. Since the accelerators are the most critical resources in the system, as long as the accelerators always have work to do, the system's throughput is maximized. Hardware-implemented finite state machines issue calls very quickly, ensuring that each thread is producing as many requests to accelerators as possible. Aggressive multithreading with hardware-implemented synchronization of accelerator calls maximizes the number of ready threads which, in turn, maximizes the number of accelerator calls **and** makes writing the domain code easy and sequential. Gorilla engines can issue a full set of accelerator calls **every** cycle, compared to tens of cycles needed for a single accelerator call on a more conventional, instruction-based network processor. Accelerators are themselves designed to maximize throughput. The end result is a system that maximizes the number of accelerator calls and, therefore, performance.

The aggressive parameterization of the templates enables rapid, extensive exploration of the implementation space, enabling the search for an optimized configuration in a reasonable time and accommodating last minute changes. Thus, templates can dramatically improve productivity.

5. VIRTUAL ROUTER CONSOLIDATION

Router virtualization is a technique to share the same packet processing hardware across different virtual routers to reduce capital and operational expenditures. FPGAs provide unique opportunity for building hardware virtual routers. Different techniques have been proposed for con-

solidating virtual routers on a single FPGA [19, 39]. A router virtualization framework should provide three important characteristics (i) isolated performance, (ii) reconfigurability, and (iii) scalability in terms of number of virtual routers.

We do not present a full fledged virtualization framework. Instead, we demonstrate how a flexible and modular network processor design using Gorilla can consolidate virtual routers efficiently. Consolidating virtual routers can be done either using isolated hardware or merged hardware. For example, when 10 10Gbps virtual routers are consolidated, 10 isolated routers can be instantiated, each handling its own traffic flow, or a single 100Gbps router that handles the merged flows. In a hardware router in which resources are allocated for worse case traffic patterns, individual routers can be merged while maintaining performance isolation for each virtual router.

Given a set of virtual routers, each supporting different protocols, it is possible to create a customized network processor that supports all the virtual routers with guaranteed throughput and minimal resource utilization. When virtual routers with the same protocol(s) are merged, area efficient techniques like multithreading and pipelining help to improve performance per area comparing to an isolated configuration. When virtual routers implementing different protocols are merged, one can support all the protocols by simply instantiating engines that support all required protocols. Although this method might save area by reusing common structures, resources are used inefficiently because the engines are more general than they need to be.

Our solution, *protocol-based consolidation*, uses a heuristic algorithm where groups of virtual routers are merged only if the virtual routers have some common protocol. Merging is only considered if the merged area is less than the area of isolated implementation of virtual routers. We demonstrate the result of such protocol-based consolidation by implementing 10*10G virtual routers using Gorilla. Each of the virtual routers supports one of the following: MPLS only, IPv4 only, or both MPLS and IPv4 protocols. Only single label MPLS processing is considered in this study. Six different virtual router configurations are considered. For example, in the 8.1.1 configuration there are eight MPLS-only virtual routers, one IPv4-only virtual router, and one IPv4/MPLS virtual router.

Figure 14 shows the LUT utilization of each of these configurations on Virtex-7 855T using isolated, merged, and protocol-based consolidation methods. When the protocol-based consolidation method is used on an eight MPLS, one IPv4, and one IPv4/MPLS virtual router configuration, the MPLS only routers are merged but not the other two virtual routers. However for the one MPLS, eight IPv4, and one IP/MPLS configuration all virtual routers are merged. This is due to the fact that for 80Gbps a single label MPLS processing requires fewer engines and fewer threads per engine than IPv4 processing. As the result, isolating a 80Gbps MPLS router from the other two routers saves a considerable area. On the other hand, when building a 80Gbps IPv4 router, the incremental area overhead for equipping engines with MPLS processing steps is smaller than adding two isolated 10G MPLS routers. Overall, for the sample virtual router configurations, protocol-based consolidation saves on average 33% of area comparing to the isolated method and 15% of area comparing to the normal merged method.

6. RELATED WORK

The architectural techniques including using accelerators, course grain pipelining, multithreading, parallelism among multiple engines, and hierarchical arbitration have been used extensively in the past. However, to the best of our knowledge, they have not been combined the Gorilla way. Also, they have not been automatically generated from stylized C or utilized FPGAs to provide programmability. The exception is our own patented work, that was not evaluated in the patent [6].

Templates (called different names in different languages including "templates" in C++ and "macros" in Lisp) have long been used in software to improve reuse. Effective parameterized hardware, though explored repeatedly in the past, has not yet achieved widespread acceptance. Recent research [24, 33] has demonstrated the power of templates, especially for high performance and/or low power designs. Gorilla differs in that it separates the functionality (application) from the parameterized template while others are more of a parameterized implementation. Others [25] have separated functionality and the micro-architecture, but do not focus on high parameterization of the micro-architecture (template.)

Requiring the user to provide a template in addition to domain code that describes functionality is Gorilla's key differentiator from traditional C-to-gates tools, such as AutoESL [2] and CatapultC [4], that attempt to infer the appropriate hardware structure. Some researchers tried to incorporate the architectural information while synthesizing the high level code but their method is limited to generating single processing engine on a pre-defined datapath [29].

Among many publications related to automatic compilation of applications into hardware, Convey [36] and Optimus [14] are perhaps the most similar to Gorilla. Convey provides predefined building blocks for a particular application personality to domain code. The tool chain compiles domain code to pipelined implementations called systolic structures. Temporal common sub-expression elimination, or loops in the pipeline, is used to reduce area overhead. Gorilla is more free formed than Convey, implementing state machines rather than application-specific processors, and focuses on balancing the performance-area trade off by sizing high level resources like engines and clusters. Optimus maps StreamIt [35] applications, that are described using streaming graph and filter kernels, to FPGAs. Gorilla targets applications with more control flow irregularity and shared resource access irregularity.

Many projects aim to improve the programmability problem for high performance FPGA-based packet processing applications. Some focus on the programmability aspect, without achieving high performance [18, 31]. NetFPGA [20] uses a set of libraries and reference designs in order to simplify the implementation of new packet processing systems. However a hardware expert is needed to design the entire hardware implementation. A Xilinx project targets 100Gbps network processing and traffic management in a pair of FPGAs, but is designed in a traditional way and is not yet available [11]. Another project [16] demonstrated 40Gbps in a FPGA for MPLS-only packet processing without any high level programmability features. Gorilla generated network processors, on the other hand, can handle 200MPPS (64B packets) for any mix of IPv4, IPv6, and MPLS on a single Virtex7 FPGA and it is programmable with C.

Routebricks [7] is a parallelized version of the Click software router and is the fastest pure software router we are aware of. Routebricks runs at 23.4M packets per second (12Gbps with 64B packets) on four systems, each containing eight Nehalem cores for a total of 32 cores. Routebricks supports four flows of 3Gbps each, a significantly easier problem than the single 100Gbps flow Gorilla supports. Gorilla also has deterministic performance and resiliency against adversarial traffic, and consumes much less power.

Packetshader [32] achieves 40Gbps on a system with two quad-core Nehalem processors, 12GB of memory, two I/O hubs and two NVIDIA GTX480 cards. Packetshader performance is dependent on intelligent NICs that balance load between cores and is currently limited by PCIe performance. The load balancers make PacketShader venerable to adversarial traffic. The partitioning introduces overhead to keep common state between cores. In addition, each GTX480 consumes up to 250W.

There are many custom network processors including those from Cavium [5], EZChip [9], Xelerated [38], Freescale [10], and LSI [21]. Based on their processor data sheets, Gorilla's performance is roughly the same as EZChip's and Xelerated's recently announced 100Gbps network processors and faster than all other vendors. A Gorilla NP is more flexible than those solutions because it is implemented entirely on an FPGA and is also at least as easy to program. From architectural perspective the difference between Gorilla and other network processors is that the hardware is specialized for the application and parallelization is expressed and managed using structures which are scalable in FPGAs.

7. CONCLUSIONS AND FUTURE WORK

We describe the Gorilla programming model, templates, canonical architectures, and tool chain. Gorilla separates the work of domain experts from the work of hardware experts, enabling each to work in their area of expertise and automatically and efficiently combining that work.

We use Gorilla to generate a family of network processors capable of handling almost all combinations of MPLS, IPv4, and IPv6 traffic at 200MPPS rate in a single FPGA. The packet processing engines, accelerators, packet splitter and reassembly are written by domain experts in a subset of sequential C, automatically compiled to hardware, and then merged with parameterized templates written by hardware experts. Domain experts only need to concentrate on their domain and can safely ignore hardware and parallelization issues. Hardware experts, on the other hand, only need to focus on hardware, rather than having to know the details of the implementation.

We studied the required FPGA resources using three modern FPGAs to demonstrate that FPGAs have enough logic, memory, and IO resources to deliver the required packet processing performance for the mentioned applications. We reported the performance results for various traffic loads and processor configurations.

Although we focused on packet processing as the proof of concept for Gorilla methodology, we believe that the same methodology can be applied on other data parallel applications like genome sequencing, or hardware simulation acceleration. We plan to port more application spaces and applications to Gorilla and to build a router using the Gorilla generated network processor described in this paper.

8. ACKNOWLEDGEMENTS

This material is based upon work supported in part by the National Science Foundation under Grants 0747438 and 0917158. We acknowledge the anonymous reviewers for their insightful comments.

9. REFERENCES

[1] Antlr compiler compiler tool. http://www.antlr.org/.

[2] AutoESL. http://www.autoesl.com/.

[3] The cooperative association for internet data analysis. www.caida.org.

[4] CatapultC. http://www.mentor.com/.

[5] Cavium network processor. http://datasheet.digchip.com/227/227-04668-0-IXP2800.pdf.

[6] DENNISON, L., AND CHIOU, D. Compilable, reconfigurable network processora simulation method. United States Patent 7,823,091, Oct. 2010.

[7] DOBRESCU, M., EGI, N., ARGYRAKI, K., CHUN, B.-G., FALL, K., IANNACCONE, G., KNIES, A., MANESH, M., AND RATNASAMY, S. Routebricks: exploiting parallelism to scale software routers. In SOSP '09: Proceedings of the ACM SIGOPS 22nd symposium on Operating systems principles (New York, NY, USA, 2009), ACM, pp. 15–28.

[8] EATHERTON, W., VARGHESE, G., AND DITTIA, Z. Tree bitmap: hardware/software ip lookups with incremental updates. SIGCOMM Comput. Commun. Rev. 34 (April 2004), 97–122.

[9] Ezchip np-4 network processor. http://www.ezchip.com/Images/pdf/NP-4_Short_Brief_online.pdf.

[10] Freescale c-5 network processor. http://www.freescale.com/webapp/sps/site/prod_summary.jsp?code=C-5.

[11] Reconfigurable computing for high performance networking applications. http://www.xilinx.com/innovation/research-labs/keynotes/Arc2011-Keynote.pdf.

[12] GUPTA, P., LIN, S., AND MCKEOWN, N. Routing Lookups in Hardware at Memory Access Speeds. In In Proceedings of INFOCOM (1998), pp. 1240–1247.

[13] HAMEED, R., QADEER, W., WACHS, M., AZIZI, O., SOLOMATNIKOV, A., LEE, B. C., RICHARDSON, S., KOZYRAKIS, C., AND HOROWITZ, M. Understanding sources of inefficiency in general-purpose chips. In Proceedings of the 37th annual international symposium on Computer architecture (New York, NY, USA, 2010), ISCA '10, ACM, pp. 37–47.

[14] HORMATI, A., KUDLUR, M., MAHLKE, S., BACON, D., AND RABBAH, R. Optimus: efficient realization of streaming applications on fpgas. In Proceedings of the 2008 International Conference on Compilers, Architectures and Synthesis for Embedded Systems (2008), pp. 41–50.

[15] JIANG, W., AND ET AL. Parallel IP Lookup using Multiple SRAM-based Pipelines, 2008.

[16] KARRAS, K., WILD, T., AND HERKERSDORF, A. A folded pipeline network processor architecture for 100 gbit/s networks. In Proceedings of the 6th ACM/IEEE Symposium on Architectures for Networking and Communications Systems (New York, NY, USA, 2010), pp. 2:1–2:11.

[17] KOHLER, E., MORRIS, R., CHEN, B., JANNOTTI, J., AND KAASHOEK, M. F. The click modular router. ACM Trans. Comput. Syst. 18, 3 (2000), 263–297.

[18] KULKARNI, C., BREBNER, G., AND SCHELLE, G. Mapping a domain specific language to a platform fpga. In Proceedings of the 41st annual Design Automation Conference (New York, NY, USA, 2004), DAC '04, ACM, pp. 924–927.

[19] LE, H., GANEGEDARA, T., AND PRASANNA, V. K. Memory-efficient and scalable virtual routers using fpga. In Proceedings of the 19th ACM/SIGDA international symposium on Field programmable gate arrays (2011), pp. 257–266.

[20] LOCKWOOD, J. W., MCKEOWN, N., WATSON, G., GIBB, G., HARTKE, P., NAOUS, J., RAGHURAMAN, R., AND LUO, J. NetFPGA - An Open Platform for Gigabit-rate Network Switching and Routing. In IEEE International Conference on Microelectronic Systems Education (June 2007).

[21] Lsi advanced payload plus network processor. http://www.lsi.com/networking_home/networking-products/network-processors/.

[22] NetFPGA 4*10G board . http://netfpga.org/foswiki/NetFPGA/TenGig/Netfpga10gInitInfoSite.

[23] Netronome flow processor. http://www.netronome.com/pages/network-flow-processors.

[24] NG, M. C., VIJAYARAGHAVAN, M., DAVE, N., ARVIND, RAGHAVAN, G., AND HICKS, J. From WiFi to WiMAX: Techniques for High-Level IP Reuse across Different OFDM Protocols. In Proceedings of MEMOCODE 2007 (2007).

[25] PATIL, N. A., BANSAL, A., AND CHIOU, D. Enforcing architectural contracts in high-level synthesis. In DAC (2011), pp. 824–829.

[26] 300Mhz Two Word Burst QDRII datasheet. http://www.cypress.com/?docID=21484.

[27] QDR plus pre-announcement by QDR consortium. http://eetimes.com/electronics-products/electronic-product-reviews/memory-products/4215361/QDR-Consortium-pre-announces-new-SRAMs.

[28] Cisco QuantumFlow architecture. http://www.cisco.com/en/US/prod/collateral/routers/ps9343/solution_overview_c22-448936.pdf.

[29] RESHADI, M., GORJIARA, B., AND GAJSKI, D. Utilizing horizontal and vertical parallelism with a no-instruction-set compiler for custom datapaths. In Proceedings of the 2005 International Conference on Computer Design (2005), pp. 69–76.

[30] RIPE Route Information Service. http://www.ripe.net/projects/ris/rawdata.html.

[31] RUBOW, E., MCGEER, R., MOGUL, J., AND VAHDAT, A. Chimpp: A click-based programming and simulation environment for reconfigurable networking hardware. In Architectures for Networking and Communications Systems (ANCS), 2010 ACM/IEEE Symposium on (Oct. 2010), pp. 1 –10.

[32] SANGJIN HAN, KEON JANG, K. P., AND MOON, S. PacketShader: a GPU-accelerated Software Router. In in Proc. of ACM SIGCOMM 2010, Delhi, India (2010).

[33] SHACHAM, O., AZIZI, O., WACHS, M., QADEER, W., ASGAR, Z., KELLEY, K., STEVENSON, J., RICHARDSON, S., HOROWITZ, M., LEE, B., SOLOMATNIKOV, A., AND FIROOZSHAHIAN, A. Rethinking digital design: Why design must change. Micro, IEEE 30, 6 (nov.-dec. 2010), 9 –24.

[34] SHAH, D., IYER, S., PRABHAKAR, B., AND MCKEOWN, N. Maintaining statistics counters in router line cards. IEEE Micro 22 (January 2002), 76–81.

[35] THIES, W., KARCZMAREK, M., AND AMARASINGHE, S. P. Streamit: A language for streaming applications. In Proceedings of the 11th International Conference on Compiler Construction (London, UK, 2002), Springer-Verlag, pp. 179–196.

[36] VILLARREAL, J., AND ET AL. Programming the Convey HC-1 with ROCCC 2.0, 2010.

[37] Xilinx ML605 reference design board. http://www.xilinx.com/products/boards-and-kits/EK-V6-ML605-G.htm.

[38] Xelerated. http://www.xelerated.com/.

[39] YIN, D., UNNIKRISHNAN, D., LIAO, Y., GAO, L., AND TESSIER, R. Customizing virtual networks with partial fpga reconfiguration. SIGCOMM Comput. Commun. Rev. 41 (2011), 125–132.

[40] ZHENG, K., HU, C., LU, H., AND LIU, B. A TCAM-based distributed parallel IP lookup scheme and performance analysis. IEEE/ACM Trans. Netw. 14 (August 2006), 863–875.

Limit Study of Energy & Delay Benefits of Component-Specific Routing

Nikil Mehta
Department of Computer
Science, 256-80
California Institute of
Technology
Pasadena, CA 91125
nikil@caltech.edu

Raphael Rubin
Department of Computer and
Information Systems
University of Pennsylvania
3330 Walnut Street
Philadelphia, PA 19104
rafi@seas.upenn.edu

André DeHon
Department of Electrical and
Systems Engineering
University of Pennsylvania
200 S. 33rd St.
Philadelphia, PA 19104
andre@acm.org

ABSTRACT

As feature sizes scale toward atomic limits, parameter variation continues to increase, leading to increased margins in both delay and energy. The possibility of very slow devices on critical paths forces designers to increase transistor sizes, reduce clock speed and operate at higher voltages than desired in order to meet timing. With post-fabrication configurability, FPGAs have the opportunity to use slow devices on non-critical paths while selecting fast devices for critical paths. To understand the potential benefit we might gain from component-specific mapping, we quantify the margins associated with parameter variation in FPGAs over a wide range of predictive technologies (45nm–12nm) and gate sizes and show how these margins can be significantly reduced by delay-aware, component-specific routing. For the Toronto 20 benchmark set, we show that component-specific routing can eliminate delay margins induced by variation and reduce energy for energy minimal designs by 1.42–1.98×. We further show that these benefits increase as technology scales.

Categories and Subject Descriptors

B.7.2 [**Integrated Circuits**]: Design Aids—*placement and routing*; B.8.1 [**Performance and Reliability**]: Reliability, Testing, and Fault-Tolerance

General Terms

Algorithms, Measurement, Reliability

Keywords

Component-Specific Mapping, Variation Tolerance, Minimum Energy

1. INTRODUCTION

As we continue to scale down feature sizes, we can no longer guarantee that all transistors will have identical parameters, nor can we determine prior to fabrication which transistors will be fast, slow, or defective. This parameter uncertainty leads to circuits that are pessimistically sized, underclocked, and overvoltaged. Because designers have no guarantee that critical paths will not contain any slow transistors, they must accept lower frequency operation. As a corollary, to maintain cycle time, V_{dd} may need to be increased in order to speed up slow transistors to meet timing, wasting energy. This energy loss is particularly detrimental for many modern designs that are energy limited. Static delay and energy margins are already significant in current technologies and will only continue to increase.

Conventional variation tolerance techniques will not be effective at reducing these margins in future technology nodes that are dominated by large random V_{th} variations [1]. For the 22nm node the ITRS predicts[1] $\sigma_{V_{th}}/\mu_{V_{th}} = 27\%$, meaning that a minimum sized transistor (L=W=22nm) with a nominal $V_{th} \approx 300$mV can have a 3σ V_{th} spread of 57mV–543mV, equating to a delay difference of 0.49–12.58 ps (Fig. 2), or nearly two orders of magnitude. The random, fine-grained nature of this delay spread means that conventional techniques that rely on pre-fabrication estimates (e.g. SSTA, statistical static timing analysis) and coarse-grained adaptations (e.g. adaptive body biasing) can only be of limited use. Sizing up transistors is an important conventional technique to reduce variation, but this comes with a cost of increased capacitance and energy—a tradeoff that must be quantified under variation (see Sec. 5.3).

FPGAs provide a unique opportunity for fine-grained control and post-fabrication configurability. If we break the one-mapping-fits-all approach and map on a per-chip basis, tools can place and route designs that avoid unreasonably slow resources while strategically selecting fast resources for critical paths. Unfortunately, there are high barriers to component-specific mapping. Individual resource delays must be extracted on a per-chip basis. Mapping must also be performed per-chip, potentially exploding CAD effort. While these challenges are great, recent work (see Sec. 2.2.1 and 2.2.2) demonstrates promising progress.

Nonetheless, the question remains as to what are the ultimate achievable benefits from component-specific mapping.

---

[1] ITRS Table DESN9 in Design Section reports 3σ variation (81% for 22nm), which we divided by 3.

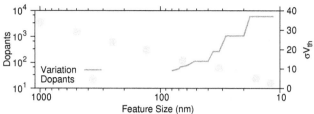

Figure 1: Decreasing dopants and increasing V_{th} variation from ITRS 2010

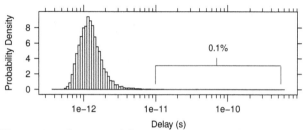

Figure 2: Inverter delay distribution under variation ($W_p=W_n=L=22$nm)

To answer this question we perform a limit study of the delay and energy benefits from component-specific mapping for future technology nodes. As a limit study, we make optimistic assumptions about the completeness and precision of delay information available and about the amount of effort we can apply to routing. We examine a router that assumes knowledge of all FPGA resource delays and measure the routed delay and energy of the Toronto 20 benchmarks [4] for predictive technology models [36] spanning the 45nm, 32nm, 22nm, 16nm, and 12nm nodes. We characterize the expected delay and energy margins from variation and by how much we expect to reduce these margins through component-specific mapping. We also explore how gate sizing affects these margins and expected benefits.

Novel contributions of this work include:

- Quantification of FPGA delay and energy margins due to random V_{th} variation in interconnect for 45nm–12nm technology nodes.
- Quantification of potential delay and energy savings for component-specific routing.
- Demonstration of the impact of interconnect buffer sizing on margins and delay and energy savings.
- Demonstration of the impact of CLB IO sparing.
- Determination of energy-optimal sizing as a function of technology and variation.

2. BACKGROUND

2.1 Variation, Delay & Energy Scaling

Parameter variation can be decomposed into lot-to-lot, wafer-to-wafer, die-to-die and within die (WID) variations. WID variation is the most significant contributer to parameter uncertainty, and can be further categorized as systematic (e.g. layout dependent), spatially correlated (e.g. distance dependent), and random. Random WID V_{th} variation, caused by effects like dopant fluctuation and channel length variance, is expected to dominate future sources of variation [1,3]. Fig. 1 shows how V_{th} variation increases due to decreases in dopant count as a function of feature size.

Increased Delay.

The first impact of variation is that gates will exhibit a large spread in delays, reducing the speed of designs as the delay of a circuit is set by its slowest path. We can express the current and delay of a gate as:

$$I_{sat} = W v_{sat} C_{ox} \left(V_{gs} - V_{th} - \frac{V_{d,sat}}{2} \right)^{\gamma} \quad (1)$$

$$I_{sub} = \frac{W}{L} \mu C_{ox} (n-1) (v_T)^2 \, e^{\frac{V_{gs} - V_{th}}{n v_T}} \left(1 - e^{\frac{-V_{ds}}{v_T}} \right) \quad (2)$$

$$\tau_p = \frac{C V_{dd}}{I_{on}} \quad (3)$$

Where $I_{on}=I_{sat}$ for $V_{dd} \geq V_{th}$, and $I_{on}=I_{sub}$ for $V_{dd} < V_{th}$. V_{th} variation can cause a large spread in gate delays: the HSPICE simulated delay of a minimum sized 22nm inverter ($W=L=22$nm) with $\sigma_{V_{th}}/\mu_{V_{th}}=27\%$ and 10,000 samples is shown in Fig. 2, and we can see delays that span several orders of magnitude. Because of Fig. 1, this spread will increase with continued technology scaling.

Increased Energy.

The second impact of variation is that it raises energy per operation, which can be expressed (ignoring short circuit currents and glitches) as follows:

$$E_{dynamic} = \sum_{i}^{n_{switch}} \frac{\alpha_i}{2} \cdot C_{load,i} \cdot (V_{dd})^2 \quad (4)$$

$$E_{static} = \sum_{i}^{n_{total}} I_{leak,i} \cdot V_{dd} \cdot \tau_p \quad (5)$$

$$E_{total} = E_{dynamic} + E_{static} \quad (6)$$

For a circuit operating at a target delay, to compensate for slower gates, designers are often forced to raise V_{dd} to increase transistor drive strength (I_{sat}, Eq. 1), increasing energy/operation.

Alternatively, many circuits may need to simply minimize energy/operation at the cost of delay. Energy/operation is the primary design constraint for many low power, embedded systems. It correlates directly with battery lifetime, and is also relevant to operating with limited thermal budgets. Even for these delay unconstrained systems, parameter variation makes energy minimization more difficult [6]. From Eq. 6 we see that the most direct way to reduce energy/operation is to lower V_{dd}. For most circuits, $E_{dynamic} > E_{static}$, so voltage scaling yields a quadratic reduction in energy. This technique has been examined in commercial FPGAs [9].

With $V_{dd} > V_{th}$ the delay of a transistor depends on Eq. 1 and is super linear; however, when $V_{dd} < V_{th}$ delay solely depends on Eq. 2 making it exponential in both V_{dd} and V_{th}. The exponential dependence in V_{dd} means that operations become significantly longer at lower voltages. As static energy/operation is expressed as leakage power times the length of an operation (Eq. 5), at low V_{dd} it will increase dramatically and eventually become the dominant source of energy dissipation. Fig. 3 shows the energy/operation of a 16-bit multiplier mapped to a 22nm FPGA. As a result, *there exists a V_{dd} at which energy is minimized.* This gives a well-defined target point of operation—we should operate at the energy optimal V_{dd} when minimizing energy/oper-

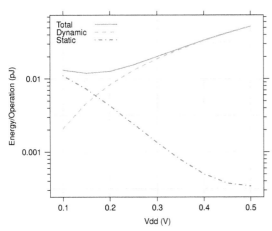

Figure 3: Minimum energy/operation for a 16-bit multiplier at 22nm (V_{th}=300mV)

ation. Furthermore, note that the minimum energy point occurs below the threshold voltage in this example, as it does for most designs. If we add V_{th} variation, the length of an operation will increase further due to delay uncertainty, increasing the minimum energy of operation [7].

Increased Failures.

The third impact of parameter variations is the increase in functional failures due to V_{th} mismatch [26]. SRAMs are commonly the first circuits to fail due to V_{th} variation which causes read upsets, write upsets, hold failures, and access time failures. However, FPGA configuration SRAMs simply hold state and are not cycled during operation, so these circuits can be fabricated with increased widths and with a separate, high V_{th} to avoid failures [34].

Static logic, however, can also fail due to V_{th} variation. We define a CMOS inverter to be defective due to variation when leakage current overpowers on current:

$$I_{PMOS,off} > I_{NMOS,on} \text{ or } I_{PMOS,on} < I_{NMOS,off} \quad (7)$$

Under these conditions the inverter can never switch; this can only happen when V_{th} variation is large enough such that a very high V_{th} device is paired with a very low V_{th} device. Furthermore, as V_{dd} decreases, the probability of a defect increases as I_{on} for both PMOS and NMOS transistors degrades (Eq. 1) [16]. Consequently, this effect is particularly acute for subthreshold operation, preventing operation near the minimum energy point.

2.2 Prior Work

Component-specific mapping for FPGAs is a simple idea: if you can determine resource delays on a chip before mapping and use that information during mapping, then you can naturally produce a more efficient design. This allows the design to avoid defects (Eq. 7) and avoid slow gates on the critical path. What is not obvious is a) how to extract resource delays, b) how to efficiently use that information to route every chip differently, and c) what is the overall benefit. This paper only attempts to quantify the third: the benefits of component-specific mapping. The ultimate feasibility of component-specific mapping is still an open question. However, several of the key challenges in measurement

and mapping have already been solved, representing partial, but realistic solutions for component-specific mapping.

2.2.1 Component-Specific Measurement

Several researchers have identified ways to quickly and accurately measure FPGA resource delays without the use of an expensive tester. One technique uses arrays of configured ring oscillators to measure aggregate delays of N=5–7 stages of LUT + interconnect for commercial 90nm and 65nm devices [29, 33]. Additionally, Wong et al. developed a technique for at-speed path delay measurements that can be configured to extract delays of N=2 stages of LUT + interconnect with 1 ps resolution [35]. The technique configures a path between a registered source LUT and a shadow registered destination LUT and sweeps clock frequency until the shadow register detects an error. Full characterization of a Altera Cyclone II EP2C35 can be achieved in 3 seconds. Using sparse sampling, Majzoobi et al. showed that spatially correlated variation can be approximated in milliseconds and characterized with only megabytes of data [23]. Remaining challenges in measurement are to resolve single element delays (e.g. individual LUTs, switches as opposed to aggregate paths) and to obtain even finer timing resolution.

2.2.2 Component-Specific Mapping

Initial work in component-specific mapping focused on defect tolerance, with HP's TERAMAC demonstrating the ability to locate and map around defective elements and tolerate defect rates of 3–10% [10]. Later ideas in defect-tolerant FPGA component-specific mapping generated multiple bitstreams and then tested each bitstream per component [24, 30, 32].

Recent work has performed mapping using delay knowledge in both placement and routing. Katsuki et al. [8] and Cheng et al. [13] perform chip-wise placement by generating a variation map per chip and using that map to place critical logic in fast regions assuming the dominant WID variation is spatially correlated. However, future technologies are dominated by random, not spatially correlated variation; placement is too coarse grained to preferentially select individual devices. Gojman et al. [11] perform fine-grained component-specific routing on an reconfigurable NanoPLA. By matching net fanout to threshold voltages, they are able to restore 100% yield in a 5nm technology with $\sigma_{V_{th}}/\mu_{V_{th}} = 38\%$.

The cost of mapping with component-specific techniques can be prohibitively expensive. Since recent FPGAs contain billions of transistors, measurement storage may be significant. More importantly, CAD must be performed per-chip. Modern CAD runtime for large designs often takes days; multiplying this CAD effort by the number of shipped parts explodes handling time.

One promising direction in reducing storage and CAD effort is Choose-Your-own-Adventure (CYA) routing [27]. In CYA, routing is performed once for all chips, but the bitstream produced by the router contains several alternative routes for each net in the design that are evaluated at load time. CYA could be conceivably extended to perform timing tests for measurement (Sec. 2.2.1) to enable delay-aware component-specific mapping.

2.2.3 Conventional Variation Tolerance

To tolerate FPGA parameter variation, researchers have employed many of the same techniques used for ASICs and

CPUs. Adaptive body biasing [25] and dual-V_{dd} assignment [5] attempt to compensate for variation at the CLB level by adjusting V_{th} and V_{dd} post fabrication. However, these techniques have insufficient granularity to deal with random variation due to circuity overhead.

Several researchers have examined using SSTA in the timing analysis steps of clustering, placement [15, 21], routing [31], and the entire CAD flow [19, 20] to better identify and optimize statistically critical paths under variation. Lin et al. demonstrate a mean delay improvement of 6.2% and a delay variance reduction of 7.5% for combined regional and random $\sigma_{V_{th}}/\mu_{V_{th}}$ of 2% and 3.3% respectively. SSTA relies on pre-fabrication models that are difficult to generate accurately, do not scale well for high V_{th} variation, and do not reflect the individual variation map of a given chip.

Device and circuit parameters can also be optimized to mitigate the impact of parameter variation. Transistor sizing is a common strategy used in ASICs to directly reduce the magnitude of variation. Larger transistors have reduced V_{th} variation as can be seen by the following relation:

$$\sigma_{V_{th}} \propto \frac{1}{\sqrt{WL}} \tag{8}$$

Increasing W in logic transistors can increase variation tolerance, at the cost of area and energy. Modern integrated circuits use few or no minimum size devices for this reason. We quantify the benefits and drawbacks of sizing both for one-mapping-fits-all and for component-specific mapping, using Eq. 8 to scale the variation of sized transistors.

3. ANALYSIS

Before looking at the results, this section briefly reviews the major effects at play to provide an intuitive basis for reasoning about the results and how they might be impacted by different designs and benchmarks.

Device-level variation does not linearly result in circuit- or application-level variation. In this section, we identify the major design and platform properties that couple with device-level variation to determine application-level performance. Note the models used in this section are simplistic, treating the primitive delays as Gaussian random variables—for the actual circuit modeling performed later in the paper, we use more primitives variables (e.g. $\sigma_{V_{th}}$) as the basis for computing delays.

3.1 Path Length

The set of device parameters will combine to define the delay of the each device, gate, or interconnect segment. For intuition, let us think about gates and interconnect segments as the unit of composition. Assume each gate or interconnect segment has delay τ_u, and further assume the delay of each unit is an identically and independently distributed (i.i.d.) random variable taken from a Gaussian distribution with mean μ_{τ_u} and standard deviation σ_{τ_u}.

Disregarding fanout and reconvergent paths, the delay along a path of length d is the sum of d gate delays. The sum of a set of Gaussians is, itself, a Gaussian.

$$\tau_{path}(d) = \tau_{u_0} + \tau_{u_1} + \ldots + \tau_{u_{d-1}} \tag{9}$$

$$\mu_{\tau_{path}(d)} = d \times \mu_{\tau_u} \quad ; \quad \sigma_{path}(d) = \sqrt{d} \times \sigma_{\tau_u} \tag{10}$$

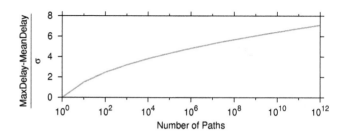

Figure 4: 50% Yield Delay vs. Number of Paths

Note that:

$$\frac{\sigma_{path}(d)}{\mu_{\tau_{path}(d)}} = \left(\frac{1}{\sqrt{d}}\right)\frac{\sigma_{\tau_u}}{\mu_{\tau_u}} \tag{11}$$

That is, as a percentage of the nominal delay, the variation decreases by a factor of the square root of the path length, d. *This means, slow circuits will see small variation, while circuits that are highly pipelined with few LUTs and interconnect segments between flops will see a larger variation as a percentage of nominal path delay.* Note that:

- It is these high-performance, short paths where we care about performance the most.
- We expect pipelining to increase in future designs.

3.2 Multiple Paths

Now, consider that we have K independent, parallel paths on a chip that have the same nominal delay; i.e. all are critical paths. Few paths on the chip are independent, but we make this simplifying assumption to develop intuitive analysis on the scaling trends. If we clock the chip synchronously, the clock cycle is limited by the longest path. This gives us:

$$T_{cycle} = \max_{\text{all paths } p_i} (\tau_{p_i}) \tag{12}$$

We would like to know the distribution of T_{cycle}, which is a max of Gaussians.

To simplify this from a distribution to a single number, we might ask what delay we can expect half of our components to meet, $T_{50\%}$. That means:

$$P(T_{cycle} \leq T_{50\%}) = 0.5 \tag{13}$$

For T_{cycle} to be $T_{50\%}$, then all K paths must have delay less than $T_{50\%}$.

$$P(T_{cycle} \leq T_{50\%}) = (P(\tau_{p_i} < T_{50\%}))^K = 0.5 \tag{14}$$

This tells us:

$$P(\tau_{p_i} < T_{50\%}) = (0.5)^{(1/K)} \tag{15}$$

When K is large, this means that $P(\tau_{p_i} < T_{50\%})$ must be a value very close to 1; for the Gaussian cumulative distribution function (Φ) to be close to one, we must allow the random variable, $T_{50\%}$, to be many σ above the mean.

$$N_\sigma(K) = \Phi^{-1}\left((0.5)^{(1/K)}\right) \tag{16}$$

This gives us:

$$T_{50\%} = \mu_{\tau_{p_i}} + N_\sigma(K) \times \sigma_{\tau_{p_i}} \tag{17}$$

Fig. 4 plots N_σ as a function of K.

This suggests: *the more critical paths we have, the slower we can expect the final circuit delay to be. As we scale to larger ICs, and hence more paths, this effect will increase.*

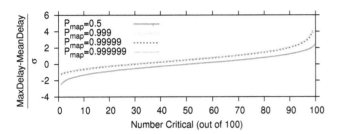

Figure 5: Achievable Delay vs. Near Critical Functions within a set of 100 Varying Resources

3.3 Choice

The path effect above results when we are forced to use every resource. We are forced to slow down the entire circuit because one or a few units are likely to be very slow. If we can choose devices to use, we can avoid these bad devices. For example, if the gates or segments are Gaussian distributed, then half of them are faster than the mean and half are slower. If we had a component with twice the resources of those needed (e.g. twice the channel width, twice the LUTs per CLB), we could avoid the half of the resources that are slower than the mean and guarantee that most ICs run at least as fast as the mean delay. When we have units with slack, the slack nodes can act as "extra" resources for the near critical resources. For example, when only 20% of the logical functions contending for resources in an interchangeable resource pool are near critical, these critical 20% effectively see an overpopulation ratio of 5.

More generally and formally, if we have equivalent sets of resource of size N and map to only use M of them, then the probability of yielding the M resources is:

$$P_{map} = \sum_{M \leq i \leq N} \left(\binom{N}{i} (P_u(\tau_{ref}))^i (1 - P_u(\tau_{ref}))^{N-i} \right)$$

Here, P_u is a Gaussian cumulative distribution function for τ_u. We define:

$$P_u(\tau_{ref}) = P(\tau_u \leq \tau_{ref}) \quad (18)$$

For fixed M and N, we can invert this and ask what P_u results in a given level of P_{map}. In turn, this tells us what τ_{ref} we can expect to achieve in order to meet the P_u bound. Fig. 5 shows speed achievable, as we vary the number of near critical units (M) contending for fast resources in a fixed pool of N=100 resources. *The more resources available for critical functions to choose from, the higher the performance we can achieve.*

3.4 Voltage for Fixed Timing Target

If, instead, we want to adjust the supply voltage (V_{dd}) to target a fixed timing target, the phenomena is similar to timing in Sec. 3.2, except the overhead σ is now in voltage. For simplicity, we assume we have K transistors that are critical and ignore the non-critical transistors. If we know we can achieve the desired timing with the nominal voltage $\mu_{V_{dd}}$, giving us a transistor on-current (I_{sat}, Eq. 1) for $\mu_{V_{on}} = \mu_{V_{dd}} - \mu_{V_{th}} - \frac{V_{d,sat}}{2}$, then we need to supply the chip with a V_{dd} large enough to provide this drive for all K critical transistors in the face of variation. Using similar reasoning to the delay case, this means that 50% of our chips

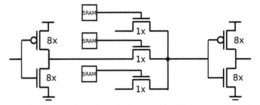

Figure 6: Directional switch circuit (3-input, 8× sized)

will need a voltage of $V_{50\%}$ or higher given by:

$$V_{50\%} = \mu_{V_{dd}} + N_\sigma(K) \times \sigma_{V_{th}} \quad (19)$$

For example, if $K=10^4$, $V_{th}=300$mV, $\mu_{V_{dd}}=700$mV, and $\sigma_{V_{th}}=20\%$ of V_{th}, or 60mV, $N_\sigma(10^4)=3.8$, and V_{dd} must be 928mV or higher 50% of the time. *The more critical transistors in a design, the higher we must expect to set the supply voltage to maintain target performance.*

4. METHODOLOGY

To evaluate the delay and energy benefits of component-specific routing, we use the predictive technology models (PTM) [36] to model delay and energy of gates as a function of variation, and employ a modified version of VPR 5.0.2 [22] to perform component-specific routing and measure total chip delay and energy.

4.1 Variation, Delay & Energy Model

We calculate individual switch delays and energies by performing extensive HSPICE simulations using the high performance PTM. We simulate at several values of V_{dd}, V_{thp} and V_{thn} and curve fit to create a continuous model of device delay and energy as a function of V_{dd} and V_{th}. This results in highly accurate models (within 1% error compared to HSPICE across $0.1V < V_{dd} < 1.2V$ and 6 $\sigma_{V_{th}}$).

To model variation for routed designs, we modify VPR such that every switch uses a randomly generated set of V_{th}'s sampled from a Gaussian distribution. Coupled with V_{dd}, we use the curve-fit models from HSPICE to compute the delay and energy of individual circuit elements. We use this method to generate 50 chips that we then keep constant across our comparisons.

To model delay and energy of a full routed design, we make additional modifications to VPR to compute both dynamic and static energy; we calculate dynamic energy by summing up all switched capacitance, and we compute static energy by summing up the leakage power of all devices. To calculate switching activity (α in Eq. 4) for dynamic energy we use the ACE 2.0 switching activity estimator [17] with random (50%) input probabilities. While we modify VPR to measure energy, we do not change VPR's cost function to target energy minimization.

An important note about our variation and energy model is that, as a simplification, we only model random WID V_{th} variation and energy dissipation in interconnect logic of FPGAs. Random variation is expected to dominate future sources of variation; interconnect switching energy is the dominant source of energy in FPGAs today [34] (e.g. Tuan shows 62% of dynamic energy in routing with the rest split evenly between logic and clocking). We assume that SRAM variation is controlled by dual V_{dd}/V_{th} processes as is standard in modern commercial FPGAs [14,34].

4.2 Architecture Model

We route on an architecture with 6-input LUTs and with 8 LUTs, 8 output pins and 27 input pins per CLB. We add additional CLB input and output pins to be utilized by delay-aware routing for additional routing flexibility (Sec. 3.3 and 5.1). All results are presented using 20% more channels than the minimum number of channels C_{min} required to route the particular benchmark design (Table 1).

The directional switch circuit we model is shown in Fig. 6. We use segment length 8, Wilton style S-Boxes and a C-Box connectivity of $F_{cin} = F_{cout} = 0.25$. When considering different gate sizes, we uniformly size up the input and output inverters of the circuit. For simplicity, we only present results for single-stage buffers. The impact of variation on multi-stage buffers in terms of slew rate effects, failure and speed tradeoffs, and energy efficiency is sufficiently complex to warrant future work. By examining single-stage buffers we build a baseline on which to understand more complex buffering schemes.

4.3 VPR Noise Reduction

VPR 5.0.2 is known to introduce experimental noise by producing inconsistent results in routing. This effect is magnified when mapping to resources that each have different delays; moreover, with high variation and low V_{dd} these delays can vary by several orders of magnitude. To minimize router noise we used the timing-targeted router [28]. We route using 200 iterations and a `-max_crit` value of 0.9999.

4.4 Experimental Setup

We compare delay-aware routing to delay-oblivious routing under variation for the Toronto 20 benchmark set [4]. We perform clustering and placement in VPR and use a single placement per benchmark for all routing experiments. Each data point is obtained by running both routers on a set of 50 Monte Carlo generated chips with V_{th} variation. The 50 chips are routed individually by the delay-aware router, while the delay-oblivious router performs a single, nominal route and evaluates that route across all chips. We report all delay and energy data at the 90% parametric yield point (i.e. we discard the 5 slowest/highest energy chips and report the max delay and energy). With 50 Bernoulli trials the 90% confidence interval for the results reported as 90% yield is 85–95%.

5. RESULTS

To quantify the delay and energy benefits of component-specific mapping for FPGAs under variation, we compare delay-aware and delay-oblivious routers through a series of experiments. First, we examine how extra resources can improve functional yield, which is necessary to operate under high V_{th} variation and low voltage. Because sizing of transistors is critical to delay, energy, and variation tolerance, we also compare both routers across a variety of switch sizes. Then we choose delay/energy optimal sizing for both the delay-aware and delay-oblivious routers to make an accurate comparison. Finally, we show how these trends scale across technologies.

5.1 CLB IO Sparing

Sec. 2.1 detailed how parameter variation can lead to functional logic failures, particularly at very low voltages. In order for delay-aware routing to avoid these defects, there must

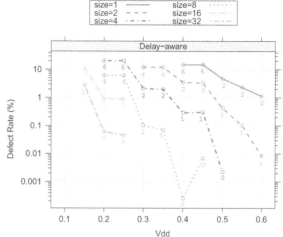

Figure 7: Tolerable Defect Rates vs V_{dd} for different extra IO pin counts (des, 22nm), all switch sizes

be enough spare resources (Sec. 3.3) for the router to find an alternative, non-defective path for a given net. FPGAs naturally have many spare resources in the form of multiple IO pins per CLB, extra channels, and flexibility in C-Box and S-Box connectivity. However, in examining the ability of delay-aware routing to avoid defects, we found that CLB IO pins are a significant bottleneck for defect avoidance. The single buffer that brings an input into the CLB, or the single buffer that fans out to C-Box switches, serve as single points of failure that can render large amounts of connectivity unusable. If a CLB is highly populated, defective IO pins can make it impossible to route.

One approach to mitigate this problem is to add spare pins to be used exclusively for defect avoidance. In Fig. 7 we examine the ability of the delay-aware router to yield by plotting tolerable defect rate as a function of voltage at 22nm for the des benchmark. As defect rate is highly dependent upon the magnitude of variation, and this magnitude is dictated by sizing (Eq. 8), we also examine routability as a function of switch size. Each point in the graph is numbered to indicate the number of extra pins required to yield at the given voltage and size. At high enough voltages (for example, above 500mV for size 4), we observe no defective switches. As we lower voltage we see defects appear and increase sharply; as expected we see higher defects rates at smaller sizes for a fixed V_{dd} due to the increased variation. We see that increasing numbers of spare CLB IO pins are required to yield as voltage is lowered. However, adding extra pins increases area, energy, and delay as accounted for in our models. For simplicity, in the remaining experiments in this paper we select 2 spare CLB pins as a configuration which generally provides defect avoidance for rates < 5% with minimal cost (< 1% energy/operation).

5.2 Delay

Fig. 8 plots parametric delay as a function of V_{dd} across a series of switch sizes, for nominal, delay-oblivious, and delay-aware routes for des at 22nm. For the nominal, no variation case, we see that, at higher voltages, size 8 switches generally provide a good tradeoff between drive strength and capacitive load, which corroborates prior work in determining delay optimal switch sizes [18]. As we reduce V_{dd}, the

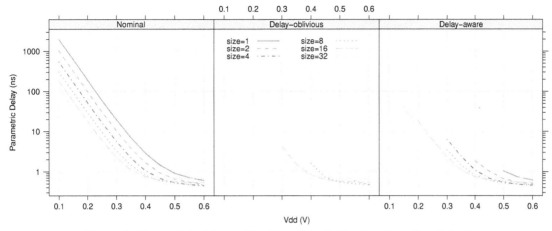

Figure 8: Parametric delay vs V_{dd} (des, 22nm), 2 extra pins, all switch sizes

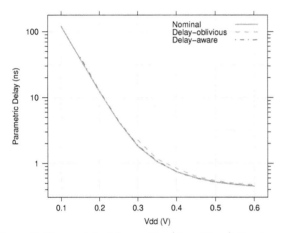

Figure 9: Parametric delay vs V_{dd} (des, 22nm), 2 extra pins, delay optimal switch sizes

increased drive strength of larger switches is necessary to reduce delay; below 400mV it is desirable to size up to 32× to minimize delay. The same basic delay-optimal sizing trends hold for delay-oblivious and delay-aware routing.

As we drop V_{dd} in the delay-oblivious case, we begin to see functional failures as described in Sec. 2.1 and shown in Fig. 7. The delay curves end at voltages where the defect rate becomes too high to achieve 90% yield. At 22nm there is sufficient variation ($\sigma_{V_{th}}/\mu_{V_{th}} = 27\%$) that enough small static CMOS inverters at low V_{dd} fail to switch which delay-oblivious cannot avoid. As we increase switch size and hence decrease the magnitude of variation, we are able to scale down V_{dd} and remain operational.

We see similar effects for the delay-aware router, where functional failures occur for small switches at low voltages. However, the delay-aware router is able to remain functional for lower voltages and smaller switch sizes through defect avoidance. For example, the delay-aware router can retain 90% yield for 32× sized gates at a V_{dd} that is 150mV lower than the delay-oblivious case (300mV vs 150mV).

Fig. 9 plots parametric delay for delay-optimal sizes across all V_{dd}'s (i.e. the composite minimum curve of Fig. 8). For

delay-optimized sizing, here we can effectively see the delay margins induced by variation as a function of V_{dd} by comparing the nominal and delay-oblivious curves. At high V_{dd} these margins are typically negligible, less than 2%. However, as we drop the supply voltage these margins increase, up to around 1.2× the nominal delay at 300mV. Delay-aware routing is able to completely eliminate variation induced delay margins, and improve delay with respect to delay-oblivious routing by 1.2×.

5.3 Energy

Fig. 10 plots parametric energy as a function of V_{dd} and switch sizes. In the no variation case we observe energy beginning to minimize around 150mV, similar to that in Fig. 3. At low voltages, delay is increased, and therefore static energy/operation increases as we spend more time leaking in a single operation. We also see that minimum sized gates always provide energy-minimal operation, a well known result in subthreshold circuit design [1, 12].

The delay-oblivious graph shows the same functional yield issues as in Fig. 8: as V_{dd} is reduced, the delay-oblivious router fails to provide 90% functional yield. In order to achieve reduced energy at lower V_{dd}'s, we must increase gate sizes in order to avoid defects. We see that 16× sized switches provide the minimal energy per operation.

For the delay-aware case we also see functional failures; however, delay-aware routing extends the range over which gates can function through defect avoidance. For example, in the 16× case, delay-aware routing enables a voltage reduction compared to delay-oblivious routing of 100mV (from 300mV to 200mV).

Fig. 11 plots parametric energy for energy optimal sizes across all V_{dd}'s (again, the composite minimum curves of Fig. 10). When comparing delay-oblivious to nominal, we see that the energy margins induced by variation are substantial, around a factor of 2 in the worst case at 300mV. These margins are significant because, while the nominal case can optimally use minimum sized devices, the delay-oblivious case must increase gate size to continue to yield, which increases switched capacitance. For example, 16× sized gates are 16 times more capacitive, but because gate capacitance contributes ≈10% of total switched capacitance in the size=1 case (wire capacitance accounts for the remain-

103

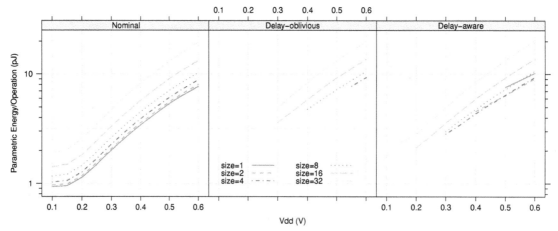

Figure 10: Parametric energy/operation vs V_{dd} (des, 22nm), 2 extra pins, all switch sizes

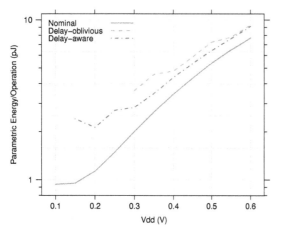

Figure 11: Parametric energy/operation vs V_{dd} (des, 22nm), 2 extra pins, energy optimal switch sizes

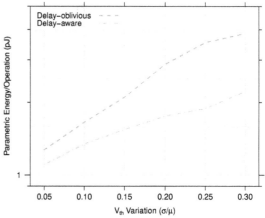

Figure 13: Minimum energy/operation vs V_{th} Variation (des, 22nm) 2 extra pins, energy optimal switch sizes

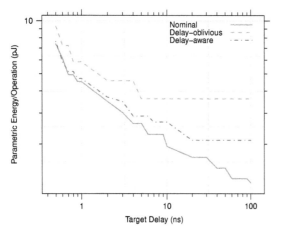

Figure 12: Parametric energy/operation vs delay target (des, 22nm), 2 extra pins, energy optimal switch sizes

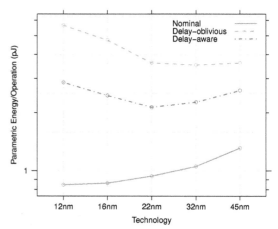

Figure 14: Minimum energy/operation vs technology (des) 2 extra pins, energy optimal switch sizes

ing 90%), increasing gate capacitance by an order of magnitude only increases switched capacitance 2×. We see that delay-aware routing is able to cut the energy margin nearly in half at 300mV, but is not able to complete eliminate it. Delay-aware routing is also able to prolong the range over which we can scale down V_{dd} and still reduce energy from 300mV to 200mV.

Table 1 shows the energy benefits of component-specific mapping as a function of technology (delay-oblivious/delay-aware). It also shows the average energy margins induced by variation across all benchmarks (delay-oblivious/nominal). On average across all benchmarks, variation increases minimum energy/operation by 2.12–6.12×, and delay-aware routing is able to improve minimum energy/operation by 1.42–1.98× as technology scales.

Fig. 12 demonstrates the energy benefits of delay-aware routing when targeting minimal energy operation under a performance constraint (as opposed to minimal energy ignoring delay). At larger delay targets, delay-aware routing achieves the required cycle times at a lower V_{dd} and hence lower energy/operation than delay-oblivious routing, with an energy savings of 1.2× for the fastest target of 2GHz and 1.8× for the slowest target of 10MHz.

5.4 Sensitivity to V_{th} Variation

The ITRS predictions for V_{th} variation are generally considered to be pessimistically high. This is exhibited by our very high defect rates at low voltages. To examine the benefits of delay-aware routing at various values of $\sigma_{V_{th}}$, Fig. 13 plots minimum energy/operation as a function of variation for des at 22nm, where the ITRS nominally predicts $\sigma_{V_{th}}/\mu_{V_{th}} = 27\%$. At 5% variation the ratio of delay-oblivious to delay-aware energy is only 1.15×. As variation increases to 30% we see the ratio scale up to 1.74×.

5.5 Technology Scaling

To analyze how our results scale with respect to feature size, we plot minimum energy/operation as a function of technology in Fig. 14. As technology scales, V_{th} variation increases substantially (we use ITRS predicted values of V_{th} as shown in Fig. 1), and we expect that delay-oblivious routing will have increased energy margins, while delay-aware routing may be able to reduce these margins. When examining the no variation case, we see energy/operation decrease with each technology generation, as expected. We also see substantial energy margins in the delay-oblivious case that delay-aware routing is able to reduce. However, we also note an interesting trend: the delay-oblivious case shows a slight net *increase* in energy beginning at the 32nm generation, due to the overhead of variation. This corresponds directly to the result from [7] that demonstrated a similar trends for ASICs, and a similar turning point at 32nm. When examining delay-aware routing, we see a similar increase, but below the 22nm generation. *This means that delay-aware routing is effectively able to allow technology scaling to continue delivering reductions in minimum operating energy for another technology generation.*

6. FUTURE WORK

This work has only explored part of the design space in evaluating component-specific routing to improve delay and energy. Perhaps most importantly, we have only examined single-stage buffers for simplicity. Multi-stage buffers

Table 1: Ratio of minimum energy per operation of delay-oblivious/delay-aware (90% yield, 2 extra IO pins)

Design	LUT	C_{min}	Technology (nm)				
			45	32	22	16	12
alu4	1492	52	1.69	1.81	1.87	2.02	2.02
apex2	1876	76	1.56	1.73	1.82	1.96	2.04
apex4	1304	78	1.29	1.46	1.57	1.87	1.90
bigkey	1816	72	1.49	1.72	1.88	2.58	2.69
clma	7808	92	1.42	1.61	1.73	2.06	2.09
des	1504	78	1.39	1.55	1.27	1.93	1.97
diffeq	1280	76	1.30	1.41	1.65	2.09	2.12
dsip	1372	64	1.58	1.82	1.90	2.01	2.05
elliptic	2784	60	1.88	2.08	2.19	2.35	2.45
ex1010	4744	114	1.17	1.27	1.48	1.88	1.91
ex5p	1092	70	1.37	1.53	1.66	1.87	1.96
frisc	2892	82	1.74	2.00	2.09	2.21	2.25
misex3	1388	68	1.63	1.78	1.81	1.78	1.76
pdc	4616	96	1.46	1.60	1.63	1.60	1.59
s298	2020	60	1.31	1.47	1.62	1.84	1.79
s38417	6232	50	1.29	1.46	1.63	2.09	2.09
s38584.1	6064	60	1.35	1.61	1.68	1.91	1.90
seq	1724	78	1.62	1.74	1.85	1.98	2.00
spla	3784	86	1.04	1.18	1.30	1.47	1.43
tseng	972	48	1.39	1.57	1.73	2.28	2.48
Geomean (benefit)			1.42	1.60	1.69	1.98	1.98
Geomean (margins[1])			2.12	3.04	4.01	5.04	6.12

[1] margin = delay-oblivious/nominal

are more realistic and representative switch circuits, but they will exhibit complex, composite effects from the results shown here (e.g. small buffers will dominate functional yield trends, but large buffers will contribute most to delay trends). Adding models for LUT variation, short-circuit power, and glitch power will also make our physical model more complete; nonetheless, we expect that their additions will not change our results significantly since interconnect delay and switching/leakage energy are dominant.

Additionally, we are limited by our selection of benchmarks in the Toronto 20. These circuits are noticeably small (see Table 1) and half are completely combinational; modern circuits are much larger and pipelined more aggressively. From the analysis in Sec. 3.1, 3.2 and 3.4, we expect delay-aware routing to show larger benefits for pipelined circuits and delay-oblivious routing to suffer greatly from the large number of near critical paths. Additionally, combinational circuits leak substantially more per operation due to the increased length of operation; pipelined circuits will use less energy per operation at low voltages, reducing the minimum energy point (Fig. 3).

Additional circuit techniques may further help to improve the energy benefits of delay-aware routing. At the minimum energy operating points, two-thirds of the energy is in leakage, and most of that leakage comes from unused devices. Power gating significantly reduces leakage energy/operation by disabling the many unused, leaky devices on a chip; delay-aware routing can further help by identifying the most leaky switches for gating. Selective sizing of gates may also help to reduce energy; instead of sizing up all gates to avoid defects, we can strategically size up gates that are most critical (such as CLB IO pins).

7. CONCLUSIONS

We show that circuits mapped using component-specific routing and delay knowledge can mitigate delay and energy induced margins from variation—**knowledge is power** [2] (or more precisely, *energy*). For a standard set of FPGA benchmark circuits mapped to 45nm, 32nm, 22nm, 16nm and 12nm predictive technologies we show that, routing oblivious to variation yields an average energy overhead of 2.12–6.12×. Routing with post-fabrication delay knowledge can eliminate delay margins, and on average, reduces minimum energy/operation relative to delay-oblivious design by 1.42–1.98×. We further show that delay-aware routing can help extend minimum energy technology scaling by an extra generation. We hope this will motivate future work in solving the significant challenges in component-specific mapping.

Acknowledgments

This research was funded in part by National Science Foundation grant CCF-0904577. Any opinions, findings, and conclusions or recommendations expressed in this material are those of the authors and do not necessarily reflect the views of the National Science Foundation.

8. REFERENCES

[1] International technology roadmap for semiconductors. <http://www.itrs.net/Links/2010ITRS/Home2010.htm> , 2010.

[2] F. Bacon. *Meditationes Sacræ. De Hæresibus.* 1597.

[3] K. Bernstein, D. J. Frank, A. E. Gattiker, W. Haensch, B. L. Ji, S. R. Nassif, E. J. Nowak, D. J. Pearson, and N. J. Rohrer. High-performance CMOS variability in the 65-nm regime and beyond. *IBM J. Res. and Dev.*, 50(4/5):433–449, July/September 2006.

[4] V. Betz and J. Rose. FPGA Place-and-Route Challenge. <http://www.eecg.toronto.edu/~vaughn/challenge/challenge.html> , 1999.

[5] S. Bijansky and A. Aziz. TuneFPGA: post-silicon tuning of dual-Vdd FPGAs. In *DAC*, 2008.

[6] D. Bol, R. Ambroise, D. Flandre, and J.-D. Legat. Interests and limitations of technology scaling for subthreshold logic. *IEEE Trans. VLSI Syst.*, 17(10):1508–1519, 2009.

[7] D. Bol, R. F. Ambroise, and J.-D. D. Legat. Impact of technology scaling on digital subthreshold circuits. In *ISVLSI*, pages 179–184, 2008.

[8] L. Cheng, J. Xiong, L. He, and M. Hutton. FPGA performance optimization via chipwise placement considering process variations. In *FPL*, pages 1–6, 2006.

[9] C. Chow, L. Tsui, P. Leong, W. Luk, and S. Wilton. Dynamic voltage scaling for commercial FPGAs. *ICFPT*, pages 173–180, Dec. 2005.

[10] W. B. Culbertson, R. Amerson, R. Carter, P. Kuekes, and G. Snider. Defect tolerance on the TERAMAC custom computer. In *FCCM*, pages 116–123, April 1997.

[11] B. Gojman and A. DeHon. VMATCH: Using Logical Variation to Counteract Physical Variation in Bottom-Up, Nanoscale Systems. In *ICFPT*, pages 78–87. IEEE, December 2009.

[12] S. Hanson, B. Zhai, K. Bernstein, D. Blaauw, A. Bryant, L. Chang, K. K. Das, W. Haensch, E. J. Nowak, and D. M. Sylvester. Ultralow-voltage, minimum-energy CMOS. *IBM J. Res. and Dev.*, 50(4–5):469–490, July/September 2006.

[13] K. Katsuki, M. Kotani, K. Kobayashi, and H. Onodera. A yield and speed enhancement scheme under within-die variations on 90nm LUT array. In *CICC*, pages 601–604, 2005.

[14] M. Klein. The Virtex-4 power play. *Xcell Journal*, (52):16–19, Spring 2005.

[15] A. Kumar and M. Anis. FPGA Design for Timing Yield Under Process Variations. *IEEE Trans. VLSI Syst.*, 18(3):423–435, March 2010.

[16] J. Kwong and A. P. Chandrakasan. Variation-Driven device sizing for minimum energy sub-threshold circuits. In *ISLPED*, pages 8–13, 2006.

[17] J. Lamoureux and S. Wilton. Activity estimation for field-programmable gate arrays. *FPL*, pages 1–8, Aug. 2006.

[18] G. Lemieux, E. Lee, M. Tom, and A. Yu. Directional and single-driver wires in fpga interconnect. In *ICFPT*, pages 41–48, December 2004.

[19] Y. Lin, L. He, and M. Hutton. Stochastic physical synthesis considering prerouting interconnect uncertainty and process variation for FPGAs. *IEEE Trans. VLSI Syst.*, 16(2):124, 2008.

[20] Y. Lin, M. Hutton, and L. He. Placement and timing for FPGAs considering variations. In *FPL*, 2006.

[21] G. Lucas, C. Dong, and D. Chen. Variation-aware placement for FPGAs with multi-cycle statistical timing analysis. In *FPGA*, pages 177–180, New York, New York, USA, 2010. ACM.

[22] J. Luu, I. Kuon, P. Jamieson, T. Campbell, A. Ye, W. M. Fang, and J. Rose. VPR 5.0: Fpga cad and architecture exploration tools with single-driver routing, heterogeneity and process scaling. In *FPGA*, pages 133–142, 2009.

[23] M. Majzoobi, E. Dyer, A. Elnably, and F. Koushanfar. Rapid FPGA delay characterization using clock synthesis and sparse sampling. In *Proc. Intl. Test Conf.*, 2010.

[24] Y. Matsumoto, M. Hioki, T. Kawanami, T. Tsutsumi, T. Nakagawa, T. Sekigawa, and H. Koike. Performance and yield enhancement of FPGAs with within-die variation using multiple configurations. In *FPGA*, pages 169–177, 2007.

[25] G. Nabaaz, N. Aziziy, and F. N. Najm. An adaptive FPGA architecture with process variation compensation and reduced leakage. In *DAC*, pages 624–629, 2006.

[26] S. R. Nassif, N. Mehta, and Y. Cao. A resilience roadmap. In *DATE*, March 2010.

[27] R. Rubin and A. DeHon. Choose-Your-Own-Adventure Routing: Lightweight Load-Time Defect Avoidance. In *FPGA*, pages 23–32, 2009.

[28] R. Rubin and A. DeHon. Timing-Driven Pathfinder Pathology and Remediation: Quantifying and Reducing Delay Noise in VPR-Pathfinder. In *FPGA*, pages 173–176, 2011.

[29] P. Sedcole and P. Y. K. Cheung. Within-die delay variability in 90nm FPGAs and beyond. In *ICFPT*, pages 97–104, 2006.

[30] P. Sedcole and P. Y. K. Cheung. Parametric yield in FPGAs due to within-die delay variations: A quantitative analysis. In *FPGA*, pages 178–187, 2007.

[31] S. Sivaswamy and K. Bazargan. Variation-aware routing for FPGAs. In *FPGA*, pages 71–79, 2007.

[32] S. M. Trimberger. Utilizing multiple test bitstreams to avoid localized defects in partially defective programmable integrated circuits. United States Patent Number: 7,424,655, September 9 2008.

[33] T. Tuan, A. Lesea, C. Kingsley, and S. Trimberger. Analysis of within-die process variation in 65nm FPGAs. In *ISQED*, pages 1 –5, March 2011.

[34] T. Tuan, A. Rahman, S. Das, S. Trimberger, and S. Kao. A 90-nm Low-Power FPGA for Battery-Powered applications. *IEEE Trans. Computer-Aided Design*, 26(2):296–300, 2007.

[35] J. S. J. Wong, P. Sedcole, and P. Y. K. Cheung. Self-measurement of combinatorial circuit delays in FPGAs. *ACM Tr. Reconfig. Tech. and Sys.*, 2(2):1–22, 2009.

[36] W. Zhao and Y. Cao. New generation of predictive technology model for sub-45 nm early design exploration. *IEEE Trans. Electron Dev.*, 53(11):2816–2823, 2006.

Analyzing and Predicting the Impact of CAD Algorithm Noise on FPGA Speed Performance and Power

Warren Shum and Jason H. Anderson
Dept. of ECE, University of Toronto, Toronto, ON, Canada
{shumwarr,janders}@eecg.toronto.edu

ABSTRACT

FPGA CAD algorithms are heuristic, and generally make use of cost functions to gauge the value of one potential circuit implementation over another. At times, such algorithms must decide between two or more implementation options of apparently equal cost. This work explores the variations in circuit quality, i.e. *noise*, that arise when CAD algorithms are altered to choose randomly when faced with such equal-cost alternatives. Noise sources are identified in logic synthesis and technology mapping algorithms, and experimental results are presented which show standard deviations of 3.3% and 3.7% from the mean in post-routed delay and power. As a means of dealing with this variation, early timing and power prediction metrics can be applied after technology mapping to find the best circuits in the presence of noise. When applied to designs with over 1.5% variation in delay and power, the best prediction models have a 40% probability of capturing the best circuit when predicting the top 10% of circuits in a group of noise-injected circuits.

Categories and Subject Descriptors

B.7 [**Integrated Circuits**]: Design Aids

Keywords

FPGAs, CAD, noise, prediction

1. INTRODUCTION

The typical FPGA CAD flow proceeds in a series of steps, with different heuristic algorithms being used in each step. In such algorithms, it is common to encounter situations where a choice must be made between two (or more) alternatives that appear to have identical quality. For example, in logic synthesis, a logic function might be implemented in multiple ways, each having the same estimated area, delay, or power. However, the choice of how that function is implemented may affect the post-routed circuit delay or power in an unpredictable way. In practice, choices between alternatives are arbitrarily made (e.g. always select the first alternative) or are controlled with the use of a random number generator. By running an algorithm multiple times using different seeds for the random number generator, one can obtain a set of circuits with different characteristics. The vari-

ation in circuit quality (area, performance, power) through seemingly neutral changes in CAD algorithms is what we call algorithmic **noise**.

The practice of executing CAD algorithms with multiple seeds in the hope of producing a higher-quality implementation is well-established for placement and routing [3, 14]. In this work, we show that noise also exists earlier in the CAD flow: in logic synthesis and technology mapping. By exposing the noise in these earlier stages, we allow seed sweeping to take place earlier, in less time-consuming stages.

This work makes the following contributions:

- An investigation of noise sources in logic synthesis and technology mapping.
- Experimental data showing the amount of power and performance noise present in CAD algorithms, using a commercial FPGA architecture.
- Timing and power prediction metrics for early identification of the best circuits in the presence of noise.

2. NOISE INJECTION IN FPGA CAD

This work focuses on the logic synthesis and technology mapping stages. In particular, we inject noise into the algorithms implemented in the academic tool, ABC [4].

2.1 Logic Synthesis

The primary data structure in ABC is an AND-Inverter Graph (AIG) which is a representation of a logic circuit using two-input AND gates and inverters. The general goal of logic synthesis algorithms is to reduce the number of nodes in the AIG and the AIG's depth (i.e. the number of logic levels from any combinational input to a combinational output). We use the *resyn2* script in ABC for technology independent logic synthesis. The script loops repeatedly through three synthesis algorithms, each of which contains an opportunity for noise injection. The algorithms are as follows:

AIG balancing: Balancing is a technique that aims to reduce the number of levels in an AIG [10]. It is done in two main steps. The **tree covering** step identifies multi-input AND gates in the AIG. The second step is **tree balancing**. For each multi-input AND gate identified by the tree covering stage, the tree balancing stage decomposes it into a balanced tree of two-input AND gates.

By default, ABC does AIG balancing in a deterministic way, making the same (arbitrary) balancing decisions every time. However, there are multiple ways to perform the balancing. To see this, consider the example in Fig. 1. With different random choices, balancing might produce either of the AIG subgraphs shown. The tree balancing stage may not have a unique optimal solution, and it is therefore a source of noise. We modified the algorithm to choose randomly in scenarios where there exist two or more equal-depth options.

AIG rewriting: AIG rewriting is an algorithm that reduces the number of nodes/logic levels in an AIG by examining subgraphs of nodes and replacing them with lower-cost

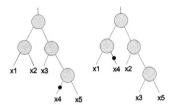

Figure 1: Examples of balanced AIGs.

substitutes [11]. The rewriting algorithm walks through the AIG and for each node n enumerates all 4-input cuts of n. A *cut* of a node n is a set of nodes (*leaves*) such that each path from a primary input to n passes through at least one node of the cut. Each cut is replaced with equivalent AIG subgraphs from a hash table of pre-computed "good" subgraphs. If the subgraph replacement candidate leads to a reduction in AIG nodes, it is accepted.

To add randomness at this stage, we modified the algorithm to allow changes even when the replacement leads to no change in the AIG node count (a *zero-cost* replacement). In our modified ABC implementation, a candidate zero-cost replacement is accepted with a 50% probability.

AIG refactoring: This technique involves computing one large cut for each AIG node, then replacing it with a factored form with fewer nodes [9]. The cuts are chosen based on how much reconvergence they contain, which is an indicator of redundancy that can be exploited by refactoring. Refactoring differs from rewriting in that it acts on a larger scale (refactoring can handle up to 16-input cuts). The noise injection in this stage is similar to the method used in AIG rewriting.

2.2 Technology Mapping

We use the priority cut-based technology mapping algorithm in ABC [12] to map the AIG logic into K-input LUTs (K-LUTs). The mapper does this by first evaluating a set of cuts for each node in an AIG, representing potential LUT implementations of that node and a subset of its predecessors in the AIG. The cuts are selected and sorted in terms of delay, number of inputs, and area. FPGA technology mapping algorithms use logic depth as a proxy for delay, and use the number of LUTs as a proxy for area.

The priority cuts for each node are evaluated based on several criteria, depending on the mapping parameters. These criteria include depth and cut size. The sorting makes use of a cut comparison function which compares two cuts. We introduce random noise in this stage when deciding between cuts with the same values for each of the specified metrics. If the cuts are tied for each of these metrics, our modification to the algorithm makes a random selection between them.

The mapping algorithm [12] makes several passes over the netlist, stitching together the best results (using depth-optimal mappings on critical paths, and area-oriented mappings elsewhere). We introduce noise in both types of passes: depth-oriented mapping and area-oriented mapping.

3. NOISE ANALYSIS

In order to evaluate noise over many random seeds, a large number of circuit compilations (over 10,000) were executed using a high-performance cloud computing system [5]. We use Altera's Quartus 10.1 to pack, place and route the circuits and perform timing and power analysis. The target FPGA family is Altera Stratix III [2]. Altera's QUIP (Quartus University Interface Program) was used to bring designs from ABC into Quartus II [1]. Modelsim 6.3e was used for simulation to get toggle rates for the power estimation, done

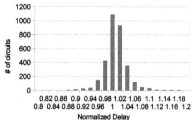

Figure 2: Number of circuits vs. normalized delay.

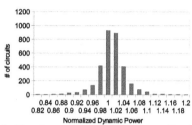

Figure 3: Number of circuits vs. normalized power.

using Quartus PowerPlay. 5,000 random input vectors were applied to each circuit. Placement and routing were performed using the *standard fit* setting (maximum effort) in Quartus II. The critical path delay was obtained using the TimeQuest timing analyzer. The benchmark set consists of 20 circuits from the MCNC benchmark set and 7 circuits from the VPR 5.0 benchmark set, in order to have data for some larger circuits. For the following sections, the word **design** will be used to refer to all circuits having the same original source file (e.g. 'alu4' is a design). A **circuit** will refer to a particular compilation of the design using certain seeds (e.g. 'alu4' compiled with synthesis seed 1 and mapping seed 2 is a circuit).

We begin by showing the amount of noise present when all noise sources are activated (balancing, rewriting, refactoring, delay and area-oriented technology mapping, and placement). Five seeds are used in each of synthesis, mapping and placement for 27 designs, making a total of $5 \cdot 5 \cdot 5 \cdot 27 = 3,375$ circuits. For these experiments, circuits with logic depth greater than the minimum depth observed for that design were removed. This is because we would like to find ways of differentiating between minimal-depth mappings, which are likely to be selected over non-minimal-depth ones.

Fig. 2 shows the number of compiled circuits vs. their critical path delay, normalized to the average for each design. Analogous power results are given in Fig. 3. The results appear as a roughly normal distribution, with the power noise being slightly greater than delay noise. The standard deviation of critical path delay is 3.3% and dynamic power is 3.7%. This is a fairly significant amount, enough to drive the use of seed sweeping to find the circuit compilations with the best results in this distribution.

We now describe our effort to isolate the effect of logic synthesis noise. 25 synthesis seeds were used for each design. All synthesis noise sources were activated (balancing, rewriting, and refactoring). For each circuit, the delay and power were averaged across five Quartus compilations using different place-and-route seeds. This was done in order to reduce the impact of placement and routing noise, in an attempt to isolate the synthesis noise.

The standard deviations of delay and power due to synthesis noise alone are 1.8% and 2.7% for delay and power, respectively. This indicates the degree to which random, zero-cost changes in the logic synthesis stage affect the overall quality of the circuit. For technology mapping, the standard deviations are 0.9% and 1.4% for delay and power, respectively, which is less than in synthesis. This is likely due to good tiebreaking mechanisms in the mapper, as well as the fact that there are fewer downstream CAD stages to be affected by noise introduced in mapping.

4. EARLY DELAY/POWER PREDICTION

While engineers commonly perform multiple placement and routing compilations using different seeds to find a high quality implementation of their design, it is a long process, taking hours or even days for the largest designs [7]. Synthesis and mapping, on the other hand, are relatively quick, leading us to question whether we can sweep seeds in the synthesis and mapping stages instead of in placement and routing. This would require early timing and power prediction metrics at the post-mapping stage, in order to find the best candidate circuits for placement and routing.

Early power and performance estimation has been done at various stages of the CAD flow. At the high-level synthesis stage, power estimation has been done to drive low-power resource allocation and binding techniques [6]. At the pre-placement stage, work has been done to predict interconnect wirelength and delay [8, 13]. The work by Manohararajah et al. [8] proposed a simple timing model based on using a single delay value for each connection depending on its source and destination node type and port (e.g. logic, I/O, memory). The work by Pandit and Akoglu [13] attempts to estimate wirelengths using structural metrics at the pre-placement stage – metrics taken from works in the ASIC domain and applied to FPGAs.

4.1 Delay Prediction

We examine ways to predict the circuit with the lowest critical path delay at the post-mapping stage, given a set of circuits compiled with different synthesis/mapping seeds. Our delay prediction model assigns each node (LUT) a certain delay, then traverses the circuit graph in topological order and computes the arrival time at each node. We examine numerous timing models, sweeping several parameters.

4.1.1 Varying Pin Delays

Due to the tree-like structure of the multiplexer within a LUT, the delay from each of its input pins to its output pin varies. In general, commercial FPGA tools automatically assign the slowest-arriving inputs to the LUT input pins with smaller delays. On this front, we investigated two timing prediction models. The **pin utilization** model bases the LUT delay of the number of used input pins on the LUT. A K-LUT with n used inputs is assigned a delay of n/K. The **pin order** model bases the LUT delay on a prediction of the ordering of the input pins, which is based on their estimated arrival times. Specifically, the model assumes that the ith-fastest input pin on a K-LUT has a delay of i/K and that the latest arriving input signal is assigned to the fastest LUT input pin; the next-latest arriving input is assigned to the second-fastest LUT input, and so on.

4.1.2 Logic, Routing and Constant Factors

The overall delay model for a LUT n is as follows:

$$Delay(n) = const + logic\_factor \cdot logic\_delay(n)$$
$$+ fanout\_factor \cdot fanout(n) \quad (1)$$

- *const*: A constant value for each LUT, which can be set to 0 or 1. If set to 1, it represents a unit delay for the LUT.

- *logic_factor*: A scaling factor for the LUT delay, *logic_delay*, calculated using one of the pin-based timing models above. We examined factors ranging from 0 to 5.
- *fanout_factor*: A scaling factor for the fanout of the node, which represents the routing delay. We examined values from 0 to 5. Fanout was capped at 10 to prevent high-fanout nodes from dominating the delay.

4.1.3 Maximum/Scaled Metrics

A logic circuit contains a single (or a few) critical path(s). Early in the CAD flow, where delay estimates are used, it is difficult to predict the paths that will be critical at the post-routing stage. Therefore, we propose a scaled model in which we take the sum of the top L maximum arrival times of delay paths in a circuit, each scaled by an exponentially decaying factor. The scaled delay model for a circuit is expressed as:

$$scaled\_delay = \sum_{i=1}^{L} max\_arrival\_time(node_i) \cdot factor^i \quad (2)$$

where $max\_arrival\_time(node_i)$ is the maximum *total* delay to reach the node with the ith latest arrival time, and $0 < factor < 1$. For our experiments, a factor of 0.95 was chosen empirically since it decays quickly enough to ignore path endpoints that are unlikely to be critical, yet not so quickly that it considers only a few.

4.2 Power Prediction

The dynamic power consumption of a circuit can be calculated as:

$$P_{dyn} = \frac{1}{2} \sum_{m=1}^{M} S(m) \cdot C(m) \cdot f \cdot V_{dd}^2 \quad (3)$$

where M is the number of nets in the circuit, $S(m)$ is the switching activity of a net m, $C(m)$ is the capacitance of net m, f is the frequency of the circuit, and V_{dd} is the supply voltage. The switching activity of a net is estimated using a fast vector simulation implemented in ABC with 1000 random input vectors. The simulator can be run in two modes:

- **Zero delay**, in which the LUTs/wires are assumed to have zero delay (i.e. on each clock cycle, all signals immediately settle into their final state).
- **Unit delay**, in which each LUT is assumed to have a delay of one. This allows for some modeling of glitches.

The overall power model is a product of switching activity and fanout over all nets:

$$P_{est} = \sum_{m=1}^{M} S(m) \cdot fanout(m) \quad (4)$$

Fanout is used as a substitute for capacitance (capped as in Section 4.1.2). V_{dd} and f from (3) are ignored since they are constant for each circuit.

5. RESULTS

We ran our prediction methods on the noise-injected circuits from Section 3. Each of the 27 benchmark designs was synthesized and mapped with different seeds to create 25 mapped candidates. Each candidate was placed and routed using Quartus using 5 different seeds. The results were averaged across the 5 placement seeds to obtain a representative average result for each mapped candidate circuit. The prediction algorithns were all implemented in ABC [4]. For each mapped candidate circuit, all prediction models were run, sweeping all parameter combinations. Designs with low swing (no circuit with more than 1.5% deviation from the average) were ignored. For the delay portion of this study, the designs were arbitrarily split into training and test sets.

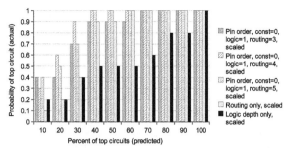

Figure 4: Probability of finding the top circuit vs. percentage of top modeled circuits considered (delay).

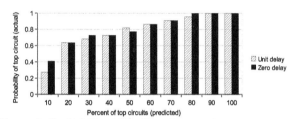

Figure 5: Probability of finding the top circuit vs. percentage of top modeled circuits considered (power).

Model parameters were chosen using data from the training design set; model predictive accuracy was evaluated using data from the test designs.

We first consider results for delay prediction. Fig. 4 shows the probability of finding the best circuit implementation in terms of delay (or one within a 0.1% margin of it). The top three models (based on the ranking of the top predicted circuit for each design) are shown, along with two simple models: "Routing only" ($fanout\_factor = 1$ in Eqn. 1, all others 0, scaled) and "Logic depth only" ($const = 1$, all others 0, scaled). The top models were chosen by taking the best *predicted* circuit for each design and summing its *actual* ranking (best=1, worst=25) across all designs. The models with the lowest sums were chosen as the best ones. The x-axis shows the percentage of the top circuits (by model score) considered, while the y-axis shows the probability of finding the actual top circuit within this group. The legend shows the timing model used.

The best model was found to be "Pin order, const=0, logic=1, routing=3, scaled" which refers to the pin order model (Section 4.1.1) with ($const = 0, logic\_factor = 1$, and $fanout\_factor = 3$) as the factor settings (Section 4.1.2), and the exponentially decaying scaling factor (Section 4.1.3). Using this model, if we take the top 10% of circuits (according to the model) we have approximately a 40% chance of selecting the actual best one (i.e. post-routing best).

Fig. 5 shows analogous results for power. "Zero delay" and "Unit delay" represent the zero delay and unit delay simulation models. Both the unit and zero delay models appear to offer similar results. Simply using a zero delay simulation and taking the top 10% of predicted circuits, the probability of capturing the top circuit is over 40%.

Table 1: Average benefit of prediction models.

Delay		
Model	% improv.	% improv. (full)
Best prediction	1.3	1.8
Logic depth only	0.6	0.3
Fanout only	0.9	0.9
Power		
Model	% improv.	
Unit delay	1.4	
Zero delay	1.1	

Table 1 shows the average delay/power savings from our prediction models if the top 10% of predicted best circuits are carried forward to placement and routing. Looking at the "% improv." column, we see that the best prediction model gives an average benefit of 1.3% in post-routed delay and 1.4% in power (compared to the average delay/power for a design) – this is more than the benefit offered by the simpler models of logic depth and fanout. It should be noted that these results were generated using relatively small training and test sets (for delay). If we instead use the entire benchmark set for *both* training and test, we can obtain improvements of up to 1.8% in delay (column "% improv. (full)").

6. CONCLUSIONS

In this paper, we studied FPGA CAD algorithm noise – variations in circuit quality due to cost-neutral changes in the compilation flow. We identified noise sources in logic synthesis and technology mapping algorithms, and we described our method of noise injection in those algorithms. Under the influence of synthesis noise, standard deviations of critical path delay and dynamic power were 1.8% and 2.7%, respectively, while the results for technology mapping were 0.9% and 1.4%, respectively. Under the influence of noise in all CAD stages, the standard deviations were 3.3% in delay and 3.7% in power.

We evaluated ways of predicting the impact of noise on delay and power at the post-technology mapping stage. Results show that the circuits predicted to be in the top 10% at the technology mapping stage in terms of delay or power, have a 40% probability of being the best circuit at the post-routing stage.

7. REFERENCES

[1] Altera Corp. Quartus university interface program. *www.altera.com/education/univ/research/quip/unv-quip.html*, 2009.
[2] Altera Corp. Stratix III device handbook. *http://www.altera.com/literature/lit-stx3.jsp*, 2011.
[3] Altera Corp. Timing closure methodology for advanced FPGA designs. *www.altera.com/literature/an/an584.pdf*, 2011.
[4] Berkeley Logic Synthesis and Verification Group. ABC: A system for sequential synthesis and verification. *www.eecs.berkeley.edu/~alanmi/abc/*, Release 00406.
[5] C. Loken et al. SciNet: Lessons learned from building a power-efficient Top-20 system and data centre. *Journal of Physics: Conference Series*, 256(1):012026, 2010.
[6] D. Chen, J. Cong, Y. Fan, and Z. Zhang. High-level power estimation and low-power design space exploration for FPGAs. In *IEEE/ACM ASP-DAC*, pages 529 –534, 2007.
[7] M. Gort and J. Anderson. Deterministic multi-core parallel routing for FPGAs. In *IEEE FPT*, pages 78 –86, 2010.
[8] V. Manohararajah, G. Chiu, D. Singh, and S. Brown. Predicting interconnect delay for physical synthesis in a FPGA CAD flow. *IEEE TVLSI*, 15(8):895 –903, Aug. 2007.
[9] A. Mishchenko and R. Brayton. Scalable logic synthesis using a simple circuit structure. In *Proc. IWLS*, pages 15 –22, 2006.
[10] A. Mishchenko, R. Brayton, S. Jang, and V. Kravets. Delay optimization using SOP balancing. In *Proc. IWLS*, pages 75–82, 2011.
[11] A. Mishchenko, S. Chatterjee, and R. Brayton. DAG-aware AIG rewriting: a fresh look at combinational logic synthesis. In *ACM/IEEE DAC*, pages 532 –535, 2006.
[12] A. Mishchenko, S. Cho, S. Chatterjee, and R. Brayton. Combinational and sequential mapping with priority cuts. In *IEEE/ACM ICCAD*, pages 354 –361, 2007.
[13] A. Pandit and A. Akoglu. Wirelength prediction for FPGAs. In *IEEE FPL*, pages 749 –752, 2007.
[14] R. Y. Rubin and A. M. DeHon. Timing-driven pathfinder pathology and remediation: quantifying and reducing delay noise in VPR-pathfinder. In *ACM FPGA*, pages 173–176, 2011.

Impact of FPGA Architecture on Resource Sharing in High-Level Synthesis

Stefan Hadjis[1], Andrew Canis[1], Jason Anderson[1], Jongsok Choi[1], Kevin Nam[1], Stephen Brown[1], and Tomasz Czajkowski[‡]

[1]ECE Department, University of Toronto, Toronto, ON, Canada
[‡]Altera Toronto Technology Centre, Toronto, ON, Canada

ABSTRACT

Resource sharing is a key area-reduction approach in high-level synthesis (HLS) in which a single hardware functional unit is used to implement multiple operations in the high-level circuit specification. We show that the utility of sharing depends on the underlying FPGA logic element architecture and that different sharing trade-offs exist when 4-LUTs vs. 6-LUTs are used. We further show that certain multi-operator patterns occur multiple times in programs, creating additional opportunities for sharing larger composite functional units comprised of patterns of interconnected operators. A sharing cost/benefit analysis is used to inform decisions made in the binding phase of an HLS tool, whose RTL output is targeted to Altera commercial FPGA families: Stratix IV (dual-output 6-LUTs) and Cyclone II (4-LUTs).

Categories and Subject Descriptors

B.7 [**Integrated Circuits**]: Design Aids

Keywords

Field-programmable gate arrays, FPGAs, high-level synthesis, resource sharing

1. INTRODUCTION

High-level synthesis (HLS) refers to the automatic compilation of a program specified in a high-level language (such as C) into a hardware circuit. There are several traditional steps in HLS. *Allocation* determines the number and types of functional units to be used in the hardware implementation. This is followed by *scheduling*, which assigns operations in the program specification to specific clock cycles and generates a corresponding finite state machine (FSM). *Binding* then assigns the operations in the program to specific functional units in a manner consistent with the allocation and scheduling results.

A well-studied area-reduction optimization in the binding step is called *resource sharing*, which involves assigning multiple operations to the same hardware unit. Consider, for example, two additions that are scheduled to execute in different clock cycles. Such additions may be implemented by

the same adder in hardware – the additions *share* the hardware adder. Resource sharing is accomplished by adding multiplexers (MUXes) to the inputs of the shared functional unit, with the FSM controlling the MUXes to steer the correct data to the adder based on the state. Since MUXes are costly to implement in FPGAs, resource sharing has generally been thought to have little value for FPGAs, except in cases where the resource being shared is large or is scarce in the target device.

In this paper, we examine the impact of the FPGA logic element architecture on the effectiveness of resource sharing. We conduct our analysis using two commercial Altera FPGA families: 1) Cyclone II [2] (4-LUTs) and 2) Stratix IV [3] (dual-output 6-LUTs). One of the contributions of this paper is to show conclusively the cases for which resource sharing is advantageous for FPGAs. Results show that certain operators (e.g. addition) that are not worth sharing in Cyclone II are indeed worth sharing in Stratix IV. This is due to the larger LUT size, which permits portions of the sharing multiplexer circuitry to be combined into the *same* LUTs that implement the operators themselves. We then show that there exist patterns of operators that occur commonly in circuits and that such patterns can be considered as *composite operators* that can be shared to provide area reductions. We use the sharing analysis results to drive decisions made in the binding phase of the LegUp open source HLS tool [13] built within the LLVM compiler [11].

2. BACKGROUND AND RELATED WORK

Recent research on resource sharing that specifically targets FPGAs includes [9, 8, 5, 14, 4]. The work of Cong and Jiang [6] bears the most similarity to our own in that it applied graph-based techniques to identify commonly occurring patterns of operators in the HLS of FPGA circuits, and then shared such patterns in binding for resource reduction. Some of the area savings achieved, however, were through the sharing of multipliers implemented using LUTs instead of using hard IP blocks. Implementing multipliers using LUTs is very costly, and thus offers substantial sharing opportunities.

The two commercial FPGAs targeted in this study have considerably different logic element architectures. In Cyclone II, combinational logic functions are implemented using 4-input LUTs. Stratix IV logic elements are referred to as adaptive logic modules (ALMs). An ALM contains a dual-output 6-LUT, which receives **8** inputs. Each of the outputs corresponds to an adaptive LUT (ALUT). The ALM can implement any single logic function of 6 variables, or alternately, can be *fractured* to implement two separate logic functions (using both outputs) – i.e. two ALUTs. The ALM can implement two functions of 4 variables, two functions with 5 and 3 variables, respectively, as well as several other combinations. In both architectures, a bypassable flip-flop is present for each LUT output.

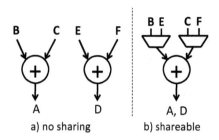

Figure 1: Illustration of sharing.

Table 1: Area data for individual 32-bit operators in the unshareable and shareable scenarios (ratios represent shareable/unshareable).

| | Cyclone II | | Stratix IV | |
	Unshareable LEs	Shareable LEs	Unshareable ALMs	Shareable ALMs
Add/Sub	32	96 (3.00)	16	25 (1.56)
Bitwise	32	64 (2.00)	32	32 (1.00)
Compare	32	96 (3.00)	24	46 (1.92)
Div	1118	1182 (1.06)	568	599 (1.05)
Mod	1119	1183 (1.06)	581	613 (1.06)
Mult	689	747 (1.08)	221	362 (1.64)
Shift	173	215 (1.24)	75	94 (1.25)

3. SHARING INDIVIDUAL OPERATORS

We first investigate the value of sharing individual operators *outside* the context of a larger circuit. For each type of operator, we wish to know whether any area savings may arise from sharing its hardware implementation vs. the case of instantiating additional instances of the operator. An example of resource sharing is illustrated in Fig. 1 for two C statements: A = B + C and D = E + F. Fig. 1(a) shows an implementation without sharing, using two adders; Fig. 1(b) depicts the *shareable* case, in which the same two adders are now implemented using the same functional unit. We wish to compare the area consumed by both possible implementations, shareable vs. unshareable, for various types of operators. The utility of sharing clearly depends on how the multiplexers are implemented.

We created two different Verilog implementations for each operator type in the LLVM intermediate representation (IR). The first implementation contains a *single* instance of the operator. The second contains the operator instance with 2-to-1 multiplexers on its inputs. We implemented these two subcircuits in FPGAs to measure their area. The subcircuits were synthesized using Quartus II ver. 11.0. To measure area, we use logic element count (LE) for Cyclone II and the number of adaptive logic modules (ALMs) for Stratix IV. We use the Quartus II INI variable `fit_pack_for_density_light` to direct the tool to minimize circuit area when packing logic into ALMs/LEs and LABs [1].

Table 1 provides area results for sharing individual 32-bit-wide operators in Cyclone II and Stratix IV FPGAs. Each row of the table corresponds to an operator type. The row labeled **bitwise** represents the AND, OR, and XOR operators, all of which consume the same area. The left side of the table gives results for Cyclone II; the right side gives results for Stratix IV. Area is shown for both the unshareable (columns labeled **unshareable**) and shareable scenarios (columns labeled **shareable**). For the data corresponding to shareable operators, values in parentheses give the ratio in area vs. the unshareable case. Sharing provides a benefit when the ratio reported is less than 2; that is, less area is consumed by sharing the operator in hardware than by instantiating two instances of the operator.

Table 1 illustrates that in both FPGA architectures, sharing is useful (from the area-reduction perspective) for mod-

ulus, division, multiplication (implemented with LUTs) and bitwise shift. Modulus and division are implemented with LUTs in both architectures and consume considerable area in comparison with multiplexers. The shift represents a barrel shift. In Stratix IV, sharing is *also* beneficial for addition, subtraction, comparison, as well as all of the bitwise operations: AND, OR, XOR. The larger LUTs in Stratix IV allow some (or all) of the sharing MUXes to be combined into the same LUTs that implement the operators, and thus, for Stratix IV, sharing is useful for a broader set of operators.

Regarding the bitwise operator data for Stratix IV in Table 1, in the unshareable case, a 32-bit bitwise logical operator uses 32 ALMs; in the shareable case, 32 ALMs are also consumed. In the unshareable case, however, each output is a function of just 2 primary inputs. Since ALMs are dual-output and can implement *any* two functions of up to 4 inputs, the unshareable case *should* have consumed just 16 ALMs. Quartus did not produce an area-minimal implementation for this case.

4. SHARING COMPOSITE OPERATORS

We now consider composite operators (patterns), which are groups of individual operators that connect to one another in specific ways. We begin by defining the key concepts used in our pattern analysis algorithm:

Pattern graph: A directed dataflow graph representing a computational pattern. Each node in the graph is a two-operand operation from the LLVM IR. Each pattern graph has a single root (output) node. The number of nodes in a pattern graph is referred to as its *size*. We require the nodes in a pattern graph to reside in the same *basic block*, where a basic block is a contiguous set of instructions with a single entry point and a single exit point.

PatternMap: A container for pattern graphs that organizes pattern graphs based on size and functionality. A key operation performed by the PatternMap is the testing of two patterns for equivalence. The equivalence checking accounts for patterns which are functionally but not topologically equivalent due to the order of operands in commutative operations. Note that pattern graphs with different schedules are not considered functionality equivalent. That is, the corresponding nodes in two equivalent pattern graphs *must* have corresponding cycle assignments in the schedule (e.g. if two operators are chained together in a single cycle in one pattern graph, the corresponding operators must be chained in the equivalent pattern graph). HLS scheduling results are used to detect such cases.

Finally, note that two pattern graphs may contain the same set of operators connected in the same way, yet corresponding operators in the graphs have different bit widths. It is undesirable to consider the two pattern graphs as equivalent if there is a large "gap" in their operator bit widths. For example, it would not be advantageous to *force* an 8-bit addition to be realized with a 32-bit adder in hardware. We developed a simple bit width analysis pass within LLVM that computes the required bit widths of operators. Two pattern graphs are not considered as equivalent to one another if their corresponding operators differ in bit width by more than 10 (determined empirically). We also consider operator bit widths in our binding phase, described below.

Valid operations for patterns: We do not allow all operations to be included in pattern graphs. We exclude operators with constant inputs, as certain area-reducing synthesis optimizations are already possible for such cases. In addition, we do not allow division and modulus to be included in pattern graphs. The FPGA implementation of such operators is so large that, where possible, they should be left as

Table 2: Area for sequential patterns of operators in the unshareable and shareable scenarios (ratios represent shareable/unshareable).

Pattern	Cyclone II Unshared LEs	Cyclone II Shareable LEs	Stratix IV Unshared ALMs	Stratix IV Shareable ALMs
Add_Add_Add_Add	128	288 (2.25)	64	73 (1.14)
Add_Sub	64	160 (2.50)	32	42 (1.31)
Add_XOR	64	128 (2.00)	33	42 (1.27)
Add_XOR_Add	96	192 (2.00)	48	58 (1.21)
OR_OR_OR	96	128 (1.33)	64	51 (0.80)
XOR_XOR	64	96 (1.50)	32	32 (1.00)

isolated operators and shared as much as possible (by wide multiplexers on their inputs).

4.1 Pattern Discovery Approach

All pattern graphs up to a maximum size S are discovered as follows: We iterate over all instructions in the program. Once a valid instruction is found, this becomes the root instruction, r, of a new pattern graph of size 1. We then perform a breadth-first search of the predecessors of r, adding all combinations of predecessors one at a time, to discover all graphs rooted at r. Each new graph is added to the PatternMap object and we stop once graph sizes exceed S or all graphs have been discovered. We then continue to the next instruction. In this work, we find all patterns up to size 10.

4.2 Pattern Sharing Analysis

We applied the pattern discovery approach described above to identify commonly occurring patterns in a suite of 13 C benchmark programs – the 12 CHStone HLS benchmarks [10], as well as dhrystone. Table 2 presents a sharing analysis for 6 patterns we found to be common. Each pattern listed occurs multiple times in at least one of the benchmarks. Our purpose here is not to exhaustively list *all* patterns that occur more than once in any benchmark; rather, our aim is to provide an illustrative analysis for the most commonly occurring patterns in these particular 13 benchmarks. The left column lists the pattern names, where each name defines the operators involved. For example, Add_Add_Add_Add is a pattern with 4 addition operators connected serially.

We follow the same analysis approach as described in Section 3. We created two Verilog modules for each pattern: one representing the unshareable case, a second having 2-to-1 MUXes on each input, representing the shareable case. The left side of Table 2 gives results for Cyclone II; the right side for Stratix IV. All operators in patterns are 32 bits wide. For the columns of the table representing area, resource sharing provides a "win" if the ratio in parentheses is less than 2 (see Section 3). The results in Table 2 are for sequential patterns, with registers on edges between operators.

For Cyclone II, we observe that sharing is beneficial in 2 of the 6 patterns (OR_OR_OR and XOR_XOR) from the area perspective. For Stratix IV, sharing is beneficial for all 6 patterns. For one of the patterns, OR_OR_OR, the shareable implementation consumed less area than the unshareable implementation. We investigated this and found that Quartus did not produce an area-minimal implementation for this pattern in the unshareable scenario, which we attribute to algorithmic noise. We also analyzed the impact of resource sharing for *combinational* patterns (i.e. patterns without registers on edges between operators) and found sharing to be less beneficial, owing to the ability of Quartus II to collapse chained operators together into LUTs, thereby reducing the opportunities for collapsing the sharing MUXes into the same LUTs as the operators.

We conclude that it is quite challenging to predict up-front when sharing will provide an area benefit, as it depends on the specific technology mapping and packing decisions made by Quartus, which appear to depend on the specific pattern implemented. However, we observe two general trends: 1) Sharing is more likely to be beneficial for composite operators that consume significant area, particularly when the MUXes that facilitate sharing can be rolled into the same LUTs as those implementing portions of the operator functionality. 2) Sharing is more advantageous when registers are present in patterns – registers prevent an efficient mapping of operators into LUTs, thereby leaving LUTs underutilized, with free inputs to accommodate MUX circuitry.

5. BINDING

For each pattern size (in descending order) we choose pairs of functionally-equivalent pattern graphs to be implemented by (bound to) a single shareable composite operator in the hardware. Any two graphs whose operations happen in non-overlapping clock cycles are candidates for sharing.

Consider two sharing candidates, patterns $P1$ and $P2$. We compute a *sharing cost* for the pair by summing the bit width differences in their corresponding operators:

$$SharingCost = \sum_{n1 \in P1, n2 \in P2} |width(n1) - width(n2)| \quad (1)$$

where $n1$ and $n2$ are corresponding operators in pattern graphs $P1$ and $P2$, respectively. The intuition behind (1) is that it is desirable for operation widths between pattern graphs sharing resources to be as closely aligned as possible.

However, two additional optimizations are possible that provide further area reductions:

1. Variable Lifetime Analysis: Our binding approach only pairs pattern graphs whose output values have non-overlapping lifetimes. Otherwise, separate output registers are required to store the values produced by each pattern. If pattern graph output value lifetimes do not overlap, a single register can be used. The LLVM compiler already has a pass to determine a variable's lifetime in terms of basic blocks spanned. We combine the results of this pass with the output of scheduling to determine the cycle-by-cycle lifetimes of each variable.

2. Shared Input Variables: If two patterns share an input variable, then adding a MUX on the input is unnecessary if the patterns are bound to the same hardware, saving MUX area. Hence, our binding algorithm prefers to pair patterns with shared input variables.

After computing the sharing cost using (1) for a pair of candidate patterns (based on their operator bitwidths), we adjust the computed cost to account for shared input variables between the patterns. Specifically, we count the number of shared input variables that feed into the two patterns and reduce the sharing cost for each such shared input variable (cost determined empirically).

Finally, we apply a greedy algorithm to bind pairs of pattern graphs to shared hardware units. Sharing candidates with the lowest cost are selected and bound to a single hardware unit. Note that owing to the costs of implementing MUXes in FPGAs, we allow a composite operator hardware unit to be shared at most twice. Once we have exhausted binding pattern graphs of a given size, we proceed to binding pattern graphs of the next smaller size. The problem is that of finding a minimum cost graph matching, and though we found that a greedy approach suffices, more sophisticated algorithms can be applied (e.g. [12]).

6. EXPERIMENTAL STUDY

We now present results for resource sharing in HLS binding for a set of 13 benchmark C programs – the 12 CHStone

Table 3: Area results for resource sharing using hard multipliers/DSP blocks.

	Cyclone II			Stratix IV		
Benchmark	No Sharing	Sharing Div/Mod	Sharing Div/Mod + Patterns	No Sharing	Sharing Div/Mod	Sharing Div/Mod + Patterns
adpcm	22541	21476 (0.95)	19049 (0.85)	8585	8064 (0.94)	7943 (0.93)
aes	18923	15418 (0.81)	15477 (0.82)	9582	8136 (0.85)	7929 (0.83)
blowfish	11571	11571 (1.00)	9306 (0.80)	6082	6082 (1.00)	5215 (0.86)
dfadd	7012	7012 (1.00)	6364 (0.91)	3327	3327 (1.00)	2966 (0.89)
dfdiv	15286	13267 (0.87)	13195 (0.86)	7043	5949 (0.84)	5915 (0.84)
dfmul	3903	3903 (1.00)	3797 (0.97)	1893	1893 (1.00)	1824 (0.96)
dfsin	27860	27982 (1.00)	26996 (0.97)	12630	11529 (0.91)	11094 (0.88)
gsm	10479	10479 (1.00)	10659 (1.02)	4914	4914 (1.00)	4537 (0.92)
jpeg	35792	34981 (0.98)	34316 (0.96)	17148	16703 (0.97)	16246 (0.95)
mips	3103	3103 (1.00)	2986 (0.96)	1610	1610 (1.00)	1493 (0.93)
motion	4049	4049 (1.00)	3897 (0.96)	1988	1988 (1.00)	1878 (0.94)
sha	11932	11932 (1.00)	12307 (1.03)	5909	5909 (1.00)	5856 (0.99)
dhrystone	5277	5277 (1.00)	5277 (1.00)	2598	2598 (1.00)	2598 (1.00)
Geomean:	10419.82	10093.65	9677.25	4980.59	4788.06	4558.11
Ratio:	1.00	0.97	0.93	1.00	0.96	0.92
Ratio:		1.00	0.96		1.00	0.95

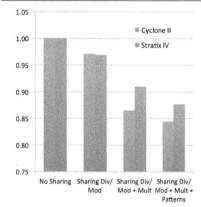

Figure 2: Normalized area results with soft (LUT-based) multipliers.

benchmarks, as well as dhrystone. For both target FPGA families, we evaluated several sharing scenarios that successively represent greater amounts of resource sharing: 1) No sharing; 2) sharing dividers and remainders (mod); 3) scenario #2 + sharing multipliers; and 4) Scenarios #2 + #3 + sharing composite operator patterns. The work in [6] implemented multipliers with LUTs instead of hard IP blocks, i.e. DSP blocks in Stratix IV and embedded multipliers in Cyclone II. To permit comparison with [6], we implemented the benchmarks in two ways: 1) with LUT-based multipliers, and 2) using hard multipliers. Scenario #3 applies only to the case of multipliers implemented with LUTs.

Table 3 gives area results for Cyclone II (left) and Stratix IV (right) when multipliers are implemented using hard IP blocks. Ratios in parentheses show the area reduction vs. the no sharing case. Observe that sharing division/modulus alone provides 3% and 4% average area reduction for Cyclone II and Stratix IV, respectively. Sharing patterns provides an additional 4% and 5% area reduction, on average, for Cyclone II and Stratix IV respectively.

Fig. 2 summarizes the average area results across all circuits for the case when multipliers are implemented with LUTs. Larger area reductions are observed, as expected, owing to the significant amount of area needed to realize multipliers with LUTs. For Cyclone II, a 16% reduction in LEs is observed when all forms of sharing are turned on (left bars for each scenario); for Stratix IV, a 12% reduction in ALMs is seen (right bars).

The pattern sharing approach introduced in this work provides a larger benefit in Stratix IV (4-5%) vs. Cyclone II (2-4%), due to the ability to exploit ALM under-utilization by combining MUX and operator functionality together into LUTs. While speed performance results are omitted for space reasons, we found that resource sharing reduced speed by 11%, on average, in both Cyclone II and Stratix IV, when all forms of sharing were turned on.

7. CONCLUSIONS AND FUTURE WORK

We investigated resource sharing for FPGAs and demonstrated that different resource sharing tradeoffs exist depending on the logic element architecture of the target FPGA. On average, resource sharing provides area reductions of 7-16% for Cyclone II, and 8-12% for Stratix IV, depending on whether multipliers are implemented using hard IP blocks or LUTs. Directions for future work include modifying the scheduling phase of HLS to encourage the generation of composite operator patterns with registers at specific points, in order to allow MUXes to be more easily combined together in LUTs with portions of the operator functionality.

8. REFERENCES

[1] Altera QUIP. *http://www.altera.com/education/univ/research/unv-quip.html*, 2009.
[2] Altera, Corp., *Cyclone II FPGA Family Data Sheet*, 2011.
[3] Altera, Corp., *Stratix IV FPGA Family Data Sheet*, 2011.
[4] E. Casseau and B. Le Gal. High-level synthesis for the design of FPGA-based signal processing systems. In *IEEE Int'l Symp. on Systems, Architectures, Modeling, and Simulation*, pages 25 – 32, 2009.
[5] D. Chen, et al. Optimality study of resource binding with multi-Vdds. In *IEEE/ACM DAC*, pages 580 – 585, 2006.
[6] J. Cong, et al. Pattern-based behavior synthesis for FPGA resource reduction. In *ACM FPGA*, pages 107–116, 2008.
[7] J. Cong, et al. High-level synthesis for FPGAs: From prototyping to deployment. *IEEE Trans. on CAD*, 30(4):473–491, 2011.
[8] J. Cong and J. Xu. Simultaneous FU and register binding based on network flow method. In *ACM/IEEE DATE*, pages 1057 – 1062, 2008.
[9] S. Cromar, et al. FPGA-targeted high-level binding algorithm for power and area reduction with glitch-estimation. In *ACM/IEEE DAC*, pages 838 – 843, 2009.
[10] Y. Hara, et al. Proposal and quantitative analysis of the CHStone benchmark program suite for practical C-based high-level synthesis. *J. of Information Processing*, 17:242–254, 2009.
[11] http://www.llvm.org. *The LLVM Compiler Infrastructure Project*, 2010.
[12] V. Kolmogorov. Blossom v: A new implementation of a minimum cost perfect matching algorithm. *Mathematical Programming Computation 1*, 1(1):43–67, 2009.
[13] A. Canis, et al. LegUp: high-level synthesis for FPGA-based processor/accelerator systems. *ACM FPGA*, pages 33–37, 2011.
[14] W. Sun, et al. FPGA pipeline synthesis design exploration using module selection and resource sharing. *IEEE Tran. on CAD*, 26(2):254 – 265, 2007.

A Fast Discrete Placement Algorithm for FPGAs

Qinghong Wu
Synopsys, Inc.
700 E. Middlefield Rd.
Mountain View, CA 94043

Kenneth S. McElvain
Department of Physics
University of California, Berkeley
Berkeley, CA 94720

ABSTRACT

Good FPGA placement is crucial to obtain the best Quality of Results (QoR) from FPGA hardware. Although many published global placement techniques place objects in a continuous ASIC-like environment, FPGAs are discrete in nature, and a continuous algorithm cannot always achieve superior QoR by itself. Therefore, discrete FPGA-specific detail placement algorithms are used to improve the global placement results. Unfortunately, most of these detail placement algorithms do not have a global view. This paper presents a discrete "middle" placer that fills the gap between the two placement steps. It works like simulated annealing, but leverages various acceleration techniques. It does not pay the runtime penalty typical of simulated annealing solutions. Experiments show that with this placer, final QoR is significantly better than with the global-detail placer approach.

Categories and Subject Descriptors

B.7.2 [**Integrated Circuits**]: Design Aids – *placement and routing*.

Keywords

Computer-aided design (CAD), individual cell temperature, window masking, dynamic window, field-programmable gate array (FPGA), placement, simulated annealing

1. INTRODUCTION

Modern FPGAs have the capacity of millions of Look-Up Tables (LUTs) and registers. Users of these large devices expect their designs to be mapped over-night. This requires the corresponding software flows to be fast and scalable. The placement stage is usually one of the most time-consuming stages. By its nature, the FPGA placement problem is an assignment problem: it assigns each cell, LUT, register, etc., to a hardware site. Ideally, assignment problems can be solved by discrete solvers, such as Integer Linear Programming (ILP) [1]. Unfortunately, the size of the problem makes it too large for an ILP solver to handle, requiring the use of heuristic solutions and leading to sub-optimal results. Traditionally, FPGA placers use simulated annealing-based [2] [3] algorithms to produce placement. By analogy with annealing in metallurgy, each step of the algorithm replaces the current solution by a random "nearby" solution, chosen with a probability that depends both on the difference between the corresponding cost function values and a global parameter, "the temperature", that is gradually decreased during the process. The current solution changes almost randomly when temperature is high, but increasingly "downhill" (for a minimization problem) as

temperature goes to zero. The allowance for "uphill" moves potentially saves the method from becoming stuck at local optima. Classic simulated annealing is slow, but can model all the hardware rules that the placer needs to obey, and it is a lot faster than ILP solvers. However, neither ILP nor the classic simulated annealing algorithm is scalable. Although simulated annealing is still popular in academic FPGA tools and smaller FPGA applications, mainstream commercial tools are switching to analytical global placement, [4]-[8] followed by a heuristic-based detail placement [9] flow. The advantage of this flow is obvious. With a large FPGA die size and comparatively insignificant sizes of logic blocks (such as Configurable Logic Blocks or CLBs on Xilinx Virtex-6), placement can be approximated as a continuous placement problem that an analytical placer handles well. The detail placer can then assign LUTs and registers to their exact physical sites. It can compute delays for each connection, and hence produce high-quality final placement results.

On closer examination of the flow, there is still a gap. Most continuous analytical placers cannot model distances smaller than 3 or 4 CLBs accurately because the problem at this level is highly discrete. Modeling is not good even at the 5 to 10 CLB level because the FPGA delay models are not continuous. A naive approach is to use the global placer locations to greedily assign the logic cells to the closest available hardware sites, and then use a detail placer to improve the placement. Detail placers are perfectly capable of handling short distances, but do not handle middle range problems well, often taking excessively long runtimes to process them. To bridge this gap, we need a high quality discrete placer that is fast enough to handle a large number of objects. With this new goal in mind, we look back again at classic simulated annealing, but limit it to low temperature. Simulated annealing is clearly discrete. Each element can go to its exact physical site. Therefore, exact point-to-point delay can be used to model timing. It is high quality, and that is why many placers use this algorithm for FPGA placement to begin with. The main problem is still runtime. Various approaches have been published to improve runtime. For example, Timber Wolf [2] limits the swap space to temperature-determined windows and [9] speeds up simulated annealing by applying directed moves. Other run-time improvement techniques such as parallelization [11] [12], clustering [13] and shorter temperature schedules [14] have also been discussed in recent literature.

Our new "middle" placer is much faster than classic simulated annealing, because it applies many new techniques tailored to low temperature. With these new techniques, it is not quite the same classical simulated annealing. Before getting into these techniques, let us first analyze the causes of long runtimes in traditional simulated annealing-based placers. Simulated annealing-based placers produce random movements. At high temperature, many of these moves are hill-climbing. We are only interested in low temperature, because the analytical global placer already does most of the work that high temperature simulated annealing previously did. At low temperature, the majority of the moves are rejected, or "wasted"; they do not directly lead to the final placement solution. An ideal revised simulated annealing placer would only take the accepted moves, both up-hill and down-hill that lead to the final placement solution. To get to this goal, "wasted" moves must be minimized to reduce runtime. Our new placer takes a few large steps towards this goal. It is able to bridge the global-detail placer gap with low run time penalty, and produces placement with much better QoR.

2. THE MIDDLE PLACER

Our placer is capable of working at different granularities, the smallest of which is the LUT and register level. At this level, many sophisticated packing rules should apply. To allow hill-climbing, we allow illegal moves, but control them using cost functions. An alternative mode is to move packed groups. In Xilinx architectures, for example, the SLICE is a natural hierarchy to consider. Pre-packed netlists are given to the placer, and the output of the placer is flattened again at the detail placement stage. The cost functions that we use model half-perimeter wire-length (HPWL), timing, and congestion. If it runs in flat mode, it also activates the legalization cost function. This is the cost function:

$$Cost = W_1 C_{HPWL} + W_2 C_T + W_3 C_C + W_4 C_L, \quad (1)$$

W_1 through W_4 are weights that are dynamically adjusted based on design characteristics and the status of cooling. C_{HPWL} is the HPWL cost, C_T is the timing cost, C_C is the congestion cost and C_L is the legalization cost.

3. ACCELERATION TECHINIQUES
3.1 Individual Cell Temperature

Temperature is a fundamental concept in simulated annealing. It represents the status of the current placement. A low temperature indicates that placement is getting very close to its final solution, while a high temperature means that there are still a lot of acceptable movements remaining. The temperature value is used in every single move of the placement process to determine whether certain up-hill move should be accepted. An up-hill move is accepted if the following is true:

$$exp \ (-\Delta Cost \ / \ Temperature) > R_f, \quad (2)$$

R_f is a positive random fraction, and $\Delta Cost$ is a positive value.

Placement results after analytical placement are not quite like placement that is cooled from high temperature simulated annealing. Cells in different areas differ in placement quality. For example, cells in areas that are not timing critical and not congested could be well-placed by the global placer. They are less likely to be moved at low temperature. Even if they are moved, the moves tend to be up-hill moves. Due to various reasons, some cells are not placed as well, and are likely to be moved. Hence, we introduced the new concept of "*individual cell temperature*" to indicate the placement status of each cell. Cells that tend to stay where they are have lower cell temperature, and cells that move more often have high temperature. The old temperature concept is retained as "*system temperature*" because it is still pertinent in the new scheme. Thus, for each cell we have the following:

$$T_C = T_S \times R, \quad (3)$$

where T_C is the Cell Temperature, T_S is the System Temperature, and R is the Temperature Ratio.

R is initialized to be 1 at the beginning of simulated annealing. At the end of each system temperature update, acceptance ratios are computed for each cell. This gives us this computation:

$$R = R_C \ / \ R_G, \quad (4)$$

where R_C is cell acceptance ratio and R_G is global acceptance ratio.

We use this R to compute cell temperature using (3); then use it in (2) to replace `Temperature`, and determine whether to accept a particular up-hill move.

The effect of cell temperature is quite significant, especially for large designs where cell temperatures vary greatly in different areas. In each system temperature, "wasted" moves are significantly reduced. To really accelerate the process, we reduce the total number of attempted moves by 15% to maintain the former QoR.

The concept of individual cell temperature can be generalized for other simulated annealing-based optimizations, where each element in the problem to be solved can have different temperatures based on their current status. Similar speed improvements can be obtained.

3.2 Window Masking

Windowing is a very common acceleration technique in simulated annealing-based placement tools [2]. The goal is to avoid attempting big moves that are unlikely to get accepted. So the tool creates a window around a cell to be moved, and only considers locations that are within the window for the new location. At higher system temperature, the window size is larger, and it gradually shrinks with lower temperatures. We extend this concept by looking inside the window.

For a given FPGA architecture, routing resources and delays are quite different at each location inside the swap window. For example, in Xilinx Virtex-6 architecture [15], there are more direct connections in adjacent CLBs in the X and Y dimensions than in diagonal CLBs. Architecture variation at each location in the window causes the acceptance ratio to be different while moving to these locations. For windows of each size, we pre-characterize the probability of each location based on the acceptance ratio and create window masks. For example, Table 1 is a window mask used for a window size 5, i.e., a move no further than 2 CLBs on a Xilinx Virtex-6 die.

Table 1. Window Mask for a Window of Size 5 (Based on Xilinx Virtex-6 Die)

		12%		
	6%	9%	6%	
5%	6%	6%	6%	5%
	6%	9%	6%	
		12%		

There are 13 filled entries in this table, and the sum of all the entries is 100%. The top entry has a value of 12%, which means that for all accepted moves, 12% are at the top of the window; i.e., they move to the 2 CLBs above. It also implies that more tries of this location will result in a higher acceptance ratio than at other locations. Higher acceptance ratio in Y dimension is largely due to the chip aspect ratio measured by CLB counts. We generate window mask tables for all window sizes that correspond to low temperature simulated annealing collecting acceptance ratio statistics from the running of a few typical large designs. These pre-computed window mask tables are loaded in memory.

After a source cell is picked and a window size given, the placer picks a random location based on the window mask table. The randomness is biased, based on the corresponding window mask.

A swap is then attempted with the cell located in that particular location. This method results in much higher acceptance ratio, when compared to uniform distribution. In other words, we reduced "wasted" moves further, and are able to reduce the total number of random trials to maintain the same QoR. Around 45% of trials are eliminated using this technique.

3.3 Dynamic Windowing

The classic windowing technique in simulated annealing ties the window size to the system temperature. However, we have noticed that this approach is inefficient. At higher temperatures, large window size causes high-cost uphill moves to occur more frequently. By looking further at (1), we found that when $R_\mathcal{E}$ is also large, these high-cost up-hill moves almost always rejected. To reduce such "wasted" moves, we tie all window sizes to $R_\mathcal{E}$, i.e., the window size for each move becomes a function of $R_\mathcal{E}$:

$$W_D = W_G \times (1 - R_\mathcal{E}), \qquad (4)$$

W_D is the dynamic window size for the current move, and W_G is the global window size at the given system temperature. When $R_\mathcal{E}$ is large, the probability of acceptance is low. Then we have a smaller W_D, which makes the up-hill move more likely to be accepted.

Using this dynamic window approach, we were able to reduce the number of trials by another 55%.

3.4 Good Placement Preservation

After analytical placement, many cells are already placed in reasonable locations. For example, if we measure HPWL, many cells are already placed inside their optimal bounding boxes. If they are not timing-critical or highly congested, there is no need to move them much, if at all. To save runtime, we characterize all the cells before each system temperature update. Cells that are well-placed have only a 10% chance of getting picked, when compared to other cells. These cells can still be moved if they are swap targets for other cells. We still move these cells because they could be in a locally optimal location. Therefore, the total number of moves is reduced again. This time, the reduction ratio is not a constant, but depends on the quality of the input placement at each system temperature. Though there is the additional runtime overhead of computing the optimal bounding-box for each cell after each system temperature update, experiments show that this method typically reduces the number of moves by 25%-30%, and results in overall speed gains of about 20%.

3.5 Other Techniques

We also incorporate other existing techniques in our approach. These techniques are applied on top of our accelerations. For example, since we have multiple cost functions, when the ones with dominant weights are already producing large positive cost values, we no longer compute the remaining cost functions. Large cells such as carry-chains, RAMs, and DSPs, have many more pins than other logic elements. We do not move them much, because computing moves for large cells consumes much more runtime. We also applied 10% greedy ripple moves in all temperatures to speed up runtime even further.

4. EXPERIMENTAL RESULTS

We performed experiments on 12 customer designs that target Xilinx Virtex-6 devices. The design names are omitted for confidentiality reasons. Because there is significant noise in the experiments, we ran all experiments three times using different

random seeds and show the average results. The netlists are packed before the middle placement stage. Table 1 compares HPWL improvement using our placement solution with and without the middle placer. The average HPWL improvement attributed by our placer is about 7.8%.

Table 1. Middle Placer Contribution to HPWL

(MP=Middle Placer, GP=Global Placer, DP=Detail Placer)

Design	# of LUTs ($\times10^3$)	# of Registers ($\times10^3$)	HWPL GP-DP ($\times10^6$)	HWPL GP-MP-DP ($\times10^6$)	Improve (%)
1	72	50	0.66	0.61	8.35
2	221	221	2.74	2.64	3.94
3	162	56	2.07	1.94	6.29
4	190	190	2.37	2.20	7.39
5	115	77	1.34	1.00	25.55
6	68	28	0.82	0.81	1.66
7	166	163	2.01	1.96	2.87
8	120	47	1.26	1.16	7.94
9	102	19	0.81	0.77	4.87
10	162	60	2.06	1.88	9.03
11	179	96	2.11	1.81	13.95
12	57	46	1.60	1.58	1.29
AVG					7.8

Table 2 compares timing QoR results with and without the middle placer. The clock period values are the final routed values measured by Xilinx ISE 13.2. The average clock period improvement with our placer is about 6.1%. The Total Negative Slack (TNS) of the final placement is about 27% better. Design 2 stands out because it is very close to positive slack and shows 95%. So do design 7 and design 12. The average TNS calculation excludes these 3 designs.

With our middle placer, the total placement time is about 20% slower. This is considered acceptable in light of the significant improvements in QoR and HPWL.

Table 2. Middle Placer Contribution to Timing QoR

Design	Period GP-DP (ns)	Period GP-MP-DP (ns)	Gain (%)	TNS GP-DP (-ns)	TNS GP-MP-DP (-ns)	Gain (%)
1	2.0	2.0	0.0	2	2	0.0
2	8.1	8.1	-1.0	11	1	95.1
3	13.7	12.7	7.2	48128	28720	40.3
4	20.5	19.3	6.1	21474	21474	0.0
5	10.6	10.6	0.4	12288	9628	21.6
6	8.6	8.5	1.0	733	827	-12.9
7	11.7	11.7	0.1	9	2	77.2
8	19.3	16.5	14.5	10826	5486	49.3
9	9.4	9.0	4.1	788	290	63.2
10	7.8	6.7	14.2	171	55	68.1
11(c1)	14.6	13.2	9.7	1852	1636	11.7
11(c2)*	14.9	12.9	13.8			
12	9.8	10.2	-3.7	11	6	43.9
AVG			6.1			26.8

* Design shows Worst Negative Slack (WNS) on different clocks for the two solutions, and thus has two entries.

Although it starts at low temperature, our middle annealing-based placer often shows 5 times more acceptance ratio than a classic simulated annealing-based placer. Table 3 compares middle placer runtimes to the enhanced simulated annealing (SA) placer without our four acceleration methods. Old acceleration techniques such as quick rejection and partial greedy moves are used in both runs. Both runs start from an analytical placer's placements. Our placer is about 2.3 times faster on average. The speed up variation in different designs is quite significant because our speed up method is highly dependent on global placement quality. Table 4 compares the final timing QoR (routed results using Xilinx ISE 13.2) and HPWL against the classic simulated annealing based placer. As shown in Table 4, average timing QoR difference is - 0.6%, and HPWL is about 1% worse.

Table 3. Middle Placer Run Time Acceleration

Design	SA Run time (hours)	Fast Mid. Placer Run time (hours)	Speed up (times)
1	1.39	0.54	2.59
2	7.67	2.71	2.83
3	8.83	3.85	2.29
4	13.03	5.60	2.32
5	9.02	4.50	2.01
6	1.83	0.85	2.15
7	3.10	1.61	1.92
8	10.93	3.92	2.79
9	2.34	1.07	2.19
10	4.69	2.24	2.09
11	4.28	2.13	2.01
12	2.42	0.81	2.98

Table 4. Middle Placer's Effect on Timing QoR and HPWL

Design	Period of SA (ns)	Period of Fast Mid. Placer (ns)	Gain (%)	HPWL of SA ($\times 10^6$)	HPWL of Fast Mid. Placer ($\times 10^6$)	Improve (%)
1	1.97	1.99	-1.1	0.62	0.61	2.38
2	7.71	8.1	-5.6	2.60	2.64	-1.37
3	12.73	12.7	0.5	1.91	1.94	-1.55
4	18.41	19.3	-4.6	2.22	2.20	0.86
5	9.95	10.6	-6.4	0.95	1.00	-4.87
6	8.45	8.5	-1.0	0.80	0.81	-1.11
7	11.48	11.7	-1.9	1.87	1.96	-4.53
8	16.46	16.5	-0.1	1.16	1.16	-0.05
9	9.07	9.0	0.4	0.78	0.77	0.64
10	6.90	6.7	3.3	1.81	1.88	-3.65
11(c1)	14.08	12.9	8.7	1.77	1.81	-2.53
11(c2)*	13.73	13.2	3.8			
12	9.68	10.2	-5.0	1.61	1.58	1.67
AVG			-0.6			-1.2

* Design shows Worst Negative Slack (WNS) on different clocks for the two solutions, and thus has two entries.

5. SUMMARY

The paper presents a fast discrete FPGA placer. This placer fills the gap between global placement and detail placement and produces high-quality FPGA placement. Four new methods that accelerate the simulated annealing algorithm are also described. Some of these methods can be generalized for other optimizations that are based on simulated annealing.

6. ACKNOWLEDGMENTS
The authors thank Saurabh Adya for many of his original ideas that evolved into the current implementation. The authors also thank Jovanka Ciric-Vujkovic, Larry McMurchie and Rita Tharakan for their valuable help on the writing of this paper.

7. REFERENCES

[1] F.A. Aloul et al, "Generic ILP versus specialized 0-1 ILP: an update," *Proc. of the International Conference on Computer-Aided Design* 2002, pp. 450 – 457.

[2] D.Braun, C. Sechen and A. Sangiovanni-Vincentelli. "The Timber Wolf placement and routing package," *IEEE Journal of Solid State Circuits*, 20(2), 1985, P 510 - 522.

[3] V. Betz and J. Rose, "VPR: A new packing, placement and routing tool for FPGA research," *Field-Programmable Logic and Applications* ,1997, pp. 213–222.

[4] P. Spindler, U. Schlichtmann, F. M. Johannes, "Kraftwerk2 - A Fast Force-Directed Quadratic Placement Approach Using an Accurate Net Model," *IEEE Trans. on Computer-Aided Design of Integrated Circuit and Systems* 27(8) 2008, pp. 1398-1411.

[5] N.Viswanathan, M.Pan, C.Chu, "FastPlace 3.0: A Fast Multilevel Quadratic Placement Algorithm with Placement Congestion Control," *ASPDAC* 2007, pp. 135-140.

[6] N. Viswanathan et al. "RQL: Global Placement via Relaxed Quadratic Spreading and Linearization," *Proc. of The Design Automation Conference* 2007, pp. 453-458.

[7] M. Kim et al. "SimPL: An Effective Placement Algorithm," *Proc. of the International Conference on Computer-Aided Design* 2010, pp. 649-656.

[8] T.-C. Chen et al.,"NTUPlace3: An Analytical Placer for Large-Scale Mixed-Size Designs With Preplaced Blocks and Density Constraints," *IEEE Trans. on Computer-Aided Design of Integrated Circuit and Systems* 27(7) 2008, pp.1228-1240.

[9] R. Ang et al, "Method and Apparatus for Placement and Routing Cells on Integrated Circuit Chips," Patent US 2008/0201678 A1.

[10] K. Vorwerk et al, "Improving Simulated Annealing-Based FPGA Placement With Directed Moves," *IEEE Trans. on Computer-Aided Design of Integrated Circuit and Systems* 28(2) 2009, pp. 179-192

[11] A. Ludwin, V. Betz, and K. Padalia, "High-quality, deterministic parallel placement for FPGAs on commodity hardware," *Proc. International Symposium on Field-Programmable Gate Arrays,* 2008, pp. 14–23.

[12] P. K. Chan and M. D. F. Schlag, "Parallel placement for fieldprogrammable gate arrays," *Proc. International Symposium on Field-Programmable Gate Arrays*, 2003, pp. 43–50.

[13] S. Mallela, and L.K. Grover, "Clustering based simulated annealing for standard cell placement, " *Proc. Design Automation Conference, 1988, pp. 312-317.*

[14] J. D. Vicente, J. Lanchares, and R. Hermida, "Annealing placement by thermodynamic combinatorial optimization," *ACM Trans. Des. Autom. Electron. Syst. (TODAES)*, vol. 9, no. 3, pp. 310–332, Jul. 2004.

[15] Xilinx Corp. "Virtex-6 FPGA Family, " http://www.xilinx.com/products/silicon-devices/fpga/virtex-6/index.htm

Rethinking FPGAs: Elude the Flexibility Excess of LUTs with And-Inverter Cones

Hadi Parandeh-Afshar
hadi.parandehafshar@epfl.ch

Hind Benbihi
hind.benbihi@epfl.ch

David Novo
david.novobruna@epfl.ch

Paolo Ienne
paolo.ienne@epfl.ch

Ecole Polytechnique Fédérale de Lausanne (EPFL)
School of Computer and Communication Sciences, 1015 Lausanne, Switzerland

ABSTRACT

Look-Up Tables (LUTs) are universally used in FPGAs as the elementary logic blocks. They can implement any logic function and thus covering a circuit is a relatively straightforward problem. Naturally, flexibility comes at a price, and increasing the number of LUT inputs to cover larger parts of a circuit has an exponential cost in the LUT complexity. Hence, rarely LUTs with more than 4–6 inputs have been used. In this paper we argue that other elementary logic blocks can provide a better compromise between hardware complexity, flexibility, delay, and input and output counts. Inspired by recent trends in synthesis and verification, we explore blocks based on *And-Inverter Graphs (AIGs)*: they have a complexity which is only linear in the number of inputs, they sport the potential for multiple independent outputs, and the delay is only logarithmic in the number of inputs. Of course, these new blocks are extremely less flexible than LUTs; yet, we show (i) that effective mapping algorithms exist, (ii) that, due to their simplicity, poor utilization is less of an issue than with LUTs, and (iii) that a few LUTs can still be used in extreme unfortunate cases. We show first results indicating that this new logic block *combined* to some LUTs in hybrid FPGAs can reduce delay up to 22–32% and area by some 16% on average. Yet, we explored only a few design points and we think that these results could still be improved by a more systematic exploration.

Categories and Subject Descriptors

B.6.1 [**Logic Design**]: Design Styles—*Logic arrays, Combinational logic*; B.7.1 [**Integrated Circuits**]: Types and Design Styles—*Gate arrays*

General Terms

Design, Performance

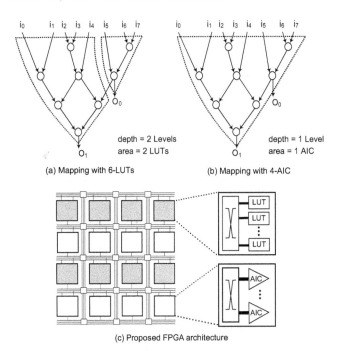

Figure 1: **Flexibility, bandwidth, cost, and delay. (a)–(b)** *And-Inverter Cones (AICs)* **can map circuits more efficiently than LUTs, because AICs are multi-output blocks and cover more logic depth due to their higher input bandwidth. (c) A possible integration of AIC clusters in an FPGA architecture.**

Keywords

FPGA Logic Block, Logic Synthesis, And-Inverter Graph, And-Inverter Cone

1. INTRODUCTION

Since their commercial introduction in the '80s, FPGAs have been essentially based on *Look-Up Tables (LUTs)*. K-input LUTs have one great virtue: they are generic blocks which can implement any logic function of K inputs, and this makes it relatively easy to perform at least some elementary technology mapping: crudely, the problem of mapping reduces to cover the circuit to map with K-input subgraphs, irrespective of the function they represent. This flexibility, and the consequent advantages, do not come for free: LUTs tend to be large (roughly, their area grows exponentially with the number of inputs) and somehow slow (equally roughly, the delay grows linearly with the number of inputs). Also, the number of outputs is intrinsically one and internal fan-out in the subgraphs used for covering is not really possible. Fig. 1(a) suggests graphically how the small number of inputs and the absence of intermediate outputs limit the usefulness of LUTs.

This seems to suggest that perhaps it would be wise to look for less versatile but more efficient logic blocks. In fact, researchers have at times looked into alternate blocks ever since FPGAs have attracted growing research and commercial interest. Yet, naturally, these alternate structures have been somehow related to the logic synthesis capabilities of the time, and thus have almost universally addressed programmable AND/OR configurations in the form of small *Programmable Array Logics (PALs)* (e.g., [14, 13, 8]). Traditionally, synthesis has been built on the sum of products representation and on algebraic transformations, but new paradigms have emerged in recent years. The one we are interested in is based on *And-Inverter Graphs (AIGs)* as implemented in the well-know academic synthesis and verification framework ABC [20]. This representation, in which all nodes are 2-input AND gates with an optional inversion at the output, is not new [11], but has received interest in recent years due to some fortunate combination when used with, for instance, *Boolean satisfiability (SAT)* solvers. Once a circuit is written and optimized in the form of an AIG, one can find very many AIG subgraphs of various depth rooted at different nodes in the circuit.

Thus, we introduce a new logic block that we call *And-Inverter Cone (AIC)*. An AIC (which is explained in detail in Fig. 3) is essentially the simplest reconfigurable circuit where arbitrary AIGs can be naturally mapped: it is a binary tree composed of AND gates with a programmable conditional inversion and a number of intermediary outputs. Compared to LUTs, AICs can be richer in terms of input and output bandwidth, because their area grows only linearly with the number of inputs. Also their delay grows only logarithimically with the input count and intermediate outputs are easier to implement. This makes it possible for AICs to cover AIG nodes more efficiently, as suggested in Fig. 1(a)-(b). In this paper, we will explore the value of AICs both as the sole components of new FPGAs as well as logic blocks for some hybrid FPGA made of both LUTs and AICs, as illustrated in Fig. 1(c). Although far from exploring comprehensively the space of AIC-based solutions, our results suggest that some hybrid solutions look particularly promising and, at the very least, deserve some further attention to refine our analysis.

The rest of the paper adapts the traditional CAD flow used on conventional FPGAs to the needs of AICs and, simultaneously, uses some of the partial results to fix the structure of our novel FPGA. Fig. 2 suggests this graphically: Section 2 addresses the design of the AIC to suit the abilities of modern AIG synthesis. Section 3 adapts traditional technology mapping to the new block. Section 4 looks at how to combine logic blocks in larger clusters with local routing, and Section 5 discusses the packing problem to complete the flow. Sections 6 and 7 then report our experimental results. We discuss related work in Section 8 and then wrap up with some conclusive remarks.

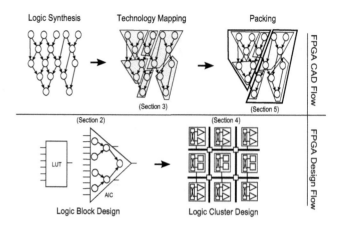

Figure 2: The paths to design and use a novel FPGA with AICs. In this paper, we alternate between adapting the traditional CAD flow to our new needs and using the results to fix our architecture. To each of the last four steps is devoted one of the sections of the paper, as indicated.

2. LOGIC BLOCK DESIGN

A new logic block is proposed in this section. This attempts to reduce the degree of generality provided by typical LUTs in order to obtain faster mappings. Unlike LUTs, our logic block is not able to implement all possible functions of its inputs. In the following, the choice of logic block is motivated and its architecture is discussed.

2.1 An AIG-inspired logic block

An *And-Inverter Graph (AIG)* is a *Directed Acyclic Graph (DAG)*, in which the logic nodes are two-input AND gates and the edges can be complemented to represent inverters at the node outputs. AIGs have been proven to be advantageous for combinational logic synthesis and optimization [20]. This graph representation format is also used for technology mapping step in both FPGA and ASIC designs [4].

Interestingly, AIGs include various cone-like subgraphs rooted at each node with different depths. Usually, the subgraphs with lower depths are more symmetric and resemble full binary trees. The frequent occurrence of such conic subgraphs serves as motivation of this work, where we propose a new logic block that can map cones with different depths more efficiently than LUTs. The basic idea is to have a symmetric and conic block with depth D, which maps arbitrary AIG subgraphs with depth $\leq D$. This logic block is called *And-Inverter Cone (AIC)*.

To illustrate the potential benefits of AICs with respect to LUTs, we refer to Fig. 1, where two levels of LUTs are required to map the same functionality that can be mapped onto a single AIC. The reason for that is twofold: on the one hand, the LUT size is limited to six inputs and the entire AIG (8 inputs) can not fit into just one 6-LUT. On the other hand, even if the size of the LUT was big enough, the mapping would still use two LUTs, as the AIG has two distinct outputs. It is worth mentioning that increasing the LUT size to accommodate more inputs would result in a huge area overhead. Instead, the proposed AIC inherently offers smaller area and propagation delay than a LUT for the same number of inputs. For example, a 4-AIC with 16 in-

Block	inputs	outputs	2:1 mux	config bits
2-AIC	4	1	3	3
3-AIC	8	3	7	7
4-AIC	16	7	15	15
5-AIC	32	15	31	31
6-AIC	64	31	63	63
6-LUT	6	1	64	64

Table 1: AICs have less configuration bits than LUTs, while they can implement circuits with a much greater number of inputs (e.g., a 6-AIC includes 8 times more inputs than a typical 6-LUT).

puts requires half the area of a 6-input LUT—using the area model of Section 6.1 with less delay. Clearly, the fact that more wires need to be connected to the AICs creates new routing congestion issues. However, as detailed in Section 4, these can largely be alleviated by packing several AICs in a limited bandwidth AIC cluster with local interconnect.

2.2 AND-Inverter Cone (AIC) Architecture

Fig. 3 shows the architecture of an *And-Inverter Cone (AIC)*, which has five levels of cells. Each cell can be configured as either a two-input NAND or AND gate. Notice that each cell has an AIC output, except for the cells belonging to the lowest level of the AIC. This provides access to intermediate nodes as in the example of Fig. 1. Moreover, these outputs enable to configure a bigger AIC as multiple smaller ones. For example, the AIC of Fig. 3, implements the AIG of Fig. 1 at the right-hand side while the left-hand side can be used to implement other functions with various combinations of 2-, 3-, and 4-AICs. Accordingly, a 5-AIC contains two 4-AICs, four 3-AICs, or eight 2-AICs.

Generalizing, each D-AIC has $2^D - 1$ cells, 2^D inputs and $2^D - 2^{D-1} - 1$ outputs. In the rest of the paper, we consider D-AICs with depths from three to six, and we will study the effect of the allowed AIC depth on the mapping solution. Depths greater than six are not considered, as they require a huge input bandwidth, which may result in major modifications of the global routing network of current FPGAs. Table 1 compares different D-AICs with the conventional 6-LUT in terms of IO bandwidth, number of configuration bits and multiplexers.

3. TECHNOLOGY MAPPING

During technology mapping, the nodes comprising the AIG are clustered into subgraphs that can be mapped onto an AIC or a LUT. This can be done in multiple ways depending on the optimization objectives including delay and area.

In this work, the primary optimization objective of technology mapping is delay minimization and consequently a mapping solution is said to be *optimal* if the mapping delay is minimum. Area reduction is also considered but just as a secondary optimization objective. Technology mapping for AICs is similar to the typical LUT technology mapping but adapted to the peculiarities of AICs, such as the fact that multiple outputs are possible. In the rest of the section, the mapping problem is first formalized and then the main four steps of the mapping algorithm are described in detail.

3.1 Definitions and Problem Formulation

A technology independent synthesized netlist (AIG format) is input to our mapping heuristic. Such netlist is automatically produced by ABC [20]. We take the input netlist and extract the combinational parts of the circuit and represent them by a DAG $G = (V(G), E(G))$. A node $v \in V(G)$ can represent an AND gate, a primary input (PI), a pseudo input (PSI, output of a flipflop), a primary output (PO), or a pseudo output (PSO, input of a flipflop). A directed edge $e \in E(G)$ represents an interconnection wire in the input netlist. The edge can have the *complemented* attribute to represent the inversion of the signal.

At a node v, the depth $depth(v)$ denotes the length of the longest path from any of the PIs or PSIs to v. The height $height(v)$ denotes the the length of the longest path from v to any of the POs or PSOs. Accordingly, the depth of a PI or PSI node and the height of a PO or PSO node are zero.

The mapping algorithm that we use in this work is a modified version of the classical depth-optimal LUT mapping algorithm [6]. It is well known that the problem of minimizing the depth can be solved optimally in polynomial time using dynamic programming [6, 15]. However, we also target area-minimization as a secondary objective, which is known to be NP-hard for LUTs of size three and greater [7, 16]. We use *area flow* heuristic [19] for area approximation during the mapping.

The mapping of a graph in LUTs requires different considerations. For a node v, there exist several subgraphs containing v as the root, which are called *cones*. Accordingly, C_v is a cone that includes node v in its root and some or all of its predecessors. For mapping C_v by a LUT, it should be K-feasible, where $inputs(C_v) \leq K$. Moreover, the cone should be *fanout-free*, meaning that the only path out of C_v is through v. If the cone is not fanout free, then the node which provides the fanout may be duplicated and will be mapped by other LUT(s), as the primary minimization objective is depth.

The AICs mapping cone candidates of v are extracted differently. In this case, rather than being K-feasible, a cone C_v, to be mappable on a D-AIC block, should be depth feasible, where $depth(C_v) \leq D$. The other constraint is that the nodes at lowest depth of C_v, should not have any path to a node outside C_v, otherwise such nodes are removed from C_v. This condition ensures that C_v to be mappable to an AIC such as the one illustrated in Fig. 3, in which no AIC output is driven by the nodes at the lowest level of the AIC.

When AICs are considered as the mapping target in addition to LUTs, the definition of the problem of mapping for depth does not change. The only difference is that the cone candidates of AICs are added to the cone candidates of LUTs for each node in the graph. Although the conditions of eligibility for LUTs and AICs are different, it is possible to have common cones between the two that are treated as separate candidates.

Next, the main steps of the mapping algorithm are described in detail.

3.2 Generating All Cones

To generate all K-feasible cones, we use the algorithm described in [9, 22], in which the cones of a node are computed by combining the cones of the input nodes in every possible way. This step of the mapping takes a significant portion of

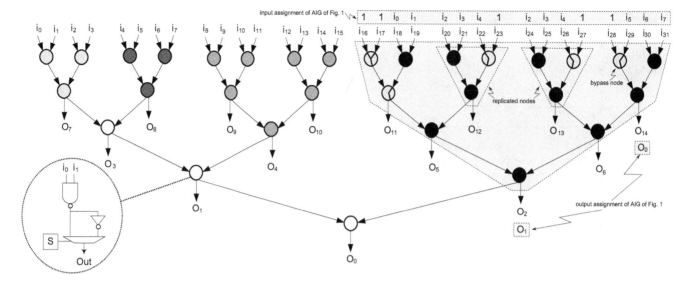

Figure 3: **Architecture of 5-AIC (AND-Inverter Cone), which has five levels of cells that are programmable to either AND or NAND gates. The 5-AIC can also be configured to 2-, 3-, and 4-AICs in many ways (highlighted cells show one possibility), without any need for extra hardware. The AIG of Fig. 1 is mapped onto the right-hand side. To propagate a signal, we can configure a cell to the bypass mode (e.g., forcing one input to 1 when this is operated as an AND). Moreover, some AIG nodes need to be replicated when the fanout of an internal value is larger than one.**

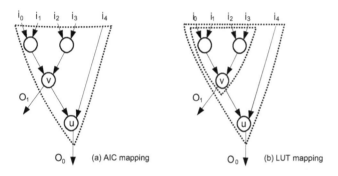

Figure 4: **Difference between LUT and AIC mapping. Since AICs are inherently multi-output blocks, the same cone rooted at u in (a) can also be a (free) mapping cone of v, while in LUT mapping, no common cone exist for any two nodes (b).**

the total execution time, specially when K is a large value such as six.

The cone generation for AICs is different from the cone generation for LUTs, as the cones of each node are produced independently from the cones of its input nodes. To generate all possible D-AIC mappable cones for a node v, the subgraphs rooted at v are examined by varying the cone depth from two to D. All possible subgraphs that meet the AIC mapping conditions described in section 3.1, are added to the cone set of node v. If a cone C_v satisfies the depth condition, but has a fanout node u at the lowest depth of the cone, u will be removed from C_v; if this still satisfies the depth condition, the cone will be added to the D-AIC mappable cone set.

The main difference between the cone generation for AICs and LUTs is having common cone candidates for different

nodes, as shown in Fig. 4. This is possible, as AICs are multi-output. In this figure, the cone that has u as its root, can be used to map both v and u. Therefore, this cone should be in the AIC cone sets of both nodes. We call this cone as a *free* cone for node v, as it maps v for free when it is selected for u mapping.

The time complexity of the D-AIC cone generation is $O(M \cdot D)$, where M is the number of nodes in the graph and D is the maximum depth of an AIC block.

3.3 Forward Traversal

Once the cones sets of both LUTs and AICS are computed for every node in the graph, the next step is to find the best cone of each node by traversing the graph in topological order. Since the primary objective in this work is to minimize the depth, the best cone of node v is the one that gives v the lowest depth. If there is more than one option, the cone which brings less area flow to v is selected (see [19] for further details). The depth and area flow of v, when mapped onto cone C_v, are dependent on the depth and area flow values of the C_v input nodes.

To compute the depth and area flow of node v, we use Equations 1 and 2, respectively. Since the FPGA blocks, including K-LUTs and D-AICs, are heterogeneous and have different depths, we should consider the interconnection wire delays for the depth computation of each node, similar to the edge-delay model [23]. Although we have both local (intra cluster) and global (inter cluster) routing wires, which have different delays, we assume that all wires have unique delay equal to the average delay of the local and global wires.

$$dp(v) = \max(dp(In(C_v)) + dp(C_v) + dp(wire)) \quad (1)$$

$$af(v) = \sum_{i=0}^{nIn(C_v)} (af(In_i(C_v)) + area(C_v) \quad (2)$$

Algorithm 1 Find the best cone for each node of the DAG

```
1:  BestC_v.dp ← ∞
2:  BestC_v.af ← ∞
3:  for i = 1 → nC_v(LUT) do
4:      v.setdp(C_v(i))
5:      v.setaf(C_v(i))
6:      cond_1 ← C_v(i).dp < BestC_v.dp
7:      cond_2 ← C_v(i).dp = BestC_v.dp
8:      cond_3 ← C_v(i).af < BestC_v.af
9:      if cond_1 || (cond_2 && cond_3) then
10:         BestC_v ← C_v(i)
11:     end if
12: end for
13: for i = 1 → nC_v(AIC) do
14:     v.setdp(C_v(i))
15:     v.setaf(C_v(i))
16:     cond_1 ← C_v(i).dp < BestC_v.dp
17:     cond_2 ← C_v(i).dp = BestC_v.dp
18:     cond_3 ← C_v(i).af < BestC_v.af
19:     if cond_1 || (cond_2 && cond_3) then
20:         BestC_v ← C_v(i)
21:     end if
22:     cond_1 ← C_v(i).dp < BestBackupC_v.dp
23:     cond_2 ← C_v(i).dp = BestBackupC_v.dp
24:     cond_3 ← C_v(i).af < BestBackupC_v.af
25:     if C_v(i).root = v then
26:         if cond_1 || (cond_2 && cond_3) then
27:             BestBackupC_v ← C_v(i)
28:         end if
29:     end if
30: end for
```

In the above equations, $dp(C_v)$ and $area(C_v)$ are the depth and area of the logic block that C_v can be mapped on. This block can be either a K-LUT or a D-AIC. If C_v is a *free* cone of node v, then $dp(C_v)$ and $dp(In(C_v))$ will refer to the depth and inputs of the sub-AIC in C_v. And for area flow computation, the term $area(C_v)$ will be removed from Equation 2.

Algorithm 1 presents the pseudo-code of the algorithm used to find the best cone of each AIG node. This function iterates over all generated cones for both LUTs and AICs of node v to find the best cone that has the lowest depth. If two cones have the same depth, the one that requires smaller area is selected. If the best cone of node v is a *free* cone, this cone will be selected for the mapping, if and only if the root of the cone—which is not v—is visible in the final mapping solution and this cone is the best cone of the root node as well. If one of these two conditions does not hold, then we need to select another cone as the best cone for v. Therefore, it is essential to maintain a non-*free* best cone—v is the root of such a cone—for v as a backup best cone.

3.4 Backward Traversal

In this step, the graph is covered by the best cones of the visible nodes in the graph, which are added to the mapping solution set S. A node is called *visible*, if it is an output or input node of a selected cone in the final mapping. Initially POs and PSOs are the only visible nodes and S is empty. The graph traversal is performed in reverse topological order from POs and PSOs to PIs and PSIs. If the visited node v is visible, then its best cone, BC_v, is selected for the mapping

and is the added to S. Then, all the input nodes of BC_v become visible and the graph traversal continues. If the BC_v is a *free* cone and it is already in S, there is no need to add it again and only the heights of the input nodes of v are updated. Otherwise, if the *free* cone is not in S, then the backup BC_v, which has v as its root, is selected for mapping and is added to S. During the backward traversal, the height of each visible node is updated. Once a BC_v is selected for mapping, the height of its input nodes are updated by adding the height of v to the depth of v within the target AIC or LUT.

3.5 Converting Cones to LUTs and AICs

The mapping solution S, which is generated during the Backward Traversal, includes all the cones that cover the graph. The next step is mapping the cones in S to either a K-LUT or a D-AIC. If the selected cone belongs to the K-feasible cone set of node v, then it should be implemented by a LUT. Otherwise, the cone is a D-AIC mappable cone, which is implemented by an AIC. The depth of the cone defines the type of the target AIC block.

4. LOGIC CLUSTER DESIGN

The proposed AICs require a much higher IO bandwidth than typical LUTs. In order to alleviate the routing problem that may result from that increase, we propose to group multiple AICs into an AIC cluster with local interconnect.

To form an AIC cluster, we integrate N D-AICs, optional flipflops at the outputs of D-AICs to support sequential circuits, and an input and an output crossbar. The input crossbar drives the inputs of the AICs in the cluster, and the output crossbar drives the outputs of the cluster. Since we do not want to change the inter-cluster routing architecture of the FPGAs, we use the same bandwidth of LUT-based clusters for AIC clusters and keep the AIC cluster area close to the area of the reference LUT cluster, which is the *Logic Array Block (LAB)* in the *Altera Stratix-III*.

To study the effect of the AIC size on the mapping results, we select different D-AICs as the base logic block in a cluster, where D varies from three to six and can be configured to implement the AIC blocks that have $depth \leq D$. However, the number of the D-AIC blocks in the cluster, N, varies for different D values such that the number of sub-AICs in the cluster remains the same and no changes occur in the cluster crossbars.

The two crossbars in the AIC cluster are the main contributors to the cluster area. Crossbars are basically constructed with multiplexers and their area depends on their density and on the number of the crossbar inputs and outputs. Since both crossbars get the outputs of N D-AICs as the input, reducing the number of the D-AIC outputs will significantly reduce the area share of the crossbars. Originally, each D-AIC has $2^D - 2^{D-1} - 1$ outputs, but in our experiments, we observed that in the extreme case only 2^{D-2} outputs are utilized and that is when a D-AIC is configured to 2^{D-2} 2-AICs. Hence, a very simple sparse crossbar is added at the output of each D-AIC to reduce the number of D-AIC outputs to 2^{D-2}.

The second technique used to reduce the crossbar area is to decrease its connectivity and make it sparse. To trade-off the crossbar density and packing efficiency in the AIC cluster, we measured the packing efficiency of the clusters having an input crossbar with 50%, 75%, and 100% connec-

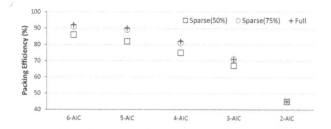

Figure 5: The packing efficiency of three crossbar connectivity scenarios: 50%, 75%, and 100%. The allowed cone depth in technology mapping is varied to study the effect of AIC size on the packing quality.

tivities. The packing efficiency is the ratio of the number of AIC clusters, assuming that each AIC cluster has unlimited bandwidth and the actual number of AIC clusters that is obtained from packing. To calculate the number of clusters in the ideal packing, we use Equation 3. In this equation, nC_i is the number of cones with depth i. Fig. 5 shows the results of this experiment for different base AIC blocks in the cluster. The reported efficiency is the average packing efficiency of the 20 biggest MCNC benchmarks.

$$nClusters_{ideal} = \sum_{i=2}^{6}(\frac{nC_i}{N \cdot 2^{6-i}}) \qquad (3)$$

One observation from Fig. 5 is that the packing efficiency is substantially reduced for all the three scenarios, when the allowed cone depth in the technology mapping is reduced. This is reasonable, as the probability of input sharing and open inputs is reduced for smaller cones. Moreover, when smaller AICs are packed to a D-AIC, a larger number of the D-AIC outputs are utilized, which increases the output bandwidth requirement. The second observation is that reducing the crossbar connectivity to 75% largely maintains the packing efficiency of the full crossbar. However, the packing efficiency for the crossbar with 50% connectivity decreases to a larger extent. Therefore, one option to reduce the crossbar area without having a sensible degradation in packing efficiency is to set the crossbar connectivity to 75%.

Exploiting the mentioned crossbar simplifications, and by using the area model of Section 6.1, the area of the AIC cluster remains close to the area of a LAB, when three 6-AICs, six 5-AICs, twelve 4-AICs, or twenty four 3-AICs are integrated in the AIC cluster. As mentioned, the input/output crossbars of the AIC cluster are fixed for all scenarios.

5. PACKING APPROACH

In the previous section, we defined the architecture of the AIC cluster. Given the AIC and LUT clusters, the next step is to pack the technology mapped netlist onto the clusters. For the packing, we use the *AAPack* [18] tool, which is an architecture-aware packing tool developed for FPGAs. The input to *AAPack* is the technology mapped netlist with unpacked blocks, as well as a description of an FPGA architecture. The output is a netlist of *packed* complex blocks that is functionally equivalent to the input netlist. Similarly, we also use *AAPack* to pack LUTs in LABs.

The packing algorithm uses an affinity metric to optimize the packing. This affinity metric defines the amount of net

Component	Area (Tr_{minW})
6-AIC block	1,512
6-AIC output Xbar	217
6-AIC FFs and muxes	1,104
AIC cluster input Xbar	22,072
AIC cluster out Xbar	2,660
AIC cluster buffers	1,447
AIC cluster with three 6-AICs	**34,678**
ALM	1,751
LAB in Xbar	16,251
LAB buffers	470
LAB with ten ALMs	**34,231**

Table 2: Areas of different components in an AIC cluster and in a LAB, measured in units of minimum-width transistor area.

sharing between p, which is a packing candidate, and B, which is a partially filled complex block. In the architecture file, the complex block should be represented as an ordered tree. Nodes in the tree correspond to physical blocks or modes. The root of tree corresponds to an entire complex block and the leaf nodes correspond to the primitives within the complex block. For the D-AIC complex block, we construct a tree similar to the DSP block multiplier tree in the original paper, by which we define different configuration modes of the D-AIC. The number of AICs in the cluster as well as the crossbars structure are also defined in the architecture file. The information is used by the packer to group the individual blocks in clusters. During the packing process, some routability checking are performed to ensure (local and global) routability of the packing solution, which considers the intra-block and the general FPGA interconnect resources.

6. EXPERIMENTAL METHODOLOGY

In this work, we use a classic area and delay model [5]: The area model is based on the transistor area in units of minimum-width transistor area; the rationale is that to a large extent the total area is determined by the transistors more than by the metal connections. For the delay model, circuits are modeled using SPICE simulations for 90-nm CMOS process technology.

6.1 Area Model

The area modeling method requires a detailed transistor-level circuit design of all the circuitry in the FPGA [5]. Fig. 6 shows an AIC cluster with three 6-AICs. Table 2 lists the area of different components in the AIC cluster and in a LAB in terms of number of minimum-width transistors. *ALM* stands for *Adaptive Logic Module*, which is the logic block in *Altera Stratix-II* and in following series. Based on this table, the area of an AIC cluster with three 6-AICs and the crossbars mentioned in Section 4 is marginally larger than a LAB with 10 ALMs. As mentioned in Section 4, the AIC cluster has almost the same area when the basic AIC block is changed.

6.2 Delay Model

The circuit level design of the AIC cluster suggested in Fig. 6 is also used for accurate modeling of the cluster delays.

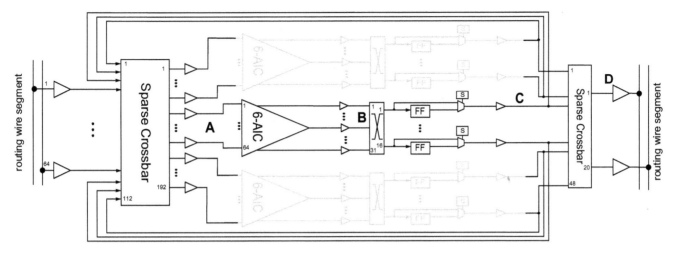

Figure 6: Structure and delay paths of an AIC cluster with three 6-AICs.

Path	Description	Delay (ps)
A → B	6-AIC main output	496
B → C	crossbar and FF-Mux	75
C → D	output crossbar of cluster	50

Table 3: Delays of different of paths in the AIC cluster of Fig. 6.

Mapping Scenario	Intra-cluster Wires
LUT	50%
6-AIC	34%
LUT/6-AIC	35%
LUT/5-AIC	37%
LUT/4-AIC	38%
LUT/3-AIC	40%

Table 4: Average ratio of intra cluster wires for the different mapping scenarios.

The crossbars in this figure are developed using multiplexers, and for these we adopted the two level hybrid multiplexer that is used in Stratix-II [17]. Hence, the critical path of each crossbar goes through two pass-gates, with buffers on the inputs and outputs of the components that include pass transistors.

We performed SPICE simulations with 90nm 1.2 V CMOS process, to determine the delay of all paths in the cluster shown in Fig. 6. The results are listed in Table 3. For the path between B and C, the delay number relates to the path that goes through the main output of the 6-AIC, which has the longest path. These delay numbers are used in the technology mapping to minimize the delay of the mapped circuit.

We also measured the delay of a LAB by SPICE simulation. Simulation results revealed that the delay of a 6-LUT in an ALM, excluding the LAB input crossbar, in 90nm CMOS process, is between 280ps and 500ps, taking into account that different LUT inputs have different delays. We use the average delay (390ps) for our experiments. Based on [1], the 6-LUT delay in 90nm process technology has a delay between 162ps to 378ps and considering the extra multiplexers that exist on the LUT output path in the ALM structure, our delay numbers appear realistic.

7. RESULTS

We contrast three architectures and various mapping strategies, using the MCNC benchmarks [24]. We consider the original FPGA, a homogeneous FPGA exclusively composed of AIC clusters, and a hybrid FPGA composed of both LUTs and AIC clusters as different experiment scenarios. In the hybrid structure, we also vary the base AIC block of the AIC-cluster from 3-AIC to 6-AIC.

Fig. 7 shows the logic delays of the benchmarks for the mentioned scenarios. The main observation is that the lowest logic delay relates to the hybrid structure, as we have both LUTs and AICs mapping options. Moreover, except for the *ex5p* and *frisc* benchmarks, the logic delay is always reduced when deeper cones are allowed, which appears predictable as a general trend. This is also visible in the number of logic-block levels on the critical path, either LUTs or AICs, as shown in Fig. 8; the graph gives an indication of the routing wires necessary to connect the logic blocks of the circuits: although some logic delays are higher for deeper cones, their total delay can be still better due to the reduced number of wires between logic blocks. Comparing LUT-only and AIC-only implementations, we see that there are circuits that have better logic delay when LUTs are used, but on average AIC-only implementation has 28% less logic delay. Moreover, except for *tseng* and *des*, the number of logic blocks on the critical path (and thus routing wires) in the AIC-only implementation is less than or equal to that of the LUT-only one.

As the current release of VPR 6.0 does not support timing driven placement and routing, we set a fixed delay value for the interconnecting wires in order to estimate the total circuit delay. This delay number is different for the different mapping scenarios and its value is specified based on the delay and used ratio of intra and inter cluster wires for each mapping scenario that is reported in Table 4. Using this wire delay, we compute the routing delay of the critical path of the circuits, using the number of logic blocks in these paths. Fig. 9 illustrates a rough estimation of the

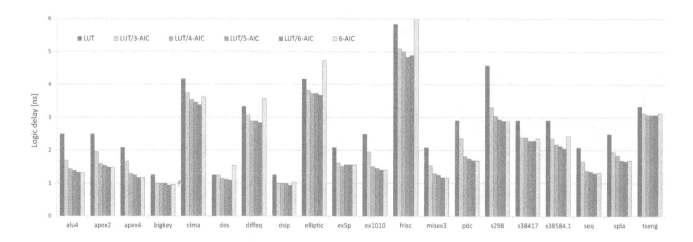

Figure 7: Logic delay of all benchmarks in the original FPGA (LUT), for the FPGA composed only of AIC (6-AIC), and for a hybrid FPGA (LUT/6-AIC).

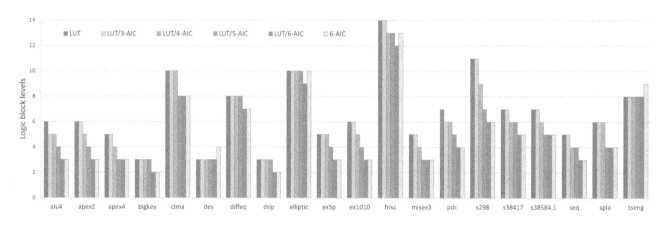

Figure 8: Number of logic blocks (both LUTs and AICs) on the critical path.

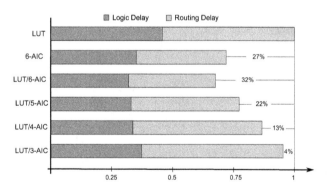

Figure 9: Geometric mean of normalized total logic and routing delays.

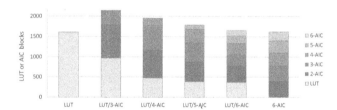

Figure 10: Number and type of logic blocks used in the various architectures and with the various mapping strategies.

total average logic and routing delays of the circuits. On average, the implementations on the pure 6-AIC architecture and on the hybrid architecture with 6-AIC and 5-AIC base blocks are 27%, 32%, and 22% faster than the baseline FPGA, respectively.

Fig. 10 presents the distribution of LUTs and AICs for the different architectures. This figure shows that when deeper

cones are allowed, less LUTs are used. Moreover, in each case the usage of each AIC type has a reverse relation with the size of the AIC. This means that the chance of mapping a node with smaller AIC is always higher. Since each of these LUTs and AICs are packed into clusters, the numbers presented there do not indicate the real logic area of the circuits. On the contrary, Fig. 11 illustrates the number of clusters after packing: this is proportional to the active area since the area of an AIC cluster is close to the area of a LAB (see Table 2) and both have the same I/O bandwidth. For some benchmarks, either the *LUT/6-AIC* hybrid architec-

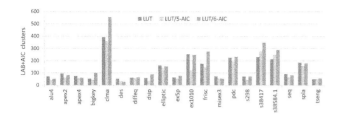

Figure 11: Area measured as the total number of clusters used, completely or partially. LABs and AIC clusters occupy approximately the same area. On average, LUT/5-AIC uses 16% less resources than LUT-only.

Benchmark	LUT	LUT/ 5-AIC	LUT/ 6-AIC
alu4	14.9	10.59	11.32
apex2	16.4	15.2	12.9
apex4	15.5	16.1	14.1
bigkey	14.3	12.6	11.6
clma	20.8	22.9	25.5
des	14.6	16.1	15.1
diffeq	10.4	13.4	13.8
dsip	18.6	17.4	12.5
elliptic	15.5	16.6	16.7
ex5p	11.2	15.9	23.2
ex1010	23.8	18.2	30.3
frisc	18.8	19.35	23.2
misex3	14	12	13
pdc	22.8	23.4	21.2
s298	13.2	9.7	15.8
s38417	12.5	18.2	19
s38584.1	11.5	18.4	17.5
seq	17.1	15.5	15.5
spla	21.5	18.8	21.1
tseng	8.3	13.1	12.5

Table 5: Average wire length in units of one CLB segments.

ture or the baseline FPGA display the lowest area; however, the *LUT/5-AIC* architecture always results in the smallest used area at a much better delay than the baseline FPGA and a slightly worse one than *LUT/6-AIC*—refer to Fig. 9. The two hybrid architectures define Pareto optimal points.

The hybrid structure of the proposed FPGA with the different cluster types needs to fix the right ratio of the two flavors of logic blocks. The packing results indicate that this ratio varies from one circuit to the other, making this problem not straightforward. We have made some preliminary experiments on this front, and we have fixed the ratio of LAB columns to AIC clusters to 1:4. The advantage of AICs is that any logic function that is mapped to a LUT is mappable to one or more AICs. The reverse is also true. Therefore, it is possible to switch to another logic block type, when we run out of one type. Moreover, considering the small size of the AIC blocks, it is quite feasible to add them as *shadow blocks* of the LUTs to the LUT clusters, by reusing the existing input crossbar. This provides the option to use either LUTs or AICs depending on the requirements.

Though, adding AICs as shadow blocks of LUTs remains as the future work.

Table 5 presents the average wire length of each benchmark, in the baseline architecture (no AIC clusters) and in the two best hybrid architectures, with the number of routing channels fixed to 180 for all the experiments. We observe that there is a fairly high variability—but averages are very similar (15.8, 16.1, and 17.3 respectively)—with a small trend against our hybrid architecture.

8. RELATED WORK

Leveraging the properties of logic synthesis netlist to simplify the logic block of FPGAs is a current research topic [2, 3]. For instance, based on the observation that circuits represented using AIGs frequently have a trimming input, a low-cost and still LUT-based logic block was designed that requires less silicon area, but it does not improve the delay [3]. Albeit somehow similar in its inspiration to modern synthesis, our work is more radical in using the AIGs to inspire the new logic cell.

Although LUT-based logic blocks dominate the architectures of commercial FPGA, PAL-like logic blocks have also been explored. In recent times, it has been shown that a fairly small PAL-like structure, with 7–10 inputs and 10–13 product terms, obtains performance gains at the price of an increase in area [8]. Much earlier, some authors have shown that K-input multiple-output PAL-style logic blocks are more area efficient than 4-input LUTs. However, the idea was abandoned because PAL-based implementations typically consumed excessive static power [14]. Our solution moves away from the typical logic block natural of traditional logic synthesis, and we have shown that it seems possible to improve both area and delay compared to LUT-based FPGAs.

There are also numerous pieces of work which have adapted or created reconfigurable logic blocks to specific needs, often by adding dedicated logic gates to existing LUTs. Among these, one can mention GARP [10] and Chimaera [25] for datapath oriented processor acceleration, macro gates [12] for implementing wide logic gates, and various sorts of fast carry chains beyond those available commercially [21]. Although they all somehow question the pure LUT as the most efficient building block, they tend to introduce modifications that are never real generic alternatives.

9. CONCLUSIONS

As several people before us, we have recognized that LUTs have many advantages but, frequently, the price to pay for these advantages is unreasonably high. We have thus explored new logic blocks inspired by recent trends in the circuits representations used in logic synthesis: we came to define AICs which are simply the natural configurable circuits homologue of the newly popular AIGs. We have explored alternate FPGAs architectures based on these AICs, essentially fitting the new logic block into a traditional FPGA architecture without changing some global parameters whose impact would be very difficult for us to master. Despite these artificial limitations, we find first results encouraging: On one hand, delay is bound to decrease as both logic delay and the number of logic blocks on the critical path reduce. With a fairly rough routing delay model we observe a delay reduction of up to 32%. On the other hand, the number of logic blocks (all of similar area) consumed by the benchmark circuits is also generally reduced; with one of our mapping approaches, the area is reduced on average by 16%. Future work will necessarily need to address placement and routing much more precisely than we had the chance to. Also, other less conservative architectures may prove more advantageous than those explored. Nevertheless, we think that our first results are sufficiently encouraging for the approach to deserve a closer inspection.

10. REFERENCES

[1] Altera Corporation. *Stratix II Device Handbook, vols. 1 and 2.* http://www.altera.com/literature/.

[2] J. H. Anderson and Q. Wang. Improving logic density through synthesis-inspired architecture. In *Proceedings of the 19th International Conference on Field-Programmable Logic and Applications*, pages 105–11, Prague, Aug. 2009.

[3] J. H. Anderson and Q. Wang. Area-efficient FPGA logic elements: Architecture and synthesis. In *Proceedings of the Asia and South Pacific Design Automation Conference*, pages 369–75, Yokohama, Japan, Jan. 2011.

[4] Berkeley Logic Synthesis and Verification Group, Berkeley, Calif. *ABC: A System for Sequential Synthesis and Verification*, Feb. 2011. Release 10216, http://www.eecs.berkeley.edu/~alanmi/abc/.

[5] V. Betz, J. Rose, and A. Marquardt. *Architecture and CAD for deep-submicron FPGAs.* Kluwer Academic, Boston, Mass., 1999.

[6] J. Cong and Y. Ding. An optimal technology mapping algorithm for delay optimization in lookup-table based FPGA designs. In *Proceedings of the International Conference on Computer Aided Design*, pages 49–53, Santa Clara, Calif., Nov. 1992.

[7] J. Cong and Y. Ding. On area/depth trade-off in LUT-based FPGA technology mapping. *IEEE Transactions on Very Large Scale Integration (VLSI) Systems*, 2(2):137–48, June 1994.

[8] J. Cong and H. Huang. Technology mapping and architecture evaluation for k/m- macrocell-based FPGAs. *ACM Transactions on Design Automation of Electronic Systems (TODAES)*, 10(1):3–23, Jan. 2005.

[9] J. Cong, C. Wu, and Y. Ding. Cut ranking and pruning: Enabling a general and efficient FPGA mapping solution. In *Proceedings of the 7th ACM/SIGDA International Symposium on Field Programmable Gate Arrays*, pages 29–35, Monterey, Calif., Feb. 1999.

[10] J. R. Hauser and J. Wawrzynek. Garp: A MIPS processor with a reconfigurable coprocessor. In *Proceedings of the 5th IEEE Symposium on Field-Programmable Custom Computing Machines*, pages 12–21, Napa Valley, Calif., Apr. 1997.

[11] L. Hellerman. A catalog of three-variable Or-Invert and And-Invert logical circuits. *IEEE Transactions on Electronic Computers*, EC-12(3):198–223, June 1963.

[12] Y. Hu, S. Das, S. Trimberger, and L. He. Design, synthesis, and evaluation of heterogeneous FPGA with mixed LUTs and macro-gates. In *Proceedings of the International Conference on Computer Aided Design*, pages 188–93, San Jose, Calif., Nov. 2007.

[13] A. Kaviani and S. D. Brown. Hybrid FPGA architecture. In *Proceedings of the 4th ACM/SIGDA International Symposium on Field Programmable Gate Arrays*, pages 3–9, Monterey, Calif., Feb. 1996.

[14] J. L. Kouloheris and A. El Gamal. PLA-based FPGA area versus cell granularity. In *Proceedings of the IEEE Custom Integrated Circuit Conference*, pages 4.3.1–4.3.4, Boston, Mass., May 1992.

[15] Y. Kukimoto, R. Brayton, and P. Sawkary. Delay-optimal technology mapping by DAG covering. In *Proceedings of the 35th Design Automation Conference*, pages 348–51, San Francisco, Calif., June 1998.

[16] I. Levin and R. Y. Pinter. Realizing expression graphs using table-lookup FPGAs. In *Proceedings of the 30th Design Automation Conference*, pages 306–11, Dallas, Tex., June 1993.

[17] D. Lewis et al. The Stratix II logic and routing architecture. In *Proceedings of the 13th ACM/SIGDA International Symposium on Field Programmable Gate Arrays*, pages 14–20, Monterey, Calif., Feb. 2005.

[18] J. Luu, J. H. Anderson, and J. Rose. Architecture description and packing for logic blocks with hierarchy, modes and complex interconnect. In *Proceedings of the 19th ACM/SIGDA International Symposium on Field Programmable Gate Arrays*, pages 227–36, Monterey, Calif., Feb. 2011.

[19] V. Manohararajah and S. Brown. Heuristics for area minimization in LUT-based FPGA technology mapping. *IEEE Transactions on Computer-Aided Design of Integrated Circuits and Systems*, 25(11):2331–40, Nov. 2006.

[20] A. Mishchenko, S. Chatterjee, and R. Brayton. DAG-aware AIG rewriting: A fresh look at combinational logic synthesis. In *Proceedings of the 43rd Design Automation Conference*, pages 532–36, San Francisco, Calif., July 2006.

[21] H. Parandeh-Afshar, P. Brisk, and P. Ienne. An FPGA logic cell and carry chain configurable as a 6:2 or 7:2 compressor. *ACM Transactions on Reconfigurable Technology and Systems (TRETS)*, 2(3):19:1–19:42, Sept. 2009.

[22] M. Schlag, J. Kong, and P. K. Chan. Routability-driven technology mapping for lookup table-based FPGAs. *IEEE Transactions on Computer-Aided Design of Integrated Circuits and Systems*, 13(1):13–26, Jan. 1994.

[23] H. Yang and D. F. Wong. Edge-map: Optimal performance driven technology mapping for iterative LUT based FPGA designs. In *Proceedings of the International Conference on Computer Aided Design*, pages 150–55, San Jose, Calif., Nov. 1994.

[24] S. Yang. Logic synthesis and optimization benchmarks user guide, version 3.0. Technical report, Microelectronics Center of North Carolina, Research Triangle Park, N.C., Jan. 1991.

[25] Z. A. Ye, A. Moshovos, S. Hauck, and P. Banerjee. CHIMAERA: A high-performance architecture with a tightly-coupled reconfigurable functional unit. In *Proceedings of the 27th Annual International Symposium on Computer Architecture*, pages 225–35, Vancouver, June 2000.

Securing Netlist-Level FPGA Design through Exploiting Process Variation and Degradation

Jason Xin Zheng, Miodrag Potkonjak
Computer Science Department
University of California, Los Angeles (UCLA)
Los Angeles, USA
{jxzheng,miodrag}@cs.ucla.edu

ABSTRACT

The continuously widening gap between the Non-Recurring Engineering (NRE) and Recurring Engineering (RE) costs of producing Integrated Circuit (IC) products in the past few decades gives high incentives to unauthorized cloning and reverse-engineering of ICs. Existing IC Digital Rights Management (DRM) schemes often demands high overhead in area, power, and performance, or require non-volatile storage. Our goal is to develop a novel Intellectual Property (IP) protection technique that offers universal protection to both Application-Specific Integrated Circuits (ASIC) and Field-Programmable Gate-Arrays (FPGAs) from unauthorized manufacturing and reverse engineering. In this paper we show a proof-of-concept implementation of the basic elements of the technique, as well as a case study of applying the anti-cloning technique to a nontrivial FPGA design.

Categories and Subject Descriptors

B.7.m [**Hardware**]: Integrated Circuits—*Miscellaneous*

General Terms

Design, Experimentation, Measurement, Security

Keywords

IP protection, active hardware metering, unclonable

1. INTRODUCTION

There exist several high-impact gaps in the design, implementation, and manufacturing of integrated circuits (ICs), including silicon capacity vs. design productivity, number of gates vs. number of pins, and the disparity between gate delays and wire delays. These gaps have been having deep and profound impacts on both IC design and manufacturing processes. In the last two decades another gap emerged that may have a far-reaching economic impact on the semiconductor industry. The gap between Non-Recurring Engineering (NRE) costs and Recurring Engineering (RE) costs has been growing exponentially as the manufacturing process continues to scale down. The numbers are truly fantastic: while in the sixties the cost of manufacturing one gate was $1, it is expected that by the end of this decade 1 trillion gates will cost only $1. Meanwhile, owing mainly to the increasing size and verification complexity of the designs that are executed, the cost of designing a modern IC product has skyrocketed. On the other hand, the cost of building a state-of-the-art semiconductor foundry has also rapidly grown to well over $1 billion. The NRE-RE cost gap provides high incentives for independent silicon foundries to recuperate setup costs by manufacturing non-authorized ICs, and for fab-less design houses to prevent manufacturing piracy.

This situation provided impetus for the initiation of active IC digital rights management (DRM) research. Several techniques have been proposed and implemented. They share at least one of two common denominators: the use of physically unclonable functions (PUFs) [2][4] or conditionally enabling through classical cryptographical one-way functions [39][7]. While these techniques address several aspects of IC intellectual property (IP) protection such as prevention of use of non-authorized ICs, they have significant limitations including rather high area, power, and frequency overheads, additional storage requirements for enabling keys, and susceptibility to operational and environmental conditions.

Most importantly, they do not offer protection of the design know-how that is often strategically important, i.e. although the attacker may not be able to produce non-authorized ICs, he can gain insight on how significant parts of the design are created.

Furthermore, the arsenal for IP protection is even more sparse for volatile SRAM-based Field-Programmable Gate-Array (FPGA) designs [20][11][15], where a locally-stored configuration bitstream is usually required. Although the stored bitstreams can be encrypted with existing cryptographical mechanisms, there exists similar overhead and storage concerns over this technique as cryptographically enabled IC designs.

In this paper, we present a novel technique for comprehensive IP protection and active device metering with very low resource and energy overhead. The technique is universally applicable to both Application-Specific Integrated Circuit (ASIC) and FPGA designs. In addition to protecting against non-authorized manufacturing, we show that reverse-engineering of the design can be prevented at user-specified levels. Key to this IP protection approach is to implement sensitive logic paths in such a way that their functionality can be altered, post-silicon, using targeted device

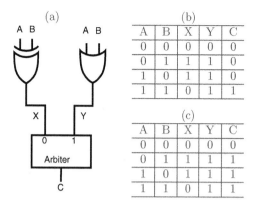

(a)

(b)

A	B	X	Y	C
0	0	0	0	0
0	1	1	1	0
1	0	1	1	0
1	1	0	1	1

(c)

A	B	X	Y	C
0	0	0	0	0
0	1	1	1	1
1	0	1	1	1
1	1	0	1	1

Figure 1: (a) Delay logic with XOR and OR gates. (b) Truth table when XOR is faster than OR. (c) Truth table when OR is faster than XOR.

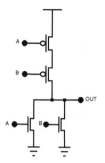

Figure 2: CMOS NOR gate.

aging. In the context of FPGA IP protection, this means that the chosen sensitive logic paths can be altered, through targeted aging, independent of the bitstream design.

The notion of sensitive logic paths reflects the understanding that not the entire silicon product requires protection against cloning or reverse engineering. For instance, it is typically not necessary to protect a generic multiplier, as such designs are easily obtainable or trivial. Sensitive logic paths refer to the portions of the design where critical know-how or functionality is embedded, such as the Finite State Machine (FSM) of a video compression engine.

1.1 Key Concepts

To understand how logic components can be configured post-silicon or post-bitstream, we introduce the concept of delay logic. Traditional binary combinational logic produces outputs that can be determined statically once the functions and the connectivities of each element, e.g. the netlist of the design, are known. Delay logic is a type of logic whose outputs depend on a third runtime factor, which is the delays of the gates in the circuit.

A small example is presented in Figure 1(a), where the outputs of an XOR gate and an OR gate are combined by an arbiter element. The output C of the arbiter is determined in the following fashion:

- If a rising edge arrives on input port 0 before input port 1, then the output is 0.

- If a rising edge arrives on input port 1 before input port 0, then the output is 1.

If the initial state of the circuit is that A and B are both 0, and A and B are changed at the same time, then it is clear that the output port C depends on the knowledge of the delay of the XOR and OR gates in the circuit. The tables in Figure 1(b) and (c) show the values of output c as a function of inputs a and b for the two possible relative speeds of the XOR and OR gates. Evidently the whole circuit behaves like an AND gate when the XOR is faster than the OR gate, and an OR gate when the OR is faster than the XOR.

The delay logic has a clear advantage to the traditional combinational logic for protection against cloning and reverse engineering, due to the fact that its output depends on a dynamic delay factor. In the context of this paper, we assume that obtaining such information from arbitrary

gates and designs is difficult. However, there are challenges to using delay logic in implementation. Many factors, such as die temperature, V_{dd}, and manufacturing variations, affect the logic delay, thus making it difficult to predict. The use of the arbiter elements relaxes the design constraints by measuring only the relative delays between two paths. Furthermore, the reliability of the delay logic can be improved through post-silicon configuration.

Negative Bias Temperature Instability, or NBTI [1][28], is an aging process that occurs to CMOS transistors when a negative bias is applied to the gate. When stressed, the breakage of hydrogen-silicon bonds creates interface traps which lead to increases in the effective threshold voltage V_{th} of the gate. Figure 2 shows a typical CMOS representation of a NOR gate. The top transistors are of PMOS type, and the bottom ones are of NMOS type. Although NBTI acts upon both the PMOS and NMOS transistors, PMOS transistors are impacted more significantly than NMOS transistors, as they are always negatively biased when turned on. Nevertheless, the overall effect of the NBTI aging is that the logic propagation delay through the gate is increased, i.e. the gate is slower.

Combining selective NBTI aging with arbiter-based delay logic yields a new approach to protecting the sensitive logic paths of an IC design. Without the knowledge of relative logic speeds, an attacker cannot understand the functionality of the circuit, even if he was equipped with the full gate-level netlist. Without proper post-silicon configuration, a non-authorized copy of the IC will not function as intended.

1.2 Related Efforts and State-of-the-Art

Security techniques for FPGA designs and implementation are a broad research area that covers a variety of issues ranging from digital rights management (DRM) and detection of malicious circuitry to reverse engineering and trusted synthesis. There are several recent surveys on FPGA security [45][30]. In our brief survey of the related work we mainly focus on directly related IC DRM and FPGA security techniques.

The first set of FPGA DRM techniques was created by John Lach and his coauthors [25][26][27]. Champagne et al. [13] discussed secure techniques for distribution of FPGA configurations.

The largest impetus for IC reverse engineering, its surprising easiness and effectiveness was created by Cambridge University researchers [5]. Consequently, several groups demonstrated that even highly security hardware security primitives such as PUFs is surprisingly easy to reverse engineer [29][40].

The IC IP protection efforts emphasized techniques that enable zero knowledge proofs that a particular hardware is designed by a specific entity. Almost all of them used some form of design watermarking and/or fingerprinting [24][19][36]. Consequently these IC fingerprinting techniques were combined with data mining techniques to form first passive metering approaches [22][3][46][47]. Passive IC metering enables counting of the number of non-authorized ICs. Since 2007, several active IC metering techniques have been developed [4][39]. These technique enable remote activation and deactivation of ICs. In many of these techniques PUF and PPUF [17][8] play a crucial role in the creation and employment of unique IC IDs. They are creative and effective solutions and advance active metering research frontier. Nevertheless, they are subject to several significant limitations such as high hardware and energy overheads, limited security protocols flexibility (e.g. no mechanisms for specifying active time intervals and only single user control), and the requirement of key storage that may be the source of security vulnerabilities. In addition, they are not amenable to quantitative security analysis and do not guarantee prevention of the IC reverse engineering.

While recovering netlists by reverse engineering actual ICs is a well-established research and business endeavor [9][33][23], there is surprisingly little reported work on IC reverse engineering to higher levels of abstraction. Notable exceptions include efforts at the University of Michigan [18], Michigan State [48][49][14], and recently the Air Force Institute of Technology [35][32][31]. However, there have been numerous reverse engineering efforts at lower levels of abstraction mainly with the goal of verifying actual implementations [10][42][12].

Recently, in a series of papers Torrance and James provided detailed description about capabilities, and limitation of state-of-the-art industrial IC reverse engineering procedures [43][44]. Even more recently, reverse engineering started to attract rapidly growing interest from academic community [41].

In addition to directly related IC structure extraction and reverse engineering techniques, there are several other related areas. They are related either because we use their techniques and tools or due to conceptual similarity. The most difficult task of IC reverse engineering is probably FSM extraction and traversal. The problem has been addressed in several communities [38][21][6]. Furthermore, identification and coverage of regular patterns has been a popular and important problem in behavioral and system synthesis [34][37].

1.3 Overview

In the rest of this paper, we present a proof-of-concept of the delay logic implemented on an FPGA platform and show that not only the delay logic responds to the slight differences in delays caused by process variation, but also to the controlled aging effects of NBTI. We introduce this delay logic element in Section 2 and explain in detail its implementation.

To show that manufacturing variation introduces observable relative delay differences that are unique for each FPGA, in Section 3 we present the experimental results of the delay logic by comparing the outputs from an array of 64 arbiters implemented on two FPGAs using identical configuration bitstreams.

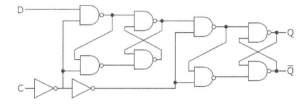

Figure 3: An example DFF design[50].

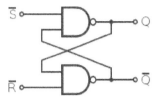

Figure 4: An SR latch[50].

We also present an preliminary aging study in Section 3 to show that short-term NBTI aging is observable on FPGA devices, and this serves as a proof-of-concept for post-silicon configuration.

In Section 4, we present a case study of incorporating the delay logic in a non-trivial logic design and proving that it works as expected.

2. THE SR ARBITER

In this section, we will introduce a delay logic element with the combined functions of an arbiter and two NAND gates that compete for the control of the arbiter output, and show how it can be implemented on an FPGA platform with repeatable results.

D-type flip-flops (DFFs) have been used as arbiters in PUF designs in the past [17], where the path differences are designed to be offset by the preceding tuning circuits. Though easy to implement, arbiters implemented using DFFs have a built-in bias due to the fact that the Clock-to-Q and D-to-Q paths are different by design. Figure 3 shows an example of an asymmetrical DFF design. In contrast, SR latches, such as the one depicted in Figure 4 are intrinsically symmetrical, making it more suitable to function as an unbiased arbiter.

Furthermore, if a trigger signal arrives at both the S and the R input ports at exactly the same time, then the output of the arbiter solely depends on the relative speeds of the two NAND gates. This is an important aspect for the FPGA platforms, as identical logic paths are almost impossible to come by. Often the minor speed differences at the transistor level are trumped by the enormous routing delay differences. Therefore an SR latch can serve both as a good arbiter and delay racing element. We will refer to this design as the SR arbiter from here on.

2.1 Implementation

We would like to highlight the necessity of nearly identical logic paths in the implementation of arbiters. Many obstacles, such as unpredictable cell placement and routing, are to be tackled to ensure that the competing paths to the arbiters are as symmetrical as possible.

Unfortunately, the target platform (Xilinx Virtex5) does

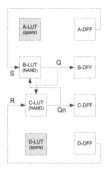

Figure 5: SR latch mapped to Virtex5 CLB Slice.

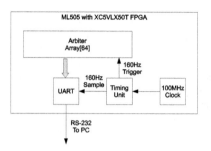

Figure 6: 64-Arbiter array test setup.

not offer any native SR latch logic cells. Therefore, an SR latch must be meticulously implemented by instantiating two lookup table (LUT) cells occupying the same logic slice (as NAND gates) and connecting them in combinational loops. By constraining the two LUT cells to the same logic slice, the combinational loop routing between the two LUTs are kept minimum and as close as possible.

Though the two NAND gates are identical in logic design, within each NAND gate, the two input-to-output paths are purposely designed to be different (as a matter of fact, the two paths cannot be designed to be identical due to the nature of SRAM-based LUT cells). While the Q-to-Qn and Qn-to-Q paths utilize the fastest path in each LUT (the highest address bit of the LUT), the S-to-Q and R-to-Qn paths utilize the slowest path in each LUT (the lowest address bit of the LUT). On the Virtex 5 FPGA where 6-input LUTs are available, the Q-Qn path only has one multiplexer, while the S/R-to-Q/Qn path has 6 layers of multiplexers. This arrangement echoes the desire to use the SR latch to measure the relative speed of the its NAND gates.

To further ensure that the signal transitions arrive as close as possible at the S and R input ports, strict relative placement constraints are used to enforce an in-slice floorplan as illustrated in Figure 5. The two DFFs in the middle of the slice (B-DFF and C-DFF) stores the Q and Qn results from the immediate NAND gates at B-LUT and C-LUT. The outer two DFFs (A-DFF and D-DFF) have very little clockskew between them, so they serve as precision triggers for the SR arbiter. To preserve local routing channels, the spare LUTs (A-LUT and D-LUT) are not allowed to be occupied by other functions. As a result, each SR arbiter occupies precisely one slice.

3. EXPERIMENT AND RESULTS

The experiment setup and results of the SR-latch based arbiters are described in this section.

3.1 Experiment Setup

The main experiment of this paper carries two objectives. The first objective is to determine whether the process variations, manifested as differences in propagation delay between two FPGA chips, can be effectively detected by the SR arbiters. To achieve this, two SR arbiters with identical placement and routing are configured on two separate FPGA chips, and a trigger pulse is sent to the S and R ports of the arbiter. The output of the arbiter indicates whether the S path is faster than the R path, or the contrary. If the S path is consistently faster than the R path on both FPGAs, then the results will agree. However, if S path is faster than

the R path on one FPGA, but slower on the other, then the arbiter results will disagree. Thus by comparing the arbiter results of two FPGAs, the propagation delay differences can be detected.

The second objective is to determine whether the effects of NBTI aging and recovery can be detected by the SR arbiters. We will exploit the frequency dependency of the NBTI aging effect by maintaining the S and R inputs of the arbiter at either logic one or zero for a prolonged period of time in the hope to slow down S or R path enough to change the outcome of the race between S and R paths. Following the aging process, the static S and R inputs are changed to toggling between one and zero in order to recover from the aging effect. The output of the SR arbiter is measured and compared after each aging and recovery cycle.

3.1.1 Environment

The target platform used in this experiment is the Xilinx ML505 reference design board. The ML505 board is equipped with a Virtex-5 FPGA (v5lx50) that can be configured via a JTAG port. For the process variation objective, two such ML505 boards are used. The two FPGAs installed on the ML505 boards will be referred to as "FPGA-A" and "FPGA-B" from here on.

A desktop PC is used to configure and collect the results from the ML505 boards via RS-232 serial ports. All tests are conducted at ambient room temperature.

3.1.2 Baseline 64-Arbiter Array Design

To facilitate the test objectives of the experiment, an FPGA design populated with an array of 64 SR arbiters is implemented. Figure 6 shows a functional diagram of the arbiter array design. Besides the array of the SR arbiters, a timing unit is used to generate the trigger pulses to the array of arbiters at a rate of 160Hz. The results of the arbiters are then captured and transmitted by a UART encoder. A chip-level floorplan of the FPGA design is shown in Figure 7.

3.1.3 Arbiter Result Scoring

An arbiter race result is represented in a binary format. A one indicates that the S path of the arbiter has won the previous race, while a zero indicates that the R path of the arbiter has won. Each result set contains 64 such binary values. A total of 1000 such result sets are collected by the host PC to compute an average score for each of the arbiters, i.e. a score of 0.0 means that the S path of the arbiter has won the race 1000 times, and a score of 0.6 means that the S path has won 600 times while the R path has won 400 times, etc. This average score is a reflection of the expected

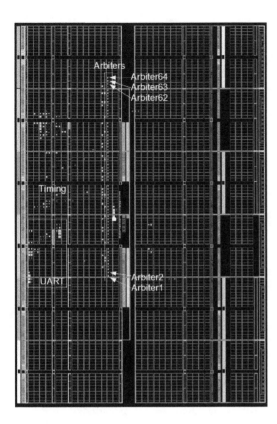

Figure 7: 64-Arbiter array floorplan.

output of the arbiter under the same condition, and will be referred to as the "arbiter score" from here on.

3.2 Results

3.2.1 Process Variation

The first objective of the experiment is to determine whether process variation between two FPGAs can be detected by the SR arbiters. The results are collected before the first NBTI aging cycle and reflect the starting state of the FP-GAs. Figure 8 plots the arbiter scores of the two FPGAs in a polar form, with the solid lines corresponding to FPGA-A, and the dashed lines corresponding to FPGA-B. The radius of each point on the plot reflects the arbiter score. For aesthetic reasons the inner circle represents a score of 0.0, and the maximum score (the tips of the longest spikes) is 1.0.

It is evident from the lack of overlaps between the solid and dashed lines in Figure 8 that the arbiter scores of the two FPGAs are significantly different. Since the two FPGAs are configured from an identical bitstream file, and the two FPGAs are exposed to the same ambient environment, we conclude that the differences in the arbiter scores observed at the same arbiter site are due to process variations at the transistor level.

Though it is possible that the minor differences in voltage and junction temperatures may contribute to the differences in arbiter scores, we believe that such differences will only cause a systematic shift of the scores. The seemingly random nature of the arbiter scores is more consistent with the effects of process variation.

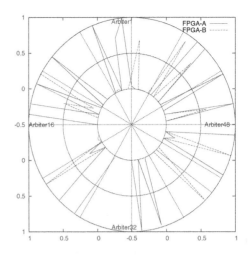

Figure 8: Arbiter score comparison.

3.2.2 NBTI Aging and Recovery

The second objective of the experiment is to determine whether NBTI aging and recovery effects can be detected by the SR arbiters. For this purpose, two variations of the baseline design are created. The first variation of the design applies a static logic pattern at the S and R ports of all arbiters to stress the the S paths of the arbiters, while the second variation stresses the R paths. During an aging session, the FPGA-A is treated with the first variation (stressing S paths), and the FPGA-B is treated with the second variation (stressing R paths).

Each aging session lasts approximately 14 hours (overnight), followed by one or two days of recovery session where the baseline design is loaded and the S and R ports are constantly toggling. The arbiter scores are recorded between aging and recovery sessions.

In Figure 9 and Figure 10, results from two aging and one recovery sessions are presented. The y (vertical) axis reflects the arbiter scores for each arbiter (x-axis). The z-axis is a series of sample times in chronological order. In Figure 9, the arbiter scores for some arbiters in FPGA-A are pushed towards 0.0 after each aging session. This is consistent with the fact that the S paths are being stressed during the aging session, and that an arbiter score closer to 0.0 reflects the higher likelihood of the R paths winning a race. After a recovery session, the arbiter scores move back towards 1.0, indicating that the NBTI stress on the S paths has receded, and that the S paths are more becoming likely to beat the R paths in a race.

The complete opposite takes place in Figure 10, where the R paths of the FPGA-B are stressed during the aging session. Note that on neither FPGA-A or FPGA-B did all arbiter scores change; in other words, the effect of a 14-hour continuous NBTI stress is not enough to change the outcome of the race between the S and R paths.

4. CASE STUDY: LEON3 PROCESSOR

The LEON3 [16] is a 32-bit general-purpose processor based on the SPARC V8 architecture. The complete VHDL source code of the LEON3 is released under the GNU Public License (GPL) for academic use. The LEON3 design is moderate in size and sophistication, therefore we chose to use

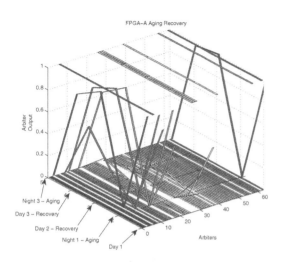

Figure 9: Aging and Recovery of FPGA-A

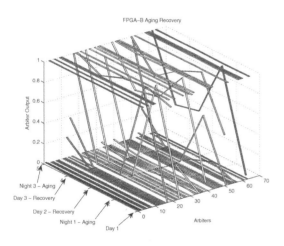

Figure 10: Aging and Recovery of FPGA-B

the LEON3 as a platform to demonstrate how arbiter delay logic can be easily incorporated into a modern design to prevent unauthorized cloning and reverse engineering. There are many techniques for delay logic to be incorporated. In this paper, we demonstrate the simplest method, which is to use arbiter outputs as an unique ID to control a shuffling network embedded in the control logic paths of the target design.

4.1 Control Logic Shuffling

A simple way to implement anti-cloning is to use physical signatures of the device to scramble control signals in a controlled way. As an analogy to cryptography, the control signals are like messages that can be encrypted, transmitted, and decrypted to be used. The encryption key is the physical signature of the chip, in this case, the arbiter scores, and the decryption key is a function of the encryption key.

One method to scramble control signals is to shuffle logic. A two-input shuffler has two data inputs, two data outputs, and a control input. When the control input is set to logic zero, the two data outputs are exact copies of the two data inputs: data input A drives data output A, and data input B drives data output B. When the control input is set to logic one, the output orders are swapped: data input A drives data output B, and data input B drives data output A. A tree of such shuffling logic controlled by arbiter outputs can ensure that the design will only function properly on FPGA devices with a matching arbiter output signature.

4.2 Selection of Control Logic

At the core of the LEON3 processor is a seven-stage (fetch, decode, register read, execute, memory, exception, and register write) integer unit. Since all instructions must go through the integer unit, the control logic in the pipeline stages is an ideal place to employ the shuffler technique to prevent unauthorized design cloning.

To minimize the performance impact on the LEON3, the selection of the control logic for shuffling is made after first sorting the control logic paths by timing slack. The control logic with the largest amount of setup slack is least likely to become the critical path after shuffling logic is inserted. Evidently the ALU control signals (ALUOP) have the most slack among all control signals in the integer unit after examining the static timing analysis. What is also interesting about the ALU control signals is that they are highly critical to the normal operation of a processor. Therefore the ALU control signals are chosen as the target of the shuffling logic.

4.3 Static Shuffling Strategy

The basic strategy to organize static shuffling is described in this section. The strategy can be summarized in three words: shuffle, rotate, and reorder. These three words represent the three types of operations that can be performed on a group of control signals.

The shuffle operation mainly makes use of two-input shufflers. Each shuffler, depending on the control port ShuffleEn (SE), can either swap the the two input signals or do nothing to them. Shown as the first and third stages in Figure 11, two adjacent control signals are sent to a shuffler. The ShuffleEn signal is controlled by the output of an SR arbiter. If there is an odd number of control signals to shuffle, the last signal is sent to a shuffler with its inverted version, i.e. the shuffler can either send out the original or the inverted ver-

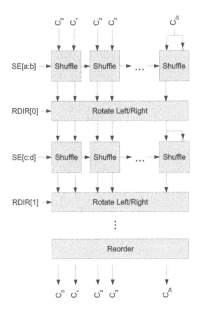

Figure 11: Shuffle, Rotate, and Reorder

sion of the signal, depending on the state of the ShuffleEn port. This type of shuffler is also referred to as an Inv-Shuffler.

The rotate operation always perform a single-bit rotation on the group of input signals. The direction of the rotation is decided by the RDIR input port, which is also driven by an SR arbiter.

The shuffle and rotation operations can be repeated as many times as possible, so long as the timing slack is sufficient. Figure 11 shows the shuffle and rotate interleaving each other to incrementally introduce entropy into the system.

The reorder operation is always the last stage before the control signals leave the shuffling tree. The purpose of having the reorder operation is to maintain the original order and polarity of the signals the same as they enter the shuffling tree. For example, if signal A and signal B are to be swapped position by a shuffler, then reorder stage must swap them back.

In a fully static shuffling tree, all the SE and RDIR inputs are controlled directly by the outputs of SR arbiters. At the compile/synthesis time, the reorder stage must know ahead of time what the SR arbiters will output to properly reorder the signals.

4.4 Metrics

The implementation result of the LEON3 with static shuffling will be judged on various metrics. The first metric is the functional correctness of the processor. The modified LEON3 processor must remain functional as before on the intended FPGA target, and any unauthorized copy must render the processor unusable. This is tested by loading the same LEON3 design on two ML505 boards. Only one of the ML505 board has the FPGA that matches the SR arbiter outputs expected by the reorder stage, i.e. this is the intended FPGA target. The other ML505 board, when loaded with the same design, should not function properly as a processor.

The second metric is the amount of area and performance

overhead incurred by the modification. The area overhead is measured in number of LUTs used, and the performance overhead is measured by the minimum clock cycle, or the maximum clock speed.

The last metric is the most difficult to quantify: how easy can the anti-cloning scheme be attacked? To answer this question, we must make several assumptions about the attacks:

- The bitstream of the design can be recovered from the EEPROM storage.

- With the proper know-how, the bitstream can be converted into a netlist.

- The SR arbiters can be readily identified from the netlist. However, the outputs of the SR arbiters cannot be determined statically, nor can they be measured dynamically.

- The attacker therefore must resort to a brute force attack by guessing the outputs of the arbiters and verify the correctness of their guesses by running a netlist-level simulation.

Using the above assumptions, the difficulty of the attack then depends exponentially on the number of arbiters used, and linearly on the number of simulation cycles required to verify the correctness of the guess. Since the number of arbiters is limited and trivial to determine, we decided to use the number of simulation cycles to verify correctness as the metric against attacks. Finally, since the netlist simulation is typically simulated with a time resolution of 1ps (10^{-12} seconds, the number of simulation cycles is essentially how many picoseconds the simulation must run before an attacker realizes that the guess is wrong.

4.5 Results

Figure 12 shows the floorplan of the modified LEON3 processor implemented on the v5lx50 FPGA. In the picture, DIV refers to the radix-2 integer divider, MUL refers to the dedicated multiplier. The integer unit, shown in orange, occupies most of the north side of the FPGA. The ALU control shufflers occupies a very small area (inside the white circle). A magnified view of the shufflers can be see in Figure 13, where the LUTs implementing the shuffle and rotate stages are pointed out. As a proof of concept, only three arbiter are used to statically control the shuffle and rotation stages. The three DFFs (highlighted in white in Figure 13) store the arbiter outputs from the selected arbiters. The two DFFs on the left are expected to hold a value of one, and the DFF on the right is expected to hold a value of zero in order for the processor to function properly.

To answer the metric of functional correctness, the modified LEON3 design is loaded on two ML505 boards. A test program that finds all prime numbers less than 1000 is then loaded to the processor to run. On the FPGA with the correct arbiter outputs, the results are returned in exactly the same way as an unmodified LEON3 processor design would. On the FPGA with the incorrect arbiter outputs, the computation did not complete. Instead, the processor quickly traps to an error state. This proves that the modified LEON3 design passes the functional correctness metric.

To evaluate the area and performance overhead of the static shuffling and arbiter array, the FPGA mapping and

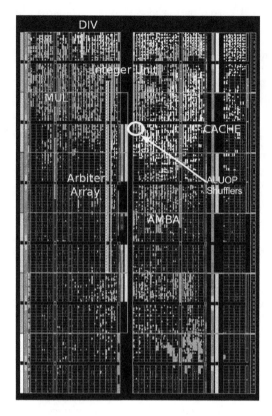

Figure 12: LEON3 Floorplan

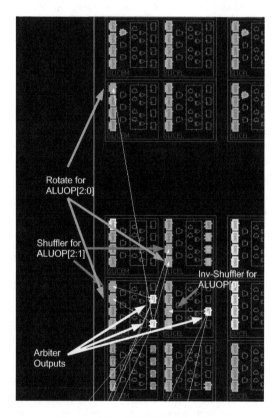

Figure 13: ALUOP Shufflers

	Unmodified LEON3	Modified LEON3	Overhead
LUT Usage	15,271	15,334	0.4%
DFF Usage	7,649	8,132	6.3%
Clock Speed	80.18MHz	80.48MHz	-0.4%

Table 1: Area and Performance Overhead

static timing reports are examined and compared against the unmodified version of the LEON3 design. As shown in Table 1, the static shufflers and the arbiter array has a very small resource overhead, using only 0.4% additional LUTs and 6.3% additional DFFs. There is no timing overhead. In fact the modified LEON3 runs slightly faster than the unmodified LEON3.

The last metric is the number simulation cycles the design must be simulated before an attacker can determine that the guessed arbiter outputs are incorrect. By simulating an instance of the modified LEON3 design with incorrect arbiter outputs until the process traps to an error state, the answer is determined to be 6.2585 milliseconds, or 6.2585 billion simulation cycles at 1 picosecond per cycle.

5. CONCLUSIONS

We have introduced a new approach to IC Digital Rights Management by combining directed NBTI aging and delay logic. As a proof of concept, we have implemented a basic form of the delay logic on an FPGA platform, and shown that the delay logic measures the relative speed differences in competing logic paths due to process variation. Furthermore, we have shown that the delay logic responds correctly to NBTI aging and recovery cycles. We also presented a case study to use static arbiter outputs and shuffling logic to add anti-cloning protection to a non-trivial FPGA design. We showed that with very little resource and no clock speed overhead, the LEON3 processor design can be made clone-proof.

6. REFERENCES

[1] M. A. Alam. A critical examination of the mechanics of dynamic NBTI for PMOSFETs. In *IEDM*, page 346. IEEE, 2003.

[2] Y. Alkabani and F. Koushanfar. Active hardware metering for intellectual property protection and security. In *USENIX Security*, pages 291–306, 2007.

[3] Y. Alkabani, F. Koushanfar, N. Kiyavash, and M. Potkonjak. Trusted integrated circuits: A nondestructive hidden characteristics extraction approach. In *Information Hiding Workshop*, pages 102–117, 2008.

[4] Y. Alkabani, F. Koushanfar, and M. Potkonjak. Remote activation of ICs for piracy prevention and digital right management. In *ICCAD*, pages 674–677, 2007.

[5] R. Anderson and M. Kuhn. Tamper resistance: A cautionary note. In *Proceedings of the 2nd USENIX Workshop Electronic Commerce*, pages 1–11, 1996.

[6] D. Angluin. Learning regular sets from queries and counterexamples. *Information and Computation*, 75(2):87–106, 1987.

[7] A. Baumgarten, A. Tyagi, and J. Zambreno. Preventing IC piracy using reconfigurable logic

barriers. *IEEE Design & Test of Computers*, 27(1):66–75, 2010.

[8] N. Beckmann and M. Potkonjak. Hardware-based public-key cryptography with public physically unclonable functions. In *Information Hiding Workshop*, pages 206–220, 2009.

[9] S. Blythe, B. Fraboni, S. Lall, H. Ahmed, and U. de Riu. Layout reconstruction of complex silicon chips. *IEEE Journal of Solid-State Circuits*, 28(2):138–145, 1993.

[10] M. Boehner. LOGEX - an automatic logic extractor form transistor to gate level for CMOS technology. In *Proceedings of the Design Automation Conference*, DAC '88, pages 517–522, 1988.

[11] L. Bossuet, G. Gogniat, and W. Burleson. Dynamically configurable security for SRAM FPGA bitstreams. In *Proceedings of the Parallel and Distributed Processing Symposium*, pages 146–153, April 2004.

[12] N. G. Bourbakis, A. Mogzadeh, and S. A. Mertoguno. Knowledge-based expert system for automatic visual VLSI reverse-engineering: VLSI layout version. *IEEE Transactions on Systems, Man and Cybernetics*, 32(3):428–436, 2002.

[13] D. Champagne, R. Elbaz, C. Gebotys, L. Torres, and R. B. Lee. Forward-secure content distribution to reconfigurable hardware. In *Reconfigurable Computing and FPGAs*, pages 450–455, 2008.

[14] T. Doom, J. White, A. Wojcik, and G. Chisholm. Identifying high-level components in combinational circuits. In *Great Lakes symposium on VLSI*, pages 313–318, 1998.

[15] S. Drimer. Volatile FPGA design security – a survey (v0.96), April 2008.

[16] J. Gaisler, E. Catovic, M. Isomaki, K. Glembo, and S. Habinc. GRLIB IP core user's manual. *Gaisler Research*, 2007.

[17] B. Gassend, D. Clarke, M. van Dijk, and S. Devadas. Silicon physical random functions. In *ACM Conference on Computer and Communications Security*, pages 148–160, 2002.

[18] M. C. Hansen, H. Yalcin, and J. P. Hayes. Unveiling the ISCAS-85 benchmarks: A case study in reverse engineering. *IEEE Design & Test*, 16(3):72–80, 1999.

[19] I. Hong and M. Potkonjak. Technique for intellectual property protection of DSP designs. In *International Conference on Acoustic, Speech, and Signal Processing*, pages 3133–3136, 1998.

[20] T. Kean. Method of using a mask programmed key to securely configure a field programmable gate array. United States Patent 7,240,218, 2001.

[21] M. Kearns and L. G. Valiant. Cryptographic limitations on learning boolean formulae and finite automata. In *ACM Symposium on Theory of computing*, pages 433–444, 1989.

[22] F. Koushanfar, G. Qu, and M. Potkonjak. Intellectual property metering. In *Information Hiding Workshop*, pages 81–95, 2001.

[23] J. Kumagai. Chip detectives. *IEEE Spectrum*, 37(11):43–48, 2000.

[24] J. Lach, W. Mangione-Smith, and M. Potkonjak. Fingerprinting digital circuits on programmable hardware. In *Information Hiding Workshop*, pages 16–31, 1998.

[25] J. Lach, W. Mangione-Smith, and M. Potkonjak. FPGA fingerprinting techniques for protecting intellectual property. In *Customom Integrated Circuits Conference*, pages 299–302, 1998.

[26] J. Lach, W. Mangione-Smith, and M. Potkonjak. Robust FPGA intellectual property protection through multiple small watermarks. In *Proceedings of the Design Automation Conference*, DAC '98, pages 831–836, 1999.

[27] J. Lach, W. Mangione-Smith, and M. Potkonjak. Fingerprinting techniques for field programmable gate array intellectual property protection. *IEEE Transactions on CAD*, 20(10):1253–1261, 2001.

[28] S. Mahapatra, M. A. Alam, P. B. Kumar, T. R. Dalei, D. Varghese, and D. Saha. Negative bias temperature instability in CMOS devices. *Microelectronic engineering*, 80:114–121, 2005.

[29] M. Majzoobi, F. Koushanfar, and M. Potkonjak. Techniques for design and implementation of secure reconfigurable PUFs. *ACM Trans. Reconfigurable Technol. Syst.*, 2:5:1–5:33, March 2009.

[30] M. Majzoobi, F. Koushanfar, and M. Potkonjak. Trusted design in FPGAs. In *Introduction to Hardware Security and Trust*, pages 195–230. Springer, 2011.

[31] J. T. McDonald, Y. C. Kim, and M. R. Grimaila. Protecting reprogrammable hardware with polymorphic circuit variation. In *Cyberspace Research Workshop*, pages 63–78, 2009.

[32] J. T. McDonald, E. D. Trias, Y. C. Kim, and M. R. Grimaila. Using logic-based reduction for adversarial component recovery. In *ACM Symposium on Applied Computing*, pages 1993–2000, 2010.

[33] R. Nakagaki, Y. Takagi, and K. Nakamae. Automatic recognition of circuit patterns on semiconductor wafers from multiple scanning electron microscope images. *Measurement Science and Technology*, 21(8):085501, 2010.

[34] R. Nijssen and J. Jess. Two-dimensional datapath regularity extraction. In *ACM/SIGDA Internation symposium on Physical Design*, pages 42–47, 1997.

[35] J. D. Parham, J. T. McDonald, M. R. Grimaila, and Y. C. Kim. A java based component identification tool for measuring the strength of circuit protections. In *Proceedings of the Sixth Annual Workshop on Cyber Security and Information Intelligence Research*, pages 26:1–26:4, 2010.

[36] G. Qu and M. Potkonjak. *Intellectual Property Protection in VLSI Design Theory and Practice*. Springer, 2003.

[37] D. S. Rao and F. J. Kurdahi. On clustering for maximal regularity extraction. *IEEE Transactions on CAD*, 12(8):1198–1208, 1993.

[38] R. L. Rivest and R. E. Schapire. Inference of finite automata using homing sequences. In *ACM Symposium on Theory of computing*, pages 411–420, 1989.

[39] J. Roy, F. Koushanfar, and I. Markov. EPIC: Ending piracy of integrated circuits. In *DATE*, pages 1069–1074, 2008.

[40] U. Rührmair, F. Sehnke, J. Sölter, G. Dror, S. Devadas, and J. Schmidhuber. Modeling attacks on physical unclonable functions. In *ACM Conference on Computer and Communications Security*, pages 237–249, 2010.

[41] D. Saab, V. Nagubadi, F. Kocan, and J. Abraham. Extraction base verification method for off the shelf integrated circuits. In *Quality Electronic Design*, pages 396–400. ASQED 1st Asia Symposium, 2009.

[42] S. Sirowy, G. Stitt, and F. Vahid. C is for circuits: capturing FPGA circuits as sequential code for portability,. In *ACM/SIGDA Symposium on Field programmable gate arrays*, pages 117–126, 2008.

[43] R. Torrance and D. James. Reverse engineering in the semiconductor industry. In *Customom Integrated Circuits Conference*, pages 429–436. IEEE, 2007.

[44] R. Torrance and D. James. The state-of-the-art in ic reverse engineering. *CHES, ser. Lecture Notes in Computer Science*, 5747:363–381, 2009.

[45] S. Trimberger. Trusted design in FPGAs. In *Proceedings of the Design Automation Conference*, DAC '07, pages 5–8, 2007.

[46] S. Wei, F. Koushanfar, and M. Potkonjak. Integrated circuit digital rights management techniques using physical level characterization. In *ACM Workshop on Digital Rights Management*, pages 3–14, 2011.

[47] S. Wei, A. Nahapetian, and M. Potkonjak. Robust passive hardware metering. In *ICCAD*, 2011.

[48] J. L. White. *Candidate subcircuit enumeration for module identification in digital circuits*. PhD thesis, Michigan State University, East Lansing, MI, 2000.

[49] J. L. White, A. S. Wojcik, M.-J. Chung, and T. E. Doom. Candidate subcircuits for functional module identification in logic circuits. In *Great Lakes Symposium on VLSI*, pages 34–38, 2000.

[50] Wikipedia. Flip-flop (electronics) — wikipedia, the free encyclopedia, 2011. [Online; accessed 26-September-2011].

Prototype and Evaluation of the CoRAM Memory Architecture for FPGA-Based Computing

Eric S. Chung, Michael K. Papamichael, Gabriel Weisz, James C. Hoe, Ken Mai

Carnegie Mellon University, Computer Architecture Lab (CALCM)

coram@ece.cmu.edu

ABSTRACT

The CoRAM memory architecture for FPGA-based computing augments traditional reconfigurable fabric with a natural and effective way for applications to interact with off-chip memory and I/O. The two central tenets of the CoRAM memory architecture are (1) the deliberate separation of concerns between computation versus data marshalling and (2) the use of a multithreaded software abstraction to replace FSM-based memory control logic. To evaluate the viability of the CoRAM memory architecture, we developed a full RTL implementation of a CoRAM microarchitecture instance that can be synthesized for standard cells or emulated on FPGAs. The results of our evaluation show that a soft emulation of the CoRAM memory architecture on current FPGAs can be impractical for memory-intensive, large-scale applications due to the high performance and area penalties incurred by the soft mechanisms. The results further show that in an envisioned FPGA built with CoRAM in mind, the introduction of hard macro blocks for data distribution can mitigate these inefficiencies—allowing applications to take advantage of the CoRAM memory architecture for ease of programmability and portability while still enjoying performance and efficiency comparable to RTL-level application development on conventional FPGAs.

ACM Categories & Subject Descriptors
C.0 [Computer System Organization]: System Architectures
General Terms: Design, standardization
Keywords: FPGA computing, memory architecture

1. INTRODUCTION

In the quest for energy-efficient computing, Field Programmable Gate Arrays (FPGAs) have emerged as a class of general-purpose accelerators to address the increasing demands for performance while reducing energy consumption. Despite their raw capabilities, today's commodity FPGAs are impractical as general-purpose computing devices. When developing an application for an FPGA, designers

are often confronted by: (1) low-level, error-prone hardware description languages (HDL), (2) "bare-bones" fabric with nothing but a sea of logic and I/O pins, and (3) low-level, vendor-specific interfaces and gateware that the application must be made compatible with.

CoRAM Memory Architecture. To address these limitations, the CoRAM memory architecture [1] is an endeavor to standardize and simplify how FPGA computing applications interact with memory and I/O, which is a critical step towards a portable FPGA abstraction. CoRAM presents a programmable, customizable view of memory that can be retargeted to different devices and platforms. The abstraction modifies the traditional FPGA's on-die SRAMs to act as in-fabric distributed portals to off-chip memory and I/O. A salient feature of CoRAM is the ability to program these customizable, on-die SRAMs using a software control thread that is portable and easy-to-use. Compared to the traditional approach where the FPGA memory hierarchy and I/O sub-system is hand-built at the RTL-level for each application, the CoRAM memory architecture can be used to efficiently support a broad range of applications.

Evaluating the Viability of CoRAM. The architectural features of CoRAM, however compelling, cannot be practical unless efficient underlying implementations are possible. In this paper, we investigate the extent to which CoRAM succeeds in serving as a performance- and cost-effective memory system replacement for FPGA-based applications. A critical aspect to be examined is whether the software-based control abstraction in CoRAM can adequately support FPGA applications with memory-intensive requirements. Our objectives require us to investigate the various ways in which the CoRAM memory architecture can be realized. At one end of the spectrum, CoRAM can be emulated on a conventional FPGA, albeit at the cost of soft logic area and performance. At the opposite end, general-purpose hard macro blocks can be embedded within conventional reconfigurable fabric to accelerate and to reduce the overhead of CoRAM operations. In this paper, we investigate both extremes and compare these implementations against traditional hand-built RTL designs on conventional FPGAs.

Prototyping Efforts and Results. Our investigation is supported by a full-featured prototype of a working CoRAM microarchitectural instance comprising: (1) a C-based language specification and compiler for software control threads, and (2) a highly parameterized RTL design of an optimized CoRAM microarchitecture retargetable to either standard cells (to model a future FPGA with hard

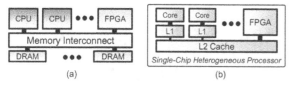

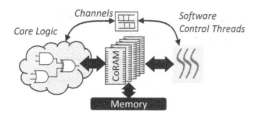

Figure 1: Assumed System Organizations.

Figure 2: CoRAM Program Model.

CoRAM support) or conventional FPGAs (as a soft logic emulation). We select three diverse applications to evaluate the hard versus soft CoRAM implementations. Our evaluation results suggest that a soft emulation of CoRAM falls short in large-scale, memory-intensive applications due to the high performance and area penalties incurred by the soft mechanisms. However, our results show that the introduction of hard macro blocks for data distribution in a future FPGA with CoRAM support can mitigate these inefficiencies—allowing applications to achieve performance and efficiency comparable to tuned applications on a conventional FPGA.

2. CORAM BACKGROUND

Assumptions. At the system level, CoRAM assumes the existence of FPGA-based accelerators co-existing with general-purpose processors on a shared memory interconnect (see Figure 1). Within an FPGA, CoRAM assumes that one or more load-store interfaces provide external memory accesses at the boundaries of the fabric. The same assumptions are similarly applicable in a single-chip hetereogeneous multicore where cores and fabric co-exist on the same die.

CoRAM Program Model. When developing an application with CoRAM support, the user perceives a simplified view of fabric as depicted in Figure 2. The core logic component is an isolated, contiguous region of fabric that preserves the hardware-centric view familiar to designers that target FPGAs today. Applications that are mapped to core logic can be programmed with any hardware synthesis language supported by contemporary tools from low-level RTL (e.g., Verilog) to high-level languages (e.g., C-to-gates, Bluespec). To create a uniform abstraction that can be made portable across different devices and platforms, the CoRAM program model restricts all communication by the core logic to the external environment through the embedded CoRAM blocks shown at the edges of core logic in Figure 2. The embedded CoRAM blocks are user-instantiated, parameterizable SRAM modules that follow a similar usage paradigm of conventional embedded SRAMs [3]. On one hand, like SRAM blocks, CoRAM blocks offer customizable high-bandwidth storage, provide deterministic access ports to independent banks with local addresses, and can be composed with flexible aspect ratios to match the requirements of the applica-

tion. On the other hand, unlike passive SRAM blocks, the contents of embedded CoRAM blocks are dynamically managed (such as loading and unloading against external main memory) using software control threads depicted in the right of Figure 2.

Software Control Threads. Software control threads form a fabric-distributed collection of logical, asynchronous control state machines for managing and mediating the data transfers between embedded CoRAM blocks and the edge memory interfaces. The software control threads and core logic are asynchronous peer entities in charge of data marshalling and computation, respectively; they communicate over bidirectional command queues. At a high level, the threads can be viewed as programmable mechanisms for prefetching an application's required data from the edge memory interface to the fabric-distributed embedded CoRAM blocks. At the lowest level, threads describe an ordered sequence of memory commands directed by control flow. The application developer relies solely on instantiated control threads to access shared memory and I/O from the beginning to end of computation.

Control threads can be used to express a rich variety of memory access patterns (e.g., random access, streaming, etc.) while maintaining portability. For example, a random-access cache controller could be implemented by combining soft logic and embedded CoRAM blocks (serving as the data array) and a control thread that implements a miss handler to memory. A stream FIFO could also be implemented by instantiating an embedded CoRAM block as a circular buffer with an associated control thread that fills or drains the buffer as needed. These portable memory building blocks can be expressed succinctly with relatively few lines of code (in most cases, under 100 lines of C) [1].

3. CORAM PROTOTYPE

There are two requirements to execute an application description in CoRAM: (1) interpreting or synthesizing high-level control threads into state machines, and (2) transporting data efficiently between memory interfaces and fabric-distributed embedded CoRAM blocks. To facilitate these, we developed a full-featured RTL prototype and control thread compiler for the CoRAM memory architecture. Our framework comprises: (1) the CoRAM Control Compiler (CORCC), which is an LLVM-based backend[1] that synthesizes C-based control thread programs into hardware finite state machines, (2) CONNECT [4], a flexible network-on-chip (NoC) generator tuned for the Virtex-6 architecture, and (3) pre-optimized macro blocks in Verilog for request handling, scoreboarding, and data distribution for the embedded CoRAM blocks. The RTL generated by our framework can be synthesized to standard cells for estimating the area, power, and performance in a hypothetical future FPGA with hardwired CoRAM support.

Figure 3 shows how a high-level CoRAM application is mapped into the synthesizable RTL generated by our framework. The embedded CoRAM blocks instantiated within the application are mapped into physical macro blocks called clusters, which aggregate up to 64 homogeneous 1024x32b SRAMs[2] into a single node. Embedded CoRAM blocks

---

[1]Low-Level Virtual Machine [2].

[2]Corresponding to the default aspect ratio of a typical FPGA BlockRAM [3].

	Max-Cfg Cluster		Mesh Router	
	Soft	Hard	Soft	Hard
LUTS+LUTRAMS	7615	-	6002	-
FFs	4741	-	1144	-
Clock (MHz)	108	840	125	610
SRAM Area (mm^2)	-	0.57	-	0.23
Die Area (mm^2)	23	0.74	18.1	0.3

Table 1: 65nm Characteristics of Single Cluster and Mesh Router.

	LUT	FF	BRAM	MHz
Microblaze (min-area-cfg)	1210	973	4	161
Stream Loop	155	118	0	345
Matrix Matrix Multiplication	2581	2802	0	192
Non-blocking Cache Miss Handler	242	316	0	354
Stream FIFO Producer Thread	544	523	0	204

Table 2: Control Thread Synthesis (Virtex-6, -2).

with aspect ratios larger than 1024x32b are constructed by spanning across multiple SRAMs within a cluster.[3] The clusters are connected to external memory links through a CONNECT-generated NoC. Figure 3 illustrates how each of the clusters are attached to one or more Control Finite State Machines (C-FSM) generated by CORCC from the control threads. At run-time, the C-FSMs issue memory commands to the NoC—the subsequent memory responses are collected at the clusters, which are responsible for steering and aligning the memory data to the destination CoRAM blocks.

Soft vs. Hard Macro Blocks. The C-FSMs, clusters, and NoC are performance- and area-critical macro blocks that can either be soft-emulated in today's FPGAs or embedded as hard logic in a future FPGA. Although we omit the details, a substantial effort was invested in optimizing the macro blocks for the Virtex-6 architecture. Table 1 compares the cost of a single maximally-configured cluster and mesh router for both hard and soft implementations.[4] Note that a maximally-configured cluster supports up to 64 SRAMs, 16 control threads, and 128 concurrent transactions to memory. In practice, a soft cluster can be configured less aggressively depending on the application—as little as 2KLUTs. When normalized to die area, both the hard cluster and mesh router achieve an order-of-magnitude improvement in area efficiency as expected. The hard macro blocks also operate at higher clock frequencies (e.g., 610MHz vs. 125MHz for the mesh router). Table 2 further shows the soft logic area of various control threads synthesized to FPGA fabric using CORCC. Across most applications, the area consumption is modest when compared to a minimally-configured Microblaze core, which suggests that control threads can be supported practically in conventional fabrics.

4. EVALUATION

Methodology. For our evaluation, we study the use of CoRAM in three diverse RTL applications reflecting

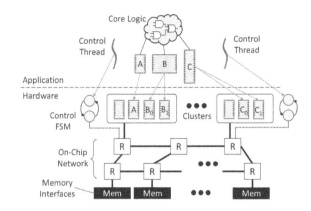

Figure 3: CoRAM Microarchitecture.

		500K	1M	2M	4M
	Technology	45nm	32nm	22nm	16nm
	Die Area (mm^2)	600	600	600	600
	LUTs (K)	500	1000	2000	4000
	Frequency (MHz)	200	200	225	250
	Bandwidth (GB/s)	25.6	51.2	102.4	204.8
	4kB CoRAM blocks	1024	2048	4096	8192
Soft	Cluster/NoC clock (GHz)	0.2	0.2	0.225	0.25
	Cluster/NoC link width	128	128	128	128
	Clusters/Nodes	Application-dependent			
Hard	Cluster/NoC clock (GHz)	0.8	0.8	0.9	1
	Cluster/NoC link width	128	128	128	128
	Clusters	16	32	64	128
	Nodes (clusters+DRAM links)	20	40	80	160
	SRAMs per Cluster	64	64	64	64
	Die Area Ovhd (%)	1.7	1.7	1.4	1.4

Table 3: Evaluation Parameters.

bandwidth-bound, compute-bound, and latency-bound characteristics [1]: (1) Black-Scholes Options Pricing (BS), (2) Matrix-Matrix Multiplication (MMM), and (3) Sparse Matrix-Vector Multiplication (SpMV). Our evaluation considers the cost and performance of the CoRAM memory architecture across multiple dimensions: (1) VLSI technology, (2) NoC topology, and (3) hard vs. soft logic implementations. Table 3 shows the parameters and characteristics of the selected FPGA configurations. All configurations (with or without hard CoRAM support) assume the existence of hard memory controllers that provide high external memory bandwidth to the fabric. All soft-logic designs are tested against different NoC topologies (ring, mesh, crossbar) to reflect the option of customization in a soft-logic implementation. Furthermore, in the soft designs, the number of clusters and SRAMs per cluster are tuned for each respective application. The hard implementation assumes a mesh topology with a fixed number of clusters and SRAMs per cluster.

Results. Figure 4 (top) shows the performance trends for all applications. On the x-axis, labels prepended with "soft" indicate designs using a soft-logic implementation on a conventional FPGA. Designs labeled "hard" indicate FPGAs with dedicated CoRAM support operating at 4X the clock rate of the soft-logic design. The right-most design point of each graph shows the results for an "ideal" application running on a conventional FPGA that can access external DRAM with no on-chip delay or soft logic area overheads. Figure 4 (bottom) shows the LUT area breakdown and LUT-

---

[3]The maximum size of the cluster can either be set automatically or configured by the user.
[4]FPGA results were obtained with XST 13.1 for the Virtex-6; ASIC results were obtained using CACTI 6.5 [5] and Synopsys Design Compiler configured with commercial 65nm standard cells.

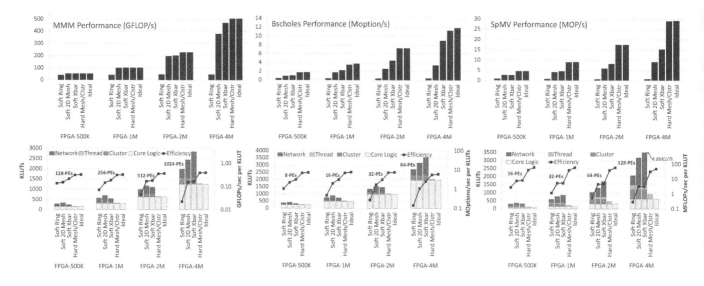

Figure 4: Performance and Area Efficiency Trends.

normalized performance for the same corresponding performance points.

What Is The Gap Between Soft versus Hard? The overall trends in Figure 4 show a gap in performance and efficiency between soft and hard implementations of CoRAM. In MMM, the performance is comparable across all design points (except for the soft ring network, which suffers from high contention). When normalized to area, however, a significant gap in efficiency (about 2X) separates soft versus hard implementations of CoRAM. The NoCs expend considerable soft logic area in buffering, which is the largest contributer to overhead. In the more memory-intensive applications such as BS and SpMV, the biggest impact on performance is the increased queuing delay as a result of higher latency and contention in the soft data distribution mechanisms. This impacts SpMV the greatest (which is latency-sensitive)—resulting in a 2X gap in performance between the soft versus hard CoRAM implementations.

Do Software Control Threads Limit Efficiency or Performance? A notable result of Figure 4 (bottom) is that the synthesized control threads constitute only a relatively small fraction of the overall area. What also stands out is that with an efficient implementation of CoRAM using hard macro blocks, the use of software-based control threads does not limit the peak performance potential of the various applications. Both SpMV and Black-Scholes, for example, were able to achieve bandwidth-limited performance even though the logic used to generate their memory accesses were described using high-level, C-based control threads. These results support the hypothesis that a high-level software-based abstraction for memory management does not fundamentally limit the performance potential of memory-intensive FPGA-based applications.

Hard CoRAM vs. Conventional FPGA. A key result of Figure 4 is that applications mapped to a hard implementation of CoRAM are capable of achieving performance and efficiency comparable to tuned applications on conventional FPGAs (labeled "ideal"). Recall, the measured results of "ideal" applications are based on simulations and synthe-

sis of core logic that do not incur any overheads in latency or area from memory control logic or data distribution between the core logic and the external memory interfaces. This suggests that hard macro blocks for CoRAM can reduce the tuning effort needed by designers when optimizing the memory accesses for an application.

5. CONCLUSIONS

This paper presented a full-featured RTL prototype of the CoRAM memory architecture for FPGA-based computing. Our prototype enabled us to investigate designs across the continuum—from soft-logic emulation on conventional FPGAs to hard macro blocks in future FPGA fabrics. Despite our best efforts, our soft implementations of CoRAM fell short of being practical in large-scale, memory-intensive applications. Our results show that the hardening of macro blocks for data distribution in CoRAM enables application development using a high-level software memory abstraction to achieve performance and efficiency that is comparable to optimized RTL development on a conventional FPGA.

6. ACKNOWLEDGEMENTS

Funding for this work was provided in part by NSF CCF-1012851 and by Altera. We thank the anonymous reviewers, members of CALCM, and Chuck Thacker for their comments and feedback. We thank Xilinx and Altera for their FPGA and tool donations. We thank Bluespec for their tool donations and support.

7. REFERENCES

[1] E. S. Chung, J. C. Hoe, and K. Mai. CoRAM: An In-Fabric Memory Architecture for FPGA-based Computing. In *Proceedings of FPGA'11*.

[2] C. Lattner and V. Adve. LLVM: A Compilation Framework for Lifelong Program Analysis & Transformation. In *Proceedings of CGO'04*.

[3] T. Ngai, J. Rose, and S. Wilton. An SRAM-programmable field-configurable memory. In *Proceedings of CICC'95*.

[4] M. K. Papamichael and J. C. Hoe. CONNECT: Re-Examining Conventional Wisdom for Designing NoCs in the Context of FPGAs. In *Proceedings of FPGA'12*.

[5] D. T. S. Thoziyoor, D. Tarjan, and S. Thoziyoor. Cacti 4.0. Technical Report HPL-2006-86, HP Labs, 2006.

A Coarse-grained Stream Architecture for Cryo-electron Microscopy Images 3D Reconstruction

Wendi Wang[†‡§], Bo Duan[†‡], Wen Tang[†‡], Chunming Zhang[†], Guangming Tan[†¶], Peiheng Zhang[†], Ninghui Sun[†¶]

†High Performance Computer Research Center, Institute of Computing Technology, CAS
¶State Key Laboratory of Computer Architecture, Institute of Computing Technology, CAS
‡Graduate University of Chinese Academy of Sciences
{wangwendi, duanbo, tangwen, zhangchunming, zph}@ncic.ac.cn, {tgm, snh}@ict.ac.cn

ABSTRACT

The wide acceptance and the data deluge in the bioinformatics and medical imaging processing require more efficient and application-specific systems to be built. Due to the recent advances in FPGAs technologies, there has been a resurgence in research aimed at the design of special-purpose accelerators for standard computer architectures. In this paper, we exploit this trend towards FPGA-based accelerator design and provide a proof-of-concept and comprehensive case study on FPGA-based accelerator design for a single-particle 3D reconstruction application in single-precision floating-point format. The proposed stream architecture is built by first offloading computing-intensive software kernels to dedicated hardware modules, which emphasizes the importance of optimizing computing dominated data access patterns. Then configurable computing streams are constructed by arranging the hardware modules and bypass channels to form a linear deep pipeline. The efficiency of the proposed stream architecture is justified by the reported 2.54 times speedup over a 4-cores CPU. In terms of power efficiency, our FPGA-based accelerator introduces a 7.33 and 3.4 times improvement over a 4-cores CPU and an up-to-date GPU device, respectively.

Categories and Subject Descriptors

C.3 [**SPECIAL-PURPOSE AND APPLICATION-BASED SYSTEMS**]: Signal processing systems

General Terms

Design, Performance

Keywords

Cryo-electron microscopy, FFT, FPGA, stream processing, memory access patterns

---

§Corresponding Author: Wendi WANG, No.6 KeXueYuan South Road, ZhongGuanCun, Beijing, P. R. China, 100190.

1. INTRODUCTION

Due to the limiting factors of the memory wall and the power wall, in recent years, the performance gains from general-purpose CPUs are diminishing. Using of alternative devices rather than general-purpose CPUs to accelerate computing-intensive applications has gained renewed interests. By introducing specialized co-processors, configurable data path and customized memory system with respect to applications' characteristics, application-specific hardware design provides a promising way to tweak performance within a given silicon budget.

However, limited by the device capacity and the high implementation cost, the typical uses of FPGAs are usually restricted to either simple and small applications, such as RSA and FIR, or key kernels of applications, such as the molecular dynamics [1]. Recently, due to the advances in modern FPGA technologies and the emergence of fast floating-point and elementary function libraries [2], there has been a resurgence in research aimed at accelerator design that leverages FPGAs to accelerate large-scale scientific applications. We exploit this trend towards FPGA-based accelerator design and propose a data path configurable stream architecture that focuses on separated design strategies for computing and memory flows. In particular, we introduce our work on accelerating a large-scale scientific application, called *EMAN* [3], which is an open source software package for single-particle 3D reconstruction from Cryo-electron microscopy images, on our customized FPGA accelerator card.

EMAN is composed of hundreds of time-consuming kernels showing diverse computing features. How to integrate kernel implementations under a unified framework determines the extent to which the application can be accelerated on FPGA. One the other side, the overall execution time of applications is often dominated by the efficiency of their data access patterns [4]. One of the main complexities for application-specific memory system design is how to support the kernel dominated data access patterns. An alternative solution to this problem is to take into account how to share data across kernels. However, the integration of multiple kernels with various functions under a unified framework and providing an efficient data exchange mechanism among them are both proven to be challenges. To address these concerns, we introduce a hybrid memory controller, which is characterized by its support for explicit pattern-based data accesses. The memory system can be viewed as both a framework for creating data access patterns and a runtime system that assists the construction of data flows.

Our stream architecture is designed by first offloading computing-intensive software kernels to dedicated hardware modules. Then configurable computing streams can be constructed by arranging the hardware modules and bypass channels sequentially. Various software functions can be implemented by a single computing stream by activating different hardware modules. For complex work, the computing flow needs to be executed and emulated in a multiple step procedure, in which the data path of each step differs. This work provides a proof-of-concept and comprehensive case study on FPGA-based accelerator design for a large-scale scientific application, which emphasizes the importance of optimizing data access behaviors. The contribution of this work is three-fold:

- We propose a coarse-grained stream architecture for accelerating a real complex application. The stream architecture is designed based on the key observations of classification of various kernels. Built upon typical computing and data access modules, the modularized design method improves coarse-grained modular reuse.

- We apply software-hardware co-design approach to build an efficient map between EMAN algorithm and FPGA architecture. Especially, we develop a novel hybrid memory controller featuring the ability to do computing dominated and pattern-based data accesses.

- Our customized accelerator is implemented as a FPGA accelerator card. The efficiency and feasibility of the proposed architecture are justified by the reported 2.54 times speedup over a 4-cores CPU. Comparing to a GPU-CUDA based version, our accelerator improves power cost by 3.4 times.

The rest of this paper is organized as follows. The background of single-particle 3D reconstruction and *EMAN* are introduced in Section II, followed by introduction to system design in Section III. Section IV explains the software and hardware co-design flow. The experimental results are discussed in Section V, whereas Section VI lists related work. Finally, we conclude our work in Section VII.

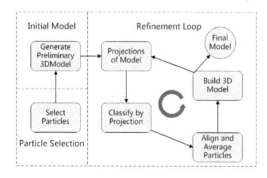

Figure 1: Flowchart of the Cryo-EM image analysis

2. SINGLE-PARTICLE 3D RECONSTRUCTION AND EMAN

EMAN is a software package designed to handle nearly all aspects of the single-particle 3D reconstruction, such

---

Algorithm 1: Reconstruction Algorithm

```
1  while model T not converged do
2      generating M projections of T;
       // particle classification
3      foreach i ∈ N particles in images do
4          for j ∈ M projections do
               // rotationally and translationally aligns
               // each particles to projected reference
5              RTFAlign(i, j);

       // class averaging
6      foreach particles i in class j do
7          RTFAlign(i, average_j);
       // generate an initial class average
8      InitialAve();
9      foreach particles i do
10         RTFAlign(i, average_initial);
       // remove particles with less similarity
11     Remove();
12     Build3D(); // build 3D model
```

Algorithm 2: RTFAlign Algorithm

```
   // step 1. rotational alignment
1  MCF (i);  // autocorrelation function
2  MCF (i'); // flip's autocorrelation function
3  unwrap (); // rectangular coordination to polar one
4  CCFX (); // cross correlation function on x-dimension
5  Rotate (); // rotates with the best rotation angle
   // step 2. translational alignment
6  CCF (); // cross correlation function with reference
7  Translate (); // translates image with the maximal CCF
   // step 3. score and select the most similarity one
8  dot (); // scores the rotational&translational alignment
```

---

as particle selection, particle alignment, 3D model projection/reconstruction and etc. The single-precision floating-point format is used to store input data and intermediate result. In single-particle 3D reconstruction for Cryo-EM images, the algorithm first generates a preliminary 3D model based on amount of particles (images), which are selected from scanned raw micrographs. The preliminary model is used as the starting point for the refinement loop. A refinement procedure for the final 3D image is a model-based iteration. True convergence is achieved when the model remains unchanged for several successive iterations. The refinement loop outlined in the right half of Figure 1.

Algorithm 1 illustrates the pseudo-code of the refinement algorithm. The iteration of refinement is a sequential process because current refinement depends on the model generated in previous iteration. However, abundant of parallelism is observed within each refinement procedure. Since the operations of classification and averaging are similar, the same parallelization can be applied. For simplicity of presentation, we only present the analysis of classification (line 3-5). Note that a common procedure of particle classification is rotational and translational alignment (RTFAlign in Algorithm 2), which occupies over 95% of total execution time. Therefore, in this paper we focus the discussion on FPGA acceleration of the particle alignment only.

Even accounts for only a small fraction of *EMAN*, particle alignment still contains hundreds of kernels and tens of parameters. However, based on partial evaluation and runtime profiling, we can classify the 23 identified kernels into

Name	Category	Description
MCF	Computing	Autocorrelation
CCFX	Computing	Cross correlation on x-dimension
CCF	Computing	Cross correlation with reference
Unwrap	Computing	Rectangular coordination to polar one
Rotate	Computing	Rotates with the best rotation angle
Translate	Computing	Translates image with the maximal CCF
DOT	Computing	Scores the rotational&translational alignment
Clip & Zero Padding	Memory	Change the size of images by clipping and padding
Rot180	Memory	Rotate image by 180°
hFlip	Memory	Flip images horizontally
Shift	Memory	Translate image by given 2D offset
Bit Reversal & Transposition	Memory	Used in FFT kernels
Matrix Transposition	Memory	Used in the row-column 2D FFT kernels
RTFAlign	Stream	Align images using RTAlign and hFlip
RTAlign	Stream	Align images using Rotate, Translate, CCFX, CCF and DOT
MakeRFP	Stream	Calculate the rotating footprint using MCF and Unwrap

3 categories: computing kernels, memory kernels and flow kernels. For simplicity of presentation, as given in Table 1, we combine some computing kernels into one kernel, thus reduce the number of computing kernels to 7.

3. MODULE-BASED SYSTEM DESIGN

A key observation underlying our stream architecture is that the kernels of *EMAN* can be broadly classified into 3 categories: computing, memory and stream.

The design of our system is geared towards architecture support that assists the implementation of each kernel category. The kernels in *EMAN* will be mapped to dedicated hardware modules on FPGA. Computing modules are used to wrap arithmetic operations, whereas memory modules are used to implement data accesses as well as related address calculation. The stream module, which incarnates concrete program functions, is composed of several computing and memory modules. In following sections, the stream architecture will be discussed in detail, which covers the topic of how to extract and classify target kernels for acceleration from *EMAN*, how to implement each kind of hardware module and how to map a complex computing flows to a stream module in hardware.

3.1 Kernel classification

There are hundreds of kernels in *EMAN*, along with computing-intensive kernels, there also exist kernels that merely containing either trivial data accesses or kernel invocations. Table 1 summarizes the data-parallel kernels in *EMAN* and the workload descriptions.

- **Memory**: Memory kernels are of limited use in their own right, by which we mean that they must be combined with computing kernels to deliver on practical functions. However, the efficiency of memory access can exert profound impacts on overall system performance. For example, FFT is the only kernel that uses the bit reversal memory access pattern, which gives poor data locality on CPU. The data access patterns, which are extracted by analyzing the memory address sequences of computing kernels, will be implemented with a unified data flow module (DFM). By introducing the DFMs, it becomes possible for computing mod-

ules to do pattern-based data accesses. The topic of how to extract data access patterns is covered in Section 3.3, whereas Section 4.3 introduces the DFM in detail.

- **Computing**: With respect to loop boundaries, computing kernels are manually composed of one or more loop statements. Instead of implementing them as heterogeneous and inflexible modules that directly work on raw memory controller interface, two additional steps are used to facilitate the implementation of computing kernels: 1) Data access patterns are expressed in the form of either fix-step counters (regular patterns) or mapping tables between iteration indices and requested memory addresses (irregular patterns); 2) All data access patterns will be implemented collectively with the DFM introduced previously. Section 3.4 discusses various issues related to kernel implantation.

- **Stream**: Stream kernels provide the specification of assembling computing and memory kernels to form large and complex computing streams. The computing flow of *EMAN* is relatively simple and regular, which makes it possible to construct computing streams as linear pipelines. Section 3.4.3 gives a concrete example of implementing the *MakeRFP* stream kernel.

3.2 System Architecture

Before we introduce the hardware implantation in detail, let us first take a glimpse of the system architecture. Figure 2 gives a broad view of our stream architecture, on which the aforementioned 3 kernel categories will be implemented. The core of the system is a hybrid memory controller, whereby on-chip and off-chip memory devices are managed with a single virtual address space. Different memory devices can be explicitly accessed by specifying exclusive memory addresses. A unique feature of the hybrid memory controller is the inclusion of 2 dedicated data flow modules (DFMs), which can be used to map various data access patterns extracted from computing kernels. Intuitively, the overhead of data transfer can be hidden by overlapping them with the computing. In order to achieve this overlap, we integrate a data pre-fetch unit in the memory controller, which can be

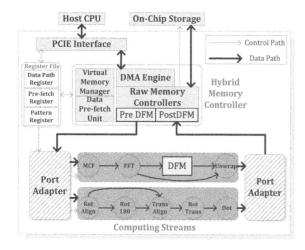

Figure 2: System architecture overview.

configured to do forward as well as backward pre-fetching at the granularity of image line. Section 4.2 gives more details about the hybrid memory controller.

The configurable computing streams are constructed by arranging the hardware modules and bypass channels to form a linear deep pipeline. Figure 4 illustrates the proposed computing stream to accelerate the RTAlign kernel. Along with two DFMs that lie in the memory controller, DFM can also be placed in the computing stream. For example, in Figure 4, there are two data re-order modules (used for data flow permutations) in the 2D FFT kernel. Multiple computing streams can coexist with each other, the port adapters are used to switch the data path between memory controller and computing streams.

Our system is configurable in two aspects by means of writing corresponding control registers: 1) The data path of the computing stream can be controlled to bypass some stages. As a result, a computing stream can be used to implement various program functions; 2) By configuring the DFMs and the data per-fetch unit, the memory controller can be controlled to do pattern-based data access.

3.3 Separating computing flow from data flow

In practice, most applications, including *EMAN*, are rarely developed to take the aforementioned 3 kernel types into consideration. It is the responsibility of the compilers and the cache system to optimize the computing and data accesses. Therefore, in most applications, it is a common phenomenon that computing and data accesses are entangled with each other.

The separation between computing flow and data flow manifests itself across two dimensions: 1) The performance of applications is deeply influenced by the ability to overlap computing with data accesses [4], by separating data flow from the computing flow, some advanced data pre-fetching and reordering strategies can be utilized; 2) A lot of data accesses of an application are highly structured, for example, data accesses within loops are often subscripted by loop indices, which results in either sequential or stride data access patterns. The cost to implement computing kernels can be greatly reduced by insulating memory read/write operations for separate consideration. In this way, computing kernels can be abstracted as black boxes with FIFO data in and FIFO results out. However, at the first place, extract-

ing data access patterns from a large computing flow is still a challenge. The good news is that, the kernel partitioning strategy (in Section 4.1.2) leads to dividing the program in loop statements. Therefore, the analysis of data access patterns can be simplified and confined within loop statements.

We developed a LLVM loop pass to facilitate the analysis process. Data blocks (images) will be clustered into arrays with the data layout optimization (in Section 4.1.3), which makes it possible to use memory addresses to identify data accesses to specific data types. By recording the memory addresses issued in loop statements, regular and irregular data access patterns can be respectively expressed as fixed-step counters and lookup tables that map iteration indices to the memory addresses in loop statement. Therefore, the memory addresses will be calculated statically, and computing kernels can dispense with the overhead of data accesses and get a simplified view of the computing flow. Computing kernels, which contain stateless loops (such as vector addition) only, can be implemented as simple arithmetic modules that operates on a continues data flow. However, it is still need to maintain registers and loop counters for stateful loops with loop-carried dependency (vector summation) and some initialization logics (image filters). In our system, data accesses and related memory address calculations will be offloaded to the data flow modules (DFMs) in memory controller, which pre-fetch, reorder and push data to each computing kernel in its expected order. Computing stream construction can be simplified as a process of instantiating computing kernels and selecting corresponding data access patterns supported by DFMs in memory controller.

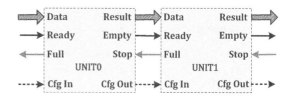

Figure 3: Interface of hardware computing modules.

3.4 Kernel implementation

Due to the complexity of *EMAN*, as explained in previous sections, it is unpractical for us to implement the entire computing flow as a single unit. On the other hand, considering the high design and debugging cost of using FPGAs, we argue that it is not a cost effective way to implement some kernels of *EMAN* on FPGAs, such as 3D-FFTs, heuristic particle selection and etc. Therefore, the performance of the FPGA-based accelerator will be dictated to a large extent by the ability to evaluate, extract and offload the most beneficial parts of the application. Runtime profiling indicates that, the *RTAlign* kernel that occupies over 95% of total execution time is the foundation, on which the process of particle classification (*Classesbymra*) and alignment (*Classalign2*) are built. In following part, we limit our discussion to FPGA acceleration of the *RTAlign* kernel only.

In follow sections, three representative kernels, mixed radix 2D FFT, matrix rotation by 180°and the RTAlign stream kernel will be used to demonstrate how to implement each kind of software kernel in our system.

3.4.1 Computing kernel

The computing kernel implementation is built upon a one-

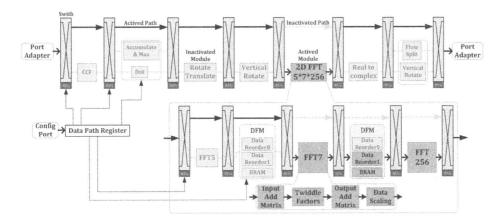

Figure 4: Computing stream for accelerating the *RTAlign* kernel.

to-one correspondence between the loop statements and the computing module (circle node) in Figure 2. In order to reduce design cost and increase modular reuse, the computing modules are implemented and wrapped in a unified interface. Illustrated in Figure 3, the unified interface is composed of standard FIFO signals, flow control signals and other signals used for kernel configuration. The advantage of partitioning and implementing the computing flow as separated kernels with common interface is that it gives us many coarse-grained building blocks that can be flexibly composed to form complex computing streams.

Based on the unified module interface, we developed an automatic HDL core generator framework with Matlab, which can be used to translate given data flow graph (DFGs) of loop statements (generated by GCC in the DOT file format) to HDL implementations. First, operators of the DFG are scheduled to different pipeline stages with respect to data dependency. Next, data flow permutations between stages will be extracted and expressed as sparse matrix, in which the number of rows and columns corresponds to the number of input and output variables of each pipeline stage, respectively. Finally, the pipelined DFG will be used to generate the Verilog HDL descriptions. The strength of the proposed core generator is justified by the ability to generate prime-radix FFT cores in our previous work [13]. There are plenty of existing work related to FFT design on FPGAs, however the issues of prime-radix FFT, 2D FFT and real FFT are rarely discussed. Therefore, it is needed to implement the FFT kernel by ourselves.

The data path of the computing stream can be configured by configuring the bypass switches in Figure 4. Therefore, different factorizations of a FFT process will be mapped to different paths flow through the computing stream. For example, the lower part of Figure 4 gives the structure of a 8960-points FFT pipeline, which is degraded to execute 1792-points FFT in current configuration. In current work, partial reconfiguration of individual hardware modules is not yet considered. With trivial overhead of data path switching, modules can be shared across configuration. We argue that with proper organization of the computing flow, the frequency of reconfiguration can be minimized or avoided.

3.4.2 Memory kernel

In order to improve alignment sensitivity, along with the reference particle, input images will be compared with up to 3 mirror images in *EMAN*. In CPU implementation, these mirror images are generated by time-consuming memory copies, which introduce misses in data cache and exert severe impact on application performance. With the support of our memory controller, the Rot180 kernel in Table 1 can be implemented in following steps: 1) Backward pre-fetch image line-by-line; 2) Buffer an entire line; and 3) Output the buffered line in reverse order. In this way, only one copy of reference image is stored in memory, other mirror images can be generated on-the-fly.

Table 2: Computing steps of the *MakeRFP* kernel.

#	Function	Activated Modules	Memory Controller[1]
1	1D FFT	2D FFT	I: interleave/ O: column write/ deinterleave
2	1D FFT	2D FFT/ Post-Vertical Rotate	O: column write/ deinterleave
6	Filter/MCF	CCF	I: interleave/2-Op.[2] O: column write
4	1D IFFT	2D FFT/ Pre-Vertical Rotate	O: column write
5	1D IFFT	R2C	I: interleave
6	1D IFFT	2D FFT/Flow Split	
7	Unwrap	RotateTranslate	I: clip/random access

[1] I for input, O for output memory access patterns.
[2] Reading two operands from two separated addresses.

3.4.3 Stream kernel

Figure 4 illustrates the computing stream built to accelerate the *RTAlign* kernel. The computing and memory modules are separated by runtime configurable bypass switches. The data flow modules (DFM) in the 2D FFT kernel are used to support data flow permutations between FFT stages. Along with the control signals for function selection (for example, the module for dot production can also be configured to do accumulation), signals for switch configuration are wrapped in runtime configurable registers. Different functions can be fulfilled by activating the required modules while bypassing the others.

If a computing flow is complex and cannot be implemented with a single data path configuration, the function can be emulated by means of activating the computing stream multiple times with a different data path each time. For example, it is needed to configure the stream pipeline 6 times to implement the *MakeRFP* kernel. Table 2 gives the function of each step, the activated modules and the involved data access patterns in detail. The 2D FFT/IFFT process is based on the row-column algorithm, which requires invoking the 1D FFT module twice.

4. HARDWARE/SOFTWARE CO-DESIGN

In order to achieve the best performance, the original computing flow of *EMAN* needs to be redesigned to be aware of the architecture features of FPGAs. We resort to a software and hardware co-design flow, which addresses the problem from both the software side (Section 4.1) and the hardware side (Section 4.2), to deliver on our performance goals.

4.1 Application redesigning

The computing flow of *EMAN* can be redesigned as a phased computing stream, which coincides with the fine-grained configurability of the FPGA and enables the possibilities to build coarse-grained deep pipeline for achieving high spatial parallelism. However, *EMAN* is largely written with C++ language, which heavily relies on templates, objects and pointers. These language features are great for high-level application design; however, they are in conflict with the semantics and syntax of Hardware Description Language (HDL), on which the FPGA designs are built. Therefore, it is needed to introduce computing flow modification that removes all high-level language features that got in the way of FPGA acceleration. On the other hand, we rely on loop fusion/fission and data layout optimizations to reduce the number of kernel stages and improve data locality, respectively.

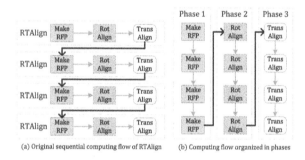

(a) Original sequential computing flow of RTAlign (b) Computing flow organized in phases

Figure 5: Modification on computing flow. The modified computing flow is organized in phases.

4.1.1 Computing flow optimization

The optimization of the computing flow consists of 3 steps: 1) Code rewriting that removes high-level language features and unreachable code (different computing paths); 2) Explicitly memory allocation that determines the input and output addresses for kernel invocations; 3) With respect to function boundaries, the original computing flow will be divided into phases. Among the motivations for computing flow partitioning, here are 2 that particularly stand out. First, a small FPGA can be used to simulate a large computing flow that otherwise would be too big to fit in. Only a part of a large computing flow needs to be presented on FPGA each time, the area consumption can be reduced considerably. Second, for a periodic computing flow (loop statements), the overhead of pipeline fill/drain as well as kernel invocation can be reduced. For example, as illustrated

in Figure 5(a), the computing flow of *RTAlign* consists of 3 functions: *MakeRFP*, *RotAlign* and *TransAlign*, which will be invoked periodically and sequentially for each image. By applying the computing flow partitioning, Figure 5(b) shows that it is possible to invoke each kernel only once. As a result, if each of the 3 kernels needs a separated configuration, the number of FPGA configuration can be reduced from 12 to 3. The importance of doing computing flow partitioning will be further justified in following section.

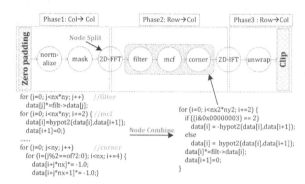

Figure 6: Coarse-grained DFG of *makeRFP* and the illustration of node split and node combine.

4.1.2 Kernel partitioning

By applying the modifications in previous section, the computing flow will be divided into separated phases that contain one or more loop statements. Each computing phase will be transformed to a data flow graph (DFG) at the granularity of loop statements, and finally mapped to a computing module in Figure 2. Illustrated in Figure 6, in order to reduce the number of stages of computing phase, we apply the well-accepted optimization of loop fusion and fission, which corresponds to node splitting and node combining of the computing flow, respectively.

Limited by the size of on-chip memory, it is important to take into account of fine-grained partitioning strategies within computing kernels. For example, we applied the row-column algorithm to implement 2D FFT/IFFT process, in which row-oriented 1D FFTs are executed before moving to column-oriented 1D FFTs. Therefore, it is need to cache the entire image on-chip (for an image of 512×512 32 bit pixels, it corresponds to $1\ MB$ storage) to implement a fully pipelined computing module, which easily exceeds the on-chip memory limitation of the FPGA and renders this approach impractical. Motivated from the fact that images are processed in batch, the overhead of splitting the 2D FFT/IFFT nodes can be amortized. Therefore, as illustrated in Figure 6, the computing flow of *MakeRFP* can be further partitioned into 3 phases. On the other hand, it is beneficial to combine adjacent loop statements to get more compact computing nodes. For example, each circle in Figure 6 represents a loop statement (nested loop is allowed) in source code. The 3 loop statements in phase 2 can be merged as a single loop with a trivial branch.

4.1.3 Data layout optimization

To facilitate the analysis of data access patterns, as illustrated in Figure 7, we propose an optimization to improve data layout and structure in *EMAN*. Along with some housekeeping parameters, the time domain data, the frequency domain data and the rotational footprint (RFP calculated on-the-fly by *MakeRFP*), are uniformly wrapped in the same *EMData* object. In original *EMAN*, different data blocks are scattered in memory space. Considering the phased computing flow introduced previously, it easier to justify spending an optimization step that clusters data by their types. By collecting separated data with the same type into continuous arrays, it becomes straightforward to map data blocks to fixed addresses in FPGA address space.

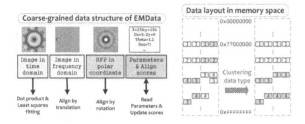

Figure 7: Data structure and layout of EMAN.

4.2 Memory system design

For applications, such as *EMAN*, that can be expressed at the granularity of loop statements and are data-intensive, investigating the possibilities of pattern-based data accesses is of paramount importance. The performance of the FPGA-based accelerator is largely dominated by the ability to support various data access patterns. The problem is compounded by the fact that data access patterns are usually entangled with the computing flow (address calculation based on loop indices). Our response to these concerns is a hybrid memory controller, which is designed to supported pattern-based data accesses. The memory controller is both a framework for creating modularized data access functionalities as well as a dynamic runtime that assist the construction of the data path.

The memory controller used in our system is a hybrid one, in which off-chip DRAM, SDRAM and parts of the on-chip BRAMs are managed with a single virtual address space. From the viewpoint of high-level users, data accesses on each kind of storage can be controlled explicitly by invoking write/read operations at the corresponding addresses in memory space. The configuration of the virtual memory space is controlled by software and can be changed online. In a typical memory space configuration, the lower address ($512\ KB$) are allocated to on-chip BRAM followed by off-chip SDRAM ($16\ MB$) and DRAM ($2\ GB$). The on-chip BRAMs are used as a scratchpad memory to store small intermediate data (scalar and vector) between consecutive pipeline stages. In addition, the BRAMs can also be used to build configurable data flow modules (DFM), which will be used to handle data flow permutations, such as the (de-)interleave operations in Table 1. The SDRAM is further divided into two parts, which are used respectively for data pre-fetching and intermediate results buffering. The final results will be offloaded to the off-chip DRAM and transferred back to host CPU via DMA through PCIe. The data bandwidth of the off-chip DRAM and SDRM is 3.2 GB and 6.4 GB, while the system bottleneck lies in the PCIe interface, which is only 2 GB.

4.3 Configurable data flow module

We rely on a data flow module, which is built upon on-

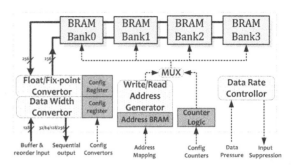

Figure 8: Data flow module with configurable address generation, data format conversion, configurable counter and flow control logic.

chip BRAMs, address generator, configurable counter and flow control logic, to support pattern-based data accesses. As illustrated in Figure 8, the 4 BRAMs are organized into 2 groups and configured to work as a Ping-Pong buffer. When one group is buffering data in current phase, the other group will be used to provide data that are buffered in previous phase. The data rate between input and output of the computing stream may become unmatched; therefore, it is necessary to provide a backward flow control mechanism, which is implemented by monitoring the back-pressure signals of the computing stream.

The extracted data access patterns are used to configure the DFMs as well as the data pre-fetching unit in memory controller. In particular, the DFMs can be controlled by varying address generation logic, data width, port number and data format converter. For example, in Figure 8, data access patterns in form of fixed-step counters (regular data access patterns) will be implemented with the counter logic by means of setting the counter range and step. In contrast, the address mapping arrays (irregular data access patterns) will be used to load the address BRAM in the write/read address generator. Various data access patterns can be achieved by configuring the address BRAM with different address sequences.

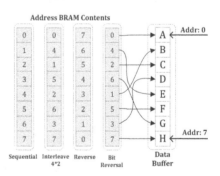

Figure 9: Address mapping in DFM. The contents are calculated off-line and used to load the address BRAM.

Take the computing steps to implement the *MakeRFP* kernel in Table 1 as an example, the DFMs in memory controller can be used to carry out various data access patterns. The process of interleaving two image lines with a length N can be achieved as follows: 1) Configuring the data pre-fetch unit in memory controller to read two lines of image each time; 2) Storing the pre-fecthed data in a BRAM group in DFM; 3) As illustrated in Figure 9, the address BRAM in the write/read address generator, which is loaded with the interleave $N \times 2$ (N is the width of an image line) content in advance, is used to generate addresses to read the data buffer. In this way, two image lines can be interleaved by reading the data buffer with respect to the content in the address BRAM. Along with the 4 address contents illustrated in Figure 9, new data access patterns can be generated on-the-fly. The primary drawback of using the indirect address mapping is that it introduces extra on-chip storage overhead. However, for this problem, the length of the address sequence is relative short, which makes the cost of additional storage affordable. For example, the longest address sequence in our system comes from the 8960 FFT module, which requires an on-chip storage as large as 15.312 KB ($\lceil \log(8960) \rceil \times 8960$).

The ability to do column-oriented writing is the major factor that contributes to the performance speedup of our system. However, the data path of the DRAM (128 bit) and SDRAM (256 bit) are both wider than current computing data path (64 bit). The DFM is used, on one side, to mitigate the data path differences by buffering 2 (4) columns before offload data to DRAM (SDRAM). On the other hand, the DFM is used to generate the column-oriented writing addresses, which can be easily mapped to the counter logic with modulo N (pixel number of an image) operation.

5. EXPERIMENT RESULTS

5.1 System and Experiment Setup

The proposed stream architecture is implemented and evaluated on our customized FPGA accelerator card. Along with various off-chip memory storage, the accelerator card, which communicates with the host CPU via PCIe gen1 interface, is composed of two FPGA chips. The computing stream is implemented on the computing node (Xilinx XC5VLX330-1), while functions related to configuration and data transferring are offloaded to the control node (Xilinx XC5VLX70T). The performance and area of the accelerator design on computing node are evaluated with Xilinx ISE 11.4 [18].

The CPU-based *EMAN* (single-threaded) is evaluated on a 4-cores 2.27GHz Intel Xeon E5520 and compiled with the latest Intel compiler. The sequential EMAM program is parallelized to 4 threads with the OpenMP. With our best effort, *EMAN* is also optimized on two GPU devices: the GeForce 8800 (G80) and the Fermi Tesla C2050 (GTX480). The G80 consists of 16 streaming multiprocessors (SMs), which contain eight streaming processors (SPs) each. The SPs within a SM run at 1.35 GHz and share a 16 KB on-chip memory, whereas all SMs share the 1 GB global off-chip memory. The GTX480 has 448 cores organized into 14 SMs, which can concurrently execute multiple warp blocks. Each SM in GTX480 has a set of registers and a 64 KB local storage, which can be configured as 16 KB L1-cache and 48 KB shared memory. All SMs share a unified 768 KB L2-cache and the 3 GB global memory.

Full system power is measured with the Fluke Norma 4000 Power Analyzer [19]. Similar as the method proposed in [16], we only consider the dynamic power, which is measured as difference between active and idle power.

5.2 Kernel performance

Figure 10 compares the execution time of the kernels listed in Table 1. Measured in terms of speedup over single-threaded CPU, the speedup of kernel implementation on FPGA varies from 2 times (*Max*) to 12 times (*CCFX*). We observe that kernels can achieve different speedup even with a similar computing flow. For example, the *Rotate* and *Unwrap* kernel in Table 1 exhibit a similar computing flow. The coordinate transformation of *Unwrap*, however, spans only one half of an image, which contributes to the reported lower speedup when compared with *Rotate*.

The performance of GPU can be greatly influenced by the data parallelism degree. For example, due to the inefficiency of parallelism degree, the CCFX kernel that executes only 1D FFTs achieves a low speedup on GPUs when compared with the MCF and CCF kernels that contain extensive 2D FFTs. Due to the advantages of frequency and bandwidth of the GDDR memory, except for 2 kernels (*CCFX* and *Translate*), the performance of using GPU is clearly better than FPGA. On the other hand, the GDDR memory on GPUs, which are optimized for sequential data accesses, incurs a high performance penalty for irregular data access patterns in kernels such as *Translate*. The performance of GTX480 is always better than the obsolete G80. The speedup of using OpenMP to parallelize EMAN on CPU ranges from 1.6 times (*Translate*) to 3.5 times (*MCF*). With the increase in image size, the speedup of using OpenMP, FPGA or G-PU will get improved marginally because of the increased data-level parallelism.

5.3 Application speedup

Table 3 shows the total execution time of a single iteration step of the 3D reconstruction process. The FPGA outperforms the 4-cores CPU by 2.54 times, while the GTX480 further outperforms the FPGA by 3.76 times. Workload in *EMAN* increases quadratically with increasing image size. However, the arithmetic intensity remains unchanged; therefore the increased workload cannot be utilized to introduce further performance improvement. For FPGA-based designs, the total execution time increases about 3 times when image size changed from 256^2 to 512^2, the improved speedup over CPU is largely attributed to the performance degradation on CPU. In contrast, large image can easily saturate the 448 cores in GTX480, which accounts for the $2x$ speedup. We consider that the FPGA and GPUs are more beneficial for Cryo-EM 3D reconstruction on large images.

Table 3: Execution time (milliseconds) and speedup (normalized to single-threaded CPU).

	Time-256	Speedup	Time-512	Speedup
CPU×1	80.86	-	377.21	-
CPU×4	30.86	2.62	124.9	3.02
G80	11.39	7.1	38.89	9.7
GTX480	5.82	13.9	13.10	28.8
FPGA	12.31	6.6	49.26	7.7

5.4 Power consumption

Table 4 lists the measured static, dynamic and working power of each device, where the static power is measured as idle power and the dynamic power is averaged across the entire execution. The working power, which is calculated as the difference between dynamic and static power, is used in following power analysis. It has long been noticed that the

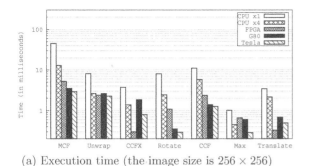

(a) Execution time (the image size is 256×256)

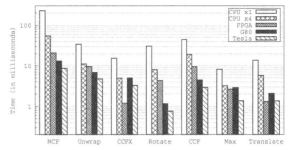

(b) Execution time (the image size is 512×512)

Figure 10: Comparison of the kernel execution time on CPU, FPGA, G80 and GTX480.

FPGA can be used to build standalone devices, which are efficient than the PCIe-based accelerator card considered in this paper. When measuring the power of system configuration with either FPGA or GPU card, in order to get an accurate evaluation, we subtract the power consumption in the host PC. And, for simplicity, we assume that the host PC remains idle during the execution of the accelerator cards. The power cost, which is computed as

$$Power\ Cost = Working\ Power \times Execution\ Time, \quad (1)$$

is defined as the energy consumed by useful computing work and measured in millijoule. The multi-threaded EMAN introduces a 3.02 times speedup at the cost of 27% increase in working power; however, with reduced total execution time, the power cost of using 4 threads is only 48% of single-threaded EMAN. The result shows that our accelerator card outperforms the CPU and GPU by 7.33 times and 3.4 times.

Moreover, it is worth to note that the GTX480 chip is built with 40nm process, while the Virtex5 FPGA used in this paper is built with the less efficient 65nm process. We argue that the performance and power cost of our FPGA-based accelerator can be further improved by upgrading to the Virtex6 and Virtex7 families.

Table 4: Power consumption analysis

Device	Static	Dyn.	Working	Power (millijoule)
GTX480	53W	147W	94W	547mJ
FPGA	7W	20W	13W	160mJ
CPU×1	89W	119W	30W	2426mJ
CPU×4	89W	127W	38W	1173mJ

Table 5: FPGA Resource Consumption

	DSP48Es	LUT-FFs	BRAMs	Freq.
MakeRFP	106(55%)	56.5K(27%)	140(48%)	180MHz
RTAlign	140(72%)	55.5K(23%)	133(46%)	180MHz

5.5 Area and frequency

Two of the most time-consuming stream kernels of *EMAN* are implemented as separated computing streams: *MakeRFP* and *RTAlign*. The area consumption and frequency of each stream are summarized in Table 5. In practice, the two streams share a lot of common blocks and can be combined to form a more compact stream. However, the two computing streams will be executed on separated FPGA accelerator card, which renders such kind of combination of little interests for now.

6. RELATED WORK

Using of 2D images to predicate and reconstruct a 3D model is a well-accepted method applied in disciplines of medical imaging, bioinformatics, multimedia and etc. In [12], the authors introduce and discuss the design of the software architecture for a 3D reconstruction system in medical imaging. The importance of using different hardware platforms to accelerate corresponding parts of the algorithm is noticed, however, no concrete accelerator design considerations are included in their work. Compared with the Cryo-electron microcopy 3D reconstruction, the computed tomography (CT) reconstruction is a much popular topic in literature. Due to the advances in the GPU technologies, accelerating the CT reconstruction on GPUs is an active research topic. In [14], the authors introduced a real-time 3D reconstruction system for x-ray CT on the GeForce 8800GTX. The results prove that GPU is very suitable for CT reconstruction. A FPGA-based accelerator is also introduced in [15], in which the authors noted that the memory accesses is a crucial point for CT reconstruction on FPGAs.

Our previous work on accelerating the *EMAN* on heterogeneous cluster described the possibilities of optimizing kernels on the G80 architecture [7]. The optimized kernels are adaptive to the new Fermi architecture [8]. The focus of this work is to exploit the power of the FPGA for further performance improvement. To our best knowledge, this is the first work on FPGA-based accelerator design for the Cryo-electron microcopy 3D reconstruction. The stream architecture on FPGAs has been widely studied in the literature [5][6]. The architecture proposed in this paper is similar to the work presented in [6], in which specialized stream units are provided to handle stream operations, the control of data stream operations can be achieved by adjusting the stream descriptor.

The background of this work is to build a heterogeneous cluster, called Chaolong-1, which contains X86 CPUs, GodsonT CPUs, GPUs and FPGAs. *EMAN* analyzed in this paper is one of the key applications in this system. Heterogeneous systems with co-processors are nothing new to the HPC community. The CUBE [10] is a massively-parallel FPGA-based cluster that consists up to 512 FPGAs. The Axel [11], which is built with the Xilinx Virtex-5 FPGAs and nVidia C1060 GPUs, demonstrates a collaborative environment for heterogeneous accelerators. The Convey HC-1^{ex} [17] contains 4 Xilinx Virtex6 LX760 FPGAs, which are connected to the host CPU thought a FSB-based interface. The memory system of the Convey HC-1^{ex} is optimized for handling concurrent random memory accesses, which renders it perfect candidate to accelerate graph-based applications.The Cray XK6 supercomputer [20] combines Cray's Gemini interconnect, AMD Opteron processors and NVIDIA Tesla 20-Series GPUs to create a heterogeneous cluster that can be upgraded to unleash more than 50 PFlops.

7. CONCLUSION

In this paper, we introduce a coarse-grained stream architecture, in which the computing data path is configurable at runtime. To facilitate the decompiling of the computing flow and the data flow, we resort to a dedicated data flow module (DFM) to provide the ability to do pattern-based data accesses. Complex functions can be emulated by configuring the data path of the computing stream multiple times, whereas both data operations as well as related address calculations are offloaded to the DFMs.

The stream architecture is evaluated by accelerating a large-scale scientific application, called *EMAN* [3], which is an open source software package for single-particle 3D reconstruction from Cryo-electron microscopy images, on our customized FPGA accelerator card. Our FPGA-based accelerator design is compared with CPU and GPU solutions. Measured in raw performance, the FPGA-based design outperforms the 4-cores CPU by 2.54 times. When compared with our previous GPU-based designs, the FPGA-based design is about $3 \sim 4$ times slower. However, we argue that it is still beneficial to use the FPGA when taking the $7 \sim 8$ times power improvement into consideration.

The data bandwidths of the PCIe interface (2GB) and off-chip memory devices ($3.2 \sim 6.4$GB) are the limiting factors that prevent our system from reaching higher performance. There are several ways to overcome this limitation: 1) upgrading current gen1 PCIe interface; 2) developing in-socket accelerator with wider memory bandwidth. The second option, which has been pioneered by the Convey HC-1^{ex} [17] system, points out our future research direction.

The overarching goal for our research is to design a heterogeneous cluster that is built with X86 CPUs as well as heterogeneous accelerators, such as GPUs and FPGAs. Although dedicated optimizations can be used to improve the performance of individual device, in order to make the heterogeneous devices working synergistically and unleash the highest aggregated computing power, a collaborative environment is still needed, in which each type of device can be used to accelerate the most appropriate work. Our previous work on job partitioning between CPUs and GPUs [8] can be considered as a primitive attempt, as an extension, we expect to include FPGAs into the job partitioning process.

8. ACKNOWLEDGMENT

This work is supported by Chinese Academy of Sciences (No.KGCX1-YW-13), 973 Program of China (NO.2012CB316502), 863 Program of China (NO.2009AA01A129), and National Natural Science Foundation of China (NO.60803030, NO.60633040, NO.60925009, NO.60921002).

9. REFERENCES

[1] Scrofano, R.; Gokhale, M.; Trouw, F.; Prasanna, V.K.; , "Hardware/Software Approach to Molecular Dynamics on Reconfigurable Computers," Field-Programmable Custom Computing Machines, 2006. FCCM'06. 14th Annual IEEE Symposium on , vol., no., pp.23-34, 24-26 April 2006.

[2] Florent de Dinechin, Cristian Klein, and Bogdan Pasca, "Generating high-performance custom floating-point pipelines," In Field Programmable Logic and Applications, IEEE, August 2009.

[3] S. J. Ludtke, P. R. Baldwin, et.al. "Eman: Semiautomated software for high-resolution single-particle reconstructions", *Journal of Structural Biology*, 128(1):82-97, 1999.

[4] Jaydeep Marathe and Frank Mueller. 2008. "PFetch: software prefetching exploiting temporal predictability of memory access streams", In Proceedings of the 9th workshop on MEmory performance: DEaling with Applications, systems and architecture (MEDEA '08). ACM, 1-8.

[5] A. Hormati, M. Kudlur, S. Mahlke, D. Bacon, and R. Rabbah, "Optimus: Efficient realization of streaming applications on FPGAs," International Conference on Compilers, Architectures and Synthesis for Embedded Systems, ACM, pp. 41-50, 2008.

[6] Nikolaos B., Sek M. Chai, Malcolm D., Dan L., Abelardo L., "Proteus: An Architectural Synthesis Toll Based on The Stream Programming Paradigm", Field Programmable Logic and Applications, 2009.

[7] G. Tan, Z. Guo, et.al. "Single-particle 3d reconstruction from cryo-electron microscopy images on GPU", In *ICS '09: Proceedings of the 23rd international conference on Supercomputing*, pages 380–389, New York, NY, USA, 2009. ACM.

[8] L. Li, X. Li, G. Tan, M. Chen, and P. Zhang. "Experience of parallelizing cryo-EM 3D reconstruction on a CPU-GPU heterogeneous system", In Proceedings of the 20th International Symposium on High performance distributed computing (HPDC '11). ACM, 195-204, 2011.

[9] D. DeRosier and A. Klug, "A reconstruction of 3-dimensional structure from electron micrographs", *Nature*, 217:130–134, 1968.

[10] Mencer, O., Tsoi, K.H., Craimer, S., Todman, T., Luk, W., Wong, M.Y., Leong, P.H.W.: "CUBE: A 512-FPGA CLUSTER", In: Proc. IEEE Southern Programmable Logic Conference (SPL 2009) (April 2009)

[11] K. H. Tsoi and W. Luk. "Axel: A heterogeneous cluster with FPGAs and GPUs", In Proc. ACM/SIGDA International Symposium on Field-Programmable Gate Arrays (FPGA), pages 115ÍC124, 2010.

[12] Holger Scherl, Stefan Hoppe, Markus Kowarschik, and Joachim Hornegger. "Design and implementation of the software architecture for a 3-D reconstruction system in medical imaging", In Proceedings of the 30th international conference on Software engineering (ICSE'08). ACM, 661-668, 2008.

[13] Duan B., Wendi W., et al. "Floating-point Mixed-radix FFT Core Generation for FPGA and Comparison with GPU and CPU", The 2011 International Conference on Field-Programmable Technology, Dec, 2011.

[14] F. Xu and K. Mueller, "Real-time 3D computed tomographic reconstruction using commodity graphics hardware", Physics in Medicine and Biology 52(12) (2007), 3405-3419.

[15] S. Coric, M. Leeser, E. Miller, and M. Trepanier. "Parallel-beam backprojection an FPGA implementation optimized for medical imaging", In Proc. ACM Int. Symp. Field-Programmable Gate Arrays (FPGAař02), pages 217-226, February 2002.

[16] Brahim Betkaoui, David B Thomas, Wayne Luk, "Comparing Performance and Energy Efficiency of FPGAs and GPUs for High Productivity Computing," The 2010 International Conference on Field-Programmable Technology (FPT'10), 8-11 Dec. 2010.

[17] Brewer, T.M.; , "Instruction Set Innovations for the Convey HC-1 Computer," Micro, IEEE , vol.30, no.2, pp.70-79, March-April 2010.

[18] Xilinx, http://www.xilinx.com.

[19] Fluke, http://www.fluke.com.

[20] http://www.cray.com/Products/XK6/KX6.aspx.

A Cycle-accurate, Cycle-reproducible multi-FPGA System for Accelerating Multi-core Processor Simulation

Sameh Asaad, Ralph Bellofatto, Bernard Brezzo, Chuck Haymes, Mohit Kapur
Benjamin Parker, Thomas Roewer, Proshanta Saha, Todd Takken, José Tierno

IBM T.J.Watson Research Center
Yorktown Heights, NY 10598
{asaad,ralphbel,brezzo,haymes,mohitk,nj,roewer,psaha,takken,tierno}@us.ibm.com

ABSTRACT

Software based tools for simulation are not keeping up with the demands for increased chip and system design complexity. In this paper, we describe a cycle-accurate and cycle-reproducible large-scale FPGA platform that is designed from the ground up to accelerate logic verification of the Bluegene/Q compute node ASIC, a multi-processor SOC implemented in IBM's 45 nm SOI CMOS technology. This paper discusses the challenges for constructing such large-scale FPGA platforms, including design partitioning, clocking & synchronization, and debugging support, as well as our approach for addressing these challenges without sacrificing cycle accuracy and cycle reproducibility. The resulting fullchip simulation of the Bluegene/Q compute node ASIC runs at a simulated processor clock speed of 4 MHz, over 100,000 times faster than the logic level software simulation of the same design. The vast increase in simulation speed provides a new capability in the design cycle that proved to be instrumental in logic verification as well as early software development and performance validation for Bluegene/Q.

Categories and Subject Descriptors

B.8.2 [**Hardware**]: Performance Reliability—*Performance Analysis Design Aids*
; C.5.0 [**Computer Systems Organization**]: Computer System Implementation—*General*

General Terms

Design, Verification, Performance

Keywords

FPGA, Multi-core, Logic Emulation

1. INTRODUCTION

Logic verification is known to be one of the major challenges of contemporary chip design. In many cases the size of the verification team surpasses that of the logic design team. Simulation based techniques remain the most common verification approaches due to their notable flexibility. Unfortunately, the algorithms used in these simulators, which are typically event-driven or cycle-based, are hard to parallelize in software due to their inherent low computation to communication ratio. Exacerbating the problem is the recent saturation of single-thread performance of general purpose computers, widening the gap between the increasing requirements for verification performance and the capabilities of software-based simulation. For example, full-chip software simulation of a Bluegene/Q compute node yields a mere 10-100 processor clocks/sec, inadequate for all but the most trivial test cases. There is a clear need for accelerating the verification task through dedicated hardware.

Existing solutions for accelerating the logic verification problem involve either generalized systems capable of addressing any digital design, or specialized solutions achieving maximum performance targeting a specific chip design effort. The first approach, although reusable for many projects, usually results in a high-cost, over-provisioned system and tends to deploy custom ASICs for acceleration. Key drawbacks of this approach include the limitations of the performance of the resulting system as well as the rising cost associated with realizing the dedicated ASIC it typically includes. The second approach usually deploys off-the-shelf field programmable gate arrays (FPGA) to directly prototype the target design. This approach usually involves designing a custom board that will deploy the minimum number of FPGA devices required to map the design at hand. The FPGA device interconnect is designed to match chip interconnect exactly and the partitioning of the chip logic onto the multiple FPGA devices is generally done by hand. Although this approach usually results in a high performance prototype with minimal footprint, it requires significant engineering effort to design, fabricate and test the prototype ahead of verifying the target chip. Scheduling the FPGA prototype build to mesh with the chip design timeline can be tricky, due to the fact that the FPGA prototype has to be ready early in the chip development cycle, a time when large changes in the chip architecture are still likely to occur, which in turn ripple through the FPGA prototype design.

FPGA based emulation has been proposed by individual research groups and EDA companies before to solve this problem. Intel has reported single-FPGA [1] and multi-FPGA [2] implementations of processors to speed up the verification process. Both implementations enable significant speed up over simulation and reach 50 MHz and 520 kHz respectively. The single-FPGA implementations reported in [1, 3] are limited in logic capacity and can only handle very small ASIC or processor designs, such as those used in the embedded space. In [2], the authors report a multi-FPGA emulator that uses 5 FPGAs on a single PCB-board and achieves 520 KHz clock speed. Our system, detailed in this paper, is capable of running at much higher speeds using more sophisticated inter-FPGA communication channels and system clocking & synchronization techniques. Other researchers also used this strategy successfully [4, 5].

The high speeds FPGA emulators are able to reach enables them to cover a wide variety of verification tasks, from cycle-accurate functional verification to architecture evaluation [6, 7, 8], HW/SW co-design [9] and soft/firmware development [5]. We will demonstrate that our platform is able to serve all these tasks simultaneously.

The approach detailed in this paper combines elements of the general-purpose and chip-specific strategies, combined with the introduction of new methodologies for connectivity implementation, design partitioning, clocking and synchronization management, and observability, to create a modular, retargetable emulation platform that achieves state-of-the art emulation performance with minimal overprovisioning. A key result of this work, then, was the creation of a set of tools, components, and methodologies from which chip-specific, high performance target emulators can be realized that operate in a cycle-accurate, cycle-reproducible manner. The critical elements driving this result include: 1) A flexible FPGA-based system architecture that enables efficient construction of simulator instances tailored to specific chip designs, 2) architected mechanisms and infrastructure logic to synchronize all FPGA devices in the system, 3) high performance flexible and modular communication links that are amenable to automation, 4) a tool for automatic partitioning of a target design onto multiple FPGA devices under user direction, and 5) techniques for efficient debugging of the target design. We developed these components and tools in the context of accelerating the verification of the Bluegene/Q compute node[1] resulting in an efficient simulation acceleration platform, codenamed *Twinstar*. While first primary use of the Twinstar platform was Blugene/Q logic simulation, the underlying tools, components, and methodology have enabled platform re-use in efficiently constructing simulators for other chip and system development projects.

The remainder of this paper will focus on the specifics of the Twinstar platform in the context of Bluegene Q compute node ASIC emulation and is organized as follows: Section 2 briefly describes the Bluegene/Q compute node ASIC architecture. Section 3 states the design goals for the Twinstar FPGA-based simulation platform. Details of the Twinstar system architecture and the methodology to map the target ASIC design are covered in sections 4 and 5, respectively. Finally, results are presented in section 6, highlighting Twinstar's impact on Bluegene/Q system development.

2. BLUEGENE/Q COMPUTE NODE ASIC

In order to appreciate the complexity of the target system, a brief overview of the Bluegene/Q compute node is presented. A more detailed description of the chip and sys-

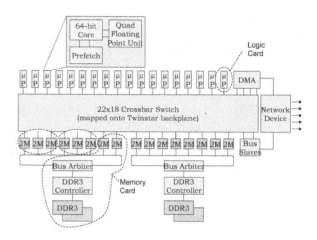

Figure 1: A simplified view of BQC architecture

tem is beyond the scope of this paper, but can be found in [10].

The 1.47 billion transistor Bluegene/Q compute chip ASIC (BQC) is a 360 mm^2 multi-core SoC implemented in IBM's 45 nm SOI CMOS technology. The chip contains 16 64-bit processor user cores[2], each equipped with a 4-way SMT in-order pipeline, 16KB of L1 data cache, a 16KB L1 instruction cache, a complex double-precision quad floating point unit, and a novel data prefetching unit. The cores are connected to a large shared 32 MB L2 cache, organized as 16 independent 2MB slices, through a central crossbar switch. The L2 slices are in turn connected through a ring architecture to 2 DDR3 memory controllers, providing support for up to 32 GB of external DDR3 memory. In addition, BQC integrates the peripheral I/O controller (PCIe) and networking hardware enabling realization of a 5-dimensional torus network in a massively parallel overall system. Figure 1 shows a high-level view of the BQC chip architecture as it pertains to partitioning and mapping onto Twinstar platform. Each of the processor cores was mapped onto a dual-FPGA node card. The central crossbar switch was partitioned into 3 functional components, namely request, response and invalidate blocks and placed onto multiple FPGA devices in the backplane. Six additional single-FPGA node cards were used to model the memory subsystem, including external DDR DRAM as shown. Figure 2 is an annotated photo of the BQC chip die.

3. TWINSTAR SYSTEM REQUIREMENTS

The main criteria for Twinstar as a simulation acceleration platform were driven by its use as a logic verification tool capable of emulating a full chip design of the size and complexity of Bluegene/Q compute node. As such, the requirements included:

• *Logic capacity:* The target for this platform is to enable full-chip verification of Bluegene/Q compute node, a multi-core high performance processor chip. Given the logic emulation capacity of the available FPGA devices available at

---

[1]This work was supported by DoE contract B554331.

[2]Extra cores for system functions

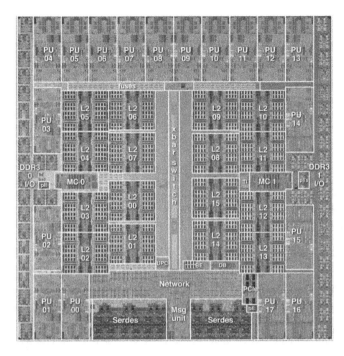

Figure 2: Annotated die photo of BQC chip

the time the project was initiated [3], this target drove the inclusion of a large number of FPGA devices in the platform.

• *Cycle accuracy:* Executing a given test on Twinstar and on a host computer running event-driven or cycle-based RTL simulation in software must yield identical traces for all visible nodes in the device-under-test (DUT). The approach used to maintain cycle accuracy for our system of multiple FPGA devices was to require that communication links passing DUT data among the various FPGA partitions do not consume any DUT cycles. Under this scheme, the links appear as mere *wires* from a DUT mapping standpoint. Another implication of this requirement is that FPGA latch and array models, which usually require modified implementations to map well on FPGA devices, must not only be functionally accurate but also timing accurate with respect to the original DUT ASIC models. The requirement also implies that every state-bearing element in the chip implementation must have a corresponding element in the FPGA model. However, note that it is permissible for the FPGA model to have extra state that is not part of the DUT state, e.g. state for instrumenting the DUT for observability, partitioning logic, etc.

• *Cycle reproducibility:* Repeated executions of the same test case must produce identical results. This requirement is critical for isolating subtle bugs in the hardware, a process that typically requires multiple runs to enable observation of all potentially relevant signals. Cycle reproducible behavior poses tight constraints on the platform design, demanding careful orchestration of all IO operations between the modeled DUT and the surrounding environment. Note that cycle reproducibility must extend even to processes like communication between the emulation platform and a host computer over an asynchronous channel (such as Ethernet). This extended capability enables, for example, the support

of a cycle-reproducible *printf*, such that sending the text to be printed from the DUT to the host computer incurs the same delay in terms of DUT clock cycles in every run. Similarly, accesses to external memory must exhibit repeatable behavior.

• *Performance:* In order to achieve target emulation throughput, a DUT frequency of at least 1 MHz was sought. Achieving this result would eliminate the need for vast amounts of simulation server farm cycles (capacity) and also make the execution of very long test cases feasible (capability), such as booting an operating system and running benchmark code on the FPGA model of the DUT.

• *Build time:* For the system to be useful in debugging logic, minimization of the time it takes from acquiring a new snapshot of the chip logic to producing the necessary configuration for the FPGA system to execute the simulation of the new logic is critical. In the case of BQC emulation,, we targeted a nightly build process, or approximately 12 hours of build time. This allocation included partitioning the logic onto the various FPGA devices, synthesizing, mapping and routing each FPGA in the system and producing the necessary configuration stream for each of the target FPGA devices.

• *Observability:* Debugging the DUT model on the FPGA can be a daunting task. To facilitate this process, a key goal of the platform is to enable observability of any DUT state-bearing latch *without* requiring a time-consuming re-build of the logic.

• *Batch processing:* A batch environment for submitting, tracking and executing simulation jobs onto Twinstar was required to enable ay multiple users to queue their jobs to maximize the utility of the system.

The above goals were used to drive the architecture and guide the implementation of Twinstar. In the following sections, we discuss the architecture and implementation approaches that enabled Twinstar system targets to be met.

4. ARCHITECTURE

The Twinstar platform consists of up to 28 FPGA node cards plugged into an active backplane forming a tightly-coupled reconfigurable system as shown in Figure 3. The node cards come in two different flavors: A *memory card* and a *logic card*, any combination of which can be used to construct a simulator instance with the desired amount of logic and memory capacity. When fully populated with logic cards, the system logic capacity reaches 60 FPGA (Xilinx Virtex5 LX330) devices; a memory-heavy configuration can model over 100 GBytes of memory using DRAM.

The *Active Backplane* provides flexible interconnect among all the node cards in the system and houses the central controller for system functions, as shown in Figure 4. The node cards are grouped into 4 quadrants (North, East, South, and West), where each quadrant includes 6 fully interconnected slots and one slot with wider buses to facilitate high bandwidth connections to the 4 FPGAs mounted on the backplane itself. The active backplane facilitates multiple connection topologies, all of which can be simultaneously supported due to the active backplane's use of dedicated point-to-point links:

• Star: All nodes are connected to the set of 4 central FPGAs on the backplane forming 4 sets of star connections. This way any node can communicate with any other node through

---

[3]we used Xilinx Virtex 5 LX330 devices for Twinstar

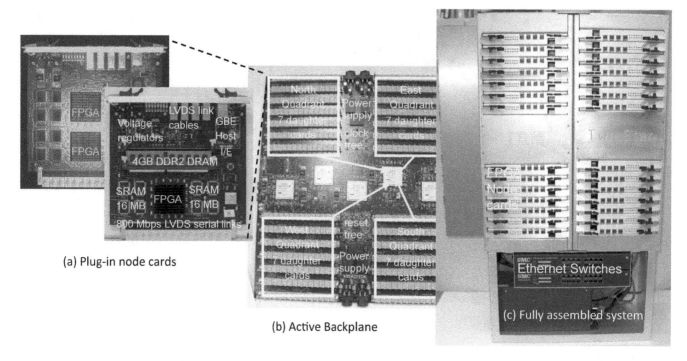

(a) Plug-in node cards

(b) Active Backplane

(c) Fully assembled system

Figure 3: Modular design approach to Twinstar platform

the central FPGAs,. The star topology is ideal for modeling the BQC crossbar switch. Moreover, it can be used for modeling various other on-chip communication topologies, such as busses or rings.
• Intra-Quadrant: Nodes in each quadrant are fully interconnected. This facilitates partitioning of logic that does not fit on one node card.
• Inter-Quadrant: Each node is connected to the corresponding node in the neighboring quadrants in the horizontal direction, adding flexibility to the system.
• Top cables provide additional point-to-point connectivity among FPGA node cards. In practice, these cables are of immense value in providing configurable bandwidth to optimize platform communication capabilities against the requirements driven by emulating a particular ASIC design..

A host computer interacts with the Twinstar FPGA system using standard Gigabit Ethernet networking gear. Each card in the system, including the active backplane and the plug-in node cards, has a unique IP address that is identifiable by the host machine. During system initialization, the host queries each slot to detect and enumerate each card in the system. A user-supplied XML-based configuration file conveys to the host software information on individual FPGA configuration image files. Figure 5 is a photograph of 2 completed Twinstar systems used to accelerate the simulation of Bluegene/Q compute ASIC.

The FPGA *memory card* is intended to model a memory element, or a special-purpose core (e.g. DMA, network, or IO peripheral). The memory card is comprised of the following components:
• A high-capacity FPGA (Xilinx V5 LX330) to model processor logic functions, including small sized (KB) arrays, such as TLBs (translation look-aside buffer), L1 caches, cache directories, etc.

• High speed SRAM (8 x 4 MB) to model large chip arrays, e.g. L2/L3 caches
• High density DDR2 SDRAM (up to 4GB) to model main memory
• A large number of high-speed LVDS-based point-to-point communication links to connect to the backplane (through its bottom edge connector) and other nodes (through top cables)
• GB-Ethernet infrastructure (including PHY, MAC, and UDP layers in hardware) to enable connection to a host machine for various host control functions, e.g. configuration, display/alter memory contents, reading internal logic state, etc.
• A control FPGA to support platform infrastructure functions including system reset sequencing, clock control, and configuration management.
• Voltage regulator modules enabling each card to be used in stand-alone mode as well as a plug-in module in the main system.

The FPGA *logic card* is a derivative design of the memory card, replacing the DDR2 subsystem with a second FPGA (also Xilinx V5 LX330). The increased logic capacity of that card makes it suitable for modeling large logic cores, such as a processor core.

5. METHODOLOGY

Starting from the ASIC RTL code, hereafter referred to as design under test or DUT, we partition the DUT logic onto multiple FPGA devices as necessary to fit the logic. DUT signals that cross partition boundaries are time multiplexed onto serial point-to-point inter-FPGA communication links since, in general, the number of physical wires between FPGA devices in the system is less than the number of DUT signals crossing a partition boundary. The DUT clock and reset networks are replaced by equivalent clock

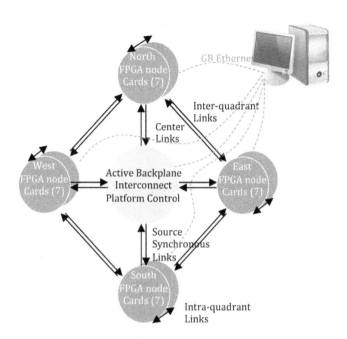

Figure 4: Twinstar System Architecure

and reset networks that are amenable to FPGA mapping. Memory arrays in the DUT are also replaced by equivalent FPGA arrays. External interfaces, such as DRAM channels, are adapted to suit the FPGA ports. Finally, execution control logic is attached to each of the FPGA partitions, facilitating interaction with the host control machine. An overarching requirement while performing these transforms is to maintain cycle accuracy and cycle reproducibility as described in section 3, while retaining the flexibility of stopping and resuming the DUT simulation at arbitrary cycles to enable efficient logic debugging of the DUT. Furthermore, it is desirable to automate these transforms as much as possible to simplify the mapping process and ensure correct-by-construction results. The following subsections present the salient components of this methodology in more detail.

5.1 Inter-FPGA communication

To maintain cycle accurate behavior, we restrict the design of the FPGA-to-FPGA communication links in the system to consume no DUT clocks; as described above, the links appear as wires in the partitioned system from a DUT standpoint. To achieve this result, we clock the internal link circuits using an infrastructure-only clock which runs at a much higher frequency than the DUT clock(s) as shown in in Figure 6. We chose LVDS-based communications to maximize overall system bandwidth and minimize link latency. At the time of Twinstar architecture definition, the total LVDS I/O pin bandwidth exceeded that of serial transceiver-based I/O for the chosen FPGA family (Xilinx Virtex5 LX). We opted for a source-synchronous, modular link design with minimal common circuits among different links so as to simplify automatic insertion by our partitioning tool. In other words, every link in the system is a self contained transmit-receive pair, with separate training pattern generator circuits at the transmitter and separate bit-, word-, and link-align circuits at the receiver side. The only common circuitry is the source

Figure 5: Photograph of 2 Twinstar FPGA systems

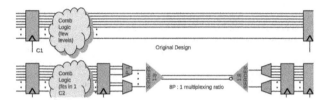

Figure 6: Cycle-accurate multiplexed communication between FPGAs

clock generation and reception which is shared among all links belonging to the same pair of communicating FPGA devices.

Figure 7 plots the maximum system frequency as a function of the number of multiplexed DUT signals carried by the LVDS link. This assumes the usage of 8:1 LVDS serdes hard macro IO blocks, 800 mbps throughput at the serial side and a 100 MHz infrastructure clock at the parallel side. The data shows that the maximum achievable system frequency is approximately 10 MHz. This is due to the end-to-end LVDS link latency, which must fit within a single emulated clock cycle. We believe this is the main factor limiting the maximum performance in our approach to cycle-accurate emulation. Taking advantage of the fact that the processor cores in Bluegene/Q run at double the frequency of the crossbar switch and memory subsystem, we developed two link designs. The first (x1) has a 32:1 mulitplexing ratio and supports connecting partitions that run at processor core frequency. The second (x2) has a capacity of 96:1 and is used to connect partitions running at half the processor frequency. With this tradeoff between capacity and latency, we are able to achieve an emulated processor clock frequency of 4 MHz.

5.2 Twinstar system clocking

Choosing the correct clocking strategy was a critical ele-

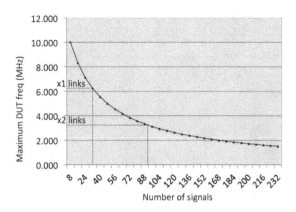

Figure 7: Maximum DUT frequency as a function of link multiplexing ratio

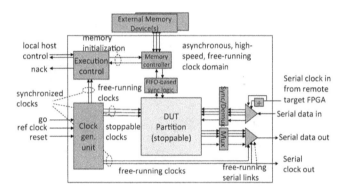

Figure 8: Cycle-reproducible execution control using free-running (infrastructure) and stoppable (DUT) clocks

ment in Twinstar system design. Due to the platform's large size, a homogeneous system-synchronous clocking approach would result in very limited performance due to clock skew across various FPGA devices in the system. At the other end of the spectrum, a per- FPGA clock domain approach with asynchronous FIFO-based communication channels between FPGAs could be implemented; although this technique may allow individual FPGA devices to run fast, the asynchronous nature of communications will yield a non-reproducible system, with repeated executions of the same workload possibly resulting in different traces of execution. Our clocking approach uses a combination of system synchronous global clocking and source synchronous point-to-point communication links. Based on link speed, link multiplexing ratio, and global clock tree skew, we can establish a maximum DUT frequency that guarantees error-free operation where all FPGA devices in the system are operating in lockstep.

Furthermore, as shown in Figure 8, we distinguish between stoppable DUT clocks and free running infrastructure clocks. For infrastructure blocks such as serial links and DRAM interfaces, stopping their clocks would result in loss of link training and loss of DRAM data due to lack of

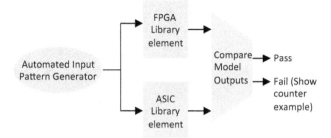

Figure 9: Array validation environment to ensure cycle accurate array mapping

refresh, and therefore these clocks need to stay free-running. On the other hand, the DUT logic itself has to be stoppable to examine/alter state of the design during debugging. Our infrastructure clock generation circuits were therefore architected to enable synchronization between the stoppable and free-running clocks in the system, allowing glitch-free stop and resume of the DUT logic at arbitrary DUT cycles.

5.3 Array mapping & validation

In general, ASIC arrays are not directly mappable to FPGA array primitives and therefore require special handling. Particularly, we address the issues of mapping multi-ported ASIC arrays as well as large array sizes, given that typical FPGA device memory is limited to dual-port memory blocks that are limited to a few MBytes.

For small arrays, to model an array with multiple read ports, we typically replicate the entire array, and write simultaneously to all arrays while exposing the replicated read ports. To model multiple write ports, we time-multiplex the single write port available in the FPGA memory block using a fast infrastructure clock that is transparent to the DUT operation. [4]

To model large arrays that exceed inherent FPGA capacity, we use external SRAM and/or DRAM memory. Again, the concept of hyper-clocking using fast, free-running, infrastructure clocks is used to hide the difference in latency and layout of the data between the emulator platform memory and the DUT target memory.

Since array mapping is predominantly a manual step in our flow, we have a rigorous validation environment to ensure that the original ASIC arrays are equivalent to the mapped Twinstar arrays in both function and cycle behavior. The environment, depicted in Figure 9, consists of a simulation testbench that instantiates both ASIC and Twinstar-mapped array models, injects random traffic across both arrays and checks that the responses coming from both are identical. Careful design of the test pattern generator is essential to ensure all corner cases are covered.

5.4 Partitioning Tool

As another key element of the tool and methodology advancement developed in conjunction with the Twinstar effort, we developed our own semi-automated partitioning tool which greatly streamlines the process of mapping a DUT such as the Bluegene/Q compute node ASIC onto Twinstar.

---

[4]This approach can also be used to model multiple read ports instead of replicating the entire array.

The partitioning tool expects 3 inputs: 1) A netlist describing the logical hierarchy of the device under test (DUT), 2) A netlist describing the physical hierarchy, i.e. the topology of the current configuration of Twinstar system, and 3) A mapping file that specifies where each instance of the logical hierarchy should land in the physical hierarchy. Both physical and logical netlists are compiled into a common intermediate format, namely, the IBM Design Automation Database (DADB2) format [5].

In terms of the physical hierarchy, it is worthwhile to note that the Verilog netlist for each board in the system is automatically generated from an in-house tool (Netlister) which is used in the actual production of the PCB boards, thereby ensuring that input to the partitioning tool faithfully represents the actual system design. The top-level physical system configuration is a hierarchical Verilog netlist that instantiates the active backplane and one or more of the node cards as needed by the desired configuration. Arbitrary configurations can be handled by the partitioning tool.

The tool analyzes the netlists and writes a complete set of VHDL files, one for each FPGA in the physical hierarchy. Each VHDL file contains the appropriate instances from the logical hierarchy embedded in a wrapper module. The wrapper instantiates serializer/deserializer macros that multiplex the logical ports onto the available (fewer) physical PCB traces along with other infrastructure components for clocking/control and memory interfaces. In addition to mapping the DUT logic onto the FPGA fabric, the tool also properly handles mapping memory macros to external discrete PCB memory. It also automatically generates timing constraint files necessary for subsequent steps in the flow. The output VHDL files and the corresponding constraints are then separately fed to vendor FPGA synthesis and routing tools to produce the configuration bitstreams for each of the FPGAs in the system. This step is done in parallel by running all FPGA synthesis jobs simultaneously on a machine farm using IBM's job queuing system (Loadleveler). Figure 10 shows a high level view of the partitioning flow.

5.5 Waveform generation

Partitioning a design across a large number of FPGAs creates its own set of challenges, a critical one being the ability to debug the target system in a single view. Exploiting the stoppable clock architecture and communication infrastructure outlined in the previous two sections along with Xilinx's configuration readback feature [11], Twinstar provides a unified waveform display, Twinstar Wevegen, allowing for cycle accurate debugging of the entire system. Unlike a software-only in-system debug, the waveform generator can take advantage of Twinstar's 4MHz cycle accurate and repeatable run to quickly arrive at the cycle of interest before beginning a detailed trace run. As shown in Table 1, debugging a Linux bringup would be prohibitive in a software only simulation. The basic idea is to run the FPGA system at full speed and stop at the region of interest. While the DUT is stopped, we scan the state of each target FPGA in the system using the Xilinx configuration readback protocol, and extract the latch state information. As the scanning is non-destructive, we then single-step the DUT clocks and repeat the scanning operation, thereby creating a cycle-by-cycle waveform or trace of the DUT execution.

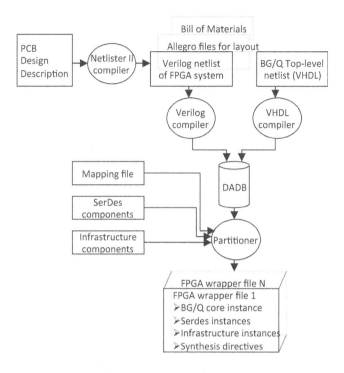

Figure 10: Array validation environment to ensure cycle accurate array mapping

According to the readback protocol, the entire configuration memory of the FPGA can be read out as a sequence of data frames [6]. Reading a frame of configuration information has the welcome side-effect of reading the current state of latches that are part of the frame. Identifying the individual latches is facilitated by a cross-reference file that is created during synthesis relating the full hierarchical design latch name to its position in the bitstream in terms of frame number and bit offset within the frame.

In "raw trace" mode of Twinstar Wavegen, the host software sends, for each cycle, the full sequence of commands necessary to scan the frames of interest and receives the entire set of response frames, then performs post-processing in software to extract the relevant latch data and display it to the user. This works fine for a limited number of cycles, but quickly becomes a performance bottlneck for larger traces. This is due to the fact that the amount of data transferred in a raw trace is fairly significant; for example in a trace run monitoring 2000 signals, each cycle transmits 5.7MB (64bytes x 2000 Frames x 45 FPGAs) of command data and receives 29.5MB (328bytes per frame x 2000 Frames x 45 FPGAs) of frame data. Collecting trace data for a 1M cycle debug can easily overwhelm memory and disk resources. To allow multiple turns for debugging, it is necessary to overcome the sluggish debug time and data overload.

Twinstar Wavegen implements a "bit filtering" mode to reduce overall traffic and increase configuration readback speed. By preloading the necessary commands for an internal capture hardware macro (ICAP) readback and the frame

---

[5]Other industry-standard EDA tools, e.g. Synopsys or Cadence, would also work.

[6]for Xilinx Virtex-5 family, a frame is 1312 bits of configuration data, which includes around 160 bits of relevant latch state. The first frame in a series of contiguous frames has to be read twice to get the updated data.

	Frequency	Time to boot Linux
Simulation	10 Hz	4.8 years
AWAN	1 KHz	17.6 days
Twinstar	4 MHz	6.3 minutes
ASIC	1600 MHz	1 second

Table 1: Time to boot Linux on various platforms

list into on-chip memory, the transmit traffic is reduced to 720KBs (8bytes x 2000 x 45 FPGAs), an 8X drop. This decrease in traffic also decreases time between frame readback requests to the ICAP dramatically. Since the host only needs to set up the configuration readback requests once, the result is a significantly higher net throughput. Increasing the throughput also implies more data transmitted back to the host, overwhelming the processing, memory, and disk resources faster. Bit filtering overcomes this by discarding redundant data and filtering out frame data by extracting relevant latch bits on the FPGA. The result is a 16X reduction in data returned to the host, or 1.8MB (20bytes x 2000 Frames x 45 FPGAs) per cycle in the 2000 signal example.

6. RESULTS

All major blocks of the Bluegene/Q compute ASIC except the network subsystem were successfully mapped onto Twinstar. Mapped elements included all processor cores with their L1 caches, quad floating-point SIMD units, memory prefetch logic, the crossbar switch, the 32 MB embedded DRAM L2 cache, the two DDR memory controllers, and 2 GB of external DDR DRAM memory. The Twinstar system was configured with 24 node cards and used 45 Xilinx Virtex-5 LX330 FPGA devices for modeling the ASIC, in addition to the control FPGA devices and discrete SRAM and DRAM components. Two such systems were built and deployed for the Bluegene/Q program. Each system had a corresponding host control machine providing two modes of operation, a batch mode and an interactive mode. For batch mode operation, the system was connected to the rest of the verification machine farm using the IBM Loadleveler batch job queuing system . This infrastructure enabled the user to easily and seamlessly direct any simulation job to either a software simulator running on a general purpose machine or to one of the Twinstar accelerator systems using a few switches. In interactive mode, the user reserved a console session through which he or she could execute on the system in real-time.

Chip-level simulation ran at 4 MHz [7] processor clock speed across the entire rack. This effective simulation frequency was over *100,000 times* faster than running on a software-based simulator at the RTL level, providing a new capability to the user that was not possible before. To put this in perspective, Table 1 shows a comparison of the time it would take to boot Linux on the various simulators available to the Bluegene/Q program, including an internal (ASIC-based) simulation acceleration tool called AWAN, as well as the actual time to boot on the target chip itself. It clearly shows the value of Twinstar, in particular for long-running simulations, such as booting an operating system and running actual code.

---

[7] 2 MHz for the memory subsystem in lock-step with the chip architecture

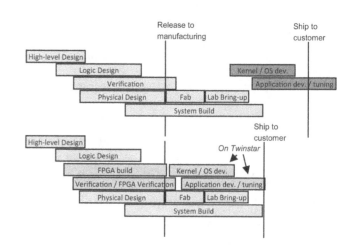

Figure 11: Shrinking time-to-market by enabling early software development

The Twinstar platform proved invaluable for the Bluegene/Q program in the following 3 areas:

1. *Logic verification*: Over 30 logic bugs at the chip-level were uncovered, fixed, and re-checked using Twinstar. Thousands of targeted tests were executed, accumulating tens of billions of simulation cycles and ultimately ensuring a high-quality tape out of the ASIC, reducing overall program risk. In fact, the first iteration of the Bluegene/Q compute ASIC was used to construct a 64-node system that outperformed many supercomputer installations.

2. *Performance validation*: Many of the Sequoia benchmark codes [12] were developed and tuned for the new Bluegene/Q architecture on Twinstar, since it possessed the cycle-accuracy and performance attributes necessary to address the developers' needs. It is worthwhile to note that a high-level software simulator would not be adequate for this task due its lack of accuracy in modeling detailed architecture and implementation characteristics, whereas a cycle-accurate simulation in software would simply be too slow for the develop-test-tune cycle, consuming days for every iteration and negatively impacting developer productivity.

3. *Early software development*: Most of the Bluegene/Q node kernel software was developed using Twinstar. Furthermore, many of the new compiler features were developed and tested on Twinstar, including compiler support for novel hardware architecture such as speculative multi-threading and the transactional memory computational paradigm. Ordinarily, these tasks could only start in earnest once the actual ASIC had been manufactured, tested, and brought up in the lab. The impact of Twinstar was to enable software development much earlier in the overall program schedule, specifically, concurrently with ASIC design, as depicted in Figure 11, thereby shrinking the overall time-to-market.

7. ACKNOWLEDGMENTS

The Bluegene/Q project, and consequently the Twinstar FPGA platform, has been supported by Argonne National Laboratory and the Lawrence Livermore National Laboratory on behalf of the United States Department of Energy, under Lawrence Livermore National Laboratory Subcontract No. B554331. The authors would like to thank the Department of Energy and the aforementioned national labs for supporting this work. The authors are also thankful to Xilinx personnel for close collaboration on the state readback using the Internal Capture (ICAP) macro for Virtex-5 devices.

8. REFERENCES

[1] P. Wang, J. Collins, C. Weaver, B. Kuttanna, S. S. G. Chinya, E. Schuchman, O. Schilling, T. Doil, S. Steibl, and H. Wang, "Intel Atom processor core made FPGA Synthesizable," in *Proceedings of the 17th Annual ACM/SIGDA International Symposium on Field Programmable Gate Arrays (FPGA'09)*, ACM, 2009.

[2] G. Schelle, J. Collins, E. Schuchman, P. Wang, X. Zou, G. Chinya, R. Plate, T. Mattner, F. Olbrich, P. Hammarlund, R. Singhal, J. Brayton, S. Steibl, and H. Wang, "Intel Nehalem Processor Core Made FPGA Synthesizable," in *Proceedings of the 18th Annual ACM/SIGDA International Symposium on Field Programmable Gate Arrays (FPGA'10)*, ACM, 2010, Monterey, California.

[3] M. Gschwind, V. Salapura, and D. Maurer, "FPGA Prototyping of a RISC Processor Core for Embedded Applications," *IEEE Transactions on Very Large Scale Integration (VLSI) Systems*, vol. 9, no. 2, pp. 241–250, 2001.

[4] J. Ributzka, Y. Hayashi, F. Chen, and G. R. Gao, "DEEP: an iterative FPGA-based many-core emulation system for chip verification and architecture research," in *Proceedings of the 19th ACM/SIGDA international symposium on Field programmable gate arrays*, pp. 115–118, ACM, 2011.

[5] P. Subramanian, J. Patil, and M. K. Saxena, "FPGA prototyping of a multi-million gate System-on-Chip (SoC) design for wireless USB applications," in *Proceedings of the 2009 International Conference on Wireless Communications and Mobile Computing: Connecting the World Wirelessly*, pp. 1355–1358, ACM, 2009.

[6] J. Wawrzynek, D. Patterson, M. Oskin, S.-L. Lu, C. Kozyrakis, J. Hoe, D. Chiou, and K. Asanovic, "RAMP: Research Accelerator for Multiple Processors," *IEEE Micro*, vol. 27, no. 2, pp. 46–57, 2007.

[7] A. Krasnov, A. Schultz, J. Wawrzynek, G. Gibeling, and P.-Y. Droz, "RAMP Blue: A Message-Passing Manycore System in FPGAs," in *International Conference on Field Programmable Logic and Applications, 2007*, pp. 54–61, 2007.

[8] C. Chang, K. Kuusilinna, B. Richards, A. Chen, R. Brodersen, and B.Nikolic, "Rapid design and analysis of communication systems using the BEE hardware emulation environment," in *Proceedings of the 14th IEEE International Workshop on Rapid Systems Prototyping, 2003*, pp. 148–154, IEEE, 2003.

[9] C.-Y. Huang, Y.-F. Yin, C.-J. Hsu, T. B. Huang, and T.-M. Chang, "SoC HW/SW verification and validation," in *Proceedings of the 16th Asia and South Pacific Design Automation Conference*, pp. 297–300, IEEE, 2011.

[10] R. Haring, "The Blue Gene/Q Compute Chip." `http://www.hotchips.org/archives/hc23/HC23-papers/HC23.18.1-manycore/HC23.18.121.BlueGene-IBM_BQC_HC23_20110818.pdf`, 2011.

[11] Xilinx, "Virtex-5 FPGA Configuration User Guide." `http://www.xilinx.com/support/documentation/user_guides/ug191.pdf`, 2011.

[12] Lawrence Livermore National Laboratory, "ASC Sequoia Benchmark Codes." `https://asc.llnl.gov/sequoia/benchmarks/`, 2009.

FPGA-Accelerated 3D Reconstruction Using Compressive Sensing

Jianwen Chen[§], Jason Cong[§], Ming Yan[†] and Yi Zou[§]

[§]Computer Science Department
University of California, Los Angeles
Los Angeles, CA 90095, USA

[†]Department of Mathematics
University of California, Los Angeles
Los Angeles, CA 90095, USA

jianwen.chen@ieee.org,{cong@cs,yanm@math,zouyi@cs}.ucla.edu

ABSTRACT

The radiation dose associated with computerized tomography (CT) is significant. Optimization-based iterative reconstruction approaches, e.g., compressive sensing provide ways to reduce the radiation exposure, without sacrificing image quality. However, the computational requirement such algorithms is much higher than that of the conventional Filtered Back Projection (FBP) reconstruction algorithm. This paper describes an FPGA implementation of one important iterative kernel called EM, which is the major computation kernel of a recent EM+TV reconstruction algorithm. We show that a hybrid approach (CPU+GPU+FPGA) can deliver a better performance and energy efficiency than GPU-only solutions, providing 13X boost of throughput than a dual-core CPU implementation.

Categories and Subject Descriptors

B.7.1 [**Integrated Circuits**]: Types and Design Styles—*Algorithms implemented in hardware*

General Terms

Algorithms, Design, Performance

1. INTRODUCTION

The industry trend of CT imaging is moving towards low-dose CT. Although it is possible to reduce the dose directly and apply image-space denoising on the noisy FBP image, a more desired approach is to reduce the number of sampling used and apply compressive sensing-based iterative reconstructions. For a review of the CT image reconstruction and optimization-based iterative schemes, please refer to the recent survey [3].

This paper presents our effort to accelerate one iterative reconstruction algorithm called EM+TV [4], which extends classic Expectation Maximization (EM) [2] algorithm by introducing Total Variation(TV) regularization terms. We implemented the EM kernel completely on virtex 6 FPGAs. Our implementation is done at C-level by using AutoESL high-level-synthesis tool [1] from Xilinx.

2. ALGORITHM OVERVIEW

Typically, image reconstruction requires the number of samples (measurements or observations) that is above the Nyquist limits. By exploiting the sparsity of the objects, the number of samples can be reduced significantly. Compressive sensing technique exploits this fact to perform the reconstruction of signals or images. In our case, suppose the image is x, we make use of the sparsity of $|\nabla x|$ in the algorithm.

EM+TV reconstruction [4] tries to solve the non-linear optimization problem:

$$\min_x \int_\Omega |\nabla x| + \alpha \sum_{i=1}^M ((Ax)_i - b_i \log(Ax)_i)$$
$$x_j \geq 0, j = 1, \cdots, N \qquad (1)$$

The first term is the TV term and the second one is the EM term. We ignore the mathematic details that can be referred in [4], but show the pseudo-code for the core computing functions instead.

2.1 Ray Tracing

EM algorithm is often implemented with a ray-driven forward-projection and a voxel-driven back-projection. To facilitate hardware sharing, we use ray-driven approach in both forward and backward projections. The code of the forward and backward projection is shown in Figure 1. The code first finds out the direction for the next voxel in the ray, then it performs multiply-and-accumulate operation to accumulate the sinogram or update the image. The tracing stops if the voxel hits the boundary of the object.

2.2 Intersection Computation

The $tracer\_precal()$ function is responsible for computing the intersection point of the ray with the object and find out the parameter required for the tracing. Given a source coordinate (s_x, s_y, s_z) and destination (d_x, d_y, d_z), the procedure finds out the intersection point with the object which is a cube $0 \leq x < N_x, 0 \leq y < N_y, 0 \leq z < N_z$. A number of divisions are used in the procedure.

3. IMPLEMENTATION & OPTIMIZATION

3.1 Parallel Backward Projection

The forward projection can be parallelized easily. A large number of parallel unit can operate on the forward ray tracers simultaneously for different source and detector pairs. For backward projection, there are dependencies among views. Moreover, even within one view, there are conflicts when two parallel units update one pixel. To resolve the data conflicts within one view, atomic functions that guarantee the mutual exclusion of an address in memory, can be used to handle such potential data conflicts. However, our target FPGA platform do not provide atomic operations

```
//EMupdate : ray−tracing algorithm
for all the views
for all the detectors
{
  tracer_precal(); // find initial ray parameters
  //   λ_x, λ_y, λ_z, λ_0,  v_x, v_y, v_z,
  //   Len_x, Len_y, Len_z,  sign_x,  sign_y,  sign_z
  if(mode==0) tempsino=0; //forward projection
  else   value= sinogram(..); //backward projection
  for  (i=0; i < N_x + N_y + N_z; i++)//(tracer_loop)
    {
    if  (λ_x <= λ_y && λ_x <= λ_z) λ = λ_x;
    else if (λ_y <= λ_z)      λ = λ_y;
    else                λ = λ_z;
      //Multiply accumulate (MAC) computation
    if(mode==0)           // forward projection
    tempsino+ = imageData(v_x, v_y, v_z) * (λ − λ_0);
    else           // backward projection
    imageData(v_x, v_y, v_z)+ = value * (λ − λ_0);
    λ_0 = λ;
      //Find the next point on the ray
    if  (λ_x <= λ_y && λ_x <= λ_z){λ_x+ = Len_x; v_x+ = sign_x;}
    else if (λ_y <= λ_z)  {λ_y+ = Len_y; v_y+ = sign_y;}
    else            {λ_z+ = Len_z; v_z+ = sign_z;}
    //Exit conditions
    if(v_x < 0||v_x > N_x − 1) break;
    if(v_y < 0||v_y > N_y − 1) break;
    if(v_z < 0||v_z > N_z − 1) break;
    }
  if(mode==0) sinogram(..)=tempsino;
}
```

Figure 1: Ray Tracing Core Engine

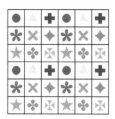

Figure 2: Ray Based Parallel Mapping

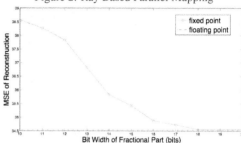

Figure 3: Fractional bit width and Reconstruction Quality

on the memory system.[1] The only way to obtain a correct design is to enforce memory requests to complete sequentially. This has substantial overhead because the memory system is designed to be weakly ordered and supports parallel data access. We instead exploit algorithm-level changes to avoid the use of atomic operations. First, we ensure the computation for different views(sources) are done in a sequential fashion. For a same view, the detectors that are far enough are set to one group. Mathematically there will be no conflicts within the group and all tracers in one group can be processed in parallel. As illustrated in Figure 2, we can choose the tracer lines of the same pattern in one group.

3.2   Fixed Point Conversion

To reduce the area of our design, we convert floating point computation into fixed point. We use standard range analysis technique to obtain the range of all the values in our datapath. Because the algorithm is iterative, static precision analysis would generate quite pessimistic results. We use dynamic analysis instead to determine the number of fractional bits.

We try different number of fractional bits and compare with the floating point reference code. As illustrated in Figure 3, the bitwidth of the fractional part will influence the reconstruction quality greatly. When 18 bits (10^{-5}) are used, the fixed point version can achieve the same reconstruction quality of the floating point version. We enlarge the bidwidth by additional 2 bits to bring in more safe margins, and use 20 bits for the fractional part. Note that it is still possible to store all those array data using 32-bit data when we use 20-bit fractional part.

3.3   Streaming Architecture

Function $tracer\_precal$ and the tracer loop computation can be executed in a task-level pipeline. We synthesize the $tracer\_precal$

and the tracer loop individually to obtain their corresponding latency reports. Because the loop bound of the tracer loop is not known, we use an average loop bound from the simulation of the test data to compute the average-case latency of the tracer loop. The throughput of the memory interfaces is also considered. Roughly the latency of the $tracer\_precal$ is around 1/4 of the latency of the tracer loop for a 128^3 test data. Because of this, we realize two $tracer\_precal$ modules and eight tracer loop module in a single FPGA. Each FPGA has 16 virtual memory channels, and each tracer loop module talks to two of them (one for read and one for write). The multi-FPGA system has 4 user FPGAs (Application Engine or AE), we distribute the work-load using SIMD fashion.

The diagram of our implementation in one FPGA is shown in Figure 4. To realize such a diagram in C level, we invoke the function $tracer\_precal$ twice and invoke the function of the tracer loop eight times. These different invocations take different FIFO channels and memory interfaces as parameters. The compiler can figure out that these function calls are independent and shall generate a parallel hardware.

The transform that converts the code in Figure 1 to a C code that calls two $tracer\_precal$ and eight $tracer\_loop$ seems counter-intuitive for software engineers. At higher-level, our manual step in this subsection can be viewed as a combination of loop unroll transform and loop distribution transform, where the distributed loops then take different unrolling factors. In practice, these decisions still need to be coded at a lower level.

The round robin distribution logic is also coded in the $tracer\_precal$ function. At the receiver side $tracer\_loop$, the control is just a simple counter to maintain the number of rays processed. Each $tracer\_loop$ would process a pre-determined number of rays. Note it is possible that the ray do not intersect with the object. In this case, the $tracer\_precal$ would send a special flag to denote that no processing is needed, but the counter should still be updated to obtain a correct exit condition.

The intersection computation we implemented is fairly generic. Currently, the control that sets the list of sources and detectors are also coded in the function, along with the lookup tables ROM for sin cos functions. Note it is very easy to change these controls to reflect another scanner machine setup.

---

[1]It is possible to realize the atomic operation within the BRAM. The off-chip memory does not support atomic updates.

164

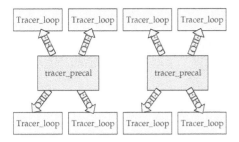

Figure 4: Overall Streaming Architecture Inside One FPGA AE

3.4 Prefetching

Generic HLS tools usually do not model board-specific IO systems. In our implementation, we model each memory access port with a request FIFO and a response FIFO. As shown in Figure 5, we need to invoke two parallel functions inside the hierarchy of $tracer\_loop$. One function is the "helper thread" $tracer\_loop\_addrGen$ which is responsible for sending memory requests for reads, and the other function is the "compute thread" $tracer\_loop\_compute$ which obtains data from response FIFO and write out the computed result into another request FIFO. This way, the helper threads can keep sending as many requests as possible (until the FIFO is full). Effectively, the helper thread is performing the prefetching of the required data, and the response FIFO serves as the prefetch buffer. Figure 5 depicts the architecture inside the $Tracer\_loop$ function.

3.5 Reducing the Data Accesses via Sparsity

The final output image of the compressive sensing algorithm is sparse. Also we know that the image voxel value is non-negative. Based on these two facts, we develop a simple heuristic to reduce the amount of data access. In the beginning of the iteration, we perform a single forward projection. If any accumulated sinogram value falls below a threshold, we conclude that any image value on that ray shall be close to zero. Based on this, we build a mask of the image called $image\_denote$. When we do the backward projection, we only update the voxels that are not masked. Note that this mask only need 1-bit data, so we merge this 1-bit data into the $imageData$ array. Through this way, we reduce the number of data access in the backward projection. Figure 6 shows the modified pseudo code.

3.6 Simultaneous Reconstruction of Two Images

After fixed point conversion, the external data accesses are all in 32-bit. The memory interface of our multi-FPGA platform supports 64-bit memory interface. Because of the data access in the tracing is somewhat random, it is hard to use the 64-bit interface to enlarge the application bandwidth. However, it is straightforward to use that to reconstruct two images simultaneously, by properly pack two 32-bit data from two images into a 64-bit data. These two images need to have exact machine setup where the $tracer\_precal$ part does not need to be changed.

We do not increase the number of MACs to support the 64-bit data. We measured that the external memory FIFO interfaces would return one data in about three cycles in the average case. [2] We simply enlarge the initiation interval (II) of the tracer loop from 1 to 2 to facilitate the sharing of MAC units.

---
[2] The peak rate is one data in every cycle. We did not reach such a high rate because our application logic is connected to a crossbar logic which performs arbitration and packet routing.

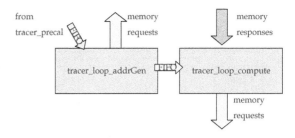

Figure 5: Streaming Architecture Inside One $Tracer\_loop$ Kernel

```
if (mode==0) // forward projection
    tempsino+ = imageData(v_x, v_y, v_z) * (λ − λ_0);
else // backward projection
    if (image_denote(v_x, v_y, v_z)==1)
        imageData(v_x, v_y, v_z)+ = value * (λ − λ_0);
```

Figure 6: Masking for Backward Projection

4. EXPERIMENTAL RESULTS

Our whole design is described in C and synthesized into verilog RTL using AutoESL HLS tool version 2011.1. The target hardware platform is Convey HC-1ex with 4 Virtex-6 LX760 user FPGAs. We designed the RTL interfaces for AutoESL tool to hook up with Convey's Personality Development Kit (PDK). Those interfaces are reused by a number of designs we implemented. PDK is the RTL-based synthesis and simulation environment for the HC-1ex platform. We synthesize the RTL generated by the AutoESL HLS tool along with the PDK infrastructure RTLs using Xilinx ISE 12.4.

Our test setup assumes a Cone-Beam CT system. Currently, we tested a phaeton data of size 128^3 which is supplied by authors of [4]. We have 36 views (sources) and the size of detector (destinations) is 301*257. According to [4], the EM+TV algorithm using 36 samples, can obtain a similar image quality that is obtained using FDK/FBP algorithm that requires 360 samples, resulting the radiation reduction by 10X.

4.1 Kernel Performance and Energy Consumption

Table 1 presents the performance and the energy consumption of the forward projection kernel and the backward projection kernel. The number is collected by averaging 1000 invocations. The performance on a dual-core CPU and many-core GPU is also reported. The CPU used is Intel Xeon 5138 with 2.13GHZ clock frequency and 35W TDP. The GPU1 column denotes Nvidia C1060 with 240 cores and 200W TDP. The GPU2 column denotes Nvidia GTX480 with 480 cores and 250W TDP. We parallelize the CPU code using OpenMP and implement the GPU kernel using Nvidia CUDA Toolkit 3.2. The throughput of the FPGA design is better than the latency because we can reconstruct two images simultaneously. The power of the FPGA application engine is measured by Xilinx xPower tool. We have 4 user FPGAs in the system. The actual system power of the Convey system is larger as the coprocessor memory, coprocessor PCB etc., also consume a lot of power.

From the Table 1 we can see that, when the latency of forward and backward is added together, our multi-FPGA engine is about 50% faster than the CUDA implementation on Tesla C1060, but about 2X slower than Fermi GTX480. When we consider the fact we can do two reconstructions simultaneously, that means our FPGA-engine is 3X faster than Tesla C1060 and in par with Fermi GTX480. The energy number is listed in the table as well. We can see that

Table 1: Performance and Energy Numbers for Computing Kernels for 128^3 data

	Power	Forward Projection		Backward Projection		Forward+Backward	
		Latency/Throughput(s)	Energy(J)	Latency/Throughput(s)	Energy(J)	Latency/Throughput(s)	Energy(J)
CPU	35W	1.81	63.4	1.67	58.4	3.48	121.8
FPGA	94W	0.305/0.153	28.7/14.4	0.308/0.154	29.0/14.5	0.613/0.307	57.7/28.9
GPU_1	200W	0.342	68.4	0.668	133.6	1.01	202
GPU_2	250W	0.085	21.3	0.276	69	0.361	90.3

Table 2: Area Results

	BRAM	DSP	LUT	FF	Slice
Consumed	79	68	113,355	104,099	36511
Total Available	720	864	474,240	948,480	118,560
Utilization	11%	7%	23%	10%	30%

Table 3: Application Performance and Energy Consumption

	Throughput(s)	Energy(J)
CPU	1189	41.6E3
GPU_1	361	72.2E3
GPU_2	114	28.5E3
Hybrid	92.0	12.7E3

the FPGA platform delivers a good performance with a much lower energy.

Note that it turns out that the execution time for backward projection is noticeably slower on GPU platforms. This is because the amount of data access is up-to 2X larger (we need to first read the voxel value and then write it back). Also we need to use more invocations (and synchronization) to avoid the conflicts and ensure the correctness. That also reduces the available parallelism. For the FPGA design, we use the same architecture for both forward and backward. Each PE is connected to two memory channels, one for read and one for write. Thus their execution times are similar. However, in the forward projection, the memory channel is somewhat under-utilized, because the number of writes is much smaller than reads. Potentially the forward projection can be made 2X faster if we separate the design for forward and backward.

Another interesting observation is that the Fermi GPU GTX480 is between 3 to 4X faster than Tesla C1060. The number of cores is 2X of C1060 and the peak off-chip bandwidth is about 1.6X (from 100GB/s to 160GB/s). So it is likely that there is an additional 2X performance benefit attributed from its cache systems. Our current FPGA design does not have a cache, but it is indeed worthwhile to investigate that possibility given the performance benefit we see from GPU.

The area results for the complete design are listed in Table 2. Note our core computing RTL consumes fewer logic slices, because the PDK infrastructure also consumes about 10% to 15% area. Most of the BRAM utilization is due to the PDK infrastructure.

4.2 Application Performance and Energy Consumption

We then test the application performance of the EM+TV algorithm on a hybrid configuration where the EM part is done by the FPGA-subsystem and the TV part is done by the GPU. In the application, the outer iteration iterates 100 times (the application calls 100 times of *EMupdate* and *TVupdate*), and the inner *EMupdate*

step iterates 3 times (each *EMupdate* calls forward and backward projection routine 3 times).

Our hybrid configuration connects Fermi GTX480 onto the Convey HC1-ex platform. After one EM iteration completes, the image data is copied into the GPU memory space and the TV CUDA kernel starts. The data transfer would not add substantial overhead in this case. We measured that a pipelined data transfer (FPGA coprocessor-side memory to PCI-e) can reach close to 1GB/s. Each EM iteration only needs to copy 128^3 or 8MB image data to GPU. And similarly we need to do the transfer backwards when one TV invocation finishes. That only adds about 0.016s for each EM+TV iteration, or about 2s for the whole EM+TV application. Because the TV kernel is highly regular stencil computation, GPU is a good choice for that application kernel. The execution time of the TV is much shorter than EM. In the energy calculation for the hybrid configuration, we assume that GPU can be powered off when it is not actively running CUDA applications. In practice, a 10% to 15% idle power may remain.

Later, we also tested a phantom with size 256^3 and $512 * 512 * 256$, and obtained a similar speedup. The algorithm is roughly linear with number of voxels if the number of iterations are unchanged.

5. CONCLUSIONS AND FUTURE WORK

In this paper, we present one FPGA-based implementation for ray-tracing EM kernels using AutoESL HLS tools. We further show that a hybrid approach provides good performance and potential energy savings. Currently, we are investigating different algorithmic or architectural approaches that can improve the data locality/reuse for the application.

6. ACKNOWLEDGEMENTS

This research is supported by the Center for Domain-Specific Computing (CDSC) which is funded by the NSF Expedition in Computing Award CCF-0926127.

7. REFERENCES

[1] J. Cong, B. Liu, S. Neuendorffer, J. Noguera, K. Vissers, and Z. Zhang. High-level synthesis for fpgas: From prototyping to deployment. *IEEE TCAD*, 30(4):473 –491, April 2011.

[2] A. P. Dempster, N. M. Laird, and D. B. Rubin. Maximum likelihood from incomplete data via the EM algorithm. *Journal of the Royal Statistical Society, Series B*, 39(1):1–38, 1977.

[3] X. Pan, E. Y. Sidky, and M. Vannier1. Why do commercial CT scanners still employ traditional, filtered back-projection for image reconstruction? *Inverse Probl*, 25(12), January 2009.

[4] M. Yan and L. A. Vese. Expectation maximization and total variation-based model for computed tomography reconstruction from undersampled data. In *Proc. SPIE Conference on Medical Imaging: Physics of Medical Imaging*, 2011.

Reconfigurable Architecture and Automated Design Flow for Rapid FPGA-based LDPC Code Emulation

Haoran Li, Youn Sung Park, Zhengya Zhang
Department of Electrical Engineering and Computer Science, University of Michigan, Ann Arbor
lihaoran@umich.edu, parkyoun@umich.edu, zhengya@eecs.umich.edu

ABSTRACT

Multitude of design freedoms of LDPC codes and practical decoders require fast simulations. FPGA emulation is attractive but inaccessible due to its design complexity. We propose a library and script based approach to automate the construction of FPGA emulations. Code parameters and design parameters are programmed either during run time or by script in design time. We demonstrate the architecture and design flow using the LDPC codes for the latest wireless communication standards: each emulation model was auto-constructed within one minute and the peak emulation throughput reached 3.8 Gb/s on a BEE3 platform.

Categories and Subject Descriptors

C.3 [**Special-purpose and application-based systems**]: Signal processing systems

General Terms

Design, Experimentation, Performance

Keywords

LDPC, decoder emulation, decoder architecture

1. INTRODUCTION

Low-density parity-check (LDPC) codes are capacity-approaching codes that can perform very close to the Shannon limit when decoded using the iterative belief propagation algorithm [1], [2]. Over the last few years, we have seen LDPC codes entering a range of important applications, from wireline [3], wireless [4]-[7], satellite [8], optical communications [9] to magnetic storage [10], to improve reliability and spectral efficiency. However, practical performance of LDPC codes can be far from their theoretical limit for two reasons: (1) a practical code's block length is limited to hundreds to thousands of bits to meet latency and complexity constraints; (2) practical decoder implementations introduce non-idealities, such as finite word length and fixed-point quantization effects [11]. It is therefore critically important to evaluate code constructions and practical decoder implementations for each new application that is brought in consideration.

Software-based code and decoder simulation are common in practice. A typical simulation setup shown in Fig. 1 consists of an encoder to produce codewords (or memory that stores known codewords), a modulator that translates bits to real values for transmission, a channel model that generates noise to corrupt the transmitted values, and a decoder that runs the belief propagation algorithm to recover the binary codeword from the real values

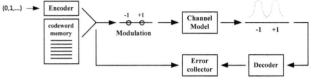

Fig. 1. Code and decoder simulation setup.

received. The decoded word is compared with the transmitted codeword to determine if an error (frame error) has occurred and, if it has, the number of bits that are wrong (bit error). Decoding errors are measured in frame error rate (FER) and bit error rate (BER). When the channel condition is poor, i.e., at low signal-to-noise ratio (SNR), decoding errors occur frequently (high FER and BER), shortening the simulation time. At high SNR, decoding errors occur infrequently (low FER and BER) and the simulation time is longer. Low FER and BER simulation is the bottleneck in code and decoder simulations.

As new generations of applications push for a higher throughput and reliability, the required FER and BER are also extended lower. For example, 10-gigabit Ethernet requires a BER of 10^{-12} or better [3]. Simulating complex decoders for these systems to extremely low BER is a challenge, as it often takes weeks or months to run a belief propagation decoder to reach a BER of 10^{-12} on a high-performance microprocessor. Recently, field-programmable gate arrays (FPGA) have been proposed to accelerate the simulations, showing three orders of magnitude speedup or more [11]-[14]. Despite the impressive speedup, FPGA emulation has not gained wide-spread use. Designing FPGA emulation is not as easy as writing C code. It requires extensive effort in creating hardware architecture and running FPGA synthesis. The barrier renders emulation inaccessible to the vast coding theory and applications community who would otherwise benefit the most in code construction and system evaluation.

In this paper, we address the challenges in creating LDPC code and decoder emulation by creating an automated design flow based on a reconfigurable hardware decoder architecture. The design flow is built upon a decoder library that consists of modules parameterized by code parameters and design parameters. Given a new LDPC code, the design flow instantiates processing elements and constructs a highly parallelized LDPC decoder. We experimented with the LDPC codes for IEEE 802.16e (WiMAX) [4], IEEE 802.11n (Wi-Fi) [5], IEEE 802.15c (wireless personal area network) [6], and IEEE 802.11ad (high-throughput wireless) [7], with block lengths ranging from 576 bits to 2,304 bits and code rates from 1/2 to 5/6. In all cases, the design flow completed decoder construction under one minute, followed by FPGA synthesis that took two hours or less. The resulting decoders operated at real-time or nearly real-time, delivering a throughput up to 3.8 Gb/s on a BEE3 platform, allowing us to reach a BER of 10^{-11} in one hour and below 10^{-12} in one day. We demonstrated the capability of the emulation platform in evaluating the functional performance of various codes. With added network interface and publically available design flow and library, the proposed FPGA

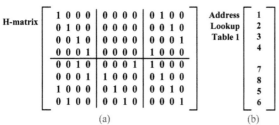

Fig. 2. An LDPC code example: (a) **H**-matrix, and (b) address lookup table for the first block column.

TABLE I
STRUCTURE OF LDPC CODES USED IN COMMUNICATION SYSTEMS

Standard	z	n_b	m_b	n	m
IEEE 802.11ad	42	16	3,4,6,8	672	126 to 336
IEEE 802.15c	21	32	4,8,16	672	84 to 336
IEEE 802.11n	27,54,81	24	4,6,8,12	648 to 1944	108 to 972
IEEE 802.16e	24,28,...,96	24	4,6,8,12	576 to 2304	96 to 1152
IEEE 802.11an	64	32	6	2048	384

TABLE II
BASIC ARCHITECTURES OF LDPC DECODERS

	Submatrix-parallel	Column-parallel	Row-parallel
Processing elements	z	n_b	m_b
Decoding time per iteration	$n_b \times m_b$	$z \times m_b$	$z \times n_b$

emulation design flow will contribute to both the theoretical and practical coding research.

2. BACKGROUND

LDPC codes are linear block codes. Each LDPC code is defined by a parity-check matrix **H** of size $m \times n$, where n is the block length and m is the number of parity checks. Almost all the latest applications have adopted LDPC codes whose **H** matrix is constructed using m_b rows and n_b columns of $z \times z$ identity matrix, its cyclic shifts, or zero matrix [3]-[7]. A simple example is given in Fig. 2, where an 8×12 **H** matrix is constructed using 2 rows and 3 columns of 4×4 submatrices and each submatrix is an identity matrix, its cyclic shift, or zero matrix. We surveyed the LDPC codes in the latest communication standards and summarized their **H** matrix structures in Table I. Each **H** matrix has a fixed n_b and often a variable m_b to control the number of parity checks (or code rate) for different channel environments. In poor channel conditions, a high m_b (low code rate) is used to introduce more redundancy for a stronger protection. The submatrix size z controls the code block length: a longer block length offers better protection at the cost of a longer latency and a higher decoding complexity.

The belief propagation decoding of LDPC codes is briefly described as follows. To begin, the received real value for each bit is used to initialize the bit's *prior* likelihood [2]. The subsequent steps are carried out iteratively in a procedure following the **H** matrix [2]. In the horizontal half iteration, we go through each row of the **H** matrix and read the prior likelihoods of the bits that participate in the parity check described by the row, followed by computing an update, known as the *extrinsic*, indicating the likelihood of each bit given the likelihoods from all other bits participating in this parity check. The horizontal half iteration is completed in m horizontal steps. In the vertical half iteration, we go through each column of the **H** matrix and read all the extrinsics corresponding to the parity checks that the bit is part of to compute an updated likelihood, known as the *posterior*. A hard decision is made based on the posterior. The vertical half iteration is completed in n vertical steps. For the second and following iterations, the horizontal half iteration is carried out in the same way as the first iteration, but instead of prior likelihoods, we use modified posterior likelihoods for the computation. More iterations improve the reliability of each bit. If hard decisions of all bits satisfy all the parity-check equations, decoding converges. For a full mathematical description of the belief propagation algorithm, we refer readers to [1], [2]. The algorithm works remarkably well in practice, and usually converges in a small number of iterations.

3. RECONFIGURABLE EMULATION

The belief propagation decoder is the most complex block of a decoder emulation platform. Many high-performance architectures have been introduced for individual codes, but they have to be customized to be useful for other codes. We design an entirely

reconfigurable architecture that is applicable to all the codes defined in Section 2.

Referring to Table I, the codes are parameterized by z, n_b, m_b, implying three natural ways of parallelizing the decoder: submatrix-parallel, column-parallel, or row-parallel. Consider a submatrix-parallel architecture, where z processing elements (PE) completes z horizontal (or vertical) steps concurrently, thus the decoding time per iteration is proportional to $n_b \times m_b$. A row-parallel architecture requires m_b processing elements for a decoding time of $z \times n_b$. Since the parameters z and m_b are variable for some applications, the submatrix-parallel and row-parallel architectures require pre-allocation of the maximum number of PEs and runtime reconfiguration. Note that the parameter n_b is fixed for each application, so the column-parallel decoder architecture supports all codes for a given application without requiring any over allocation of PEs. The decoding time however varies with m_b. The three basic architectures are listed in Table II for comparison. Additional architectures can be created by parallelizing or serializing the three basic architectures or mixing them. More parallel architectures demand more hardware resources. To achieve the maximum throughput on a given FPGA platform, we can create multiple decoders to run parallel emulations.

3.1 Emulation System Design

We choose the column-parallel architecture for the code and decoder emulation platform shown in Fig. 3, where n_b PEs are allocated and connected to a parity-check node for the horizontal step computation, the output of which is sent to each PE. The vertical step is completed within the PE. The PE datapath is entirely data-driven. The controller only needs to generate the address counter for each PE to access the correct data and write to the correct location. The address sequence is determined by the **H** matrix and stored in an address lookup table. An example of the address lookup table is shown in Fig. 2.

An input generator consists of a Gaussian noise generator and a set of valid codewords stored in a memory. Gaussian noise is added to codewords to emulate the effect of an additive white Gaussian noise channel (AWGN). The channel SNR is adjusted by noise variance.

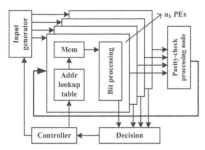

Fig. 3. Column-parallel decoder architecture.

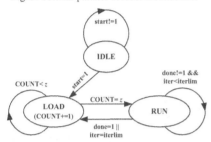

Fig. 4. Emulation control state machine.

A controller orchestrates the emulation. Its simplified state machine is shown in Fig. 4. System starts up in IDLE state. Upon receiving a "start" signal, it transitions to LOAD state, where n_b PEs load inputs from the input generator concurrently. As each PE is assigned to a block column of z bits, the loading will take z clock cycles. A persistent state variable COUNT keeps track of the loading state. COUNT increments by 1 each clock cycle until it reaches z, when the system transitions to RUN state.

The belief propagation algorithm operates by the **H** matrix row-by-row for the horizontal half iteration, and then column-by-column for the vertical half iteration. Here we merge the vertical step with the horizontal step by performing vertical steps following each horizontal step. The interleaved processing lengthens the pipeline of each horizontal step, but completes one decoding iteration in approximately m clock cycles.

Towards the end of RUN state when the posterior likelihood of each bit is finalized, the controller enables the decision block to make hard decisions. If decoding converges or the iteration limit is reached, a "done" signal is generated to trigger the transition to LOAD state for loading another input vector; if decoding fails to converge when the iteration limit is reached, a frame error and number of bit errors are recorded before moving to LOAD state; if decoding fails while the iteration limit is not yet reached, it will remain in RUN state by starting another decoding iteration.

3.2 Design- and Run-Time Reconfiguration

The emulation system is entirely parameterized. Given an LDPC code, the number of block columns, n_b, determines the number PEs in the system. The submatrix size z and the number of block rows m_b determine the control schedule, including number of loading cycles, address counter, memory write enable, and decision enable. The **H** matrix structure, i.e., the locations of '1' entries in the **H** matrix of each block column, decides the address lookup table entries. The PE and parity-check node complexities are also determined by these parameters: the depth of the extrinsic memory is $z \times m_b$, the depth of the posterior memory is z, the parity-check node implements a n_b:1 adder tree (for the sum-product algorithm [2]) or a n_b:1 compare-select tree (for the min-sum algorithm [15]).

The parameterized emulation system enables convenient reconfiguration. The number of PE blocks, extrinsic and posterior memory depth, parity-check node topology are design-time reconfigurable or limited run-time reconfigurable by over allocation and selective enabling. Control constants and address lookup table are run-time reconfigurable. Hence it is possible to design one IEEE 802.16e-compatible LDPC decoder to be reconfigured for all 19 LDPC codes specified by the standard [4] by setting control constants and address lookup tables.

We consider additional parameters that are important in code and decoder designs: word length and quantization of prior, extrinsic and posterior likelihoods, limit on the number of decoding iterations, channel SNR, and algorithm control knobs (such as the offset in the min-sum algorithm [15]). Word length and quantization are design-time reconfigurable, but it is expensive to change in run time. Decoding iteration limit, channel SNR and some algorithm controls can be easily reconfigured in run time.

To sum up, many parameters are in consideration for code and decoder designs and the interplays among the parameters are of great interest. To speed up these evaluations, we need a reconfigurable architecture to minimize the number of redesigns, such that one design can be reused for many different evaluations. However, many parameters cannot be made run-time reconfigurable easily, which necessitates redesigns. A fast and automated design flow will greatly facilitate this effort.

4. DESIGN FLOW

We propose a design flow based on the BEEcube Platform Studio (BPS) targeting BEEcube BEE3 multi-FPGA platform [16]. The steps of the design flow are illustrated in Fig. 5. The first step is to establish a Simulink design library that consists of the modules that make up an emulation system: PE, parity-check node, noise generator, decision block, and controller. The modules are designed using Xilinx blockset to be readily synthesized. Design parameters including memory size, word length, and quantization are coded as parameters in the modules. The small number of modules are quick to design and easily reusable.

In the second step, a Matlab script is used to perform four tasks: (1) initialize code structure parameters, z, n_b, and m_b, and design parameters, word length and quantization, that are being referenced in the design library; (2) parse the given LDPC code to build address lookup table for each PE; (3) instantiate n_b PEs, a parity-check node, an input vector generator, a decision block, a controller and connect these modules into a complete decoder in Simulink; and (4) create an interface wrapper using BPS blockset to provide configuration registers for run-time reconfigurable parameters: control schedule, decoding iteration limit, channel SNR, and algorithm knobs, as well as output registers that capture BER and FER. This step is fully automated and can be completed in well under 1 minute.

The third step involves BPS compilation, which takes less than 2 hours based on all the experiments we carried out. The resulting bit file is programmed on the BEE3 FPGA platform for emulation experiments.

The proposed design flow integrates with the BPS flow and simplifies the design process. The library and script will be made available online. We take advantage of Virtex 5 FPGAs' Ethernet functionality by connecting them to the network, each with its own IP address. Remote users can control emulations through function calls in Matlab or C code. We expect this work to contribute to the coding research community and encourage collaborations among researchers.

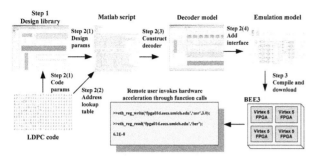

Fig. 5 BEE3-based automated design flow.

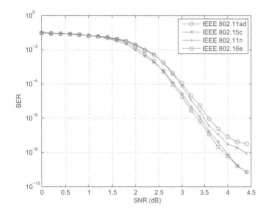

Fig. 6. BER plots of 1/2 LDPC codes used in four communication standards: IEEE 802.11ad, IEEE 802.15c, IEEE 802.11n, IEEE 802.16e. The results are obtained using 10 decoding iterations and 5-bit Q5.0 fixed-point quantization using offset min-sum algorithm.

5. RESULTS

We designed a common design library and applied the automated design flow to LDPC decoders for four different applications, IEEE 802.11ad, IEEE 802.15c, IEEE 802.11n, and IEEE 802.16e. Excluding the initial effort in making the common design library, the flows including compilation completed within two hours. For a better utilization of the target Xilinx Virtex-5 XC5VLX155T device, we created multiple decoder copies, each with its own input generator, decision block and controller, on a single FPGA to run parallel emulations. The device utilization details are listed in Table III based on 5-bit fixed-point quantization and offset min-sum algorithm [15]. The utilization includes fixed overhead created by BPS to handle interfaces and controls. Note that the reported level of parallelism was not limited by the resources available on FPGA, but by the runtime memory of the 32-bit operation system on which the compilation was done.

The designs in Table III have all been successfully compiled and they meet a minimum clock frequency of 100 MHz and deliver throughputs from 380 to 950 Mb/s (in decoding the 1/2 rate code of the longest block length in each standard). With four such FPGAs available on the BEE3 platform [16], we can achieve an emulation throughput up to 3.8 Gb/s, allowing us to reach a BER of 10^{-11} in one hour and below 10^{-12} in one day (with at least 100 bit errors observed for statistical significance).

Fig. 6 shows the performance of four 1/2-rate LDPC codes used in the four communication standards. The 10^{-9} BER point was captured within one minute. Evaluation of code construction and decoder design can be made quickly, e.g., selection of submatrices, code block length, code rate, word length and quantization, iteration limit, algorithm tuning and error floor studies. The rapid

TABLE III
DEVICE UTILIZATION AND THROUGHPUT OF LDPC EMULATION PLATFORMS (BASED ON XILINX VIRTEX-5 XC5VLX155T)

	802.11ad	802.15c	802.11n	802.16e
# of parallel emulators	6	3	4	4
Slice registers	22,323 (23%)	19,971 (21%)	22,221 (23%)	22,689 (23%)
Slice LUTs	35,723 (37%)	31,497 (32%)	38,883 (40%)	39,214 (40%)
Occupied slices	12,568 (52%)	11,819 (49%)	13,565 (56%)	13,219 (54%)
BRAMs	131 (62%)	130 (61%)	130 (61%)	178 (84%)
Throughput (at 100MHz)	950 Mb/s	380 Mb/s	695 Mb/s	710 Mb/s

FPGA-based code emulation will contribute to future coding theory and applications research.

6. ACKNOWLEDGMENTS

This work is supported by National Science Foundation under grant CCF-1054270. We acknowledge the donations made by BEEcube, Xilinx, Intel, and the technical advice by Dr. C. Chang and Dr. K. Camera from BEEcube.

7. REFERENCES

[1] R. G. Gallager, *Low-Density Parity-Check Codes*. Cambridge, MA: MIT Press, 1963.

[2] D. J. C. MacKay, "Good error-correcting codes based on very sparse matrices," *IEEE Trans. Inf. Theory*, vol. 45, pp. 399-431, Mar. 1999.

[3] *IEEE Standard for Local and Metropolitan Area Networks – Specific Requirements Part 3*, Sep. 2006, IEEE Std. 802.3an.

[4] *IEEE Standard for Local and Metropolitan Area Networks Part 16*, Feb. 2006, IEEE Std. 802.16e.

[5] *IEEE Standard for Local and Metropolitan Area Networks – Specific Requirements Part 11*, Feb. 2007, IEEE Std. 802.11n.

[6] *IEEE Standard for Local and Metropolitan Area Networks – Specific Requirements Part 15.3*, Oct. 2009, IEEE Std. 802.15c.

[7] IEEE P802.11 – Task Group AD. (2010). *PHY/MAC Complete Proposal Specification* [Online]. Available: http://www.ieee802.org/11/Reports/tgad_update.htm.

[8] *ETSI Standard TR 102 376 V1.1.1: Digital Video Broadcasting (DVB)*, ETSI Std. TR 102 376, Feb. 2005.

[9] F. Chang, K. Onohara, and T. Mizuochi, "Forward error correction for 100 G Transport Networks," *IEEE Communications Mag.*, vol. 48, no. 3, pp. S48-S55, Mar. 2010.

[10] A. Kavcic and A. Patapoutian, "The read channel," *Proc. IEEE*, vol. 96, no. 11, pp. 1761-1774, Nov. 2008.

[11] Z. Zhang, L. Dolecek, B. Nikolic, V. Anantharam, and M. J. Wainwright, "Design of LDPC decoders for improved low error rate performance: quantization and algorithm choices," *IEEE Trans. Communications*, vol. 57, no. 11, pp. 3258-3268, Nov. 2009.

[12] ———, "Investigation of error floors of structured low-density parity-check codes by hardware emulation," in *IEEE Global Communications Conf.*, Nov. 2006.

[13] Y. Cai, S. Jeon, K. Mai, and B.V.K.V. Kumar, "Highly parallel FPGA emulation for LDPC error floor characterization in perpendicular magnetic recording channel," *IEEE Trans. Magnetics*, vol. 45, no. 10, pp. 3761-3764, Oct. 2009.

[14] X. Chen, J. Kang, S. Lin, and V. Akella, "Accelerating FPGA-based emulation of quasi-cyclic LDPC codes with vector processing," in *Conf. Design, Automation and Test in Europe*, Mar. 2009, pp. 1530-1535.

[15] J. Chen, A. Dholakia, El Eleftheriou, M. P. C. Fossorier, and X. Hu, "Reduced-complexity decoding of LDPC codes," *IEEE Trans. Communications*, vol. 53, no. 8, pp. 1288-1299, Aug. 2005.

[16] BEEcube. (2011). *BEEcube Products* [Online]. Available: http://www.beecube.com/product.

Reliability of a Softcore Processor in a Commercial SRAM-Based FPGA[*]

Nathaniel H. Rollins
NSF Center for High-Performance
Reconfigurable Computing (CHREC)
Brigham Young University
Provo, UT
nhrollins@byu.edu

Michael J. Wirthlin
NSF Center for High-Performance
Reconfigurable Computing (CHREC)
Brigham Young University
Provo, UT
wirthlin@ee.byu.edu

ABSTRACT

Softcore processors are an attractive alternative to using radiation-hardened processors in space-based applications. Unlike traditional processors however, the logic and routing of a softcore processor are vulnerable to the effects of single-event upsets (SEUs). This paper applies two common SEU mitigation techniques, TMR with checkpointing and DWC with checkpointing, to the LEON3 softcore processor. The improvement in reliabilty over an unmitigated version of the processor is measured using three metrics: the architectural vulnerability factor (AVF), mean time to failure (MTTF), and mean useful instructions to failure (MuITF). Using configuration memory fault injection, we found that DWC with checkpointing improves the MTTF and MuITF by over $35\times$, and that TMR with triplicated input and outputs improves the MTTF and MITF by over $6000\times$.

Categories and Subject Descriptors

B.8.1 [**Performance and Reliability**]: Reliability, Performance, and Fault-Tolerance

Keywords

Reliability, Softcore processors, AVF, MTTF, MuITF

1. INTRODUCTION

Microprocessors used in space-based applications must be protected from the effects of high-energy particles. They are usually protected through a very expensive process called radiation-hardening. Radiation-hardened processors are built with special design libraries and fabrication processes that are more resilient to high-energy radiation [1]. These larger

transistors resist high-energy particles by requiring more energy to switch. Although tolerant to radiation, rad-hard processors are larger, slower, consume more power, and can cost hundreds of thousands of dollars [1,2] when compared to a commercial processor. Additionally, rad-hard processors used in space are often one to two decades old [2].

As an alternative to an older expensive rad-hard processor, processors can be implemented in the logic of a field-programmable gate array (FPGA). A processor implemented in an FPGA is called a *softcore* processor because the logic is reconfigurable. In contrast to rad-hard processors, softcore processors are fast, flexible, less expensive, and reconfigurable. If a softcore processor can be adequately protected from soft-errors, softcore processors can be an attractive alternative for space-based applications.

The flexibility and reprogramability that FPGAs provide for space-based applications comes at a price – SRAM-based FPGAs are inherently sensitive to the effects of faults caused by high-energy particles. These single event upsets (SEUs) can occur not only in FPGA user memory bits but also in the FPGA configuration bits that define the logic and routing of the softcore processor. SEUs in the FPGA configuration memory remain until repaired with configuration scrubbing.

Although FPGA user memory is also sensitive to the effects of SEUs, there are far more configuration bits in an FPGA device than user memory bits. For the device used in this study (Xilinx Virtex4 FX60), there are almost 21 million configuration bits, which is over $4\times$ more bits than user memory bits. Configuration bits control slices (including flip-flops, LUT-RAMs, and SRLs), IOBs, DCMs, DSPs, BRAM interconnect, all instance attributes, and all routing. User memory bits include BRAM data bits and user flip-flops. Thus SEUs are more likely to occur in the logic and routing (configuration memory) than in user memory.

This paper characterizes the sensitivity of a mitigated softcore processor in the presence of configuration SEUs using a novel fault injection technique. The mitigation techniques include triple modular redundancy (TMR), and duplication with compare (DWC) with checkpointing. TMR is used since it is one of the most popular FPGA design mitigation techniques [3], and DWC with checkpointing is used since it is one of the most popular processor mitigation techniques [4]. The metrics used to characterize the reliability of the softcore processor designs include architectural vulnerability factor (AVF) [5], mean-time to failure (MTTF), and mean useful instructions to failure (MuITF) [6].

---

[*]This work was supported in part by the I/UCRC Program of the National Science Foundation under the NSF Center for High-Performance Reconfigurable Computing (CHREC) under Grant No.0801876.

2. SOFTCORE PROCESSOR DESIGNS

The processor used in this study is Aeroflex Gaisler's 32-bit LEON3 processor [7]. This processor is chosen for this study because of its popularity in the space community, and because it is open-source. Only the core microarchitectural components of the LEON3 processor are included in this study. Figure 1 shows that the core units include the 7-stage integer pipeline, a 12 window register file, the hardware multiplier and divider, 1 Kbyte direct-mapped instruction and data caches, interrupt controller, and on-chip main memory.

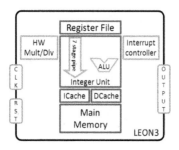

Figure 1: The LEON3 core processor units used in this study.

The first mitigation technique used in this study to protect the LEON3 processor uses duplication with compare (DWC) and rollback checkpointing [8]. DWC is a reliability technique that uses a duplicate processor to detect upsets. This study includes two versions of the DWC LEON3 processor: one with duplicated clock and reset inputs, and one with single clock and reset signals. Duplication is a popular processor reliability technique that can either be used in lockstep [8] or dual-core execution (DCE) [9]. This study uses duplicated processors in lockstep.

Softcore processors have the luxury of using strict lockstep since *any* processor signal can be exposed on a cycle-by-cycle basis. The softcore processor architecture can be changed at any time to use any given signal for lockstep comparison. The lockstep DWC processors compare register file outputs, program counters, instruction registers, processor output, and main memory input in a strict lockstep fashion.

The second LEON3 processor mitigation technique uses TMR and roll-forward checkpointing [8]. This study uses two versions of the TMR LEON3 processor: one with triplicated inputs and outputs (with off-chip voting), and one with untriplicated inputs and outputs (with on-chip voting). Similar to the duplicated processor design, the TMR design keeps the triplicated processors in strict lockstep execution using the register file outputs, program counters, instruction registers, processor outputs, and main memory inputs. But unlike DWC design, when one of the processors falls out of lockstep, a rollback is not required. Instead, the state of the two processors still in lockstep is used to correct and resynchronize the one that is out of lockstep.

3. RELIABILITY METRICS

Traditional processor reliability is measured in terms of mean time to failure (MTTF) or more recently, mean instructions to failure (MITF) [6]. This section shows how these metrics are modified for softcore processors. The metrics introduced in this section can be applied to the processor as a whole or to individual components of the processor to compare component reliability.

Not every configuration upset in a softcore processor will lead to erroneous processor output. The traditional metric for expressing how likely it is for an upset to lead to a processor error is called the *architectural vulnerability factor* (**AVF**) [5]. When applied to softcore processors, AVF is used to represent the percentage of configuration memory bits that, when upset, cause the processor to execute incorrectly.

Each of the 21 million bits in the configuration memory is classified as being either required for architecturally correct execution (**ACE**) or unnecessary for architecturally correct execution (**unACE**) [5]. Upsetting an ACE bit causes a program running on the softcore processor to produce an incorrect output, while upsetting an unACE bit does not hinder correct program execution. In this study the terms ACE bits and *sensitive bits* are used synonymously. Likewise, unACE is a synonym for *insensitive*.

In keeping with the meaning of a sensitive bit, upsets to ACE bits are classified as being either detected, recoverable upsets (**DRU**), detected, unrecoverable errors (**DUE**), or silent data corruption (**SDC**) bits. DUE upsets are those which are detected by the processor, but from which the processor cannot recover. SDC upsets are upsets which are never detected, and which cause erroneous output. In order to have a reliable processor, there should be as few SDCs and DUEs as possible. SDCs are especially bad since they are not even detected. When mitigation techniques are used, the number of DRU bits indicate how many upsets are detected and prevented from causing erroneous output. Mitigation techniques attempt to reduce the number DUEs and SDCs and increase the number of DRUs.

Processor SEU sensitivity is measured using AVF. The AVF of a processor is the percentage of the processor area that is sensitive to DUEs and SDCs [5]. Equation 1 shows how AVF is computed for softcore processors. AVF is the fraction of the configuration bits (CFGbits) that contain SDC or DUE bits.

$$\text{AVF} = \frac{\#\ \text{SDCs} + \#\ \text{DUEs}}{\text{CFGbits}} \quad (1)$$

AVF can be used to estimate the reliability in terms of mean time to failure (MTTF). MTTF is computing using the configuration upset rate (λ) with AVF: $\text{MTTF} = \frac{1}{\lambda} \cdot \frac{1}{\text{AVF}}$. The upset rate of the processor is equal to λ_{bit} multiplied by the number of configuration bits used by the processor ($\lambda_{proc} = \lambda_{bit} \cdot \text{CFGbits}$). λ_{bit} depends upon the spacecraft orbit, space environment, and device properties. For this paper, λ_{bit} is estimated using a galactic cosmic ray (GCR) environment at solar minimum. The estimated upset rate is 1E-10 upsets/bit-day [10] or 1.16E-15 upsets/bit-s. The MTTF for the processor or a component is:

$$\begin{aligned} \text{MTTF} &= \frac{1}{\lambda_{bit}} \cdot \frac{1}{\text{CFGbits}} \cdot \frac{1}{\text{AVF}} \\ &= \frac{1}{\lambda_{bit}} \cdot \frac{1}{\#\ \text{SDCs} + \#\ \text{DUEs}}. \end{aligned} \quad (2)$$

Although MTTF provides a reasonable reliability measure, it does not account for the processor performance costs. Instead of measuring the time between two errors, mean instructions to failure (MITF) measures the amount of work accomplished between two errors [6]. MITF expresses how many instructions a processor commits, on average, between two errors. When mitigation techniques are used, MITF must also account for any performance costs (ρ) incurred from the mitigation techniques. The performance cost represents the execution of additional *unuseful* instructions. Equation 3 shows how the mean *useful* instructions to failure (MuITF) is computed for a processor:

$$
\begin{aligned}
\text{MuITF} &= \frac{\text{frequency}}{\lambda_{bit}} \cdot \frac{1}{\text{CFGbits}_{mit}} \cdot \frac{\text{IPC}}{\text{AVF} \cdot \rho} \\
&= \frac{\text{frequency}}{\lambda_{bit}} \cdot \frac{\text{IPC}/\rho}{\# \text{ SDCs} + \# \text{ DUEs}}.
\end{aligned} \quad (3)
$$

4. HARDWARE FAULT-INJECTION

To evaluate the reliability of the mitigated and unprotected LEON3 processor designs, hardware fault-injection is used to identify ACE bits in the configuration memory. Traditionally, ACE (and unACE bits) bits only occur in user memories and registers. ACE bits are normally identified by tracking them in the pipeline or with the use of models [11]. This section describes the hardware fault-injection procedure that identifies ACE bits in the configuration memory.

Hardware fault-injection is a well-known way of evaluating the impact of SEUs on commercial SRAM-based FPGA devices [12]. Fault-injection is performed by upsetting each and every bit in the FPGA configuration memory, one at a time, while a program executes on the LEON3 processor. For the Xilinx Virtex4 FX60 FPGA used in this study, almost 21 million upsets and program executions are required to test every bit in the entire configuration memory.

For each of the 21 million configuration bits in a mitigated LEON3 processor, a novel fault-injection procedure is followed. Overall, the program running in the LEON3 processor on the DUT FPGA runs four times for each of the 21 million configuration memory bits. The program runs once to ensure that at least one checkpoint is taken. In the next program iteration, the program runs for a random number of clock cycles before the configuration bit is upset. After the upset is inserted, the second program execution finishes and then runs a third time. This gives the LEON3 ample time to detect the upset. If the upset is detected, the upset is immediately repaired by the fault-injector. If the upset is detected, the DUT also attempts to recover from the upset with a checkpoint rollback. If the upset goes undetected by the end of the third program execution, it is repaired. Finally, the program runs for a fourth time to ensure that any checkpoint recovery attempted by the DUT is successful. A similar process is followed to test the configuration bits in the unmitigated LEON3, but only the middle two program runs are required.

The full fault-injection procedure is run for all 21 million configuration bits for eight micro-benchmarks used in this study. The benchmarks used in this work are limited by the constraints of the fault injector and by the number of BRAMs available on the FPGA. Running the fault-injection procedure on a LEON3 design with one of these micro-benchmark programs takes up to 60 hours to complete. The benchmark programs in this study test worst-case behavior of the processor structures and instruction set architecture. Since these benchmarks are meant to be stress tests, it is expected that they will be pessimistic compared to a typical workload.

5. PROCESSOR COMPARISONS

This section discusses the reliability-cost trade-off for applying mitigation to the LEON3 softcore processor. The costs are measured in terms of area and performance with respect to an unmitigated processor. Reliability is measured using AVF, MTTF, and MuITF.

The area costs of the unmitigated LEON3 processor are compared with the mitigated LEON3 designs in Table 1. Area is compared in terms of slices, BRAMs, DSPs, and configuration memory bits (CFGbits). The values in braces show the area increase (in terms of configuration bits) compared to the unmitigated processor. It shows that the area cost of DWC is significantly more than double and that the area cost of TMR is more than triple.

Area Costs				
LEON3 Design	Slices	BRAM	DSP	Config Bits (CFGbits)
Unmitigated	3058	12	4	677,065 (1.00×)
DWC (1 clock)	8628	33	8	1,834,002 (2.71×)
DWC (2 clocks)	8569	33	8	1,837,719 (2.71×)
TMR (1 in/out)	14,623	36	12	3,102,921 (4.58×)
TMR (3 in/out)	14,286	36	12	3,075,399 (4.54×)

Table 1: Area cost comparison for the LEON3 processor designs.

The addition of checkpointing to the LEON3 processor introduces a small performance cost (ρ). The value of this cost is proportional to the frequency with which checkpointing occurs. In this study, checkpointing occurs once per program execution. The performance cost incurred is reported in terms of the additional number of clock cycles needed, on average, to run a program compared to when running on an unmitigated processor. On average, programs on the DWC processors run 1.01× longer than when run on an unmitigated processor. The time to record checkpoint information accounts for the small performance penalty incurred by the DWC designs. The performance cost for the TMR processor is negligible since the only incurred performance cost comes when roll-forward checkpointing corrects and resynchronizes one of the three processors.

Hardware fault-injection is used to measure the total number of ACE and unACE bits for each of the LEON3 softcore processor designs and with each of the benchmark programs (Table 2). The percentage in the unACE column indicates the percentage of used configuration bits that, when upset, do not hinder correct program execution. The percentages reported in the DRE, DUE, and SDC columns are with respect to only the ACE bits for the given processor.

Table 2 shows that both mitigation techniques significantly reduce the number of SDCs and DUEs of the unmitigated processor, but that TMR with triplicated inputs and outputs almost eliminates SDCs and DUEs. In the TMR design with untriplicated inputs and outputs, 66% of the SDCs and DUEs occur in the untriplicated output, 21% occur in the clock and reset signals, and 12% occur in the voters. In the DWC designs, a large majority of the SDCs and DUEs occur in the top-level outputs, clock and reset signals, or in the comparator unit.

Table 2 also shows that some of upsets are detected by the unmitigated LEON3 processor (as shown in the DUE

HW Fault-Injection Results				
LEON3 Design	unACE	ACE		
		DRU	DUE	SDC
Unmitigated	537,825 (79.4%)	0 (0.0%)	11,637 (8.36%)	127,603 (91.64%)
DWC & Check (1 clock)	1,378,813 (75.2%)	450,527 (98.98%)	762 (0.16%)	3900 (0.86%)
DWC & Check (2 clocks)	1,369,864 (74.5%)	463,957 (99.17%)	376 (0.08%)	3522 (0.75%)
TMR (no triplicated in/out)	2,458,337 (79.2%)	642,523 (99.68%)	73 (0.01%)	2061 (0.31%)
TMR (triplicated in/out)	2,421,376 (78.7%)	654,013 (99.996%)	13 (0.002%)	10 (0.002%)

Table 2: Full fault-injection results for the LEON3 processors.

column). Although no upset detection techniques have been explicitly applied, the LEON3 processor pipeline has some built-in error detection. The processor throws an error signal when interrupts occur in an unexpected way.

The measured number of ACE bits are used to calculate the AVF, MTTF, and MuITF of each LEON3 processor. The reliability of each of the LEON3 processors is shown in Table 3. The average number of instructions per clock cycle (IPC) – required by MuITF (Equation 3) is 0.62. The MuITF frequency value used in Equation 3 is 33 MHz, since that is the frequency at which our fault-injection hadware runs. The AVF values in the table show that the unmitigated LEON3 is almost 100× more vulnerable than the LEON3 protected with DWC and checkpointing, and over 27,000× more vulnerable than the LEON3 protected with full TMR and roll-forward checkpointing. The reliability results show that although full TMR provides the best protection, DWC and checkpointing may be an acceptable lower-cost alternative.

LEON3 Processor Reliability			
LEON3 Design	AVF	MTTF (years)	MuITF $\times 10^{20}$ (instructions)
Unmitigated	20.6%	0.83 (1.00×)	5.36 (1.00×)
DWC (1 clock)	0.25%	24.8 (29.9×)	158.6 (29.6×)
DWC (2 clocks)	0.22%	29.7 (35.7×)	189.7 (35.4×)
TMR (1 in/out)	0.07%	55.2 (66.4×)	356.0 (66.4×)
TMR (3 in/out)	0.00075%	5032 (6058×)	32,469 (6058×)

Table 3: Comparison of AVF, MTTF, and MuITF of an unmitigated LEON3 against the mitigated LEON3 processors.

6. CONCLUSION

This study demonstrates the improvements in reliability by applying DWC and TMR with checkpointing to a softcore processor. Three metrics were used to compare an unmitigated softcore processor with these two mitigated designs: architectural vulnerability factor (AVF), mean time to failure (MTTF), and mean useful instructions to failure (MuITF). The AVF, MTTF, and MuITF were measured through hardware fault-injection. The reliability of this unmitigated processor is improved by applying DWC with checkpointing and TMR with checkpointing to the processor. DWC with checkpointing is shown to improve the MTTF and MuITF by over 35×. TMR with checkpointing improves the MTTF and MuITF by over 6000×.

This study shows that although softcore processors are sensitive to the effects of SEUs, the faults can be characterized and protected with appropriate mitigation techniques. An adequately protected softcore processor is an attractive alternative to a rad-hard processor for space-based applications. Compared to rad-hard processors, softcore processors are faster, flexible, less expensive, and reconfigurable.

7. REFERENCES

[1] Q. Zhou, K. Mohanram, Gate sizing to radiation harden combinational logic, Computer-Aided Design of Integrated Circuits and Systems, IEEE Transactions on 25 (1) (2006) 155 – 166.

[2] J. F. Bell, et al., Mars reconnaissance orbiter mars color imager (MARCI): Instrument description, calibration, and performance, Journal of Geophysical Research 114.

[3] L. Sterpone, M. S. Reorda, M. Violante, F. L. Kastensmidt, L. Carro, Evaluating different solutions to design fault tolerant systems with SRAM-based FPGAs, Journal of Electronic Testing: Theory and Applications 23 (2007) 47–54.

[4] A. Ziv, J. Bruck, Analysis of checkpointing schemes with task duplication, Computers, IEEE Transactions on 47 (2) (1998) 222 –227.

[5] S. S. Mukherjee, et al., A systematic methodology to compute the architectural vulnerability factors for a high-performance microprocessor, Microarchitecture, IEEE/ACM International Symposium on 0 (2003) 29.

[6] C. Weaver, et al., Techniques to reduce the soft error rate of a high-performance microprocessor, SIGARCH Comput. Archit. News 32 (2004) 264–.

[7] J. Gaisler, E. Catovic, Multi-Core Processor Based on LEON3-FT IP Core (LEON3-FT-MP), in: DASIA 2006 - Data Systems in Aerospace, Vol. 630 of ESA Special Publication, 2006.

[8] D. Pradhan, N. Vaidya, Roll-forward and rollback recovery: performance-reliability trade-off, in: Fault-Tolerant Computing, 1994. FTCS-24. Digest of Papers., Twenty-Fourth International Symposium on, 1994, pp. 186 –195.

[9] H. Zhou, A case for fault tolerance and performance enhancement using chip multi-processors, Computer Architecture Letters 5 (1) (2006) 22 –25.

[10] R. Hillman, et al., Space processor radiation mitigation and validation techniques for an 1,800 MIPS processor board, in: Radiation and Its Effects on Components and Systems, 2003. RADECS 2003. Proceedings of the 7th European Conference on, 2003, pp. 347 – 352.

[11] S. S. Mukherjee, M. Kontz, S. K. Reinhardt, Detailed design and evaluation of redundant multithreading alternatives, Computer Architecture, International Symposium on 0 (2002) 0099.

[12] E. Johnson, M. Caffrey, P. Graham, N. Rollins, M. Wirthlin, Accelerator validation of an FPGA SEU simulator, Nuclear Science, IEEE Transactions on 50 (6) (2003) 2147–2157.

Leveraging Latency-Insensitivity to Ease Multiple FPGA Design

Kermin Fleming¶   Michael Adler†   Michael Pellauer†   Angshuman Parashar†
Arvind¶   Joel Emer†¶

†Intel Corporation
VSSAD Group
{michael.adler, michael.i.pellauer,
angshuman.parashar, joel.emer}
@intel.com

¶Massachusetts Institute of Technology
Computer Science and A.I. Laboratory
{kfleming, arvind,
emer}
@csail.mit.edu

ABSTRACT

Traditionally, hardware designs partitioned across multiple FPGAs have had low performance due to the inefficiency of maintaining cycle-by-cycle timing among discrete FPGAs. In this paper, we present a mechanism by which complex designs may be efficiently and automatically partitioned among multiple FPGAs using explicitly programmed latency-insensitive links. We describe the automatic synthesis of an area efficient, high performance network for routing these inter-FPGA links. By mapping a diverse set of large research prototypes onto a multiple FPGA platform, we demonstrate that our tool obtains significant gains in design feasibility, compilation time, and even wall-clock performance.

Categories and Subject Descriptors

B.5.2 [**Design Aids**]: Automatic Synthesis

General Terms

Design, Performance

Keywords

FPGA, compiler, design automation, high-level synthesis, switch architecture, DSP, programming languages

1. INTRODUCTION

FPGAs are an extremely valuable substrate for prototyping and modeling hardware systems. However, some interesting designs may not fit in the limited area of a single FPGA. If a design cannot fit onto a given FPGA, the designer is faced with a handful of choices. The designer may use a larger single FPGA or refine the design to reduce area, neither of which may be possible. A third possibility is to partition the design among multiple FPGAs. This option is typically feasible from an implementation perspective, but has some serious drawbacks. Manual partitioning may obtain high performance, but represents a time consuming design effort. Tool-based partitioning, while automatic, may suffer performance degradation.

Consider a processor model manually partitioned across two FPGAs in which there are two channels sharing a multiplexed physical link between the FPGAs. One channel is information from the processor decode stage. It has a narrow bit-width, but occurs in nearly every processor cycle. The other channel is the memory interface between the last-level cache and backing main memory. This link is exceptionally wide, with perhaps as many as 1000 bits. However, because programs generally have good locality this link is infrequently used. Exploiting this knowledge produces a high performance partitioned design: data for each link is sent as it becomes available and only when it is available. This implementation recognizes two high-level properties of the underlying design; first, although physical wires are driven to specific values every cycle, not all of those values impact behavior, and second, that links may have different semantic properties, such as priority.

Automatic partitioning tools avoid the engineering overhead of manual partitioning at the cost of throughput. In the previous example the tool needs to automatically understand when the last-level cache is making requests in order to achieve high throughput. If the tool cannot derive this high level meaning, then it must conservatively transport all or nearly all values between FPGAs on each cycle to maintain functional correctness. Because extracting this high-level meaning is difficult, existing tools take the conservative approach in partitioning. Thus, pin bandwidth, serialization, and latency become major sources of performance degradation in partitioned designs, even if the values transported between FPGAs ultimately do not impact design behavior on the majority of cycles. The difficulty in extracting high-level knowledge from RTL lies in automatically differentiating the cycle-by-cycle timing behavior of RTL from the *functional* behavior of a design.

In this paper, we explore the application of latency-insensitive design [3] [4] [25] to multiple FPGA implementation. We view hardware designs as sets of modules connected by latency-insensitive FIFOs, as shown in Figure 1(a). Because inter-module communication is restricted to latency-insensitive FIFOs, we have broad freedom in implementing the network between modules. We leverage this freedom in our compiler to automatically generate design implementations that span multiple FPGAs. A key technical contribution of this paper is the automatic synthesis of a high-performance and deadlock-free network optimized for design-specific, latency-insensitive communications.

By partitioning designs among multiple FPGAs at latency-insensitive FIFOs, we gain efficiency over traditional tools in two ways. First, only data explicitly enqueued into the FIFOs needs to be transported. Second, we have more options in transporting data over multiplexed physical links. By compiling several large research prototypes to a multiple FPGA platform, we will demonstrate that our tool obtains significant gains in design feasibility, compilation

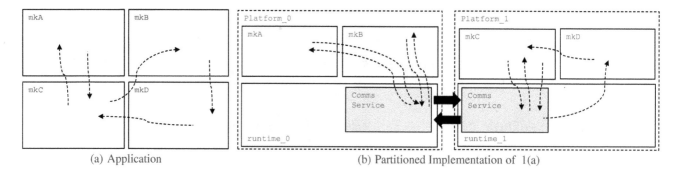

(a) Application   (b) Partitioned Implementation of 1(a)

Figure 1: A sample application and its mapping to a two FPGA platforms. Links to modules on the same FPGA are directly tied together by a FIFO, while inter-FPGA links are tied to a synthesized communications complex produced during compilation.

time, and, in some cases, wall-clock performance as compared to conventional single FPGA implementations.

2. BACKGROUND

2.1 Model of Computation

Recently, a number of highly modular research prototypes [7] [16] have been developed using latency-insensitive design [6]. In latency-insensitive design, the goal is to maintain the *functional* correctness of the design in response to variations in data availability. Latency-insensitive designs are typically composed of modules connected by FIFO links; modules are designed to compute when data is available, as opposed to when some clock ticks. Latency-insensitive design offers a number of practical benefits, including improved modularity and simplified design exploration.

The usual framing of latency-insensitive design focuses on the properties of and composition of modules, but we observe that an implication of this style of design is that the FIFOs connecting the modules may take an arbitrarily long time to transport data without affecting the functional correctness of the original design. As a result, the send and receive ends of these special FIFOs can be spatially far apart, even spanning multiple FPGAs. By breaking designs at latency-insensitive boundaries, it is possible to automatically produce efficient and high performance implementations that span multiple FPGAs.

In this work, we constrain our designs to be composed of modules, abstract entities that interface to each other through asynchronous, latency-insensitive FIFO communications channels. The communications channels have two critical properties. First, they have arbitrary transport latency, and second, they have arbitrary, but finite size. Internally modules may have whatever behavior the designer sees fit, provided that the interface properties are honored. For example, once data has been obtained from an interface FIFO, it might pass through a traditional latency-sensitive hardware pipeline. Alternatively, the module could be implemented in software on a soft-core, or even on an attached CPU. Module internals do not matter. Rather, it is the latency-insensitive interface that is fundamental. Modules alone describe an abstract model of computation. To achieve a physical implementation, modules are mapped on to platforms, entities which can execute modules. Figure 1 shows a stylized representation of a design with four modules mapped on to two platforms. Cross-platform links transit special communications services, providing longer latency, point-to-point communications.

We will solve the following problems related to the mapping of modules to platforms. First, we introduce a means of describing latency insensitivity in an HDL. Second, we develop a portable technique for interfacing latency-insensitive modules with external, latency-sensitive wired devices. Third, we present a high-

performance, low latency architecture for an automatically synthesizable deadlock-free inter-FPGA communication network. We tie the preceding technologies together in a compiler capable of automatically partitioning latency-insensitive designs across multiple FPGAs.

2.2 Related Work

There exist a number of commercially available tools [11] [14] capable of mapping RTL designs across multiple FPGAs. These tools generate emulators intended primarily for ASIC verification in preparation for silicon implementation. As such, they are required to maintain the cycle accurate behavior of all signals in the design. Existing partitioning tools are differentiated by whether they provide dedicated [26] or multiplexed [2] [12] chip-to-chip wires.

Dedicated wire partitioning tools include inter-FPGA link delays in the circuit-level timing equations for determining setup and hold times. The result is that the emulation clock is greatly slowed, since delays on chip-to-chip wires are much longer than on-chip delays. However, board-level wires have physical meaning that may be useful in certain debugging regimes.

Multiplexed-wire partitioner operate by first running a single clock cycle of the base design on each FPGA, then propagating values across multiplexed inter-device links, and finally running another model cycle once all value from the previous cycle become available. As with dedicated wires, multiplexed wires incur performance overhead. To maintain cycle accuracy, the emulator must conservatively transport all inter-FPGA values every cycle, whether or not they impact the behavior of the succeeding cycle. As a result, these partitioning tools do not typically exhibit high performance, achieving cycle-accurate operating speeds of a few megahertz [23] [24].

Our tool is fundamentally different than either emulator approach: it is not required to maintain the cycle behavior of an unpartitioned design. Our language primitives allow designers to explicitly annotate locations in which it is safe to change cycle-by-cycle behavior of the design. As a result, partitions are free to run independently and operate on data as soon as it becomes available. This allows us to take advantage of the natural pipeline parallelism of hardware designs at a much finer grain than existing partitioners [22]. Furthermore, because we permit only FIFOs to cross between FPGAs, we transport only useful, explicitly enqueued values.

Our compiler operates on latency-insensitive designs, and the examples the we will present in Section 7 were originally written in this style. However, it is not necessary to write latency-insensitive designs to make use of our compiler. Methodologies [4] [20] [25] exist that transform existing latency-sensitive RTL designs into a functionally equivalent set of latency-insensitive modules *while preserving the timing behavior of the original design*. These methodologies seek to preserve the cycle-accurate behavior of some design signals, while permitting some parts of the original design to be re-written to

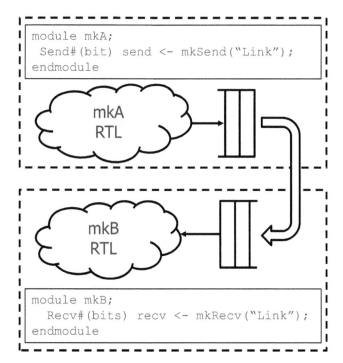

```
module mkA;
  Send#(bit) send <- mkSend("Link");
endmodule
```

```
module mkB;
  Recv#(bits) recv <- mkRecv("Link");
endmodule
```

Figure 2: A pair of modules connected by a latency-insensitive link. The arrow represents an automatically generated connection. Users may specify a minimum buffering as an argument to the link constructor. Here, "Link" is a tag used to match connections during the compilation flow.

more efficiently map onto an FPGA. Although our tool provides no inherent guarantees relating to cycle-accuracy, it can be composed with these transformation tools should the cycle-accuracy of some signals be required. Indeed, one of our example codes in Section 7 uses the A-Ports technique. In association with these tools, our compiler can be used to verify any synchronous design, including those designs written in a latency-insensitive style. However, there is no free lunch: as the number of cycle-accurate signals increases, our synthesized implementations will degrade in performance until they reach parity with traditional cycle-accurate partitioning tools.

3. LANGUAGE LEVEL SUPPORT

Synthesizable HDLs provide a very basic hardware abstraction: register, logic, and, to a lesser extent, memory. The behavior of these elements are tied to a specific clock cycle. Traditional HDLs reflect the physical reality of high-performance hardware, but do not offer the compiler much room to change cycle-to-cycle behavior. For example, if a programmer instantiates a two element FIFO in an HDL, this FIFO will have some well defined cycle-by-cycle behavior. Even changing the depth of this FIFO is difficult to automate, because changing the depth will almost assuredly perturb cycle behavior and potentially break the design. Even if the HDL design can tolerate such perturbations, it is difficult for the HDL compiler to prove that this is the case. Inter-FPGA communications, which are almost guaranteed to take many FPGA cycles, do not map well into the existing HDL model.

We avoid the generally undecidable problem of reasoning about cycle accurate behavior in the compiler by introducing a primitive syntax for latency-insensitive, point-to-point FIFO communication. By using these constructs, the programmer asserts that the compiler is free to modify the transport latency and buffer depth of the FIFO in

question. This is a departure from previous multi-FPGA compilation efforts in that we permit the compiler to modify not only the cycle behavior of the communications link, but also the resulting behavior of the user design. Users must ensure that their designs can tolerate these changes in behavior, but in practice this is little different than interfacing to a normal, fixed-behavior FIFO. By providing a latency insensitive primitive, we push the burden of reasoning about high-level design properties to the designer, simplifying the task of the compiler to generating high-speed, deadlock-free interconnects.

An example of our link syntax [18] is shown in Figure 2. mkSend and mkRecv operate as expected, with the send endpoint injecting messages into a logical FIFO with arbitrary latency and receive draining those messages. This syntax is convenient because it provides a simple way for logically separate modules to share links while abstracting and encapsulating the physical interconnect. In our compiler, only these special FIFOs are treated as latency-insensitive. Other FIFOs, for example those provided in the basic language, retain their original fixed-cycle behavior.

In addition to point-to-point communications, we also provide a ring interconnect primitive. Like the point-to-point links, rings are named, but may have many ring stops. Ring stops are logically connected in sequence, with messages flowing around the ring. The ring primitive is useful in scaling and sharing runtime services across FPGAs, since rings can have an arbitrary number of stops.

4. PLATFORM SERVICES

Our model of computation consists of modules communicating over latency-insensitive FIFO links. However, practical FPGA systems must have some interaction with some platform-specific, wired devices, like memory. The difficulty with these physical interfaces is that we require the freedom to move modules to any FPGA, but physical interfaces are fundamentally unportable. Therefore, we introduce an abstraction layer [17], mapping physical devices into platform services using our latency-insensitive primitives. Services are tied to specific FPGA, but can be used by any module in a design. Common services in a platform runtime include memory and inter-FPGA communications, but application-specific platforms may include network, wireless, or video interfaces.

Platform services differ from user modules in two ways. First, platform services are permitted to have an arbitrary wire interface. Second, platform services are shared among user modules. In addition to a wired, external interface, platform services must also expose a set of latency-insensitive links, typically some form of request-response. User modules may then interface to a platform service as if it was a regular module.

Since memory is perhaps the most important service, we will use our memory hierarchy, shown in Figure 3, as an example of the design of a platform service. This hierarchy is an extension of the scratchpads [1] memory abstraction. In the scratchpads hierarchy, modules are presented the abstraction of disjoint, non-coherent, private memory spaces with simple read-write interfaces. Backing this simple interface is a complex, high-performance cache hierarchy.

Platform services may be shared among several different clients. In this case, each client instantiates an interface provided by the service. The service interface internally connects the clients, typically using a ring, and handles multiplexing among clients. In the case of scratchpads, each client instantiating a memory link gets a private cache, connected to the rest of the memory hierarchy by a ring. At runtime, the clients are dynamically discovered as they begin sending memory traffic. The ring permits a single, parametric service implementation to automatically scale to an arbitrary number of clients and FPGAs.

Scratchpads also introduces a shared chip-level L2 cache to improve memory performance. In the context of multiple FPGAs, there are two interesting cases for a chip-level resource: the case in which a resource is located on a remote FPGA and the case in which several FPGAs provide the same resource.

Several FPGAs may provide the same service. If useful, a chip-level ring can also be introduced connecting the services and allowing them to communicate with each other dynamically. In the case of scratchpads, a secondary ring is introduced on top of the chip-level caches allowing all of them to share a single interface to host virtual memory.

In the case that a platform does not provide a service required by a module mapped to the platform, the compiler will automatically connect the module to an instance of the service provided on a different FPGA. This flexibility in module mapping is valuable in cases where the runtime service is used infrequently. In the case of scratchpads, the local ring will automatically be mapped across FPGA boundaries, resulting in a correct, but lower performance implementation. In the case that there are several remote services available, modules have the option of specifying the FPGA whose service should be used.

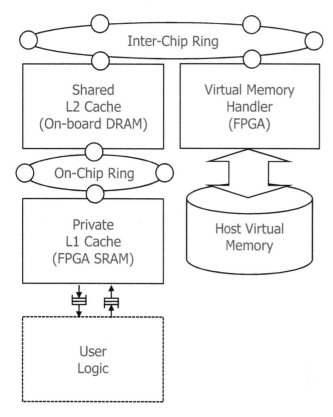

Figure 3: A view of a scalable multiple FPGA memory hierarchy. Fast, private local caches are backed by a shared last-level cache, which in turn is backed by global virtual memory. The structure of the hierarchy is automatically inferred at compilation time.

5. COMPILATION

Given a design consisting of modules connected by latency-insensitive links, the goal of our compiler is to automatically connect all the exposed links by synthesizing the most efficient network interconnect possible. Modules mapped to the same FPGA may have an interconnect as simple a FIFO, but distant modules may require an interconnection network transiting several FPGAs.

Our compilation currently assumes a static, user provided mapping of modules to FPGAs. Initially, source modules are decorated with tags representing the specific FPGA to which the module has been mapped. Synthesis proceeds per FPGA in two passes. During the first pass, source for the entire system is compiled in order to discover latency-insensitive endpoints. As endpoints are discovered within modules, they are tagged with the FPGA to which they belong. At the end of the pass, each FPGA has a set of dangling send and receive connections associated with it. Some of these represent local connections and some represent off-chip links. Sends and receives are then matched by name on a per FPGA basis. Matched links represent connections local to the specific FPGA and are connected by a simple FIFO, while unmatched names are propagated to a global matching stage.

The global matching stage synthesizes a router for each specified communication link between FPGAs. Connections are matched based on name and are routed to their destination, which may require inserting links across several intervening FPGAs. The routers generated in this stage are parameterized instances of the communications stack discussed in Section 6. Each connection is assigned a virtual channel on each link that it traverses. At the source and sink of each path, the inverse latency-insensitive primitive is inserted into the router, for example, a dangling send will have its corresponding receive inserted. During the second compilation pass, the inserted connections will be matched with existing connections, completing the inter-chip routing. For intermediate hops, a new, unique link is introduced between ingress and egress. Figure 4 depicts the result of the global compilation for a single FPGA, in which the user logic has three inter-chip links, and one link transits the FPGA.

The generated routers are then injected into the compilation for each FPGA, and compilation proceeds as in the first pass. However, at the end of this pass, during the local match step, all formerly dangling connections are matched to local endpoints at the synthesized routers. The Verilog generated by this final step can be simulated or passed to back-end tools to produce bit-files.

6. INTER-FPGA COMMUNICATION

There are two issues in synthesizing an inter-FPGA communications network for latency-insensitive links: performance and correctness. Partitioning designs at latency-insensitive FIFOs allows us to transport only explicitly enqueued data. However, to achieve high performance a network must be able exploit the pipeline parallelism inherent in the partitioned design. In practice this means that many messages must be in-flight at any given time.

For our model of computation, network correctness means the in-order delivery of messages, a relatively simple requirement. However, because many links can cross between FPGAs, there is a need to multiplex the physical links between the FPGAs. This multiplexing can introduce deadlocks, but we will show that our synthesized networks are *deadlock-free*.

Deadlocks arise in shared interconnect when dependent packets are forced to share the same routing paths, which can cause the packets to block each other. To get around this issue, virtual channels are introduced to break dependence cycles [5]. In traditional computer architectures, this is a tractable problem since the communications protocols are known statically and dependencies can be explicitly broken at design time. However, reasoning about the communications dependencies of an arbitrary hardware design is difficult. Therefore, we simply allocate a virtual channel to each link crossing between FPGAs. Virtual channel allocation alone is not sufficient to ensure deadlock freedom, because full virtual channels can still cause head-of-line blocking across the shared physical links. To resolve this issue we must also introduce flow control across each virtual channel. Together, universal virtual channel allocation and flow control are sufficient to guarantee that our compiler does not introduce deadlocks into previously deadlock-free latency-insensitive designs. This property is an easy corollary of the Dally-Seitz theo-

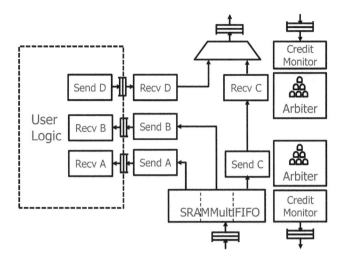

Figure 4: An example of a synthesized router connecting an FPGA with to two other FPGAs (not shown). Link "Recv C" is routed through the FPGA.

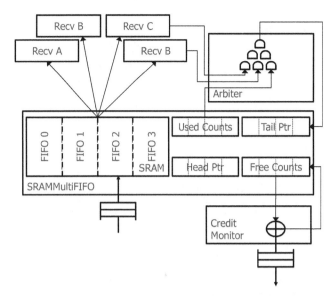

Figure 5: FIFOs are folded onto a single logical SRAM resource. Each FIFO in the SRAM represents a buffer for a single virtual channel

rem [5], wherein we insert a virtual channel for each communication link, trivially preventing dependent packets from blocking one another.

We require flow control per link to guarantee functional correctness in our partitioned designs. This is a seemingly costly proposition, particularly since flow-control packets incur a high latency round-trip between FPGAs. This latency appears to create a cost-performance tradeoff between buffering per virtual channel and the performance and area of the router, since too little buffering can cause the sender to stall before the receiver even begins to receive packets, while too much buffering reduces the area available to the user design. A naive register-based flow control implementation with buffering sufficient to cover a round-trip latency of 16 cycles requires half the area of a large FPGA. Clearly, this kind of implementation does not scale beyond a pair of FPGAs.

The problem with the register-based design is that it is too parallel and therefore needlessly wasteful of resources. In any cycle, any of the registers in any of the buffers can potentially supply a data value to transmit. However, we observe that the inter-chip bandwidth between FPGAs is limited to a single, potentially wide, data per cycle. This bandwidth limitation means that to sustain the maximum rate across the link, we need to enqueue and dequeue exactly one FIFO in any given cycle. Therefore, a structure with low parallelism, but high storage density is sufficient to sustain nearly the maximum throughput of the physical channel.

Most modern FPGAs are rich in SRAM, with a single chip containing megabytes of storage. Although large amounts of memory are available, the bandwidth to each slice of this memory is limited to a single four to eight byte word per cycle. Because inter-FPGA communication is similarly constrained, we can map many virtual channels with relatively large buffers onto the resource-efficient SRAM without significant performance loss. We call this optimized storage structure, depicted in Figure 5, the SRAMMultiFIFO (SMF) [8]. Because the SMF maps many FIFOs onto a single SRAM with a small number of ports, it must introduce an arbiter to choose which FIFO will use the SRAM port in a given cycle. SMF FIFOs have uniform and constant size which simplifies control logic at the cost of storage space. The SMF is fully pipelined, and each mapped FIFO can utilize the full bandwidth of the SRAM.

Area usage for SMF and a functionally similar register-based FIFO implementation are shown in Figure 6. The SMF scales in BRAM usage as FIFO depth increases, consuming only around 2% of slices on a Virtex-5 LX-330T. The low area usage of the SMF-based switch makes it amenable to FPGA platforms with a high-degree of inter-platform interconnection. On the other hand, the registered buffer schemes can quickly exhaust large amounts of area. The largest implementable registered FIFO switch has no better performance than a more resource efficient, but deeper SMF switch, despite the inherent parallelism of the registered implementation.

The density of the SMF fundamentally changes the way that communication networks between FPGAs are designed. Unlike processor network on chips, which multiplex virtual channels and offer extremely limited buffering in network to conserve area, SMF based switches can liberally allocate virtual channels to each connection traversing the inter-FPGA link without significant area penalty. As a result, concerns involving shared virtual channels [15] do not apply to our switches and routing scheme. Because SMF provides deep buffers, each flow-controlled inter-chip channel can sustain full bandwidth across high-latency physical links. Deep buffers also permit us to send control messages relatively infrequently, minimizing throughput loss.

In our switches, we use a simple block-update control control scheme. The virtual channel source keeps a conservative count of the number of free buffer spaces available at the virtual channel sink. Each time a packet it sent, this count is decremented. The virtual channel sink maintains a count of the free buffer space available, which is updated as user logic drains data out of the virtual channel. When this free space counter passes a threshold, it is set to zero and a bulk credit message is sent to the virtual channel source. These credit messages are given priority over the data message to improve throughput.

Although the SMF is the core of our router, the router architecture consists of three pipelined layers: packetization, virtual channel, and physical channel. At compilation time the compiler generates an FPGA-specific router using parametric components from a library.

The physical channel layer consists of specially annotated FIFOs provided by the platform runtime. The backing implementation of these FIFOs is irrelevant from the user's perspective and could range from LVDS to Ethernet. Given the specific name of an inter-FPGA communications link provided by the platform runtime, the

	LUTS	Registers	BRAM	Relative Performance
Registered FIFOs, depth 8	10001	22248	0	1
Registered FIFOs, depth 32	25494	68813	0	1.11
SRAM MultiFIFOs, depth 32	4996	4778	2	1.09
SRAM MultiFIFOs, depth 128	5225	4850	8	1.11

Figure 6: Synthesis and performance metrics for various switch architectures. Results were produced by mapping a simple HAsim dual core processor model to two FPGAs. In this design, 29 individual links and 1312 bits cross the inter-FPGA boundary.

compiler simply instantiates a connection to the link and ties it in to the synthesized communication hierarchy.

We introduce packetization into our communications hierarchy to simplify both the virtual-channel hardware layer and to handle the presence of wide links. Since all communications links and link widths are statically determined at compile time, our compiler can infer bit-optimal packet protocol for each link. These protocols are specific instantiations of a header-body packet schema in which the header contains information about the packet length, type, and virtual channel. The parameterized packetization and de-packetization hardware then infer an efficient implementation based on the data width to be transported. In the case that the data width is wider than the physical link, marshalling and de-marshalling logic is automatically inserted. However, if the data width is sufficiently small, the packet header and body will be bit-packed together. Since the data communicated between FPGAs tends to be narrow, this is a significant performance optimization.

7. EVALUATION

To evaluate the quality of our compiler, we partition three large research prototypes. These prototypes already used latency insensitive links to obtain better modularity and we were able to partition and run these designs *without source modification*. We tested our designs on two platforms: the ACP [13], consisting of two Virtex-5 LX330 chips mounted on Intel's front-side bus, and a multiple FPGA software simulator, which can model an arbitrary number and interconnection of FPGAs.

Partitioning a design using our compiler has four potential benefits. First, wall-clock runtime of the design can decrease, due to improved clock frequency and increased access to resources. Second, some designs can be scaled to handle larger problem sizes, again due to increased access to resources. Third, synthesis times are reduced due to the smaller size of design partitions. Fourth, partial recompilation is available in earnest because only those FPGAs that have changing logic need to be rebuilt. Different designs will experience different combinations of these salutary effects. On the other hand, because we are partitioning a design between chips, any communication between the chips will have increased latency. Our experiments will show that this negative effect is minimal for typical designs; the natural pipeline parallelism of hardware and improved operating frequency together compensate for increased latency.

Wireless Processing: Airblue is a highly parametric library for implementing OFDM baseband processors such as WiFi and WiMAX. A typical baseband pipeline implemented in Airblue, shown in Figure 7 has relatively little feedback, although the main data path has high bandwidth and low latency requirements. Typical wireless protocols implemented using Airblue have protocol-level latency requirements on the order of tens of microseconds. Based on these requirements, our compiler presents an ideal mechanism for scaling Airblue protocol implementations to multiple FPGAs, because our partitioned implementations can be made to favor high-bandwidth links even in the presence of inter-FPGA traffic on non-critical links. The latency introduced by inter-FPGA hops is small, approximately

100 nanoseconds, and well within the timing requirements of the high-level protocols.

To evaluate our compiler, we partition a micro-architectural simulator for SoftPHY [9], a recently proposed cross-layer protocol which extends commercially deployed forward error correction schemes to improve wireless throughput. Partitioning benefits the micro-architectural simulator in two ways. First, because only the microarchitecture of the error correction algorithm varies, by partitioning the simulator at the error correction algorithm, the bulk of the hardware simulator needs to be compiled only once. To test a different algorithm, a relatively small logical change, only one bit-file needs to be rebuilt. Second, because the clock frequencies of the FPGAs can be scaled, the wall-clock performance of the simulator improves.

Figure 12(b) shows the normalized performance of two experiments: one using a complex software channel model and the other using a simpler hardware channel model. In the first experiment, the software channel model is the performance bottleneck and limits the throughput of both the single and multiple FPGA implementations. In this case the multiple FPGA implementation achieves near performance parity with the single FPGA implementation, even though it has a much higher clock frequency. For one data point, QAM-64, the multiple FPGA implementation slightly outperforms single FPGA implementation. This is because QAM-64 produces more bits per software communication and begins to overwhelm the serial portions of the slower single FPGA implementation.

When a simpler channel model is implemented in hardware, the multiple FPGA implementation outperforms the single FPGA implementation. In this case, the normalized performance is tied to the clock frequency ratios of the two designs. For BPSK, which stresses the FFT, the ratio is highest, since the FFT is located on FPGA 0 in the partitioned implementation. For higher bit-rate modulation schemes, the bit-wise error correction, located on FPGA 1, is the bottleneck. Since the ratio of the clocks of the single FPGA implementation and FPGA 1 is smaller, the performance gap narrows.

Processor Modeling: HAsim is a framework for constructing high speed, cycle-accurate simulators of multi-core processors. Like many FPGA-based processor models, HAsim uses multiple FPGA cycles to simulate one model cycle [19]. HAsim uses a technique called A-Ports [20] to allow different modules in the processor to simulate at different and runtime variable FPGA-cycle-to-Model-cycle Ratios (FMR). This makes HAsim amenable to our multi-FPGA implementation technique, as the A-Ports protocol can be layered on top of our latency-insensitive links without affecting the ability of A-Ports to resolve the cycle-by-cycle behavior of the original design. HAsim is written in a highly parameterized fashion, both in terms of the structure and the number of the cores modeled. HAsim models can scale to hundreds or thousands of cores by changing a handful of parameters, an important feature for modeling future processors. The difficulty in modeling such large processors is that, even though describing the models using HAsim is straightforward, the models themselves do not fit in a single FPGA.

HAsim is divided into a functional partition and a timing partition, which separates the calculation of simulation results from the amount of time that those results take in the modeled processor [21]. This partitioning creates a high degree of feedback. For example,

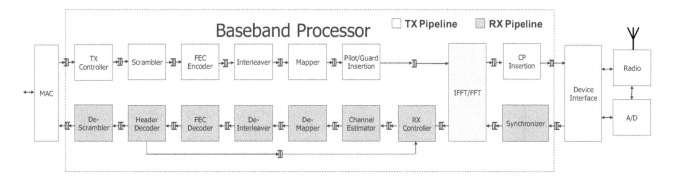

Figure 7: An Airblue 802.11g-compatible transceiver. In the SoftPHY experiment, only the forward error correction (FEC) decoder block is modified.

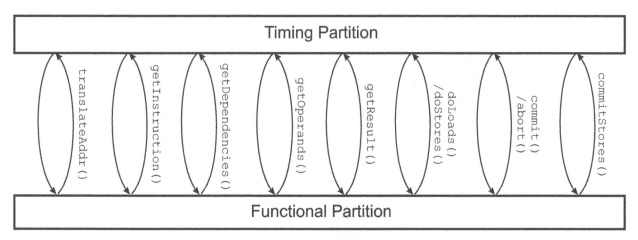

Figure 8: HAsim partitioned processor simulator. The timing partition relies on the functional partition for all computation related tasks, for example, instruction decoding.

the timing partition must query the functional partition to decode an instruction and wait for a response before proceeding. Similar feedback loops arise in the other processor stages and in the cache model. Despite this level of feedback, a natural mapping of HAsim to two FPGAs is placing the timing and functional partitions on separate FPGAs. This partition, shown in Figure 8, is attractive because all HAsim timing models share a common functional partition, enabling our compiler to compile the functional partition once and reuse it among many timing models.

The timing-functional partitioning works well because in practice HAsim is latency tolerant by design. In order to scale to multi-core configurations without using large numbers of FPGAs, HAsim uses time-multiplexing to map several virtual processors onto a limited number of physical processors. This multiplexing means that individual logical cores can wait dozens of cycles for responses from the functional model without reducing overall model throughput. Moreover, this tolerance scales as the number of simulated cores increases.

Although HAsim gracefully degrades its performance in the presence of limited resources, introducing more resources both speeds simulation and enables HAsim to scale to simulations of larger numbers of more complex cores. In particular, large HAsim models need large amounts of fast memory. Partitioning HAsim designs among multiple FPGAs automatically introduces new chip-level resources, like DRAM, into the synthesized implementation, increasing cache capacity and memory bandwidth.

On a single FPGA, HAsim scales to 16 cores before the FPGA runs out of resources. By mapping HAsim to two FPGAs, we are able to build a partitioned model capable of supporting up to 128 cores and give them access to approximately twice the memory capacity and bandwidth of a single FPGA implementation. We achieve super linear scaling in problem size because many structures in HAsim are either time-multiplexed among all cores or scale logarithmically with the number of cores.

We evaluate the throughput of the models by running a mix of SPEC2000 integer and floating point applications in parallel on the modeled cores. Figure 12(c) shows the normalized performance of the multiple-FPGA simulator relative to the single FPGA simulator. For small numbers of cores, the gap between the single FPGA and multiple FPGA simulator is large, due to the request-response latency between the timing and functional partitions. However, as the number of simulated processors scales, models become more latency tolerant, and the performance gap closes.

The raw performance of various HAsim implementations is shown in Figure 9. Single FPGA performance decreases from 8 to 16 cores due to increased cache pressure on the simulator's internal memory hierarchy. The partitioned processor model achieves about 75% the aggregate throughput of the 16-core single FPGA implementation, due to the latency of communication between chips. As we scale the number of cores in the partitioned model, throughput increases until 36 cores. The reason for this throughput improvement is that larger numbers of cores in a multiplexed model are more resilient to inter-link latency. Balancing this improved latency tolerance is cache pressure. In the multiple FPGA implementation cache pres-

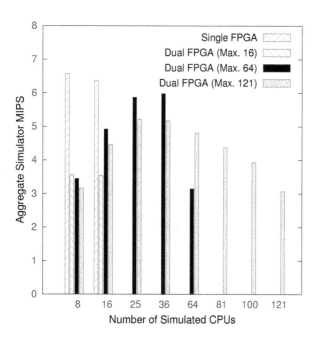

Figure 9: Performance results for various HAsim simulation configurations. Simulated cores run a combination of wupwise, applu, gcc, mesa, mcf, parser,perlbmk, and ammp from the SPEC2000 suite.

sure effects emerge dominate only in larger simulations because both FPGAs have chip-level memory caches, doubling the cache size and bandwidth available to the model. As the maximum number of simulated cores increases, the FPGA becomes more crowded, reducing operating frequency. As a result, partitioned models supporting more cores have lower performance even when simulating the same number of cores as a smaller model.

Video Decoder: H.264 [10], shown in Figure 11, is a state of the art video decoder, which has seen broad deployment both in custom hardware and in software. H.264 has several potential levels of implementation with widely varying feature sets and performance requirements. When implementing these various feature sets, it is useful to have a platform for rapidly evaluating the performance of different micro-architectures and memory organizations. The lower compile times offered by our compiler are useful in this kind of architectural exploration.

H.264 is naturally decomposed into a bit-serial front-end and a data parallel back-end. The front-end handles decompression and packet decoding, while the back-end applies a series of pixel-parallel transformations and filters to reconstruct the video. H.264 has limited feedback between blocks in the main pipeline. The pipeline synchronizes only at frame boundaries, which occur at the granularity of millions of cycles. Intraprediction does require some feedback from interprediction, but this feedback is somewhat coarse-grained, occurring on blocks of sixty-four pixels.

Because H.264 generally lacks tight coupling among processor modules, many high performance partitionings are possible. We choose to partition the bit-serial fronted because the front-end computation does not parallelize efficiently. As such, its performance can only be increased by raising operating frequency. The front-end also contains a number of difficult feedback paths, which end up limiting frequency in a single FPGA implementation.

Figure 12(a) shows the performance of a partitioned implementation of H.264 relative to a single FPGA implementation. In the case of the low resolution, the multiple FPGA implementation outperforms the single FPGA implementation by 20%. This performance

gain comes from increasing the clock frequency of the partitioned implementation relative to the single FPGA implementation. However, at higher resolution, interprediction memory traffic becomes more significant which has the effect of frequently stalling the processing pipeline. As a result some part of the latency of inter-chip communications is exposed and the multiple FPGA performance degrades slightly.

Compilation Time: To this point, we have focused on the wall clock performance and design scaling that our system provides. However, our compiler also provides another important performance benefit: reduced compilation time. FPGAs have notoriously long tool run times, primarily due to their need to solve several intractable problems to produce a functional design. In practice, these run times represent a serious impediment both to experimentation and to debugging. Our compiler helps alleviate the compilation problem in two ways. First, partitioned designs are fundamentally easier to implement; in the context of nonlinear run times, even a small decrease in design size can reduce compilation time significantly. Second, by partitioning we obtain a degree of modular compilation. If the design is modified, but the gate-ware of a partition has not changed, then that partition does not need to be recompiled. This savings is significant in two contexts: debugging and micro-architectural experimentation. In the case of debugging, the utility of the shortened recompilation cycle is obvious. However, the need to compile a partition only once per set of experiments is perhaps more beneficial. In this case, effort can be spent tweaking the tools to produce the best possible implementation of the shared infrastructure, in order to accelerate all experiments. In the case of HAsim, a single functional partition can be used in conjunction with all timing partitions.

Figure 13 shows selected compilation times for single and partitioned designs. It is important to note that the numbers reported for multiple FPGA designs reflect parallel compilation on the same machine, although to maximize speed, compilation should be distributed. By partitioning we achieve reduced compilation time, even though, in aggregate, we are building more complex, higher frequency designs. For Airblue and HAsim, the two examples in which modular recompilation of FPGA 1 is a useful, our recompilation facilities represent a substantial time savings.

8. CONCLUSION

In this paper, we present a language extension and compiler that leverages latency-insensitive design to produce high-performance implementations spanning multiple FPGAs. Our language and compiler permit us to build larger research prototypes, improve compilation time, and, in some cases, gain performance over single FPGA implementations.

Our compiler performs best in partitioning digital signal processing applications. These applications usually feature high bandwidth and computation requirements, but very little global control or feedback. As a result they are more resilient to the latency introduced in chip-to-chip communication and have the potential for super-linear performance increases when scaling to systems with multiple FPGAs. Applications with larger amounts of feedback, like processor prototypes, may experience performance degradations relative to a single FPGA due to latency. However, these applications still benefit from improved access to resources, design scaling, and reduced compile times.

The compiler that we have presented in this paper is promising, and we see four areas of exploration moving forward. The first is hardware-software co-design. Sequential languages are intrinsically latency-insensitive and so can be easily integrated into our model of computation. We believe that our proposed syntax provides a convenient mechanism for bridging the gap between both host PC and soft-cores instantiated on the FPGA itself. Second, we see compiler optimization as an area for exploration. Unlike traditional

	LUTS	Registers	BRAM	fMax(MHz)
Airblue, SOVA, Single	115780	67975	46	25
Airblue, SOVA, FPGA 0	77982	56499	34	65
Airblue, SOVA, FPGA 1	46852	21707	39	45
HAsim, 16 cores, Single	185002	153906	127	70
HAsim, 16 cores, FPGA 0	119231	102161	136	75
HAsim, 16 cores, FPGA 1	123892	99066	88	80
HAsim, 64 cores, FPGA 0	148107	108617	220	65
HAsim, 64 cores, FPGA 1	164920	111145	133	70
H.264, Single	79839	59212	63	55
H.264 FPGA 0	66893	52860	65	65
H.264 FPGA 1	13998	9493	19	85

Figure 10: Synthesis metrics for single and multiple FPGA implementations of our sample designs. Xilinx 12.1 was used to produce bit-files. To limit compile times, we stepped fMax at increments of 5MHz.

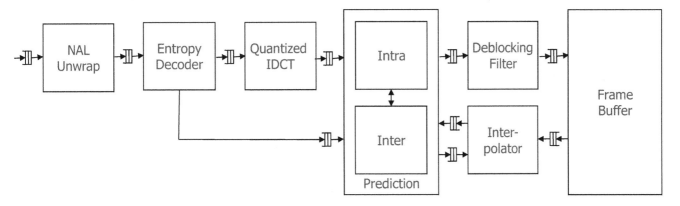

Figure 11: An H.264 decoder.

circuits, for which communication is statically determined by wire connection, our designs exhibit complex phase behavior. We believe that static, dynamic, and feedback-driven optimization techniques may be applied with great effect. Third, we see an opportunity to introduce quality of service (QoS) to improve design reliability. Programmer visible QoS is necessary to maintain program correctness for those workloads with high-level latency and throughput requirements, such as wireless protocols. Fourth, we see our model of computation as a means of alleviating the place and route problem, even on a single FPGA. Reuse of pre-routed components is difficult in current FPGA design because there is no guarantee that the tool can make timing on inter-component interface wires. However, our paradigm breaks these long paths with registered buffer stages as needed. For many designs, including those presented in this paper, small delays are negligible, especially during debugging and design exploration. By enabling large-scale component reuse, the synthesis back-end might reduce to a simple and fast linking step, dramatically reducing compile times.

Acknowledgements: During the course of this work, Kermin Fleming was supported by the Intel Graduate Fellowship.

9. REFERENCES

[1] Michael Adler, Kermin Fleming, Angshuman Parashar, Michael Pellauer, and Joel S. Emer. LEAP Scratchpads: Automatic Memory and Cache Management For Reconfigurable Logic. In *FPGA*, pages 25–28, 2011.

[2] Jonathan Babb, Russell Tessier, Matthew Dahl, Silvina Hanono, David M. Hoki, and Anant Agarwal. Logic Emulation With Virtual Wires. *IEEE Trans. on CAD of Integrated Circuits and Systems*, 16(6):609–626, 1997.

[3] Luca P. Carloni, Kenneth McMillan, and Alberto L. Sangiovanni-Vincentelli. Theory of Latency-Insensitive Design. *IEEE TRANSACTIONS on Computer-Aided Design of Integrated Circuits and Systems*, 20(9), September 2001.

[4] Josep Carmona, Jordi Cortadella, Mike Kishinevsky, and Alexander Taubin. Elastic Circuits. *IEEE Transactions on Computer-Aided Design*, 28(10):1437–1455, October 2009.

[5] W. J. Dally and C. L. Seitz. Deadlock-Free Message Routing in Multiprocessor Interconnection Networks. *IEEE Trans. Comput.*, 36:547–553, May 1987.

[6] Nirav Dave, Man Cheuk Ng, Michael Pellauer, and Arvind. Modular Refinement and Unit Testing. In *MEMOCODE'10*.

[7] K. Fleming, Chun-Chieh Lin, N. Dave, Arvind, G. Raghavan, and J. Hicks. H.264 Decoder: A Case Study in Multiple Design Points. In *Formal Methods and Models for Co-Design, 2008. MEMOCODE 2008. 6th ACM/IEEE International Conference on*, pages 165 –174, Jun. 2008.

[8] Kermin Fleming, Myron King, Man Cheuk Ng, Asif Khan, and Muralidaran Vijayaraghavan. High-throughput Pipelined Mergesort. In *MEMOCODE*, pages 155–158, 2008.

[9] Kermin Elliott Fleming, Man Cheuk Ng, Samuel Gross, and Arvind. WiLIS: Architectural Modeling of Wireless Systems. In *ISPASS*, pages 197–206, 2011.

[10] ITU-T Video Coding Experts Group. Draft ITU-T Recommendation and Final Draft International Standard of Joint Video Specification, May, 2003.

[11] http://www.cadence.com/products/sd/palladium_series/pages/default.aspx. "Cadence Palladium".

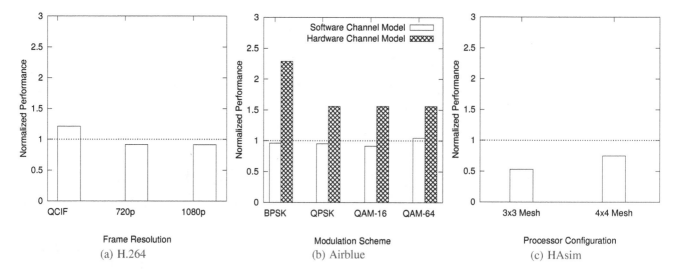

Figure 12: Performance results for various two FPGA partitioned workloads. Performance is normalized to a single FPGA implementation of the same hardware.

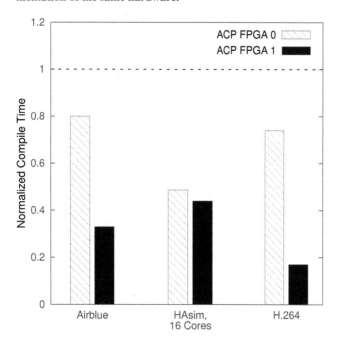

Figure 13: Compilation time relative to a single FPGA build. Xilinx 12.1 was used to produce bit-files. Compile times were collected on an unloaded Core i7 960 with 12 GB of RAM. Note that for multiple FPGA builds, the FPGA builds can proceed in parallel.

[12] http://www.eda.org/itc/scemi.pdf. Standard Co-Emulation Modelling Interface (SCE-MI): Reference Manual.

[13] http://www.nallatech.com. Nallatech ACP module.

[14] "http://www.synopsys.com/Systems/ FPGABasedPrototyping/pages/certify.aspx". "Synopsys Certify".

[15] Michel A. Kinsy, Myong Hyon Cho, Tina Wen, G. Edward Suh, Marten van Dijk, and Srinivas Devadas. Application-aware Deadlock-free Oblivious Routing. In *ISCA*, pages 208–219, 2009.

[16] M. C. Ng, K. Fleming, M. Vutukuru, S. Gross, Arvind, and H. Balakrishnan. Airblue: A System for Cross-Layer Wireless Protocol Development. In *ANCS'10*, San Diego, CA, 2010.

[17] Angshuman Parashar, Michael Adler, Kermin Fleming, Michael Pellauer, and Joel Emer. LEAP: A Virtual Platform Architecture for FPGAs. In *CARL '10: The 1st Workshop on the Intersections of Computer Architecture and Reconfigurable Logic*, 2010.

[18] M. Pellauer, M. Adler, D. Chiou, and J. Emer. Soft Connections: Addressing the Hardware-Design Modularity Problem. In *DAC '09: Proceedings of the 46th Annual Design Automation Conference*, pages 276–281. ACM, 2009.

[19] M. Pellauer, M. Adler, M. Kinsy, A. Parashar, and J. Emer. HAsim: FPGA-Based High-Detail Multicore Simulation Using Time-Division Multiplexing. In *The 17th International Symposium on High-Performance Computer Architecture (HPCA)*, February 2011.

[20] M. Pellauer, M. Vijayaraghavan, M. Adler, Arvind, and J. Emer. A-Ports: An Efficient Abstraction for Cycle-Accurate Performance Models on FPGAs. In *Proceedings of the International Symposium on Field-Programmable Gate Arrays (FPGA)*, February 2008.

[21] M. Pellauer, M. Vijayaraghavan, M. Adler, Arvind, and J. Emer. Quick Performance Models Quickly: Closely-Coupled Timing-Directed Simulation on FPGAs. In *IEEE International Symposium on Performance Analysis of Systems and Software (ISPASS)*, April 2008.

[22] Charles Selvidge, Anant Agarwal, Matthew Dahl, and Jonathan Babb. TIERS: Topology Independent Pipelined Routing and Scheduling for Virtual Wire Compilation. In *FPGA*, pages 25–31, 1995.

[23] Todd Snyder. Multiple FPGA Partitioning Tools and Their Performance. Private communication, 2011.

[24] Russel Tessier. Multi-FPGA Systems: Logic Emulation. *Reconfigurable Computing*, pages 637–669, 2008.

[25] Muralidran Vijayaraghavan and Arvind. Bounded Dataflow Networks and Latency-Insensitive Circuits. In *MEMOCODE'09*, Cambridge, MA, 2009.

[26] Nam Sung Woo and Jaeseok Kim. An Efficient Method of Partitioning Circuits for Multiple-FPGA Implementation. In *Proceedings of the 30th international Design Automation Conference*, DAC '93, pages 202–207, New York, NY, USA, 1993. ACM.

A Scalable Approach for Automated Precision Analysis

David Boland and George A. Constantinides
Department of Electrical and Electronic Engineering
Imperial College London
London, UK
{david.boland03, g.constantinides}@imperial.ac.uk

ABSTRACT

The freedom over the choice of numerical precision is one of the key factors that can only be exploited throughout the datapath of an FPGA accelerator, providing the ability to trade the accuracy of the final computational result with the silicon area, power, operating frequency, and latency. However, in order to tune the precision used throughout hardware accelerators automatically, a tool is required to verify that the hardware will meet an error or range specification for a given precision. Existing tools to perform this task typically suffer either from a lack of tightness of bounds or require a large execution time when applied to large scale algorithms; in this work, we propose an approach that can both scale to larger examples and obtain tighter bounds, within a smaller execution time, than the existing methods. The approach we describe also provides a user with the ability to trade the quality of bounds with execution time of the procedure, making it suitable within a word-length optimization framework for both small and large-scale algorithms.

We demonstrate the use of our approach on instances of iterative algorithms to solve a system of linear equations. We show that because our approach can track how the relative error decreases with increasing precision, unlike the existing methods, we can use it to create smaller hardware with guaranteed numerical properties. This results in a saving of 25% of the area in comparison to optimizing the precision using competing analytical techniques, whilst requiring a smaller execution time than the these methods, and saving almost 80% of area in comparison to adopting IEEE double precision arithmetic.

Categories and Subject Descriptors

G.1.0 [**General**]: Error analysis

Keywords

Range analysis, Precision analysis, Word-length Optimisation

1. INTRODUCTION

When designing custom accelerators for an algorithm, the majority of effort is placed into the architecture design, with the aim of maximizing the performance obtained from a fixed silicon or power budget. However, if one were to also to attempt to optimize the precision used throughout the hardware design to meet an algorithm designer's output specification, it is possible to save a significant amount of silicon area; this in turn could affect the performance achievable. Such a specification may take the form of ensuring the result lies within some desired output range or meeting a threshold on the maximum error introduced by the use of finite precision arithmetic (floating point or fixed point). While performance improvements are also available for general purpose processors and GPUs by moving from double to single precision, provided a proof can be created that ensures this move is valid, these devices are largely restricted to these two IEEE standards, unlike FPGAs which have the freedom to implement any precision. Unfortunately, it is hard to choose a number system which can guarantee any range or error specification can be met. As a result, simulation is sometimes used to justify that a customized precision is valid [1, 2], but in this case the resulting implementation comes with no guarantees, or a standard precision is selected to make a 'fair' comparison against a software implementation [3, 4], ignoring the potential performance improvements. To create a custom accelerator that benefits from this freedom with guaranteed numerical properties, we require a tool to verify such an algorithm specification is met. While no tool exists which can provide such verification for an arbitrary algorithm over a set of input ranges and precision in a tractable time, there remains an interest in developing analytical tools that can guarantee a certain precision is sufficient to satisfy the specification.

There are two important metrics for any such tool: quality of bounds and scalability. A tool that can find tight bounds will generally be able to guarantee that hardware will satisfy the chosen output criterion with a lower precision than a tool which calculates wide bounds. We note that any approach that does not calculate a bound, such as simulation, cannot be used to make hardware with any such guarantees. Scalability is important because rounding errors arise from every operation and the tool must be able to keep track of all the potential errors that can arise throughout an algorithm and be able to calculate bounds, taking all these errors into account, in a reasonable time to be of any use. Furthermore, a scalable and fast tool is important given that it is likely to be used repeatedly within some compilation and synthesis framework to find the best precision for a custom hardware design.

Existing work in this field has typically focused on either one of these goals, meaning the methods that can calculate tight bounds are restricted to trivial computational kernels [5–7], or the restricted LTI domain [8], and the methods which can scale to larger problems do not find tight bounds and consequently limit the potential hardware optimizations [9, 10]. This work describes a tool to calculate bounds for any algorithm consisting of algebraic operations or elementary functions and can be translated into straight-line single static assignment (SSA) form by performing a static analysis of input code. This means our approach can support algorithms consisting of loops with known bounds at compile-time, by loop unrolling, and conditional statements by analyzing branches separately. The bounds our approach finds approach the quality of the time consuming methods in a significantly shorter execution time, and our method is capable of scaling to larger examples than all but the simplest existing method. Furthermore, unlike existing approaches, our approach also allows a user a finer level of control over the trade-off between execution time and quality of bounds.

We examine the use of our method on two iterative algorithms to solve a system of linear equations: successive-over-relaxation and MINRES [11]. We demonstrate how our approach obtains better bounds than existing analytical techniques, and this provides the ability to tune the precision used in a hardware acceleration of an algorithm, which we show to translate into a significant reduction in silicon area. Furthermore, we highlight that our approach is much more scalable than most of the existing approaches, enabling the analysis of much larger algorithms than these examples.

This paper first describes background into the main existing approaches to calculate bounds in Section 2 and how to model floating point round-off error in Section 3. We follow this with an analysis of how this model of error affects the scalability of the methods to find bounds in Section 4, before we describe how our new method addresses problems in the existing methods in order to obtain a control over the execution time in Section 5. We follow this with a description of our algorithm to compute tight bounds in Section 6, after which we demonstrate the benefits of our algorithm relative to alternate methods in Section 7 and draw conclusions in Section 8.

2. BACKGROUND

While there is a large amount of literature in the field of word-length optimization focused on developing tools to optimize the potential area, frequency and power benefits to satisfy a design specification [12], the number of basic analysis techniques is limited. In this section, we elaborate on these methods: simulation, interval arithmetic, Taylor forms, and Handelman representations, so as to compare these with our new suggested approach in later sections.

The most straightforward way to estimate an error is through simulation. The aim of any simulation-based approach is to find the inputs that will cause the extreme ranges of the data set. Unfortunately, the size of the search space for the inputs will generally be too large to explore exhaustively, meaning that even advanced simulation methods, for example statistical profiling [13], can only estimate the error because corner cases can be missed.

To calculate true bounds for general algorithms, the most well known analytical approach is interval arithmetic (IA) [14]. Interval arithmetic represents every value as lying within some interval: $[x_1, x_2]$, where x_1 and x_2 are the lower and upper bounds respectively. The intervals are then propagated through the computation according to basic rules, given in (1) (provided the range of a divisor does not include zero), which calculate at each stage the new worst case bound.

$$[x_1, x_2] \odot [y_1, y_2] = [\min(x_1 \odot y_1, x_2 \odot y_2, x_1 \odot y_2, x_2 \odot y_1), \quad (1)$$
$$\max(x_1 \odot y_1, x_2 \odot y_2, x_1 \odot y_2, x_2 \odot y_1)],$$
$$\odot \in \{+, -, \bullet, \div\}.$$

However, interval arithmetic suffers from the so-called dependency problem, where if the same variable is used twice, information is lost. A trivial example is the following: for a variable x which lies in the interval $[0, 1]$, perform the operation $x - x$. The interval should be $[0, 0]$, but the result using interval arithmetic

would be $[-1, 1]$. Several simple examples can demonstrate how this problem may cause bounds that are significantly wider than the tightest bounds [14]. As a result of these problems, there is an active community of researchers in *robust computing* who have developed ways to mitigate this problem [15].

One simple method to reduce the effect of the dependency problem is to split intervals into the union of much smaller intervals (2), and evaluate each of these independently, because dependencies between smaller intervals have a reduced effect of widening bounds (3). While effective, the number of intervals that must be evaluated grows as of $\Theta(n_o n_s^{n_{sv}})$, where n_o is the number of operations, n_s is the maximum number of interval splits of the n_{sv} variables that are split. This means this technique is not computationally scalable. Even though some methods have been proposed to choose these splits more wisely, such as the GAPPA tool which uses a set of in-built and user-defined 'hints' [16], or the work by Nicolici et al. which uses Satisfiability-Modulo Theories (SMT) [5], the quality of bounds these approaches can obtain are heavily limited by the run-time. As a result of the slow run-time, the authors of the latter paper have attempted to improve scalability using vector approximations [17], and by adding additional constraints informing the solver to ignore certain regions [18], but both sacrifice the tightness of bounds and because they use interval splitting, they will always suffer if there are many input variables, as we will see in Section 7.

$$[x_{lower}, x_{upper}] = \bigcup_{i=1}^{n} [x_{ilower}, x_{iupper}]. \quad (2)$$

$$f([x_{lower}, x_{upper}]) \supseteq \bigcup_{i=1}^{n} f([x_{ilower}, x_{iupper}]). \quad (3)$$

Alternatively, more recently a set of methods that can be loosely grouped together under the name of Taylor forms, analyzed in detail [19], have gained popularity. These use a polynomial representation of the error terms with the intuition that this allows cancellation of dependencies; in the case of the earlier example, use of these approaches would result in $x - x = [0, 0]$ as desired. Unfortunately, a polynomial with second order terms or higher contains dependencies within the polynomial, and finding optimal bounds for a multivariate polynomial has been shown to be NP-hard [20].

The most well-known of these Taylor forms, affine arithmetic (AA) [21], avoids this problem by restricting polynomials to first order, ensuring the polynomial contains no dependencies, meaning applying interval arithmetic to the final polynomial can find the ideal bounds. However, to ensure the polynomial only contains first order terms and still bounds the potential range, it must approximate bounds on any higher order terms which are created, using a new variable. Unfortunately, any difference between the true range of the higher order terms and their approximation will result in wider bounds, whilst the dependency information between the higher order and lower order terms is lost. In addition, the added variables can affect the scalability, as will be seen in Section 7.

To minimize both these effects, the more general method by Berz et al., named Taylor methods with Interval Remainder bounds (TwIR) [22] represents range using the form (T_ρ, I_ρ), where T_ρ is a polynomial consisting of all the terms that are less than or equal to an order, ρ, chosen by a user, and I_ρ is an interval which bounds the remaining higher order terms. Using a single interval I_ρ avoids continually introducing new variables to bound higher order terms whenever they are created, as performed by affine arithmetic. Furthermore, the choice of maximum order gives a user some form of control over the trade-off between execution time and quality of bounds, because by retaining higher orders, the higher order dependencies can cancel. Unfortunately, using the single interval I_ρ

means any operations involving I_ρ suffer from the same dependency problem as interval arithmetic, and in addition, finding the final bounds of the polynomial T_ρ still relies on interval arithmetic, and this is a problem once again because dependencies will exist between the high and lower order terms of the polynomial.

Recently, a novel approach has been suggested which is capable of reducing the effect of widening of bounds due to dependencies in a multivariate polynomial by bounding the polynomial using Handelman representations [6]. This approach obtained superior bounds to those achievable by interval arithmetic, and also showed significant advantages in the case of division, by using a rational expression instead of a polynomial. Unfortunately, the approach was limited to small problems because it made no effort to control the size of the polynomial, as will be discussed in Section 4. Furthermore, it was limited to algorithms consisting of the basic algebraic operations $\{+, -, \bullet, \div\}$, unlike Taylor forms which have the added advantage of being applicable to more complex functions by applying polynomial approximation techniques [23].

In this work, we aim to develop a new approach which can find bounds approaching those of the method using Handelman representations, whilst scaling to much larger problems. We also demonstrate how to allow a user to have a strong control over the trade-off between execution time and the tightness of bounds, and show how our new method can be used to create superior hardware designs.

3. FLOATING POINT MODEL OF ERROR

In this work, we have elected to use the multiplicative model of floating point error used throughout numerical analysis literature. This represents the closest radix-2 floating-point approximation $\hat{x}$ to any real value x as (4) [24], where η represents the number of mantissa bits used and δ represents the small unknown roundoff error. It is similarly possible to specify that the radix-2 floating-point result of any scalar operation ($\odot$) is bounded as in (5), provided the exponent is sufficiently large to span the range of the result. This allows us to create a polynomial representing the potential range of a variable, as shown for a simple example in Table 1. This is the only general model that is useful for worst-case floating-point error modelling, with the exception of a bit-blasting of the floating-point logic. More sophisticated models, such as detecting if one operand is a power of two and setting the error to zero in this case will have minimal impact and severely complicate modelling [25]. We note, however, that provided the chosen model, or a similar model for fixed-point error, can be expressed using polynomials or rational functions, the algorithms described in this work are still applicable.

$$\hat{x} = x(1 + \delta) \quad (|\delta| \leq \Delta, \text{where } \Delta = 2^{-\eta}). \quad (4)$$

$$\widehat{x \odot y} = (x \odot y)(1 + \delta). \quad (5)$$

Table 1: Construction of polynomials

x, y are inputs, Δ is the error bound determined by the precision, so that $|\delta_i| \leq \Delta$	
Input Code	Polynomial Representation of Variable Value
$a = x \bullet y;$	$a = xy(1 + \delta_1)$
$b = a \bullet a;$	$b = (xy(1 + \delta_1))^2(1 + \delta_2)$
$c = b - a;$	$c = [(xy(1 + \delta_1))^2(1 + \delta_2) - xy(1 + \delta_1)](1 + \delta_3)$

The value of this representation is that it allows us to apply the symbolic analysis techniques described in the Section 2 to calculate bounds on the range or relative error of any variable in the input code. In the following section, we analyze the limitations of these techniques when using this model of error, before describing how we overcome them in Section 5.

4. SCALABILITY OF EXISTING APPROACHES

In this section, we analyze the worst case execution time of these approaches in terms of the number of operations in the code, denoted n_o, and the number of input variables n. Throughout this paper, we define a term as a product of the variables raised to some integer powers, e.g. $\delta_1\delta_2\delta_3^2$, and a monomial as a term multiplied by some real coefficient, e.g. $10\delta_1\delta_2\delta_3^2$.

Using interval arithmetic, an interval evaluation is performed to find an initial bound on the result of every floating point operation in the code, after which an extra interval evaluation is used to find the bounds taking into account the floating point model of error, where the latter is computed by replacing each of the variables δ_i with an interval. Consequently, the worst case execution time is proportional to the number of operations, or of $O(n_o)$, or as we have mentioned in Section 2, in the case of applying interval splitting, it will then scale as of $O(n_o n_s^{n_{sv}})$

At the other extreme lies the approach using Handelman representations. In this approach, the polynomials representing the variable value, as in Table 1, are first expressed as a sum of monomials in which each term appears at most once. The number of monomials in the polynomials quickly become large, as an example, the floating-point error model for an algorithm consisting of a series of floating-point multiplications (Figure 1), will have a polynomial representation as in (6); the number of monomials in this polynomial, when expanded into canonical form, is exponential in n_o.

```
y = 1
for i = 1; i ≤ n_o; i + + do
    y = y • x[i]
end for
```

Figure 1: Code to calculate the product of vector elements.

$$x_1 x_2 \ldots x_{n_o}(1 + \delta_1)(1 + \delta_2)\ldots(1 + \delta_{n_o}) \quad (6)$$

As we indicated in Section 2, Taylor forms control this growth in the polynomial representing the variable value by only retaining monomials up to a given order. However, the lack of similar control over the number of variables ensures that the size of this polynomial remains unbounded. This is a problem when using the floating point model of error of Section 3, because a variable (δ_i) bounding the finite precision error will be added after every operation, and hence the total number of variables for an intermediate polynomial bounding the range of a variable in an algorithm, when including the n input variables, is of $O(n_o + n)$. This means that for a Taylor form limited to a maximum order ρ, the number of monomials in any polynomial representing the range of an intermediate variable can grow as of $O\binom{n_o+n+\rho}{\rho}$, so bounding the higher order terms will require of $O\binom{n_o+n+\rho}{\rho}$ interval evaluations. Altogether, the execution time to do this for n_o operations grows as $O(n_o \times \binom{n_o+n+\rho}{\rho})$.

This analysis makes it clear that only interval arithmetic has an execution time that scales well with the number of floating point operations. However, as shown in [6, 21, 22], interval arithmetic is unable to find tight bounds due to dependencies between variables. The use of interval splitting in conjunction with any of these approaches allows some trade-off between quality of bounds and execution time, but this scales poorly in the number of variables. Our aim is to create an approach where the execution time of grows in proportion with the code size $O(n_o)$ and provides a user a very flexible level of control over the trade-off between execution time per floating point operation and can still obtain bounds approaching the tightness of the approach described in [6]. We discuss the main method to obtain a control over the execution time in Section 5, and our overall approach which uses this heuristic to calculate bounds on the range of any variable in an algorithm in Section 6.

5. CONTROLLING EXECUTION TIME

The basic concept we employ to obtain control over the execution time needed to bound the result of an operation is to directly control the number of monomials in every polynomial to a user chosen value, N, and hence the worst case execution time to create any intermediate polynomial will be some function of N. Since the worst case execution time to create any intermediate polynomial becomes constant, the overall execution time of our algorithm grows as $O(n_o)$, and the choice of N provides the user the ability to trade potential tightness of bounds with execution time. We note that this is a much finer level of control than Taylor forms.

To create the intermediate polynomial of only N monomials that still bound the correct result, we apply the algorithm described in Figure 2. This algorithm retains the monomials that have the greatest potential contribution to the final bounds, as calculated by computing the bounds of each monomial using interval arithmetic, and then representing the worst case bounds of the remaining monomials, again computed using interval arithmetic, using a new polynomial that consists of a constant C_i a single monomial ζ_i. The constant C_i is chosen to center the monomial ζ_i over the desired range, for this has previously been shown to obtain the best error properties [26]. The rationale for choosing monomials with the greatest potential contribution is that many monomials within a polynomial represent a small contribution towards the final bounds of the function, and hence if the dependency information of these monomials is lost, it has little impact on the final result. Table 3 demonstrates the potential contributions to the final bounds for all the individual monomials when computing bounds for a simple problem. We note that unlike Taylor forms, our approach may retain higher order monomials at the expense of lower order monomials, for example, in Table 3, the second order monomial $x_1 y_1$ has a higher contribution than the first order monomial δ_1. However, because some input variables may have much wider ranges than other input variables, and often wider ranges than variables bounding finite precision errors, this approach is logical as it is most important to retain the dependency information for the variables with the widest bounds whereas the dependency information for small perturbations can be sacrificed in favor of a reduced execution time.

$(\hat{p}, k) = $ **Simplify Polynomial** (p, N, k)
1: $\hat{p} = N$ monomials from p with the largest magnitude of potential contribution to final bounds, calculated by IA
2: $C_k + \zeta_k$ = new polynomial bounding potential contribution of other monomials in p
3: $\hat{p} = \hat{p} + C_k + \zeta_k$
4: $k = k + 1$

Figure 2: Algorithm to control the size of the polynomial.

6. TRADING QUALITY FOR EXECUTION TIME

While the algorithm given in Figure 2 would be sufficient to control the execution time if integrated into either affine arithmetic or Taylor series with interval remainder bounds, a combined approach would still suffer from dependencies within a polynomial. This was addressed by the use of Handelman representations [6], which also had added benefits by retaining correlation between a numerator and denominator polynomial in the case of division. As such, in this section, we describe how we combine this algorithm with the technique to bound rational functions using Handelman represen-

Table 3: Potential contribution of each monomial in $(1 + x_1)(1 + y_1)(1 + \delta_1)$.

Compute $a = x \bullet y$, where $x = [0.8; 1.2]$, $y = [0.9; 1.1]$ in 6 bit floating point			
let $\|x_1\| \leq 0.2$, $\|y_1\| \leq 0.1$, $\|\delta_i\| \leq 2^{-6} \Rightarrow x \in 10(1 + x_1)$, $y \in 10(1 + y_1)$			
$a = 10(1 + x_1)10(1 + y_1)(1 + \delta_1)$			
$= (100 + 100x_1 + 100y_1 + 100\delta_1 + 100x_1y_1 + 100x_1\delta_1$			
$\quad + 100y_1\delta_1 + 100x_1y_1\delta_1)$			
Monomial	Potential Contribution	Monomial	Potential Contribution
100	100	$100x_1$	± 20
$100\delta_1$	± 0.09765625	$100x_1\delta_1$	± 0.01953125
$100y_1$	± 10	$100x_1y_1$	$\pm$ -2
$100y_1\delta_1$	± 0.009765625	$100x_1y_1\delta_1$	± 0.001953125

tations to create a scalable framework to find tight bounds for the range or relative error for any variable in an algorithm.

6.1 Representing the range of a variable

One of the problems with using the polynomial simplification algorithm described in Figure 2 is that when we replace all of the monomials with small contributions to the final bounds with a new monomial, we lose information on whether those monomials with small contributions were a function of only the input variables, or a function of both input variables and finite precision errors. This is an issue when computing bounds on the relative error. To compute the relative error, if we have a polynomial p representing the range in infinite precision, and a polynomial $\hat{p}$ representing the range in the presence of finite precision errors, the bound on the relative error is found by maximizing the function $\left|\frac{p - \hat{p}}{p}\right|$. However, if we were to control the size of the polynomials p and $\hat{p}$ using the algorithm described in Figure 2, we would lose any correlation between the added monomials bounding small contributions in p and $\hat{p}$.

To demonstrate how this can become a problem, we use a simple example shown in Table 2(a). In this example, we attempt find bounds on the relative error of the computation $(x \bullet y) \bullet z$, where $x \in [0.8; 1.2]$, $y \in [0.9; 1.1]$, $z \in [9.9; 10.1]$ using a 6-bit precision, where we limit the maximum number of monomials in a polynomial to be 6. If we were to compute the relative error of this operation, according to Table 2(a), we must compute bounds of the function $\left|\frac{z_1 + 2.048\zeta_2 - 10\delta_1 - 25.3203\zeta_3}{10 + 10x_1 + 10y_1 + z_1 + 10x_1y_1 + 2.048\zeta_2}\right|$. The problem with this is that there is correlation between the monomials z_1, ζ_2 and ζ_3 which is lost due to the simplification algorithm. This leads to much wider bounds on the relative error.

In order to avoid this problem, we separate a polynomial $\hat{p}$ into the sum of two polynomials $p + p_\epsilon$, where the polynomial p consists of monomials that are only a function of the input variables, and the polynomial p_ϵ store the additional monomials resulting from the introduction of finite precision errors. We note now that even if we apply the polynomial simplification algorithm, the polynomial p will bound the result in infinite precision, and by keeping these polynomials separate, we can now compute bounds on the relative error by finding bounds of the rational function $\left|\frac{p_\epsilon}{p}\right|$; This allows us to find much tighter bounds, as shown in Table 2(b).

This technique is an effective method to describe polynomials, however, to take advantage of correlation between numerator and denominator polynomials using the Handelman representations approach, we bound the range of any intermediate variable in the code using a rational function of the form $\frac{n + n_\epsilon}{d + d_\epsilon}$, where n and d are the numerator and denominator polynomials that contribute to the bounds in infinite precision, and n_ϵ and d_ϵ store the additional monomials resulting from the introduction of finite precision errors.

6.2 Bounding the range of variables in finite precision arithmetic for a user algorithm

In order to compute bounds on the range or relative error for any variable within an algorithm, we first compile the target al-

(a) Using the algorithm defined in Figure 2 to control polynomial size with $N = 6$.

Calculate the relative error of the computation $(x \bullet y) \bullet z$, where $x \in [0.8; 1.2]$, $y \in [0.9; 1.1]$, $z \in [9.9; 10.1]$ in floating point with a 6-bit mantissa.			
let $|x_1| \le 0.2$, $|y_1| \le 0.1$, $|z_1| \le 0.1$ $\Rightarrow$ $x = (1 + x_1)$, $y = (1 + y_1)$, $z = (10 + z_1)$. Also let $\forall_i$, $|\delta_i| \le 2^{-6}$, $|\zeta_i| \le 2^{-6}$			
Create polynomials to bound the range of every intermediate variable			
Code	Polynomial bounding variable range	Polynomial in canonical form	Simplified polynomial
$a = x \bullet y$	$a = (1 + x_1)(1 + y_1)$	$1 + x_1 + y_1 + x_1 y_1$	$1 + x_1 + y_1 + x_1 y_1$
	$\hat{a} = (1 + x_1)(1 + y_1)(1 + \delta_1)$	$1 + x_1 + y_1 + x_1 y_1 + \delta_1 + x_1\delta_1 + y_1\delta_1 + x_1 y_1 \delta_1$	$1 + x_1 + y_1 + x_1 y_1 + \delta_1$ $+0.32\zeta_1$
$b = a \bullet z$	$b = (1 + x_1 + y_1 + x_1 y_1)(10$ $+ z_1)$	$10 + 10x_1 + 10y_1 + 10x_1 y_1 + z_1 + x_1 z_1 + y_1 z_1 + x_1 y_1 z_1$	$10 + 10x_1 + 10y_1 + z_1$ $+10x_1 y_1 + 2.048\zeta_2$
	$\hat{b} = (1 + x_1 + y_1 + x_1 y_1 + \delta_1$ $+0.32\zeta_1)(10 + z_1)(1 + \delta_2)$	$10 + 10x_1 + 10y_1 + 10x_1 y_1 + 10\delta_1 + 3.2\zeta_1 + z_1 + x_1 z_1 + y_1 z_1$ $+x_1 y_1 z_1 + z_1\delta_1 + 0.32 z_1\zeta_1 + 10\delta_2 + 10x_1\delta_2 + 10y_1\delta_2$ $+10x_1 y_1\delta_2 + 10\delta_1\delta_2 + 3.2\zeta_1\delta_2 + z_1\delta_2 + x_1 z_1\delta_2 + y_1 z_1\delta_2$ $+x_1 y_1 z_1\delta_2 + z_1\delta_1\delta_2 + 0.32 z_1\zeta_1\delta_2$	$10 + 10x_1 + 10y_1 + 10\delta_1$ $+10x_1 y_1 + 25.3203\zeta_3$
Find relative error of every intermediate variable			
Variable	Rational function bounding relative error		Bound on relative error using IA
$\left|\frac{a - \hat{a}}{a}\right|$	$\left|\frac{\delta_1 + 0.32\zeta_1}{1 + x_1 + y_1 + x_1 y_1}\right|$		$\|0.0303\|$
$\left|\frac{b - \hat{b}}{b}\right|$	$\left|\frac{z_1 + 2.048\zeta_2 - 10\delta_1 - 25.3203\zeta_3}{10 + 10x_1 + 10y_1 + z_1 + 10x_1 y_1 + 2.048\zeta_2}\right|$		$\|0.1026\|$

(b) Using the algorithm defined in Figure 2 with $N = 3$ to control separate polynomials for range in infinite precision and the additional monomials resulting from finite precision errors.

Create polynomials to bound the range of every intermediate variable			
Code	Polynomial bounding range of variable	Simplified polynomial, p, bounding range in infinite precision	Simplified polynomial, p_ϵ, bounding finite precision errors
$a = x \bullet y$	$(1 + x_1)(1 + y_1)(1 + \delta_1)$	$1 + x_1 + 7.68\zeta_1$	$\delta_1 + x_1\delta_1 + 0.12\zeta_2$
$b = a \bullet z$	$(1 + x_1 + 7.68\zeta_1 + \delta_1 + x_1\delta_1 + 0.12\zeta_2)(10 + z_1)(1 + \delta_2)$	$10 + 10x_1 + 83.9680\zeta_3$	$1.2\zeta_2 + \delta_2 + 15.6723\zeta_4$
Find relative error of every intermediate variable			
Variable	Rational function bounding relative error		Bound on relative error using IA
$\left|\frac{a_\epsilon}{a}\right|$	$\left|\frac{\delta_1 + x_1\delta_1 + 1.12\zeta_2}{1 + x_1 + 7.68\zeta_1}\right|$		$\|0.0244\|$
$\left|\frac{b_\epsilon}{b}\right|$	$\left|\frac{1.2\zeta_2 + \delta_2 + 15.6723\zeta_4}{10 + 10x_1 + 83.9680\zeta_3}\right|$		$\|0.0363\|$

gorithm into a 2-input static single assignment (SSA) intermediate representation consisting of vector operations, with the aid of techniques such as loop unrolling. For our tests, we performed this by hand, but for more complex examples we could make use of front end compilation tools such as GCC. Throughout our algorithms, we prefer to operate on vectors so as to take advantage of the fact that every element in a vector will typically share the same denominator and hence our algorithms are designed to retain this correlation so as to improve the tightness of bounds. We then proceed to calculate bounds on this intermediate representation using a set of simple algorithms summarized in Figures 3 and 4. In the rest of this section, we explain the rationale behind each of these algorithms.

6.2.1 Bound variable in code

In the main algorithm, we first create a set V containing all the input variables in the algorithm, stored as vectors wherever this is applicable. Our algorithm proceeds by sequentially examining each operation in the intermediate representation and creates new rational functions which bound the range of every output element from this operation. The operations we support are scalar multiplication, scalar division, scalar addition and subtraction, vector addition and subtraction, dot products, and any other function to which a polynomial approximation can be computed. The reason we prefer to perform vector operations is because after creating each new rational function, the number of monomials in the polynomials n and d are controlled according to a user choice N_1, and the number of monomials in the polynomials n_ϵ and d_ϵ are controlled according to a user choice N_2, where the choice of N_1 and N_2 is left to a user to trade the execution time against the potential quality of bounds. If we were to perform each operation on scalars instead of performing a single vector operation, then if the number of monomials in the polynomials d and d_ϵ created by the operation are greater than N_1 or N_2, then the denominator polynomials would be simplified

with the use of a different variable ζ_k for each element in the vector, meaning that none of the denominators for the vector would be the same. By performing a single vector operation, we only need to simplify the denominator polynomial for the entire vector once, retaining correlation for all the vector elements and improving the overall bounds. We note that because each numerator polynomial for a vector will in general be different, each of these are simplified individually, and in order to capture round-off uncertainty, we first apply the model error described in Section 3 to the polynomial n_ϵ. Finally, once we have created a rational function to bound the range of the desired output variable, we calculate bounds using Handelman representations [6] to try to improve the bounds by taking into account any dependencies in the rational function.

6.2.2 Compute rational function

Figure 4 describes how we create a rational function representing the value of every intermediate variable in the SSA version of the code. In general, a rational function representing a result of the basic algebraic operations ($\odot \in \{+, -, \bullet, \div\}$) applied to two other rational functions can easily be computed symbolically. For example, equation (7) shows how to perform multiplication of two rational functions $v_a = \frac{n_a + n_{a\epsilon}}{d_a + d_{a\epsilon}}$ and $v_b = \frac{n_b + n_{b\epsilon}}{d_b + d_{b\epsilon}}$; in this equation, we have used brackets to separate the polynomials which consist of both input variables and finite precision errors. However, for addition or subtraction when the denominator polynomials are different for the two operands, we apply a different approach. The reason for this exception is that this operation can result in a very large numerator polynomial which has lots of correlation with the denominator polynomial, as shown in equation (8). However, when we subsequently simplify the numerator and denominator polynomials, according to the algorithm in Figure 3, we lose correlation between these polynomials, and because the number of monomials

$$\frac{n_a + n_{a\epsilon}}{d_a + d_{a\epsilon}} \times \frac{n_b + n_{b\epsilon}}{d_b + d_{b\epsilon}} = \frac{n_a n_b + (n_a n_{b\epsilon} + n_b n_{a\epsilon} + n_{a\epsilon} n_{b\epsilon})}{d_a d_b + (d_a d_{b\epsilon} + d_b d_{a\epsilon} + d_{a\epsilon} d_{b\epsilon})} \tag{7}$$

$$\frac{n_a + n_{a\epsilon}}{d_a + d_{a\epsilon}} + \frac{n_b + n_{b\epsilon}}{d_b + d_{b\epsilon}} = \frac{n_a d_b + n_b d_a + (n_a d_{b\epsilon} + n_b d_{a\epsilon} + n_{a\epsilon} d_{b\epsilon} + n_{b\epsilon} d_{a\epsilon} + n_{a\epsilon} d_{b\epsilon} + n_{b\epsilon} d_{a\epsilon})}{d_a d_b + (d_a d_{b\epsilon} + d_b d_{a\epsilon} + d_{a\epsilon} d_{b\epsilon})} \tag{8}$$

Bound variable in code (N_1, N_2. code).
// N_1, N_2 are user chosen variables to control the maximum polynomial sizes
// We denote vectors of rational functions bounding variables with v where
the ith rational function of this vector is indexed v^i. The total number of
elements in a vector is given by $|v|$. As a scalar variable is a vector consisting
of only one element so we omit the superscript i.
// Number of variables δ_j, ζ_k are determined at run time.
1: Create set V of all input variables as vectors of the form:
$$v_a = \left[\frac{n_a^1 + n_{a\epsilon}^1}{d_a^1 + d_{a\epsilon}^1}, ..., \frac{n_a^{|v_a|} + n_{a\epsilon}^{|v_a|}}{d_a^{|v_a|} + d_{a\epsilon}^{|v_a|}} \right].$$
2: $(j, k) = (1, 1)$.
3: **for** every operation $v_a \odot v_b$ in intermediate representation **do**
4:   $(v_\star, j)$ = Compute rational function $(v_a, \odot, v_b, N_1, N_2, j)$
5:   **for** $i = 1$ to $|v_\star|$ **do**
6:     $(n_\star^i, n_{\star\epsilon}^i, k)$ = Simplify Polynomials$(n_\star^i, n_\star^i \delta_j + n_{\star\epsilon}^i (1 + \delta_j)$
    , $N_1, N_2, k)$
7:     $j = j + 1$
8:   **end for**
9:   $(d_\star, d_{\star\epsilon}, k)$ = Simplify Polynomials$(d_\star, d_{\star\epsilon}, N_1, N_2, k)$
10:   $j = j + 1$
11:   Add $v_\star$ to V
12: **end for**
13: Bound desired variable v in V using Handelman representations

---

$(\hat{p}, \hat{p}_\epsilon, k)$ = **Simplify Polynomials** $(p, p_\epsilon, N_1, N_2, k)$
1: $(\hat{p}, k)$ = Simplify Polynomial (p, N_1, k)
2: $(\hat{p}_\epsilon, k)$ = Simplify Polynomial (p_ϵ, N_2, k)

Figure 3: Overall algorithm to find bounds on the range or relative error of a variable from a user input code.

$(v_\star, j)$ = **Compute rational function** $(v_a, \odot, v_b, N_1, N_2, j)$
1: **if** $(\odot == \bullet)$ **then**
2:   $(d_\star, d_{\star\epsilon}) = (d_a d_b, d_a d_{b\epsilon} + d_{a\epsilon} d_b + d_{a\epsilon} d_{b\epsilon})$
3:   **if** $(|v_b| == 1)$ **then**
4:     **for** $i = 1$ to $|v_a|$ **do**
5:       $(n_\star^i, n_{\star\epsilon}^i) = (\hat{n} + n_a^i n_b, \hat{n}_\epsilon + n_{a\epsilon}^i n_b + n_a^i n_{b\epsilon} + n_{a\epsilon}^i n_{b\epsilon})$
6:     **end for**
7:   **else**
8:     $(n_\star, n_{\star\epsilon}) = (0, 0)$
9:     **for** $i = 1$ to $|v_a|$ **do**
10:       $(n_\star, n_{\star\epsilon}) = (n_\star + n_a^i n_b^i, n_{\star\epsilon} + n_a^i n_{b\epsilon}^i + n_{a\epsilon}^i n_b^i + n_{a\epsilon}^i n_{b\epsilon}^i$
        $+ \delta_j (n_a^i n_b^i + n_a^i n_{b\epsilon}^i + n_{a\epsilon}^i n_b^i + n_{a\epsilon}^i n_{b\epsilon}^i))$
11:       $n_{\star\epsilon} = \hat{n}\delta_j + \hat{n}_\epsilon (1 + \delta_{j+1})$
12:       $j = j + 2$
13:     **end for**
14:   **end if**
15: **else if** $((\odot == +)$ **or** $(\odot == -))$ **then**
16:   **if** $(d_a, d_{a\epsilon}) == (d_b, d_{b\epsilon})$ **then**
17:     **for** $i = 1$ to $|v_a|$ **do**
18:       **if** $(|v_b| == 1)$ **then**
19:         $(\hat{n}^i, \hat{n}_\epsilon^i) = (n_a^i \odot n_b, n_{a\epsilon}^i \odot n_{b\epsilon})$
20:       **else**
21:         $(\hat{n}^i, \hat{n}_\epsilon^i) = (n_a^i \odot n_b^i, n_{a\epsilon}^i \odot n_{b\epsilon}^i)$
22:       **end if**
23:     **end for**
24:     $(d_\star, d_{\star\epsilon}) = (d_a, d_{a\epsilon})$
25:   **else**
26:     $v_1 = \frac{d_a + d_{a\epsilon}}{1}, v_2 = \frac{d_b + d_{b\epsilon}}{1}$
27:     v_1 = Compute Polynomial Approximation$(v_1.\lambda x.x^{-1})$
28:     v_2 = Compute Polynomial Approximation$(v_2.\lambda x.x^{-1})$
29:     (v_1, j) = Compute rational function $(v_a, \bullet, v_1, N_1, N_2, j)$
30:     (v_2, j) = Compute rational function $(v_b, \bullet, v_2, N_1, N_2, j)$
31:     $(v_\star, j)$ = Compute rational function $(v_1, \odot, v_2, N_1, N_2, j)$
32:   **end if**
33: **else if** $(\odot == \div)$ **then**
34:   $v_b = \frac{d_b + d_{b\epsilon}}{n_b + n_{b\epsilon}}$
35:   $(v_\star, j)$ = Compute rational function $(v_a, \bullet, v_b, N_1, N_2, j)$
36: **else**
37:   $v_\star$ = Compute Polynomial Approximation$(v_a, \odot)$
38: **end if**

Figure 4: Algorithm to create rational functions bounding intermediate variables.

---

in the numerator is substantially larger than in the denominator, this will result in wider bounds.

As a result, in Figure 4, we instead compute a rational function to bound the result by first normalising the denominators of the two input rational functions to 1 by multiplying their numerator polynomials by a polynomial approximation of the reciprocal of their denominator polynomials. To do this, we could use any of the well known techniques in approximation theory such as Taylor approximations, Chebyshev approximations, or the Remez algorithm [23]. Though we note that this can lead to wider bounds due to the errors in the approximation, experimentally these errors have in general been found to be smaller.

To implement other elementary functions, such as $(\odot \in \{\sqrt{}, \sin(), \exp()\})$, we apply polynomial or rational function approximations, as for the reciprocal. In general, the choice of approximation will trade trade quality of bounds with execution time, and due to the wealth of research in this area, this is left to a user to choose the optimum approximation. However, we note that this flexibility is unavailable when using AA and TwIR, for these require the polynomial approximation to be of a specific form. Furthermore, because if we approximate a function over a smaller range, the approximation will generally have less worst case error, we use Handelman representations to find bounds on the range of a variable before performing any polynomial approximation.

7. RESULTS

We have created two examples, shown in Figures 5 and 6 to help demonstrate the benefits of our proposed approach in terms of scalability and quality of bounds. In these figures, we present the original pseudo code alongside a breakdown of this pseudo code into vector operations, because in the original code, no order is specified, but the order of operations affects the accumulation of errors,

and hence this information is required to calculate any bound on the range or relative error. Using these examples, we compare against all the main competing methods that are capable of bounding error: IA, TwIR, AA, and Handelman representations. Furthermore, we also examine the impact of using interval splitting on the diagonals of the matrix A and the elements of the vector b when finding bounds using IA, as splitting these intervals will have the greatest impact on the final bounds. We do not perform this for the second example because it is unclear which variables would be best to split, whilst splitting all variables, as we shall see in Section 7.1, would require too large an execution time. These examples are large compared to similar publications in the field, consisting of approximately 20-30 input variables and 400-500 floating point operations, each of which add a new variable bounding the floating point roundoff error; in contrast, the examples of [5, 6] consist of up to 10 input variables and 30 floating point operations, with the former paper not taking finite precision errors into account.

In the first test, because there is little error in the first order approximation of the reciprocation of $a_{i,i}$, and all the non-affine operations are multiplications of polynomials or rational functions by input variables $a_{i,j}$, the majority of the information regarding the

190

$$A = \begin{pmatrix} 100 & -10 & -15 & -4 & 16 \\ -10 & 105 & -13 & 4 & 14 \\ -14 & -13 & 90 & 19 & -11 \\ 12 & 4 & 14 & 110 & 15 \\ 16 & 14 & -10 & 8 & 95 \end{pmatrix} \pm 1\% \quad b = \begin{pmatrix} 200 \\ -120 \\ -160 \\ 180 \\ -100 \end{pmatrix} \pm 1,$$

```
// The i^th element of a vector v is indexed v^i as before
// The j^th row of a matrix A is indexed A(j)
1: for k = 1; k ≤ 8; k + + do
2:    for j = 1; j ≤ 5; j + + do
3:       x^j = (1 − w)x^j + (w/A(j)^j)(b_j − Σ_{i=1,i≠j}^5 A(j)^i x^i)
4:    end for
5: end for
```

```
1:  // Initialisations                        11: for k = 1; k ≤ 8; k + + do
2:  for i = 1; i ≤ 5; i + + do                12:    for j = 1; j ≤ 5; j + + do
3:     wDIVa_i = w/A(i)^i                      13:       ax = ASUBdiagA^j • x
4:     for j = 1; j ≤ 5; j + + do             14:       bSUBax = b^j − ax
5:        ASUBdiagA_i^j = A(j)^i               15:       rhs = wDIVa^j • bSUBax
6:     end for                                 16:       lhs = U_SUBw • x^j
7:     ASUBdiagA_i^i = 0                       17:       x^j = lhs − rhs
8:  end for                                    18:    end for
9:  U_SUBw = 1 − w                             19: end for
10: // Algorithm
```

Figure 5: 5x5 Successive over relaxation benchmark: inputs, original code and code expressed using vector operations.

final range of the x values is in the first order terms. This implies IA, AA and TwIR should be able to calculate tight bounds, so in this test, we wish to demonstrate that our approach will perform well even where the existing methods ought to perform well. In contrast, the second test involves products of polynomials or rational functions bounding intermediate variables, and the division and square root of a multivariate polynomial, so we use this test to show that our approach can perform well in a more complex algorithm.

When performing polynomial approximations, for a fair comparison we use the same polynomial approximation method for affine arithmetic and our approach, with the exception of using Handelman representations to find the input range for this approximation, as mentioned in Section 6.2.2.

7.1   Test 1: Successive over relaxation

Scalability: Figures 7(a) and 7(b) demonstrate how the execution time on an Intel Xeon E5345 of each of the methods grows with the number of operations when computing the range or relative error for intermediate variables over the course of the successive over relation algorithm.

For the range analysis case, seen in Figure 7(a), IA, 1st order TwIR, AA and our approach initially scale well with the number of operations, whereas TwIR of orders greater than 1 and Handelman representations scale poorly due to exponential growth, as mentioned in Section 4. However, it is also clear that only IA and our approach have an execution time proportional to the number of operations, as expected from our analysis in Section 4, and this means that as the number of operations gets large, our approach can run faster than AA and TwIR. As we have previously mentioned, this is because our approach directly controls the size of the polynomials bounding the range of every intermediate variable, unlike Taylor forms where these polynomials are proportional to the number of operations because a variable bounding the roundoff error is added after every operation. Furthermore, as expected from our analysis in Section 4, AA scales worse than 1st order TwIR because it gains an extra variable for every operation as a result of bounding the error of the higher order terms created by the multiplicative model of error. Finally, this graph shows that when applying IA splitting,

$$A = \begin{pmatrix} 50 & -60 & 70 & -80 \\ -60 & -60 & 40 & 70 \\ 70 & 40 & 40 & -30 \\ -80 & 70 & -30 & -40 \end{pmatrix} \pm 0.25, \quad x_0 = \begin{pmatrix} 0 \\ 0 \\ 0 \\ 0 \end{pmatrix}, \quad b = \begin{pmatrix} 80 \\ 60 \\ 40 \\ 70 \end{pmatrix} \pm 0.25$$

```
1:  v_1 = b − Ax_0;                          11:  ρ_1 = √(δ² + β_{i+1}²)
2:  β_1, η = ||v_1||_2                        12:  ρ_2 = σ_i α_i + γ_{i−1} γ_i β_i
3:  v_0, w_0, w_{−1} = [0000]'               13:  ρ_3 = σ_{i−1} β_i
4:  σ_0, σ_1 = 0 ; γ_0, γ_1 = 1              14:  γ_{i+1} = δ/ρ_1
5:  for i = 1; i ≤ 3; i + + do                15:  σ_{i+1} = β/ρ_1
6:     v_i = v_i/β_i                          16:  w_i = (v_i − ρ_3 w_{i−2} − ρ_2 w_{i−1})/ρ_1
7:     α = v_i^T A v_i                        17:  x_i = x_{i−1} + σ_{i+1} η w_i
8:     v_{i+1} = Av_i − αv_i − βv_{i−1}       18:  η = −σ_{i+1} η
9:     β_{i+1} = ||v_{i+1}||_2                19: end for
10:    δ = γ_i α_i − γ_{i−1} σ_i β_i
```

```
1:  // Initialisations                      25:  tmp1_δ = γ_i • α_i
2:  for i = 1; i ≤ 4; i + + do              26:  tmp2_δ = γ_{i−1} • σ_i
3:     Ax^i = A(i) • x_0                     27:  tmp2_δ = tmp2_δ • β_i
4:     v_0^i = 0                             28:  δ = tmp1_δ − tmp2_δ
5:     w_0^i = 0                             29:  tmp_ρ1 = δ • δ
6:     w_{−1}^i = 0                          30:  tmp_ρ1 = tmp1_ρ1 + tmp_β
7:  end for                                  31:  ρ_1 = √tmp_ρ1
8:  v_1 = b − Ax;                            32:  tmp1_ρ2 = σ_i • α_i
9:  tmp_β = v_1^T • v_1                      33:  tmp2_ρ2 = γ_{i−1} • γ_i
10: β_1 = √tmp_β                             34:  tmp2_ρ2 = tmp2_ρ2 • β_i
11: σ_0, σ_1 = 0 ; γ_0, γ_1 = 1             35:  ρ_2 = tmp1_ρ2 + tmp2_ρ2
12: // Algorithm                             36:  ρ_3 = σ_{i−1} • β_i
13: for i = 1; i ≤ 3; i + + do              37:  γ_{i+1} = δ/ρ_1
14:    v_i = v_i/β_i                         38:  σ_{i+1} = β_i/ρ_1
15:    for j = 1; j ≤ 4; j + + do           39:  tmp1_w = ρ_3 • w_{i−2}
16:       Av^j = A(j) • v_i                  40:  tmp2_w = ρ_2 • w_{i−1}
17:    end for                               41:  tmp2_w = tmp1_w − tmp2_w
18:    α = v_i • Av                          42:  tmp1_w = v_i − tmp2_w
19:    αV = α • v_i                          43:  w_i = tmp1_w/ρ_1
20:    βV_{−1} = β_i • v_{i−1}               44:  tmp_x = η • w_i
21:    tmp_v_{i+1} = αV − βV_{−1}            45:  tmp_x = γ_{i+1} • tmp_x
22:    v_{i+1} = Av − tmp_v_{i+1}            46:  x_i = x_{i−1} + tmp_x
23:    tmp_β = v_1 • v_1                     47:  η = −σ_{i+1} • η
24:    β_{i+1} = √tmp_β                      48: end for
```

Figure 6: MINRES algorithm benchmark: inputs, original code and code expressed using vector operations.

while the execution time is still proportional to the number of operations, there is a large difference between the execution time depending on the number of splits. We have calculated bounds using IA without splitting, IA where the chosen variables are split into two regions (IA with splitting v1), and IA where the chosen variables are split into three regions (IA with splitting v2); there is a significant difference in execution time between these approaches, indeed this difference can be several orders of magnitude, as seen in Table 4. This is because IA with splitting v1 requires 2^{10} separate interval evaluations, IA with splitting v2 requires 3^{10} interval evaluations. Clearly, performing any further splits is not scalable, and we note that this is with an intelligent selection of 10 variables to split; a naïve approach of splitting every variable would scale far worse and is unlikely to find much tighter bounds.

When bounding the relative error of intermediate variables, as seen in Figure 7(b), the execution time for AA and TwIR grows much faster whereas it remains similar for our approach. The cause of this is that to compute the relative error using AA or TwIR, one must first generate two polynomials, a polynomial p representing the range of the desired variable in the absence of finite precision errors, and a polynomial $\hat{p}$ representing the range in the presence of these errors, then compute bounds of the function $\left|\frac{p−\hat{p}}{p}\right|$. To bound this using AA or TwIR, one must first compute a polynomial approximation of $\tilde{p} = 1/p$ then bound the result of the computation $(p − \hat{p}) \times \tilde{p}$, and because p, $\hat{p}$ and $\tilde{p}$ are large polynomials where

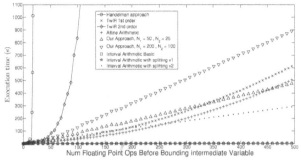

(a) Execution time of various methods to bound the range of the intermediate variable after a given number of operations within a 5x5 Successive Over Relaxation.

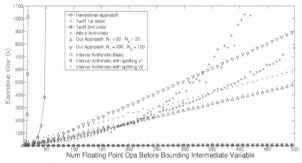

(b) Execution time of various methods to bound the relative error of the intermediate variable after a given number of operations within a 5x5 Successive Over Relaxation.

Figure 7: Range and relative error of various methods to bound error applied to a 5x5 successive over relaxation.

Table 4: Average bound on range for x vector of various methods to calculate bounds applied to a successive over relaxation of a 5x5 matrix for precision of 20 bits.

Method	Range	Execution time (s)
IA	0.2452	2.279100e-02
IA with Splitting v1	0.2014	10
IA with Splitting v2	0.1866	580
1st Order TwIR	0.2492	612
Affine Arithmetic	0.1369	555
Our Approach $N_1 = 50, N_2 = 25$	0.1366	430
Our Approach $N_1 = 200, N_2 = 50$	0.1330	767

the number of monomials in these polynomials are proportional to the number of operations, the result will be a very large polynomial with many monomials that must be bounded.

Quality of bounds: Table 4 shows the computed bounds on ranges for the first example for the methods that could calculate bounds in a tractable time. It is clear that our approach can compute tighter bounds than the existing approaches, in a reduced execution time, illustrating our claim that our approach can perform well where the existing methods perform well. However, for relative error analysis, shown in Figure 8, only our approach is able to track how relative error decreases with increasing precision. This is because monomials representing floating point error are a function of roundoff variables and input variables, meaning they are second order or greater, and hence first order methods such as AA and 1st order TwIR can only approximate these errors, whereas by retaining 'most significant terms', our approach can retain these higher order terms. In addition, our approach calculates bounds on the worst case relative error that are close to the estimates found by random simulation, implying our bounds are tight.

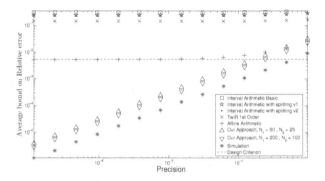

Figure 8: Average bound on relative error for x vector of various methods to calculate bounds applied to a successive over relaxation of a 5x5 matrix.

Table 5: Slice use and max frequency of 5x5 successive over relaxation required according to analytical tools to guarantee the relative error is less than 5.5×10^{-3}, or using IEEE standards.

Method	Exponent (# bits)	Mantissa (# bits)	Slice Regs	Slice LUTs	Frequency (MHz)
Simulation	8	11	3562	3012	330
Our Approach	8	13	4261	3647	330
Affine Arithmetic	8	18	6606	5368	300
IA/TwIR	∞	∞	∞	∞	N/A
IEEE Single Precision	8	24	8407	6815	280
IEEE Double Precision	11	53	27200	22066	251

In order to demonstrate how this can be used to improve a hardware design, we created a hardware implementation where every floating point operation in line (3) of Figure 5 has a dedicated floating point operator, and the dot product is performed using a row of parallel multiplication units and an adder reduction tree, as in [27]. Table 5 shows the resources required, post place and route on a Virtex 5 LX 330T, to create such a hardware implementation using the minimum precision necessary to ensure the relative error lies below a threshold of 5.5×10^{-3} according to the various methods. Using our approach, we can create a hardware design with 25% less silicon area than by using AA. The other methods cannot prove bounds, meaning that one could either attempt a simulation based approach, which in these examples would use less hardware at the cost of sacrificing guarantees that it will satisfy the desired bounds, or implement it using IEEE double precision floating point units, but our approach could save up almost 80% of the silicon area in comparison. Our proofs could also be used to show that IEEE single precision would be sufficient to meet the design criterion, for use on other hardware platforms, but this precision would be much greater than is necessary. Furthermore, because our approach can track how relative error decreases with increasing precision we could use it to tune hardware to satisfy a tighter bound on relative error, to which even AA would be unable to find a proof.

Execution time vs quality of bounds performance trade-off: Figure 9 compares the ability of our approach to trade quality of bounds with scalability with the existing approaches. In this figure, we have chosen to compare bounds on the range, because as we have shown earlier, only our approach is capable of tracking how relative error decreases with precision. All the values in this figure are based on a mantissa of 20 bits. Furthermore, in order to compare more approaches, in Figure 9(a), we have restricted the successive over relaxation example to only two iterations, allowing us also to see the quality of bounds and execution time trade-off for increasing the order of TwIR to 2nd order, while in Figure 9(b) we plot the same after the conclusion of the example. For our approach, we have varied N_1 in 10 equal size increments from 20 to

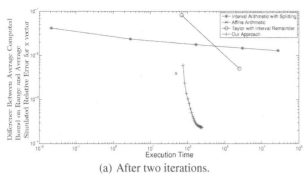

(a) After two iterations.

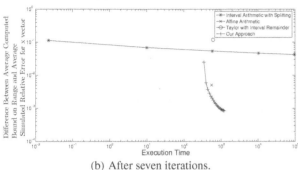

(b) After seven iterations.

Figure 9: Trade off between bound on range and execution time of various implementations of our approach to find the average bound on the range of x vector of a successive over relaxation of a 5x5 matrix for various approaches.

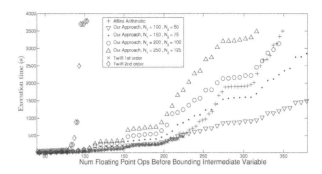

Figure 10: Execution time vs number of operations using various methods to bound range applied to a MINRES algorithm of a 4x4 matrix.

Table 6: Slice use and max frequency of MINRES implementation required according to analytical tools to guarantee the relative error is less than 1×10^{-3}, or using IEEE standards.

Method	Exponent (# bits)	Mantissa (# bits)	Slice Regs	Slice LUTs	Frequency (MHz)
Our Approach, $N_1{=}150, N_2{=}75$	8	21	25221	20793	225
Our Approach, $N_1{=}100, N_2{=}50$	8	22	26216	21519	225
AA/IA/TwIR	∞	∞	∞	∞	N/A
IEEE Single Precision	8	24	30121	24797	150
IEEE Double Precision	11	53	74883	89063	120

400 and set $N_2 = \frac{1}{2}N_1$. Finally, to obtain a measure of the quality of the bounds, we plotted the difference between the calculated and simulated bounds. This graph demonstrates how our method provides a much finer grain level of control in comparison to existing methods: interval splitting has a large growth in execution time for every extra split of the desired variables, AA has no control over this trade-off, and TwIR has a large difference in execution time between 1st and 2nd order. In addition, this graph shows once again that our approach is capable of computing the tightest bounds of all these methods, and because these bounds approach those found by simulation, it implies the bounds we compute are tight. While we note that in the case of Figure 9(a), due to the additional overheads of our algorithm in creating rational functions representing the value for intermediate variables, AA can run quicker than our approach, as the number of operations becomes larger as in Figure 9(b), because AA cannot trade execution time for quality of bounds, our approach has the flexibility to run quicker than AA at a cost of bounds, or for longer than AA to obtain tighter bounds.

7.2 Test 2: MINRES

Scalability: Figure 10 demonstrates how the execution time of each of the methods grow with the number of operations when computing the range for intermediate variables over the course of the MINRES algorithm. We note that 2nd order TwIR once again scales poorly with the number of operations, while IA and 1st order TwIR fail to compute bounds after 228 operations because they are unable to prove the input to the $\sqrt{}$ function is non-negative. For our approach and AA, in comparison to Figures 7(a) and 7(b), in this experiment the execution time grows much faster with the number of operations. This is because, as mentioned when introducing these examples, this algorithm contains the multiplication of two polynomials or rational functions which may be large, as

opposed to the multiplication of a rational functions by a single input variable. However, we comment that as the maximum number of monomials in the product of two polynomials is given by the product of the number of monomials in each polynomial and the size of the AA polynomial for intermediate variables can be much greater than $N_1 + N_2$. This ensures AA suffers much more than our approach, and is the reason our approach becomes faster than AA after much fewer operations than in Figures 7(a) and 7(b). We also comment that this graph grows in steps, unlike Figures 7(a) and 7(b), this is because operations which are products of two polynomials create many more monomials and take longer than operations which are sums of two polynomials, and hence the execution time for sums is below the worst case execution time for any operation.

Quality of bounds: When attempting to calculate bounds on the relative error that would enable one to create an optimized hardware design, once again, only our approach was capable of computing bounds for this example that track how relative error decreases with increasing precision. This was similarly because errors arising from the use of finite precision arithmetic will be second order or greater, meaning first order methods can only approximate these errors. Table 6 shows the resource use for a parallel implementation of the MINRES algorithm, as described in [28], that would be required to satisfy a bound on relative error of less than 1×10^{-3} using the various methods to calculate bounds, including our approach with different run-times, as well as using IEEE single and double precision. This table again demonstrates significant savings can be made in comparison with IEEE double precision. Furthermore, by using an increasing run-time, we obtain tighter bounds that result in a smaller hardware implementation, highlighting the trade-off we aim to achieve. Finally, we note that in this instance our software has the potential to obtain a proof that IEEE single precision is sufficient to satisfy the desired precision, and this illustrates how our tool could also be used to obtain performance improvements on other hardware platforms.

Altogether, we have shown that our approach can compute bounds that are much tighter than competing approaches in a smaller execution time, and has the ability to scale to much larger examples because its execution time scales worst case linearly with the number of operations. Furthermore, by retaining higher order information, our technique can also track the effect of finite precision errors and compute tight bounds on the relative error introduced by the use of finite precision arithmetic, and this enables us to design hardware that meets a given relative error specification with less silicon area; in the case of the successive over relaxation example, saving over 80% of the slices in comparison to IEEE 754 double precision, and in the case of the MINRES example, saving over 65% of the slices in comparison to IEEE 754 double precision.

8. CONCLUSION

This paper has presented a new algorithm to calculate bounds on finite precision errors. We have demonstrated that this algorithm is more scalable than the existing approaches, meaning that it has the potential to be applied to examples that even the most advanced methods previously could not realistically handle, and can also calculate tighter bounds than the existing methods. In addition, we have shown that our algorithm has significantly greater control over the trade-off between execution time and quality of bounds making it useful for both small and large examples within an optimization framework. Finally, we have shown that because our tool can not only find tight bounds on the range, but also track how relative error decreases with increasing precision, we can use it to create hardware designs with guaranteed error properties that obtain significant silicon area savings over hardware implementations that adhere to IEEE standard single or double precision.

9. REFERENCES

[1] G. Chow, K. Kwok, W. Luk, and P. Leong, "Mixed precision processing in reconfigurable systems," in *Proc. IEEE. Symp. on Field-Programmable Custom Computing Machines*, 2011, pp. 17–24.

[2] A. R. Lopes and G. A. Constantinides, "A fused hybrid floating-point and fixed-point dot-product for FPGAs." in *Proc. Int. Symp. on Applied Reconfigurable Recomputing*, 2010, pp. 157–168.

[3] M. deLorimier and A. DeHon, "Floating-point sparse matrix-vector multiply for FPGAs," in *Proc. Int. Symp. on Field-Programmable Gate Arrays*, 2005, pp. 75–85.

[4] L. Zhuo, G. R. Morris, and V. K. Prasanna, "High-performance reduction circuits using deeply pipelined operators on FPGAs," *IEEE Trans. Parallel Distrib. Syst.*, vol. 18, no. 10, pp. 1377–1392, 2007.

[5] A. Kinsman and N. Nicolici, "Bit-width allocation for hardware accelerators for scientific computing using SAT-modulo theory," *IEEE Trans. Comp.-Aided Des. Integ. Cir. Sys.*, vol. 29, pp. 405–413, 2010.

[6] D. Boland and G. Constantinides, "Bounding variable values and round-off effects using handelman representations," *IEEE Trans. Comp.-Aided Des. Integ. Cir. Sys.*, vol. 30, no. 11, pp. 1691–1704, 2011.

[7] Y. Pang, K. Radecka, and Z. Zilic, "Optimization of imprecise circuits represented by Taylor series and real-valued polynomials," *IEEE Trans. Comp.-Aided Des. Integ. Cir. Sys.*, vol. 29, pp. 1177–1190, August 2010.

[8] G. Constantinides, P. Cheung, and W. Luk, "Optimum wordlength allocation," *Proc. Int. Symp. Field-Programmable Custom Computing Machines*, pp. 219–228, 2002.

[9] D.-U. Lee, A. Gaffar, R. Cheung, O. Mencer, W. Luk, and G. Constantinides, "Accuracy-guaranteed bit-width optimization," *IEEE Trans. Comp.-Aided Des. Integ. Cir. Sys.*, vol. 25, no. 10, pp. 1990–2000, 2006.

[10] M. L. Chang and S. Hauck, "Automated least-significant bit datapath optimization for FPGAs," *Proc. IEEE Symp. on Field-Programmable Custom Computing Machines*, pp. 59–67, 2004.

[11] R. Barrett, M. Berry, T. F. Chan, J. Demmel, J. Donato, J. Dongarra, V. Eijkhout, R. Pozo, C. Romine, and H. V. der Vorst, *Templates for the Solution of Linear Systems: Building Blocks for Iterative Methods*. Philadelphia: SIAM, 1994.

[12] G. Constantinides, A. Kinsman, and N. Nicolici, "Numerical data representations for FPGA-based scientific computing," *Design & Test of Computers*, vol. 28, no. 4, pp. 8–17, 2011.

[13] Z. Zhao and M. Leeser, "Precision modeling and bit-width optimization of floating-point applications," in *High Performance Embedded Computing*, 2003, pp. 141–142.

[14] R. E. Moore, *Interval Analysis*. Englewood Cliff, NJ: Prentice-Hall, 1966.

[15] B. Einarsson, *Handbook on Accuracy and Reliability in Scientific Computation*. Soc for Industrial & Applied Math, 2005, ch. 10, pp. 195 – 240.

[16] F. de Dinechin, C. Q. Lauter, and G. Melquiond, "Assisted verification of elementary functions using gappa," in *Proc. Symp. Applied computing*, 2006, pp. 1318–1322.

[17] A. Kinsman and N. Nicolici, "Computational bit-width allocation for operations in vector calculus," in *IEEE Int. Conf. on Computer Design*, oct. 2009, pp. 433 –438.

[18] ——, "Robust design methods for hardware accelerators for iterative algorithms in scientific computing," in *Proc. Design Automation Conference*, 2010, pp. 254–257.

[19] A. Neumaier, "Taylor forms - use and limits," *Reliable Computing*, vol. 9, pp. 43–79, 2003.

[20] N. Courtois, A. Klimov, J. Patarin, and A. Shamir, "Efficient algorithms for solving overdefined systems of multivariate polynomial equations," in *Proc. Int. Conf. on Theory and application of cryptographic techniques*, 2000, pp. 392–407.

[21] L. H. de Figueiredo and J. Stolfi, *Self-Validated Numerical Methods and Applications*. Rio de Janeiro: IMPA/CNPq, 1997.

[22] K. Makino and M. Berz, "Taylor models and other validated functional inclusion methods," *International Journal of Pure and Applied Mathematics*, vol. 4, pp. 379–456, 2003.

[23] J.-M. Muller, *Elementary Functions: Algorithms and Implementation*. Birkhauser, 2005.

[24] N. J. Higham, *Accuracy and Stability of Numerical Algorithms*, 2nd ed. Philadelphia, PA, USA: Soc for Industrial & Applied Math, 2002.

[25] W. S. Brown, "A simple but realistic model of floating-point computation," *ACM Trans. Math. Softw.*, vol. 7, pp. 445–480, December 1981.

[26] H. Ratschek, "Centered forms," *SIAM Journal on Numerical Analysis*, vol. 17, no. 5, pp. pp. 656–662, 1980.

[27] D. Boland and G. Constantinides, "Optimising memory bandwidth use and performance for matrix-vector multiplication in iterative methods," *ACM Trans. Reconfigurable Technol. Syst.*, vol. 4, pp. 22:1–22:14, 2011.

[28] ——, "An FPGA-based implementation of the MINRES algorithm," in *Proc. Int. Conf. Field Programmable Logic and Applications*, Sept. 2008, pp. 379–384.

Optimizing SDRAM Bandwidth for Custom FPGA Loop Accelerators

Samuel Bayliss and George A. Constantinides
Department of Electrical and Electronic Engineering
Imperial College London
London SW7 2AZ, United Kingdom
{s.bayliss08, g.constantinides}@imperial.ac.uk

ABSTRACT

Memory bandwidth is critical to achieving high performance in many FPGA applications. The bandwidth of SDRAM memories is, however, highly dependent upon the order in which addresses are presented on the SDRAM interface. We present an automated tool for constructing an application specific on-chip memory address sequencer which presents requests to the external memory with an ordering that optimizes off-chip memory bandwidth for fixed on-chip memory resource. Within a class of algorithms described by affine loop nests, this approach can be shown to reduce both the number of requests made to external memory and the overhead associated with those requests. Data presented shows a trade off between the use of on-chip resources and achievable off-chip memory bandwidth where a range of improvements from 3.6× to 4× gain in efficiency on the external memory interface can be gained at a cost of up to a 1.4× increase in the ALUTs dedicated to address generation circuits in an Altera Stratix III device.

Categories and Subject Descriptors

B.5.2 [**Design Aids**]: Automatic Synthesis

Keywords

FPGA, Memory, SDRAM, Loop Transformations

1. INTRODUCTION

The number of pins available on semiconductor devices, and the data rate available on such pins, has not scaled with transistor density [1]. This, among other factors has contributed to a 'memory wall' in which the parallelism achievable with computing devices is limited by the speed in which large off-chip memories can be accessed. On-chip memory hierarchies improve the performance of the memory system by exploiting *reuse* and *reordering*. Reusing data retained in an on-chip memory hierarchy reduces the number of requests made on the external memory interface. Where those data requests can be safely reordered, often the control overhead of servicing the requests can be reduced and hence bandwidth on the external interface is more efficiently used.

The impact of this reordering can be seen when Dynamic Random Access Memories (DRAMs) are used as the external off-chip memory. DRAM devices have a hierarchical structure in which a small number of independent memory arrays (banks) are themselves composed of rows and columns. A row within a bank must be explicitly opened ('activated') before columns within it can be accessed, and closed ('precharged') before another row can be selected. While DRAM capacity has scaled in line with Moore's observations [19], increased cell density has been used to deliver increased capacity for a given silicon die area. This means the wirelengths upon which 'precharge' and 'activation' times are dependent are not significantly shorter than 10 years ago. With smaller transistors delivering shorter clock periods, the overall impact is that timing delays associated with 'precharge' and 'activation' *cycles* take up an ever increasing proportion of clock cycles on the external memory interface. The net effect is that the bandwidth efficiency of external memory interfaces has reduced with each generation of SDRAM device.

On general purpose computers, the sequence of memory requests generated by the CPU is typically unknown at design time, so the processor cache is designed to be able to reuse and reorder data without a static compile time analysis of the program to be executed. Complex dynamic memory controllers in modern CPUs buffer and dynamically reorder cache line-fill requests to external memory. Both the cache and memory controllers therefore contain memory and associative logic to buffer and dynamically select and service memory requests [15]. As well as the direct area cost this imposes, caches and dynamic memory controllers make it very difficult to predict memory performance at compile-time.

In this paper we ask the question: "What can be done in the memory controller to improve bandwidth efficiency without sacrificing predictability?". In essence, we aim to trade logic resources (whose availability increases with process scaling in line with Moore's observations [19]) for increased memory interface bandwidth (which does not). While this area has been explored in the context of dynamic memory controllers and caches which operate on random streams of data, we focus instead on opportunities for a *statically* scheduled reordering of memory requests. Such an approach means we can exactly determine memory access latency and bandwidth at compile time for each memory access. This knowledge might be exploited in a high level synthesis flow to improve resource scheduling algorithms.

For a given application targeted for hardware acceleration on an FPGA, the structure of computation kernels often makes static analysis tractable. More specifically, for such kernels, the sequence of memory requests can often be determined at compile time. In this paper, we give a methodology grounded in a formal computation framework (the Polyhedral Model [16, 20]) for reordering memory requests for such kernels. We present results which are specific to hardware implementation, in that they exploit the reconfigurability of the device by creating a custom memory architecture for a specific application, and also take advantage of the abundance of arithmetic logic components available on modern FPGAs to build sophisticated controller structures.

The key contributions are :

- A parametric representation of the set of SDRAM rows accessed by a memory reference within a loop nest as the integer points enclosed in a convex polytope, avoiding the need for enumeration.
- An efficient procedure to optimize the size of such a parametric representation, corresponding to optimizing the FPGA resources dedicated to constructing an address generator.
- A method of code generation from the optimized set description which minimizes the number of row swaps performed by, and the number of commands issued to, the SDRAM memory controller.
- A method for producing an efficient and flexible pipelined memory address sequencer.
- An evaluation of the SDRAM bandwidth efficiency improvements achieved by such an approach, considering separately the improvements attributable to the reuse of data and the reordering of data requests to external memory.
- An evaluation of the cost of that transformation in terms of additional logic required in address generation and on-chip data storage.

We have implemented our method as a complete design flow that reads in a subset of C, extracts a polyhedral representation and generates a Verilog implementations of an optimized SDRAM address sequencer. We would encourage readers to try the flow via the web interface, available at http://cas.ee.ic.ac.uk/AddrGen.

2. BACKGROUND

2.1 SDRAM Memory

DRAMs are designed for high yield and low cost in manufacturing and have densities which exceed competing technologies. However, the design trade-offs needed to achieve these characteristics means much of the burden of controlling DRAM memory access falls on an external memory controller.

SDRAM memories store data as charged nodes in a dense array of memory cells. An explicit 'activate' command sent to the memory selects a row within the array before any reads or writes can take place. This is followed by the assertion of a sequence of memory 'read' or 'write' commands to columns within the selected row followed by a 'precharge' command which must be asserted before any further rows can be activated. The physical structure of the memory device determines the minimum time which must elapse between the issuing of 'precharge', 'activate' and 'read'/'write' operations. The on-chip memory controller is responsible for ensuring these timing constraints are met. When reading consecutive columns within a row, DDR SDRAM memories can sustain two words-per-clock-cycle data-rates. With a 200MHz DDR2 device (with 8-bit wide data bus), this means 400MBytes/s. However the overhead of 'precharge' and 'activate' commands and the delays associated with their respective timing constraints means that achieving this peak bandwidth is rare. In the worst case, where single bursts are requested from different rows within the same bank, memory bandwidth is reduced to 20% of its peak rate.

2.2 Previous Work

High performance memory systems are a goal across the spectrum of computing equipment and a large body of work exists which seeks to improve cache performance ([13] provides a comprehensive review). Most of this work assumes the sequence of addresses is randomly (but not necessarily uniformly) distributed and describes optimizations of dynamic on-chip structures to exploit data locality. Where scratchpad memories have been used within a memory hierarchy, there are examples of static analysis to determine which specific memory elements are reused. Of particular note is the work of Darte et al. [11] and Liu et al. [18]. These two works both explore data-reuse using a polytope model. One develops a mathematical framework to study the storage reuse problem, and the other is an application of the technique within the context of designing a custom memory system implemented on an FPGA, but without considering the impact that ordering has on performance.

The 'Connected RAM' (CoRAM) methodology presented in [10] is notable in that, in common with our approach, it seeks to decouple communication and computation threads within an application. Their methodology lacks a compilation framework to extract communication threads from high level descriptions and optimally schedule those threads to access external memory, and our work presented here is a step towards the realization of that goal.

Prior work focused more specifically on SDRAM controllers can be divided into work that seeks to optimize the performance of random data streams using runtime on-chip structures, and that which uses static analysis techniques on kernels of application code. A review of different dynamic scheduling policies for SDRAM controllers can be found in [23]; the results show that none of the fixed scheduling policies are optimal across all benchmarks, providing a motivation for our approach of using FPGA programmability to pursue an application-specific approach. The method presented in [2] describes a memory controller that guarantees an allocation of bandwidth and bounded access latency to many requestors in a complex SOC. A set of short templates are defined which optimize bursts of data and time-slots are allocated using a dynamic credit-controlled priority arbiter. The static approach we demonstrate in this paper enables fine-grained scheduling of datapath operations that is difficult using a dynamic approach.

Other static compile time approaches to improving SDRAM efficiency can be found in [17] where different data layouts are used to improve efficiency in a image processing application. A block-based layout of image data is proposed rather than a traditional row-major or column-major layouts and a Presberger arithmetic model is used estimate the number of 'precharge' / 'activate' operations required in the execution of a video benchmark. Their results show a 70-80% accuracy compared to simulation results and achieve up to 50% energy savings. In [9], a strategy is proposed for allocating arrays to different memory banks to hide the latency of row activation. Their heuristic approach assumes each logical row of an allocated array fits within an SDRAM row; an assumption is likely to be restrictive in handling large data-sets. While our proposed methodology does not consider bank allocation directly, we believe it complimentary to the concept demonstrated in [9], since by reordering memory accesses to cluster together accesses to the same row, SDRAM rows which are accessed consecutively can be allocated to different banks with a simple permutation of address bits.

To the best of our knowledge, our work is the first to propose static analysis of loop nests for developing hardware address generators for application-specific SDRAM-optimized memory reordering *and* data reuse.

```
char A[56];
for (x₁ = 0 ; x₁ <= 2 ; x₁++) {
  for (x₂ = 2 − x₁ ; x₂ <= 2 ; x₂++) {
    for (x₃ = x₁; x₃ <= x₂ ; x₃++) {
      A[7 * x₁ + 8 * x₂ + 9 * x₃] = func (...);
    }
  }
}
```

Figure 1: C source code for 3-level nested loop example.

Figure 2: Memory address and associated fields.

3. MOTIVATING EXAMPLE

The example shown in Figure 1 provides a motivation for our methodology, and will be used as a running example throughout the paper to illustrate the algorithms. This is a 'toy' loop nest with three levels ($n = 3$) and a memory array within the innermost loop. The array A is assumed to reside in external SDRAM memory and, for didactic reasons, we assume that each row in the memory has length of 16 bytes and each individual memory request is made of bursts of 4 bytes. In real SDRAMs, of course the the size of each row is much greater (as is the number of accesses made in a real application), but we have tried to keep this example as simple as possible to illustrate the key features and novelties of our approach.

For simplicity of exposition, we assume the array A originates at address 0 in memory. Thus element $A[i]$ of the array resides at memory address i. Every memory request to DDR2 SDRAM is made with the selection of a unique bank and row within the memory followed by a burst request (of 4 or 8 words). Considering only a single bank within the device, each memory address within that bank can be divided into three bit fields, corresponding to SDRAM Row, Burst, and Byte Within Burst, as shown in Fig. 2. These three fields can be represented as vectors in $\mathbb{Z}^3$ where the three dimensions denote the Row, Burst and Byte Within Burst respectively.

When running the code in Figure 1 without transformation, a sequence of seven memory requests is generated, as shown in order in Table 1. This sequence exhibits several features. Firstly, the ordering of requests means that both the first and second rows of the SDRAM are opened more than once. SDRAM timing constraints mean a significant penalty is incurred when 'activating' and 'precharging' SDRAM rows, hence this is an inefficient order to access the memory; a more efficient order would activate each row only once. Secondly, in some rows, there are multiple accesses to the same burst, for example Row 2 Burst 0 is accessed both by $(x_1, x_2, x_3) = (0, 2, 2)$ and by $(x_1, x_2, x_3) = (1, 2, 1)$. If we can store the data from the burst in on-chip memory and reuse it later in the computation, we can reduce the number of external memory transactions. The final important feature of this sequence is the presence of 'holes': not all bursts are accessed within each row; indeed, burst number 1 is never accessed for any row. Careful attention to these holes is important in ensuring code is correct (since spurious write operations corrupt data) and efficient (since non-essential read operations reduce bandwidth efficiency).

Our aim, therefore, is to establish an automatic methodology for deriving an efficient memory subsystem capable of addressing

Table 1: Sequence of memory accesses generated by example in Figure 1.

Order	x_1	x_2	x_3	Array Index	Row	Burst
1	0	2	0	16	1	0
2	0	2	1	25	1	2
3	0	2	2	34	2	0
4	1	1	1	24	1	2
5	1	2	1	32	2	0
6	1	2	2	41	2	2
7	2	2	2	48	3	0

these three features, by reordering external memory accesses when appropriate, by storing reused data on-chip when possible, and by ensuring only those memory locations accessed by the original code are accessed by the derived memory subsystem.

4. METHODOLOGY

In our methodology, we restrict ourselves to kernels of code for which the set of associated memory accesses can be determined at compile-time and are independent of any inputs to the program. This is referred to as the static control portion of a program. We also restrict ourselves to code that can be expressed as a perfect loop nest using normalised loop indices. Loops in this form are easily expressed in a mathematical form referred to as the Polyhedral Model, which enables reasoning about code transformations. Examples naturally arise in most DSP and Scientific Computing applications [20, 21, 22]. However, we are not restricted to these domains, recent advances show that we can convert imperfect loop nests into canonical forms [8] and demonstrate that static control code kernels can be automatically extracted from intermediate compiler representations [12] further broadening the applicability of our technique.

The steps in our compilation flow are described in the flowchart in Figure 3, with each step enumerated below.

1. Parse the 'C' kernel code and construct a polytope representation.

2. For each memory reference, augment the polytope description with a variable representing the SDRAM row (r) and burst (u) accessed.

3. For each memory reference, find a unimodular matrix and change of variables such that the maximum number of variables can be eliminated from the polytope by the sufficient conditions in [24].

4. Check the necessary conditions for elimination in [24] for the remaining variables and use code generation tools to generate transformed code with a reordered loop structure.

5. Generate pipelined hardware which implements the loop indexing function

In the sections which follow, we formulate an initial problem description and describe each of these steps, demonstrating each transformation using our example code from Figure 1.

4.1 Decoupling Memory Access from Execution using On-Chip Memory

In any useful program, there is a mix of read accesses and write accesses. In general, freedom to reorder the statements executing in a program is restricted by data dependencies within the program: the true read-after-write dependencies within the code prevent arbitrary reordering of loop structures. We tackle this problem by

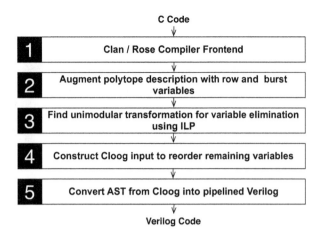

C Code
↓

1	Clan / Rose Compiler Frontend
2	Augment polytope description with row and burst variables
3	Find unimodular transformation for variable elimination using ILP
4	Construct Cloog input to reorder remaining variables
5	Convert AST from Cloog into pipelined Verilog

↓
Verilog Code

Figure 3: Flowchart showing steps in methodology.

separating a program into separate communication and execution threads. The simplest legal schedule sees all the data required for program execution loaded into on-chip memory before execution begins, and then the data items written during execution are buffered and written back after execution completes.

In many cases, such a simple communication/execution split is infeasible, because the on-chip memory is not large enough to simultaneously store all the data read and written within a loop nest. In such a case, the set of memory items corresponding to some specific iteration of an outer loop (or outer loops) can be loaded into on-chip memory, execution of the inner loops can progress and when complete, data written during execution of those inner loops can be written back to memory before repeating for the next iteration of the outer loops.

We can introduce on-chip memory buffers for decoupling memory access from execution at any level of the loop nest and use the parameter t to denote the level at which a buffer is introduced. Where $t = 1$, this denotes we indicate the introduction of a buffer at the outermost level of the loop nest and the requirement to prefetch all the data required for execution of the loop nest before execution begins. At the opposite extreme, parameterisation where $(t = n + 1)$ indicates introduction of a buffer large enough to contain just the elements accessed in a single iteration of the innermost loop. While not considered within this paper, standard loop-tiling transformations can be used to give the user even finer control over the size of the required on-chip data buffer.

For some chosen level of parameterisation, the set of read accesses generated by the inner loop nest for a specific outer loop iteration vector can be arbitrarily reordered, as can the set of write accesses, and the data-dependencies are guaranteed to be satisfied under the condition that all the read accesses required to fill the data buffer occur before execution and the write accesses required to commit results are scheduled after execution and before data is fetched for the next iteration.

In the sections that follow, we present a methodology for exploiting data reuse and reordering memory transactions using the abstract notation of the Polyhedral Model, for a fixed value of t.

4.2 Representing Memory Accesses in the Polyhedral Model

The Polyhedral Model is a mathematical description of a sequence of computations. For code which can be expressed within its restrictions, the Polyhedral Model provides a basis for reasoning about loop transformations, data dependencies and memory access patterns.

Consider the example in Figure 1 which contains three nested loops ($n = 3$). There is an loop variable for each level in the loop (labelled x_1, x_2 and x_3). In the general case, these loop variables are $x_1, x_2, \ldots, x_n$ with the inner-most loop arbitrarily labelled with the highest index. Each execution of the statement(s) within the innermost loop is associated with a unique iteration vector $x \in \mathbb{Z}^n$. The bounds of each loop are affine functions of the loop (induction) variables of outer levels in the loop nest. The set S_E of integer vectors executed by a given loop nest, which we refer to as the *iteration space*, can be represented parametrically as the integer points contained within a polytope described as a set of linear inequalities $Ax \le b$, where A is an $m \times n$ matrix, b is an m-vector, and the vector inequality is interpreted as $x \le y$ iff $x_i \le y_i$ for all i.

For the loop in Figure 1, the polytope representing the specified loop bounds is given in (1) :

$$\begin{pmatrix} -1 & 0 & 0 \\ 1 & 0 & 0 \\ -1 & -1 & 0 \\ 0 & 1 & 0 \\ 1 & 0 & -1 \\ 0 & -1 & 1 \end{pmatrix} \begin{pmatrix} x_1 \\ x_2 \\ x_3 \end{pmatrix} \le \begin{pmatrix} 0 \\ 2 \\ -2 \\ 2 \\ 0 \\ 0 \end{pmatrix} \qquad (1)$$

The loops contain memory references within the innermost loop, where the array indexing functions are themselves affine functions of the loop variables, *i.e.* of the form $A[fx + h]$ where f is an n-dimensional row vector and h is a scalar. For the example code, $f = [7\ 8\ 9]$, $h = 0$. We can describe the set S_M of memory addresses accessed within a loop nest as in (2). For our example code, if we were to enumerate the elements of this set, we would obtain $S_M = \{16, 24, 25, 32, 34, 41, 48\}$, as illustrated in Table 1. Crucially, for each memory reference, the number of elements in the set S_M is always less than or equal to the number of iteration vectors in S_E. This is because, while each memory reference accesses only a single memory element, multiple iteration vectors can access the same memory element. Exploitation of this is referred to as data *reuse* since elements in S_M could be stored in on-chip memory and reused on more than one iteration in S_E.

$$S_M = \{fx + h \mid \exists x \in S_E\} \qquad (2)$$

Each of the memory accesses in S_M corresponds to a specific row and aligned burst in external SDRAM memory. Beyond data-reuse, we can achieve higher off-chip memory bandwidth by reordering accesses so that accesses to the same row in external memory are grouped together and thus the number of row-swaps (and associated 'precharge' and 'activate' commands) is minimized. We represent SDRAM rows and bursts *explicitly* in the Polytope Model to help us reorder accesses for improved bandwidth efficiency.

4.3 Explicit Representation of SDRAM rows in the Polytope Model

The first step of our procedure is to explicitly represent the rows and bursts of SDRAM access by introducing new variables into the polytope representing the iteration space.

If the size of each SDRAM row is R words, the row accessed by memory address fx is given by $r = fx$ div $R = \lfloor fx/R \rfloor$, where $\lfloor \cdot \rfloor$ represents the *floor* function. If the size of each SDRAM burst is B words, then the burst number is similarly given by $u = \lfloor (fx - rR)/B \rfloor$. Unfortunately, neither of these representations is amenable to linear algebraic manipulation, due to the floor functions.

However, we may note that from the properties of the floor function:

$$\lfloor \frac{fx}{R} \rfloor - 1 < r \leq \lfloor \frac{fx}{R} \rfloor \quad (3)$$

and

$$\lfloor \frac{fx - rR}{B} \rfloor - 1 < u \leq \lfloor \frac{fx - rR}{B} \rfloor \quad (4)$$

which we can write as the linear equalities below, without loss of information

$$fx - R + 1 \leq Rr \leq fx \quad (5)$$

$$fx - rR - B + 1 \leq Bu \leq fx - rR \quad (6)$$

We then add these 4 extra inequalities to those already present defining the loop bounds, to form an augmented system of linear inequalities that completely describe not only the iteration space, but the SDRAM rows and bursts accessed within the innermost loop:

$$\begin{pmatrix} A & 0 & 0 \\ f & -R & 0 \\ -f & R & 0 \\ f & -R & -B \\ -f & R & B \end{pmatrix} \begin{pmatrix} x \\ r \\ u \end{pmatrix} \leq \begin{pmatrix} b \\ R-1 \\ 0 \\ B-1 \\ 0 \end{pmatrix}. \quad (7)$$

The corresponding augmented system for our Figure 1 is shown below

$$\begin{pmatrix} -1 & 0 & 0 & 0 & 0 \\ 1 & 0 & 0 & 0 & 0 \\ -1 & -1 & 0 & 0 & 0 \\ 0 & 1 & 0 & 0 & 0 \\ 1 & 0 & -1 & 0 & 0 \\ 0 & -1 & 1 & 0 & 0 \\ 7 & 8 & 9 & -16 & 0 \\ -7 & -8 & -9 & 16 & 0 \\ 7 & 8 & 9 & -16 & -4 \\ -7 & -8 & -9 & 16 & 4 \end{pmatrix} \begin{pmatrix} x_1 \\ x_2 \\ x_3 \\ r \\ u \end{pmatrix} \leq \begin{pmatrix} 0 \\ 2 \\ -2 \\ 2 \\ 0 \\ 0 \\ 15 \\ 0 \\ 3 \\ 0 \end{pmatrix} \quad (8)$$

In principle, we can use this augmented definition of the polyhedral loop bounds to rearrange the loops in our loop nest to move the variables r and u which iterate over SDRAM rows and burst accesses respectively to the outer levels of the loop body. This transformation gathers together the memory accesses to a specific SDRAM row, and reduces the number of row swaps ('activation' and 'precharging' of rows) incurred.

An example code where this transformation results in an optimal ordering of accesses is shown in Fig. 4. The original code for this example is shown in Fig. 4(a). Interpreting the addition of explicit row and burst variables as the introduction of new loop iteration variables, the augmented polytope description corresponds to Fig. 4(b), where each of the two innermost loops iterates exactly once, by construction. By itself, this transformation has only made the loop body more complex, however, it now allows us to move the r and u variables to the outermost loops using standard loop transformation techniques [5, 6], and add a buffer following [18] resulting in Fig. 4(c). Note now that the x_2 loop only iterates once and can thus be eliminated giving the end result shown in Fig. 4(d). This code is far preferable, as it accesses each row only once, streaming the data into a buffer, coalescing data reads into bursts where possible.

```
char A[256];
for (x1 = 0; x1 <= 15; x1++) {
  for (x2 = 0; x2 <= 15; x2++) {
    .. = f( A[x1 + 16 * x2] );
  }
}
```

(a) Original Code

```
char A[256];
for (x1 = 0; x1 <= 15; x1++) {
  for (x2 = 0; x2 <= 15; x2++) {
    // Note : / is integer division.
    for ( r = (x1+16*x2)/16; r <= (x1+16*x2)/16;
          r++ ) {
      for ( u = (x1 + 16 * x2 - 16 * r)/4;
            u <= (x1+16*x2-16*r)/4; u++ ) {
        ... = f( A[x1 + 16 * x2] )
      }
    }
  }
}
```

(b) Augmented Code

```
char A[256];
char buff[16][4][4];
for (r = 0; r <= 15; r++) {
  for (u = 0; u <= 4; u++) {
    buff[r][u][0..3] = burstread(r,u);
    for ( x2 = r; x2<=r; x2++ ) {
      for ( x1 = 4*u; x1<=4*u+3; x1++ ) {
        ... = f( buff[x2][x1/4][x1%4] );
      }
    }
  }
}
```

(c) Intermediate Code

```
char A[256];
char buff[16][4][4];
for (r = 0; r <= 15 ; r++) {
  for (u = 0; u <= 4; u++) {
    buff[r][u][0..3] = burstread(r,u);
    for (x1 = 4*u; x1 <= 4*u+3; x1++ ) {
      ... = f( buff[r][u][x1-4*u] );
    }
  }
}
```

(d) Transformed Code

Figure 4: C source code for 2-level nested loop example.

In general, however, such a direct transformation may not be possible. As already noted, the earlier example in Fig. 1 contains 'holes'. Moving row and column accesses to the outermost loop levels will, in this case, fill in the holes, introducing superfluous reads/writes and/or requiring complex guard statements to skip the holes. Thus our transformation engine aims to determine when such transformations can be safely applied, and manipulates the loop structure to allow their application. The first question to address, therefore, is when a loop variable, e.g. x_2 in Fig. 4, can be eliminated from the augmented code without changing the set of memory locations accessed.

199

4.4 Variable Elimination

We may formalise the question: Is the set $\{fx \mid \exists x \in \mathbb{Z}^n, Ax \leq b\}$ equal to another set $\{f'y \mid \exists y \in \mathbb{Z}^m, A'y \leq b'\}$ for some choice of f' (representing the new array indexing function), A' and b' (representing the new loop bounds), with $m < n$? If the answer is 'yes', this tells us that we may eliminate a variable, resulting in a lower complexity addressing sequencer.

A related problem has been studied in the context of operational research by Williams [24], who looked at the specific case $y = (x_1 \, x_2 \, \ldots x_{q-1} \, x_{q+1} \, x_n)^T$, i.e. the loop iterators are kept the same, but one variable is deleted (as in Fig. 4). Williams gives the following sufficient conditions for this special case:

- The q th column in matrix A has at least one entry with the value +1 with corresponding entries in all other rows being 0, negative or +1 *or*

- The q th column in matrix A has at least one entry with the value -1 with corresponding entries in all other rows being 0, positive or -1

We generalise Williams' result by trying to transform the loop body such that the above conditions are satisfied. We draw on the theory of *unimodular loop transformations* [5, 20, 25] to write $\{fx \mid \exists x \in \mathbb{Z}^n, Ax \leq b\} = \{fUz \mid \exists z \in \mathbb{Z}^n, AUz \leq b\}$ for an arbitrary unimodular matrix U, allowing us to apply Williams' elimination procedure to the matrix AU rather than to the original matrix A. We may therefore expose further opportunities for variable elimination.

For our specific example from Fig. 1, we have added dimensions r and u to our polytope description, alongside the original loop variables $[x_1, x_2, x_3]$. Adding r and u does not change the number of items in the set S_M, since each memory reference addresses data in exactly one row and one burst. Applying the unimodular transformation in (9), we can transform our matrix describing loop bounds into those shown in (10).

$$U = \begin{pmatrix} 1 & 0 & 0 & 0 & 0 \\ -1 & 1 & 0 & 0 & 0 \\ 0 & -1 & 1 & 0 & 0 \\ 0 & 0 & 0 & 1 & 0 \\ 0 & 0 & 0 & 0 & 1 \end{pmatrix}, \begin{pmatrix} x_1 \\ x_2 \\ x_3 \\ r \\ u \end{pmatrix} = Uz \, , \; AUz \leq b \quad (9)$$

$$AU = \begin{pmatrix} -1 & 0 & 0 & 0 & 0 \\ 1 & 0 & 0 & 0 & 0 \\ 0 & -1 & 0 & 0 & 0 \\ -1 & 1 & 0 & 0 & 0 \\ 1 & 1 & -1 & 0 & 0 \\ 1 & -2 & 1 & 0 & 0 \\ -1 & -1 & 9 & -16 & 0 \\ 1 & 1 & -9 & 16 & 0 \\ -1 & -1 & 9 & -16 & -4 \\ 1 & 1 & -9 & 16 & 4 \end{pmatrix} \quad (10)$$

From (10), we can see that Williams' sufficient conditions can be applied to eliminate the z_1 and z_2 variables. The set of integer points enclosed by the polytope with the z_1 and z_2 projected out has been reduced (from 7 to 5) by the transformation, but crucially, the number of unique rows and bursts accessed is the same. The difference here is that we have exploited the explicit representation of rows and bursts to enable data reuse. Where a burst within a specific row was activated more than once in the original code, in the transformed code, here it is only accessed once. The problem remaining is to find an appropriate matrix U to enable variable elimination, which we address below.

4.5 Integer Linear Program to Maximise Eliminable Variables

We wish to find a unimodular matrix U that enables a change of variable ($x = Uz$) such that the maximum number of variables can be eliminated from a polytope by Williams' conditions. To do this, we construct the formulation in Figure 5 and solve using CPLEX[14]. The formulation expresses a matrix multiplication of the input matrix A whose coefficients are known, with the unknown matrix U made up of decision variables. The elements of the resulting matrix AU are labelled $P_{i,j}$ in our formulation. The decision variables D_i form the diagonal elements of the unimodular matrix U we are trying to find, and $N_{i,j}$ are the lower triangular elements. The upper triangular elements of the unimodular matrix U are all zero. This ensures that the resulting matrix is unimodular since all lower triangular matrices with diagonal elements of -1 or +1 are unimodular.

Integer Linear Program for finding unimodular matrix to maximize variable elimination by Williams' conditions[24], $A_{k,m}$ is input integer matrix, Sz is a bound on size of any entry in the unimodular matrix.

max: $\displaystyle\sum_{i=1}^{n} sum_i$

subject to:

% *restricts lower triangular elements to [-Sz Sz]*

$1 \leq i \leq n, \, 1 \leq j < i \quad N_{i,j} \leq Sz$

$1 \leq i \leq n, \, 1 \leq j < i \quad N_{i,j} \geq -Sz$

$1 \leq i \leq n, \, i \leq j \leq n \quad N_{i,j} = 0$

% sum_i *is 0 if pos_i and neg_i are both zero*

$1 \leq i \leq n \quad pos_i + neg_i - sum_i \geq 0$

$1 \leq i \leq n, \, 1 \leq j \leq n \quad P_{i,j} = A_{i,j}D_j + \displaystyle\sum_{g=j+1}^{n} A_{i,g}N_{g,j}$

% $M_{i,j}$ *is precomputed value guaranteed to be larger than $P_{i,j}$*

$1 \leq i \leq n, \, 1 \leq j \leq n \quad P_{i,j} + (M_{i,j} - 1)neg_i \leq M_{i,j}$

$1 \leq i \leq n, \, 1 \leq j \leq n \quad P_{i,j} - (M_{i,j} - 1)pos_i \geq -M_{i,j}$

$1 \leq i \leq n \quad D_i \in \{-1, 1\}$

$1 \leq i \leq n \quad sum_i \in \{0, 1\}$

$1 \leq i \leq n \quad pos_i \in \{0, 1\}$

$1 \leq i \leq n \quad neg_i \in \{0, 1\}$

$1 \leq i \leq n, \, 1 \leq j \leq n \quad P_{i,j} \in \mathbb{Z}$

$1 \leq i \leq n, \, 1 \leq j \leq n \quad N_{i,j} \in \mathbb{Z}$

Figure 5: Finding a unimodular matrix which maximises the number of eliminable columns.

The formulation presented finds the optimal unimodular matrix for eliminating variables by Williams' conditions subject to the constraint that each of the lower triangular coefficients $N_{i,j}$ is bounded in the range [-Sz, Sz]. This is done to ensure that we can always calculate a constant value $M_{i,j}$ which is guaranteed to be bigger than $P_{i,j}$, as required by the constraints. The constraints containing $M_{i,j}$ are trivially satisfied if the associated binary variable (neg_i or pos_i) is zero. This binary variable is only allowed to become 1 if all the variables in a column of P are less-than-or-equal-to 1 or greater-than-or-equal-to -1. If either the neg_i or the pos_i variable for a particular column is non-zero, the sum variable becomes non-zero. Since the optimization procedure is attempting to maximize the sum of the sum_i variables, the optimization

```
char buff [56];
for (r = 1 ; r <= 3 ; r++) {
    for ( z3  = ceil( (16r+2) / 9) ;
              z3 <= min (6, 2r+1) ; z3++) {
        for ( u   = -4*r+2*z3 ;
                  u <= -4*r+2*z3 ; u++) {
            burstwrite(r, u) = buff [...];
        }
    }
}
```

Figure 6: Transformed source code for memory accesses in example code from Figure 1.

procedure effectively pulls up the sum variables, finding optimal values for the decision variables which form the unimodular matrix D_i for the diagonal elements, and $N_{i,j}$ for the lower triangular elements. The binary values sum_i declare whether under the change of variables, $x = Uz$, the variable z_i is eliminable.

Having found a unimodular function which gives a change of variables and allows elimination of variables, we apply that unimodular transformation to the original polytope and eliminate the appropriate variables by a Fourier-Motzkin projection [4]. In our example code, the unimodular matrix in (9) is generated, which when multiplied by the bounds in (8) gives (10) which allows for the elimination of the z_1 and z_2 indices by the sufficient conditions in [24]. In our experiments, all the ILP formulations generated complete within sub-second timing on a desktop PC.

After applying the optimal unimodular transformation, further necessary conditions from [24] are checked to see if those variables not identified as eliminable in the ILP formulation (which quickly checks for sufficient conditions) can be eliminated without creating holes. The interested reader is referred to [24] for further explanation of these conditions, with the note that the complexity of checking these conditions is dependent on the coefficients of the loop bounds, which increase with the size of the data-set to be processed. Our ILP approach scales instead with the number of nested loops which is independent of the size of the input data.

For our example code in Figure 1, checking these necessary conditions shows that the remaining variable (z_3) can be eliminated from the row dimension without creating holes, but cannot be eliminated from the burst dimension without creating holes. This is consistent with our sequence of memory accesses shown in Table 1. We use this information to reorder the loops.

Since all the variables can be shown to be eliminable from the 'r' dimension, that dimension is traversed at the outermost level of the generated loops, followed by the only remaining existential variable shown not to be eliminable in our ILP formulation, z_3. This loop variable is nested inside the 'r' variable but outside the 'u' variable in the resulting code, because it can be eliminated without causing holes in the 'r' dimension, but cannot be eliminated without causing holes in the 'u' dimension. All the other dimensions can be safely projected out. When this projection and 'C' code generation is performed using Cloog [6], the data transfer code in Figure 6 is produced.

In this generated code, we observe that the sequence of rows accessed is monotonic, bursts within each row are accessed only once and the set of rows and bursts accessed is exactly the set in the original code description (i.e. the holes in the original set are preserved). The access pattern of this transformed code contains two fewer memory accesses and two fewer row activations, as a result, its usage of scarce external memory bandwidth is more efficient than the code in the original example. The final stage of our procedure is to generate an efficient hardware address generation function to implement this transformed loop structure.

4.6 Code Generation

The tool Cloog [6] is used to generate nested loop structures by traversing the integer points within an input polytope in a specified order. Cloog generates an abstract syntax tree which can be directly translated into 'C' code. For our example, the generated code is given in Figure 6. We choose to work with this abstract syntax tree to produce pipelined streaming hardware which implements the loop index generation.

Statements which may occur in the abstract syntax tree include 'for' statements, 'assignment' statements and 'compound' statements describing the serial composition of more than one statement. The expressions within those statements include integer division, multiplication by a scalar, reduction of a vector using min, max and summation functions and modulus, floor and ceiling functions.

The expressions contained with the upper and lower bounds of the 'for' statements and within the right hand side of assignments statements can be quite complex, and without pipelining, negatively impact the achievable clock frequency. However, because Cloog derives nested loop structures in which the inner loop indices only depend on indices in outer loops, we can arrange the logic as a feed-forward pipeline with distributed control, adding arbitrary pipeline stages and using the auto-pipelining features of a logic synthesis tool (Altera Quartus II) to distribute them in a manner which minimizes the length of the critical timing path. This ensures that our hardware implementation of address generation is scalable to meet future requirements for high clock-speeds.

The synthesis and transformation of the Cloog AST format into hardware address generators produces Verilog code which can be synthesized for implementation in an FPGA. Results are reported in Section 5 showing the efficiency of each of the generators produced using our tool.

5. RESULTS

We show the effectiveness of our approach with three benchmarks :

Matrix-Matrix Multiply (50x50) This benchmark must access two matrices simultaneously using the columns of one and the rows of the other. The large strides in memory this implies means that row-swaps occur frequently within the inner loop of the benchmark.

Sobel Edge Detection (96x64) This benchmark reuses data as a sliding window perform a convolutions over an image. The input and output image row sizes do not align with SDRAM row boundaries which makes manual optimization of this benchmark difficult.

Triangular Backsubstitution (72x72) This benchmark demonstrates the applicability of our technique to non-rectangular loop nests, it demonstrates a non-constant stride over the blocks in the input matrix. Only the necessary upper triangular elements of the matrix are loaded into on-chip memory buffers.

Each benchmark is expressed as C code and passed through our automatic flow. Static control portions of the input code are marked with #pragma preprocessor directives and a polyhedral description is automatically extracted using [7]. After transformation using the methodology in Section 4, synthesizable address generators expressed in Verilog are generated as the tool output. The address generators produced were connected to the Altera High Performance SDRAM Controller II [3] in a testbench environment which recorded the SDRAM interface usage at each cycle and the overall benchmark runtime. These results are reported in Table 2.

In these results we can see the total cycles required to fetch data in each benchmark decreases as we decrease the parameterisation level (t). This is in part due to data *reuse*. When data reuse buffers are inserted at the outermost levels of the loop ($t = 1$), all accessed data is preloaded into on-chip memory at the start of execution, and fetched from on-chip memory during execution. We would expect to see a significant reduction in the number of 'read' / 'write' cycles on the external interface as the parameterisation level is reduced and more data is buffered on-chip. If we compare the original code ($t = n+1$) with the ($t = 1$) parameterisation, we see results consistent with this expectation, with a $400\times$, $94\times$ and $33\times$ reduction in the number of 'read' / 'write' cycles in each respective benchmark.

Alongside this evidence of data reuse, Table 2 also shows a breakdown of the total benchmark time into the cycles in which the interface performs 'reads' and 'writes', the cycles in which it is idle due to bus turnaround time (transition from read-to-write and vice-versa) and the 'precharge' / 'activate' and 'refresh' cycles lumped together with their respective delay cycles. This information is also presented visually in Figure 7. From this we can see that the *reordering* of memory transactions through our loop transformations increases the *efficiency* of the memory interface usage. In the original code in each of the three benchmarks, ~75% of memory interface cycles are used for the control overhead of changing SDRAM rows and bus turnaround cycles. Since our static analysis approach groups together the memory requests which occur in rows and bursts, it significantly increases the efficiency of the external memory interface. If we compare the original code with the parameterisations which have reuse buffers inserted outside the innermost loop level ($t = 3$ for MMM, $t = 4$ for SOB and $t = 2$ for GBS), we see a reduction in the proportion of memory cycles used for 'precharge' and 'activate' commands from 73.24% to 21.27%, from 70.80% to 59.35% and from 57.12% to 23.45% in the MMM, SOB and GBS benchmarks respectively. The gains in efficiency (the proportion of 'read' / 'write' cycles as compared to the original code) vary from $3.6\times$ to $4\times$ across the benchmarks and their associated parameterisations.

In order to show that the efficiency gains arising from this reordering are achievable at a reasonable cost, we report synthesis and F_{max} results from the slow 1100mV corner of the static analysis tool in Quartus 10.1 with physical synthesis and register retiming options enabled. Registers were inserted by our code-generation flow to ensure the address generator met a 133MHz clock frequency. This corresponds to half the command frequency of our external DDR2 Memory since the minimum burst size of DDR2 memory (4 words) means two clock cycle periods are needed to process consecutive back-to-back memory requests. It should be noted that since our address sequence generators can be pipelined to an arbitrary depth, they are scalable to future memory speeds at the cost of increased register count and initial latency. The post place-and-route maximum frequency of our address generator designs and their resource requirements (ALUTs and registers) are reported in Table 3. We show the number of on-chip memory words needed to implement our three benchmarks, inserting reuse buffers at different levels in the loop nest. This is reported in words rather than an absolute number of bytes to reflect the fact that external SDRAM interfaces typically bundle together multiple parallel data-channels with commands issued by a single set of control signals (which allows scaling of our benchmark runs from 8-bit to 64-bit data types). The synthesis results show that our backend generation tool will scale to useful clock-frequencies. The logic resource utilization of the address generators at different parameterisation levels varies from a reduction of $0.5\times$ to an increase of $1.4\times$ when compared to the original code in the three benchmarks. It should be noted however that even the largest address generator presented uses less than 4% of the smallest available Stratix III device (EP3SL50). From the synthesis results reported, we can conclude that our address generators achieve their reordering at a very reasonable logic cost, and will scale to useful clock frequencies in modern devices.

Table 3: Synthesis results for benchmark codes.

Benchmark	Level	Req. on-chip mem. words	ALUTs	Regs	Frequency
MMM	t=1	61200	575	764	296 MHz
MMM	t=2	21200	1050	1666	174 MHz
MMM	t=3	416	1346	2098	179 MHz
MMM	Orig.	0	1003	2740	184 MHz
SOB	t=1	11411	592	717	300 MHz
SOB	t=2	579	1551	2251	182 MHz
SOB	t=3	19	1355	1907	144 MHz
SOB	t=4	7	1200	2566	153 MHz
SOB	Orig.	0	1107	3607	148 MHz
GBS	t=1	2772	833	1156	242 MHz
GBS	t=2	288	952	1366	211 MHz
GBS	Orig.	0	804	2263	186 MHz

To explore the trade-off between on-chip *memory* resources and external memory interface performance, Figure 8 shows how the overall number of memory access cycles scales with the amount of on-chip memory dedicated to buffering data for each of the benchmarks. From this we can see that if all the data in the MMM benchmark can be stored on-chip, $1500\times$ fewer memory access cycles are needed to transfer data from external memory. What is more significant about this plot however, is that it shows that one can, using an automatic tool, explore more reasonable trade-offs such as the $t = 3$ parameterisation for the SOB benchmark, which achieves a $6.6\times$ reduction in memory access cycles at a cost of 20 additional words of on-chip memory. In each benchmark, our pareto-optimal fronts show multiple feasible points for on-chip memory usage and automatic generation of address generators using our methodology allows evaluation of the performance trade-off each embodies early in the design cycle.

While other work, such as [18], demonstrates similar trade-offs between on-chip memory usage and performance due to data *reuse*, our *explicit* representation of SDRAM rows and bursts and static reordering of memory transactions achieves additional performance gains from the efficient utilization of the memory interface; the SOB $t = 3$ parameterisation achieves $6.6\times$ better performance than the original code, despite having only a $4.7\times$ reduction in 'read' / 'write' cycles due to data reuse. This is because by *reordering* transactions we have reduced the proportion of 'precharge' / 'activate' cycles in that benchmark from 70.8% of the total interface cycles to 52.5%. Hence we can conclude that both the data reuse uncovered using our methodology *and* the transaction reordering achieved through loop transformations contribute to the memory performance improvements we demonstrate.

In Table 4, we report the tool runtime. The runtimes are aggregated across all parameterisations of the benchmarks. We report separately the time spent checking *sufficient* conditions for variable elimination using our ILP formulation and the time spent exhaustively checking the *necessary* conditions for elimination of a variable in the event that the sufficient conditions are not met. From Table 4, we note that for each benchmark, the mean time for checking the sufficient conditions for variable elimination using our ILP formulation is less than a second, with a narrow standard deviation. In

Table 2: Simulation results for benchmark codes.

Benchmark	Level	Read/Write Cycles	Bus Turnaround Cycles	Precharge/Activate Cycles	Refresh Cycles	Total Cycles
MMM	t=1	5012 (94.00%)	8 (0.15%)	202 (3.79%)	110 (2.06%)	5332
MMM	t=2	66804 (91.54%)	976 (1.34%)	3638 (4.98%)	1560 (2.14%)	72978
MMM	t=3	590000 (73.86%)	21428 (2.68%)	169870 (21.27%)	17506 (2.19%)	798804
MMM	Orig.	2000004 (23.78%)	66904 (0.80%)	6159756 (73.24%))	183596 (2.18%)	8410260
SOB	t=1	8920 (94.93%)	8 (0.09%)	300 (3.19%)	168 (1.79%)	9396
SOB	t=2	77328 (71.18%)	1868 (1.72%)	27136 (24.98%)	2300 (2.12%)	108632
SOB	t=3	180300 (32.43%)	57188 (10.92%)	274766 (52.47%)	11392 (2.18%)	523646
SOB	t=4	442932 (32.58%)	80348 (5.91%)	806860 (59.35%)	29334 (2.16%)	1359474
SOB	Orig.	839236 (24.18%)	100768 (2.90%)	2457648 (70.80%)	73632 (2.12%)	3471284
GBS	t=1	1832 (91.88%)	8 (0.40%)	154 (7.72%)	0 (0.0%)	1994
GBS	t=2	13888 (67.90%)	1368 (6.69%)	4798 (23.45%)	400 (1.96%)	20454
GBS	Orig.	61348 (25.55%)	36640 (15.26%)	137146 (57.12%)	4972 (2.07%)	240106

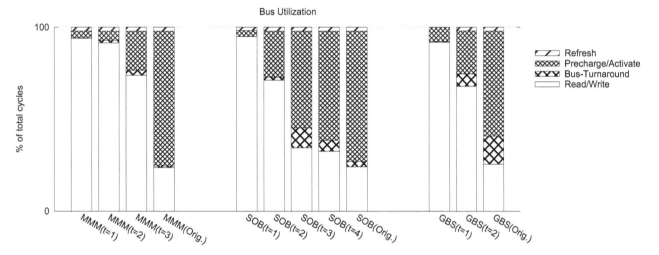

Figure 7: SDRAM Memory Interface Utilization : Breakdown by Command Type.

comparison, the mean time taken to check the *necessary* conditions for variable elimination (for those variables which cannot be eliminated by the sufficient conditions) is much greater, and varies much more significantly between the different parameterisations of each benchmark. This is because the runtime of our ILP formulation depends on the number of loop variables (n) present in the source code while checking the necessary conditions takes time proportional to the number of points in the iteration space S_E. In practice this means the time taken to check the *necessary* conditions for variable elimination scales poorly with large benchmarks, but runtimes are greatly improved if we first eliminate the variables which meet the sufficient conditions using our ILP formulation. Together

Table 4: Tool Runtime.

Benchmark	Time Taken (Suf. Cond.)		Time Taken (Nec. Cond.)	
MMM	$\mu(0.24s)$	$\sigma(0.03s)$	$\mu(25.62s)$	$\sigma(30.36s)$
SOB	$\mu(0.44s)$	$\sigma(0.20s)$	$\mu(518.63s)$	$\sigma(1009.05s)$
GBS	$\mu(0.26s)$	$\sigma(0.01s)$	$\mu(0.41s)$	$\sigma(0.57s)$

these results show that our ILP formulation, and the checking of sufficient conditions for variable elimination using Williams' results [24] allow us to produce safe performance enhancing loop transformations with reasonable compile-time. The methodology and tool built around these insights allows the automatic production of address generators whose logic cost is reasonable in modern devices and whose frequency scales to useful clock speeds. Pa-

rameterisation allows the trade-off of on-chip memory resources for performance by both a reduction in the amount of data transferred on the external memory interface and an improvement in the efficiency of that interface through reduction in the proportion of interface cycles required for 'precharge' and 'activation'.

6. CONCLUSION

In this work, we have described an analytical framework for hardware compilation which allows a parameterised trade-off between the usage of on-chip memory and arithmetic resources and external memory bandwidth. We show the applicability of our technique to a range of benchmarks and demonstrate scalability to support high clock frequencies demanded by future memory technologies.

Furthermore the tool (available at http://cas.ee.ac.uk/AddrGen) provides a starting point for those wishing to experiment with polyhedral compilation techniques within a hardware compilation flow.

In future work, we propose applying recent advances in integer point counting techniques to produce optimized static schedules for SDRAM memory access which allow exact calculation of execution time at compile time and safe overlapping of parallel execution threads and serialized external memory accesses. Such an approach will allow us to use external SDRAM memory within a general synthesis flow, exploiting knowledge of memory delays to enable efficient hardware implementations through multi-cycle logic evaluation and resource sharing.

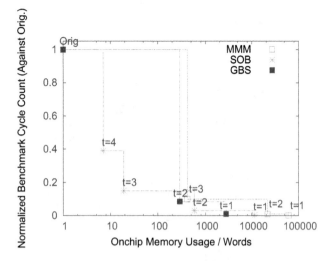

Figure 8: Pareto-optimal fronts showing designs parameterised at different levels.

7. REFERENCES

[1] Assembly and Packaging. *International Technology Roadmap for Semiconductors*, 2010.

[2] B. Akesson, K. Goossens, and M. Ringhofer. Predator : A Predictable SDRAM Memory Controller. In *CODES+ISSS '07 : Proceedings of the 5th IEEE/ACM International Conference on Hardware/Software Codesign and System Synthesis*, pages 251–256, Salzburg, Austria, 2007.

[3] Altera. DDR2 and DDR3 SDRAM Controller with UniPHY User Guide. http://www.altera.com/literature/hb/external-memory/emi_ddr3up_ug.pdf, June 2011.

[4] C. Ancourt and F. Irigoin. Scanning Polyhedra with DO Loops. In *PPOPP '91 : Proceedings of the 3rd ACM SIGPLAN Symposium on Principles and Practice of Parallel Programming*, pages 39–50, Williamsburg, United States, 1991.

[5] U. K. Banerjee. *Loop Transformations for Restructuring Compilers: The Foundations*. Kluwer Academic Publishers, Norwell, MA, USA, 1993.

[6] C. Bastoul. Code Generation in the Polyhedral Model is Easier than you Think. In *PACT '13 : IEEE International Conference on Parallel Architecture and Compilation Techniques*, pages 7–16, Juan-les-Pins, France, 2004.

[7] C. Bastoul. Extracting Polyhedral Representation from High Level Languages. Technical report, Paris-Sud University, 2008.

[8] M. Benabderrahmane, L. Pouchet, A. Cohen, and C. Bastoul. The Polyhedral Model is More Widely Applicable than you Think. In *ETAPS CC'10 : Proceedings of the International Conference on Compiler Construction*, pages 283–303, Paphos, Cyprus, 2010.

[9] H.-K. Chang and Y.-L. Lin. Array Allocation Taking into Account SDRAM Characteristics. In *ASP-DAC '00: Proceedings of the 2000 Asia and South Pacific Design Automation Conference*, pages 497–502, New York, NY, USA, 2000.

[10] E. S. Chung, J. C. Hoe, and K. Mai. CoRAM: An In-Fabric Memory Architecture for FPGA-based computing. In *FPGA '11 : Proceedings of the 19th Annual International Symposium on Field Programmable Gate Arrays*, pages 97–106, Monterey, CA, USA, 2011.

[11] A. Darte, R. Schreiber, and G. Villard. Lattice-Based Memory Allocation. *IEEE Transactions on Computers*, 54(10):1242–1257, 2005.

[12] T. Grosser, H. Zheng, R. A, A. Simbürger, A. Grosslinger, and L.-N. Pouchet. Polly - Polyhedral Optimization in LLVM. In *IMPACT'11 : First International Workshop on Polyhedral Compilation Techniques*, Chamonix, France, 2011.

[13] J. Hennessey and D. Patterson. *Computer Architecture : A Quantitative Approach*. Morgan Kaufmann, 6th edition, 2006.

[14] IBM. Introduction to CPLEX Optimization Studio. http://www-01.ibm.com/software/integration/optimization/cplex-optimizer/, June 2010.

[15] L. Johnson. Improving DDR SDRAM Efficiency with a Reordering Controller. *Xcell Journal*, 3:38–41, 2009.

[16] W. Kelly and W. Pugh. A Framework for Unifying Reordering Transformations. Technical Report CS-TR-3193, Dept. Of Computer Science, University of Maryland, 1993.

[17] H. S. Kim, N. Vijaykrishnan, M. Kandemir, E. Brockmeyer, F. Catthoor, and M. J. Irwin. Estimating Influence of Data Layout Optimizations on SDRAM Energy Consumption. In *ISLPED '03 : Proceedings of the 2003 International Symposium on Low Power Electronics and Design*, pages 40–43, Seoul, South Korea, 2003.

[18] Q. Liu, G. A. Constantinides, K. Masselos, and P. Y. K. Cheung. Automatic On-chip Memory Minimization for Data Reuse. In *FCCM '07 : Proceedings of the 15th Annual IEEE Symposium on Field-Programmable Custom Computing Machines*, pages 251–260, Napa Valley, CA, USA, 2007.

[19] G. E. Moore. Cramming More Components onto Integrated Circuits. *Proceedings of the IEEE*, 86(1):82–85, 1998.

[20] L.-N. Pouchet, C. Bastoul, A. Cohen, and N. Vasilache. Iterative optimization in the Polyhedral Model: Part I, One-Dimensional Time. In *Proceedings of the International Symposium on Code Generation and Optimization*, CGO '07, pages 144–156, Washington, DC, USA, 2007. IEEE Computer Society.

[21] F. Quilleré, S. Rajopadhye, and D. Wilde. Generation of Efficient Nested Loops from Polyhedra. *International Journal of Parallel Programming*, 28:469–498, 2000.

[22] P. Quinton and V. V. Dongen. The Mapping of Linear Recurrence Equations on Regular Arrays. *Journal of VLSI Signal Processing*, 1(2):95–113, 1989.

[23] S. Rixner, W. J. Dally, U. J. Kapasi, P. Mattson, and J. D. Owens. Memory Access Scheduling. In *ISCA '00 : Proceedings of the 27th Annual International Symposium on Computer Architecture*, volume 28, pages 128–138, Vancouver, BC, Canada, 2000.

[24] H. P. Williams. The Elimination of Integer Variables. *The Journal of the Operational Research Society*, 43(5):pp. 387–393, 1992.

[25] M. E. Wolf and M. S. Lam. A Data Locality Optimizing Algorithm. In *Proceedings of the ACM SIGPLAN 1991 conference on Programming language design and implementation*, PLDI '91, pages 30–44, New York, NY, USA, 1991. ACM.

VirtualRC: A Virtual FPGA Platform for Applications and Tools Portability

Robert Kirchgessner, Greg Stitt, Alan George, Herman Lam
NSF Center for High-Performance Reconfigurable Computing (CHREC)
Department of Electrical and Computer Engineering
University of Florida
{kirchgessner, gstitt, george, hlam}@chrec.org

ABSTRACT

Numerous studies have shown significant performance and power benefits of field-programmable gate arrays (FPGAs). Despite these benefits, FPGA usage has been limited by application design complexity caused largely by the lack of code and tool portability across different FPGA platforms, which prevents design reuse. This paper addresses the portability challenge by introducing a framework of architecture and middleware for virtualization of FPGA platforms, collectively named VirtualRC. Experiments show modest overhead of 5-6% in performance and 1% in area, while enabling portability of 11 applications and two high-level synthesis tools across three physical platforms.

Categories and Subject Descriptors

C.3 [**Special-purpose and Application-based Systems**]: Real-time and embedded systems.

General Terms

Performance, Design.

Keywords

FPGA, portability, virtual architectures.

1. INTRODUCTION

Field-programmable gate arrays (FPGAs) have been widely shown to often achieve significant performance improvements compared to microprocessors [6], graphics-processing units (GPUs) [10], et al., while also reducing power consumption [10]. Despite such advantages, many application designers have avoided FPGAs due to significantly lower design productivity as compared to other devices [3].

Although numerous factors lead to low productivity [4], a major contributor is the lack of application *portability* [4] across FPGA boards and systems, herein referred to as *platforms*. Differences in platform architectures prevent developers from exploiting common design reuse techniques, forcing them to redesign significant portions of an application and write platform-specific register-transfer-level (RTL) code. This problem also extends to debugging, performance analysis, and high-level synthesis (HLS)

tools [4] which could ideally support any platform architecture. Existing tools, however, require a platform-support package for each individual platform, making it infeasible to support the numerous available platforms.

To address these problems, we introduce a framework for FPGA platform virtualization called VirtualRC (Virtual Reconfigurable Computing). VirtualRC enables application portability by providing a configurable virtual platform architecture and corresponding software middleware that the framework can potentially map onto any physical platform. With VirtualRC, application designers target a user-customizable virtual platform, which simplifies development and enables the same RTL code to execute on any supported physical platform. In this paper, we evaluate VirtualRC on three PCIe and PCI-X FPGA platforms from GiDEL, Pico Computing, and Nallatech, demonstrating a modest performance overhead of 5-6% and an area overhead of less than 1% using application case studies and benchmarks. We showcase application portability across three platforms with 11 different RTL applications that required no coding changes. We similarly demonstrate the portability of RTL synthesized from two HLS tools, ROCCC [11] and AutoESL [6].

2. RELATED WORK

Previous works have addressed portability via application-specialized platform interfaces. Saldaña et al. [8] proposed a method of enabling the MPI programming model across FPGA platforms via HW/SW middleware. Reves et al. [7] presented a portable virtual architecture specific to software-defined radio applications. VirtualRC is conceptually similar, but also enables virtual FPGA platforms where designers can configure any application-specialized platform architecture. Coole et al. [1]

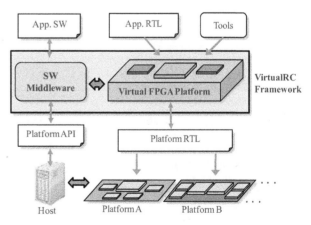

Figure 1: Overview of the VirtualRC framework for FPGA platform virtualization, which enables application and tool portability across multiple physical FPGA platforms.

introduced virtual FPGA devices for fast placement and routing, which is complementary to VirtualRC platform virtualization.

Standardized APIs such as OpenFPGA's GenAPI [5] and Intel's Acceleration Abstraction Layer (AAL) [2] address portability by providing a standardized software API for communicating with platform resources. Similarly, OpenCL [9] provides communication between heterogeneous devices. VirtualRC provides a unique API, but could potentially use any interface.

3. VirtualRC

As shown in Figure 1, VirtualRC provides a configurable *virtual FPGA platform* and a *software middleware* API for communication with the virtual platform.

To use VirtualRC, an application designer or tool first analyzes application characteristics and then requests a corresponding virtual platform architecture based upon provided configuration options. For example, a designer or tool could request one external memory with a 32-bit read port for the streaming of floating-point inputs, and another external memory with a 16-bit write port for writing fixed-point results. Given this request, VirtualRC generates a virtual platform, represented by an empty RTL entity, whose interface matches the requested configuration of resources. For the previous example, the virtual platform interface would have a 32-bit input corresponding to the read port of one virtual memory, and a 16-bit output corresponding to the write port of the second memory, in addition to control signals. The application designer then writes their application RTL code using the virtual platform as a top-level interface. Alternatively, an HLS tool could generate an application circuit that connects to the virtual platform interface. Finally, a set of platform RTL, ideally provided by the physical platform vendor, implements the virtual platform architecture on the physical platform by converting the interfaces and protocols into those used by the physical platform. Although the exact structure of platform RTL depends upon the virtual and physical platforms, for the platforms we evaluated most of this RTL consisted of simple control logic and specialized buffers for changing streaming data widths.

In creating the virtual platform architecture, we analyzed numerous FPGA platforms from GiDEL, Nallatech, DRC, Pico, and XtremeData, and identified several architectural features common to all platforms: 1) one or more FPGAs; 2) a platform bus for communicating between the host and the FPGA platform; 3) an FPGA communication controller that allows software to access on-chip resources such as block RAM and registers; and 4) one or more external memories. The virtual platform architecture provides these same four resources, as shown in Figure 2, with a unified interface and communication protocol that allows designers and tools to use the same RTL code on any supported platform. Details of the interfaces and communication protocols are omitted here for brevity.

The virtual platform supports the following configuration options. For the external memories, the virtual platform allows designers to specify the number of virtual memories, the size, the number of ports (read or write), and the data width of each port. To enable an arbitrary number of ports per virtual memory, the virtual memory interface contains an arbiter using a round-robin policy for concurrent accesses. Virtual memories may either be mapped externally to physical memory or internally to on-chip memory, depending upon application requirements and available resources. Currently, designers are required to manually perform this mapping, which we will automate in future work. For the FPGA

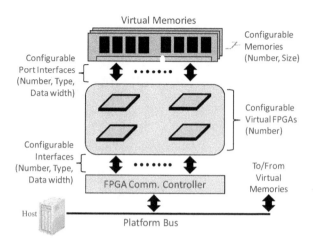

Figure 2: Virtual platform architecture overview.

communication controller, the platform allows designers to configure different data widths and numbers of controllers. The platform also supports a configurable number of virtual FPGA devices, although at present each virtual device simply acts as a top-level entity for a corresponding physical FPGA.

To enable software code portability, VirtualRC provides a simple C++ middleware API with overloaded read and write primitives that enable transparent communication of a variable or array to and from virtual resources. The API directly translates VirtualRC communication routines to virtual components into native API calls to the physical platform. The goal of the current API is to show proof of concept for communication with the virtual platform; the middleware could potentially use any API.

4. EXPERIMENTS

In this section, we describe the experimental setup (4.1) and then analyze performance and area overhead (4.2). We then evaluate the portability of applications and high-level synthesis tools (4.3) using VirtualRC across multiple physical platforms.

4.1 Experimental Setup

In our experiments, we evaluate VirtualRC on three different FPGA platforms: the GiDEL PROCStar III; Nallatech H101; and Pico Computing M501. Each platform has a significantly different platform architecture and API, and all use a variety of FPGAs from different vendors. To support each platform, we manually created the platform RTL shown in Figure 1, which required several days to several weeks. However, vendors could add this support in much less time due to familiarity with their platforms.

Bitstreams for the GiDEL PROCStar III were generated using Altera Quartus 9.1 SP2. Bitstreams for the Pico M501 and Nallatech H101 were generated using Xilinx ISE 12.3. Driver versions used were 8.8 (GiDEL), FUSE 1.5 (Nallatech) and a pre-release version of the M501 drivers (Pico). All software was compiled used g++ version 4.4.3 with –O3 optimizations.

4.2 Overhead Analysis

This section analyzes VirtualRC overhead. Section 4.2.1 presents case studies that evaluate application performance overhead. We then analyze FPGA memory bandwidth overhead (4.2.2), software middleware overhead (4.2.3), and resource overhead (4.2.4).

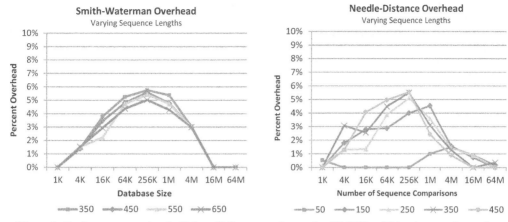

Figure 3: Performance overhead of VirtualRC compared to native PROCStar III implementations of Smith-Waterman and Needle-Distance applications.

4.2.1 Application Case Studies

This section evaluates VirtualRC performance overhead by comparing application performance of native implementations for a physical platform with implementations targeting VirtualRC on the same platform. We obtained source code for previously published implementations of Smith-Waterman and Needle-Distance [6]. We chose these two bioinformatics applications due their existing implementation on the GiDEL PROCStar III and substantial speedup over an optimized software baseline.

To evaluate overhead, we used VirtualRC to create the virtual platform interface required by the applications. We then manually mapped the application interfaces to the virtual resource interfaces. Porting the application required no modifications to the application code, and took approximately half an hour per application (not including compilation). We also created software based upon the original application software using the VirtualRC API. We then executed these applications on the PROCStar III, both with and without VirtualRC, for varying input sizes.

Figure 3 illustrates the performance overhead of the Smith-Waterman and Needle-Distance applications on VirtualRC for the PROCStar III. Across all input sizes, VirtualRC had a peak performance overhead of 6% for Smith-Waterman and 5% for Needle-Distance, with the overhead approaching 0% for large inputs sizes in both cases. This overhead was due to differences in the software required to interface VirtualRC with the original FIFO interface. In future work, we will add a configurable FIFO interface option to VirtualRC that will minimize this overhead. VirtualRC had an area overhead of approximately 1% for both applications due to the configurable virtual resources.

4.2.2 External Memory Bandwidth Overhead

For each platform (PROCStar III, M501, and H101), we measured the effective memory bandwidth with and without VirtualRC for varying transfer sizes. For VirtualRC, we configured a single virtual memory with a width equal to the native memory width. For the case without VirtualRC, we used a state machine and FIFO buffer to characterize a streaming application without blocking. We then measured the number of clock cycles required to transfer a specified amount of data.

Table 1 evaluates the FPGA to external memory overhead. Overhead was smallest on the PROCStar III due to similarities between the GiDEL IP and VirtualRC interfaces, with a maximum overhead of 5% for small transfers and negligible overhead for large transfers. For the H101, maximum overhead was 8% for small transfers and again negligible for large transfers. The VirtualRC overhead was largest on the M501 due to significant differences in interface and features from VirtualRC. The M501 IP limits transfer from 32 to 4096 bytes. Although VirtualRC adds additional overhead, it also enables transfers of any size. We measured a peak overhead of 25% for small transfers, which decreased substantially for transfers over 1KB. Since FPGA applications commonly use large data streams, the overhead of VirtualRC in these cases would be negligible.

4.2.3 Software Middleware Overhead

In order to analyze the software overhead for memory accesses, we measured the effective bandwidth for each platform using the VirtualRC API versus physical API for varying transfer sizes.

Table 2 evaluates the host to external memory overhead. The PROCStar III had the smallest overhead of 6% due to similarities between GiDEL and VirtualRC APIs. Maximum overhead on the H101 was 27% for small transfers, and became negligible for transfers above 32 KB. The largest overhead of 46% was measured for the M501, but only for transfers of 16 to 64 bytes. Since data transfers between host and FPGAs are costly, applications tend to avoid small transfers, making the overhead of VirtualRC insignificant for common situations.

Table 1: Read (write) overhead from FPGA to external memory due to VirtualRC memory virtualization.

	16B	1KB	16KB	256KB	1MB
PROCStar III	5%(4%)	3%(2%)	1%(1%)	0%(0%)	0%(0%)
H101	4%(8%)	0%(0%)	0%(0%)	0%(0%)	0%(0%)
M501	25%(19%)	5%(4%)	2%(2%)	NA	NA

Table 2: Read (write) overhead from host to FPGA memory due to VirtualRC middleware API.

	16B	1KB	16KB	256KB	1MB
PROCStar III	5%(0%)	6%(0%)	0%(1%)	0%(0%)	0%(0%)
H101	16%(27%)	13%(27%)	12%(17%)	0%(0%)	0%(0%)
M501	44%(46%)	1%(0%)	1%(0%)	0%(1%)	0%(0%)

4.2.4 FPGA Resource Overhead

Resource overhead from VirtualRC stems from three major components: the FPGA communication controller; the virtual memory read interface; and the virtual memory write interface. We determined overhead for each component using the appropriate vendor tool (Quartus, ISE). Each component was found to require less than 1% of total device resources in terms of LUTs, registers, and block RAMs. The exact resource usage on each platform varied based on how similar the native IP interface was to the VirtualRC interface specification.

Table 3: Demonstration of VirtualRC application portability.

	PROCStar III		M501		H101	
	Freq. (MHz)	Time (ms)	Freq. (MHz)	Time (ms)	Freq. (MHz)	Time (ms)
1D Convolution FP	125	39.29	125	247.90	100	91.06
2D Convolution FP	106	13.18	106	15.18	100	43.25
Option Pricing	125	12.15 s	125	14.40 s	-	-
Sum Abs. Differences	98	14.72	98	15.62	98	86.71
Needle Distance	125	194.00	125	116.20	100	199.51
Smith Waterman	125	116.00	125	133.00	100	225.00
Image Segmentation	125	12.40	125	16.39	100	4.81
OpenCores SHA256	125	64.05	125	120.49	100	25.97
OpenCores FIR	125	24.51	125	413.80	100	106.16
OpenCores AES128	125	25.33	125	503.78	100	126.18
OpenCores JPEG Enc.	125	15.29	125	23.93	100	21.24

Table 4: Demonstration of VirtualRC tool portability.

	PROCStar III		M501		H101	
	Freq. (MHz)	Time (ms)	Freq. (MHz)	Time (ms)	Freq. (MHz)	Time (ms)
ROCCC 8pt FFT	125	15.66	125	16.61	100	39.91
ROCCC 5-tap FIR	125	17.78	125	18.73	100	40.57
AutoESL Convolution	125	4.29	125	7.31	100	2.49

4.3 Portability Analysis

In this section we demonstrate application and tool portability provided by the VirtualRC framework. To evaluate portability, we created a variety of applications for VirtualRC and also obtained examples from OpenCores (www.opencores.org) that we modified to use VirtualRC. We then executed each application and tool on the three aforementioned vendor platforms, using the exact same application code. Each application consists of RTL code in addition to C++ code using the VirtualRC middleware that transfers data to and from the virtual platform. Since performance was not the goal of these experiments, a frequency of 125 MHz was used except in cases where the estimated design operating frequency was lower.

Table 3 summarizes a list of applications used to demonstrate portability on all three platforms. VirtualRC successfully executed all applications on each platform, with the exception of Option Pricing, which would not fit on the H101. This limitation was not a consequence of VirtualRC, but was caused by non-parameterized RTL code that we could not reduce in size to fit on the older Virtex-4 FPGA on the H101. It is important to note that this table is *not* intended as a performance comparison due to significant platform differences. Instead, the purpose here is to show characteristics of each application on each system.

Table 4 demonstrates tool portability. As with Table 3, these results are *not* intended to be compared across devices or between tools. Instead, these results verify that VirtualRC enables identical HLS code synthesized into RTL to work seamlessly across multiple platforms that are not directly supported by the tools. Therefore, VirtualRC potentially enables HLS tools to support any platform without effort from the tool vendors.

5. LIMITATIONS AND FUTURE WORK

There are several limitations that we plan to address as future work which include expanding VirtualRC to enable embedded processor architectures, further reducing overhead for different use cases, estimating virtual platform performance for design exploration, and integrating alternative software APIs.

6. CONCLUSIONS

To address challenges in FPGA design productivity owing to lack of code portability, we introduced VirtualRC. VirtualRC provides developers with a user configurable virtual platform interface, enabling portability across any supported platform. We evaluated VirtualRC using applications and benchmarks, and show a performance overhead of 5-6% with an area overhead of less than 1%. We also demonstrated that VirtualRC enables application and tool portability by executing the same code for 11 different applications and two tools across three physical platforms.

7. ACKNOWLEDGEMENTS

This work was supported in part by the I/UCRC Program of the National Science Foundation under Grant No. EEC-0642422. The authors gratefully acknowledge vendor equipment and tools provided by Altera, GiDEL, Nallatech, Pico Computing, and Xilinx that helped make this work possible.

8. REFERENCES

[1] Coole, J., and Stitt, G. Intermediate fabrics: Virtual architectures for circuit portability and fast placement and routing. In *Hardware/Software Codesign and System Synthesis (CODES+ISSS), 2010 IEEE/ACM/IFIP Int. Conf. on* (2010).

[2] Intel. Intel QuickAssist Technology AAL (White Paper).

[3] Jones, D., Powell, A., Bouganis, C.-S., and Cheung, P. GPU versus FPGA for high productivity computing. In *Field Programmable Logic and Applications (FPL), 2010 Int. Conf. on* (2010).

[4] Merchant, S., Holland, B., Reardon, C., George, A., Lam, H., Stitt, G., Smith, M., Alam, N., Gonzalez, I., El-Araby, E., Saha, P., El-Ghazawi, T., and Simmler, H. Strategic challenges for application development productivity in reconfigurable computing. In *Aerospace and Electronics Conference, 2008. IEEE National* (2008).

[5] OpenFPGA. OpenFPGA GenAPI version 0.4 Draft for Comment.

[6] Pascoe, C., Lawande, A., Lam, H., George, A., Sun, Y., Farmerie, W., and M., H. Reconfigurable supercomputing with scalable systolic arrays and in-stream control for wavefront genomics processing. In *Proc. of Symposium on Application Accelerators in High-Performance Computing* (2010).

[7] Reves, X., Marojevic, V., Ferrus, R., and Gelonch, A. FPGA's middleware for software defined radio applications. In *Field Programmable Logic and Applications, 2005. Int. Conf. on* (2005).

[8] Saldaña, M., Patel, A., Madill, C., Nunes, D., Wang, D., Chow, P., Wittig, R., Styles, H., and Putnam, A. MPI as a programming model for high-performance reconfigurable computers. *ACM Trans. Reconfigurable Technol. Syst. 3* (November 2010), 22:1–22:29.

[9] Stone, J., Gohara, D., and Shi, G. OpenCL: A parallel programming standard for heterogeneous computing systems. *Computing in Science Engineering 12*, 3 (may-june 2010), 66 –73.

[10] Tian, X., and Benkrid, K. High-performance quasi-monte carlo financial simulation: FPGA vs. GPP vs. GPU. *ACM Trans. Reconfigurable Technol. Syst. 3* (November 2010), 26:1–26:22.

[11] Villarreal, J., Park, A., Najjar, W., and Halstead, R. Designing modular hardware accelerators in c with ROCCC 2.0. In *Field-Programmable Custom Computing Machines (FCCM), 2010 18th IEEE Annual Int. Sym. on* (may 2010), pp. 127 –134.

Multi-Ported Memories for FPGAs via XOR

Charles Eric LaForest, Ming G. Liu, Emma Rae Rapati, and J. Gregory Steffan
Department of Electrical and Computer Engineering
University of Toronto, Canada
{laforest,steffan}@eecg.toronto.edu, {emma.rapati,minggang.liu}@utoronto.ca

ABSTRACT

Multi-ported memories are challenging to implement with FPGAs since the block RAMs included in the fabric typically have only two ports. Any design that requires a memory with more than two ports must therefore be built out of logic elements or by combining multiple block RAMs. The recently-proposed Live Value Table (LVT) [8] design provides a significant operating frequency improvement over conventional approaches. In this paper we present an alternative approach based on the XOR operation that provides multi-ported memories that use far less logic but more block RAMs than LVT designs, and are often smaller and faster for memories that are more than 512 entries deep. We show that (i) both designs can exploit multipumping to trade speed for area savings, (ii) that multipumped XOR designs are significantly smaller but moderately slower than their LVT counterparts, and (iii) that both the LVT and XOR approaches are valuable and useful in different situations, depending on the constraints and resource utilization of the enclosing design.

Categories and Subject Descriptors

B.3.2 [**Memory Structures**]: Design Style—*Shared Memory*

General Terms

Design Performance Measurement

Keywords

FPGA, memory, multi-port, parallel, XOR

1. INTRODUCTION

FPGAs are increasingly used to implement complex systems-on-chip that require frequent communication, sharing, queuing, and synchronization among distributed functional units and compute nodes. These high-contention storage mechanisms are often implemented using *multi-ported memories* that allow multiple reads and writes to occur simultaneously. A good example is the register file of an FPGA-based *soft processor*, for which even a simple in-order RISC processor requires one write port and two read ports, while processors that issue instructions more aggressively require

even more ports. The challenge is that constructing a multi-ported memory out of FPGA logic elements is inefficient [8]. Furthermore, FPGA substrates typically provide block RAMs (BRAMs) that provide only two ports, and hence memories with more than two ports must be "soft", i.e., constructed using FPGA logic and/or hard BRAMs. However, the ability to construct efficient soft multi-ported memories is important as it frees FPGA vendors from having to include hard BRAMs with more than two ports in their fabrics.

1.1 Prior Approaches

Implementations for FPGA-based multi-ported memories have only recently been formally described and studied [8]; we summarize the conventional approaches here. A straightforward approach is to construct a multi-ported memory using logic elements—for example Altera's adaptive logic modules (ALMs)—enjoying flexibility but at a heavy cost in area and performance. **Replication** enables constructing a memory with any number of external read ports, but can support only a single external write port that must be connected to one of the two ports of each replicated BRAM. **Banking** divides the read and write ports across multiple separate BRAMs, supporting concurrent read and writes but fragmenting and isolating the data across banks. The **Live Value Table** (LVT) approach [8] augments a banked approach with a table that uses output multiplexers to steer reads to the most recently-updated bank for each memory address. The LVT approach improves significantly on the area and speed of comparable designs built using ALMs, although the internal LVT table itself scales somewhat poorly, can consume a lot of area, and usually becomes the critical path. Finally, **Multipumping** can be applied to any memory design to multiply its read and write ports by operating that memory at a multiple of the external clock frequency. Multipumping reduces the area required for a memory with a certain number of ports, but also reduces its maximum achievable external operating frequency.

1.2 An XOR-Based Approach

The XOR operation ($\oplus$) has interesting and useful properties, particularly that $A \oplus B \oplus B = A$. For example, XOR can be used in RAID systems [11] to implement parity and provide data recovery capability should one hard-drive of an array of drives fail. In this paper we present an alternative to the LVT approach that is based on XOR. Similar to the LVT approach, the XOR approach internally uses banking and replication. However, the XOR design avoids the need for a Live Value Table to direct reads and also avoids the corresponding output multiplexing, instead allowing the logic of each read port to consist solely of an XOR of values read from a bank of BRAMs. We demonstrate that XOR designs consume far fewer ALMs but more BRAMs than corresponding LVT designs, and that XOR designs can also be faster and consume less total area than LVT designs for some configurations.

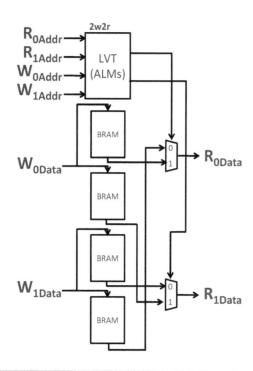

Figure 1: A 2W/2R Live Value Table (LVT) design.

1.3 Contributions

This paper makes the following contributions:

1. we present a novel design for implementing multi-ported memories based on the XOR operation;

2. we describe methods to improve the speed of the original LVT design, at an area cost for some designs but an area savings for others;

3. we thoroughly evaluate and compare XOR and improved LVT approaches on an Altera Stratix IV device across a broad range of depths and numbers of ports, as well as multiple factors of multipumping;

4. we demonstrate that the XOR and LVT designs have significant resource diversity, as XOR designs use far fewer ALMs but more BRAMs than LVT designs;

5. we show that both XOR and LVT designs can be the smallest or fastest option, depending on the depth and number of ports of the desired memory.

2. THE LIVE VALUE TABLE DESIGN

The live value table (LVT) multi-ported memory design allows the implementation of a memory with more than one write port to be based on BRAMs, as opposed to being limited to building such a memory solely from logic elements. The basic idea of an LVT design is to augment a banked design with the ability to connect each read port to the most-recently written bank for a given memory location. In this section we briefly summarize the construction and operation of the LVT design; a full treatment is provided by LaForest and Steffan [8].

As a simple example, Figure 1 shows a two-write-two-read (2W/2R) LVT-based memory. Each write port requires its own bank of BRAMs, and each bank requires two BRAMs to provide

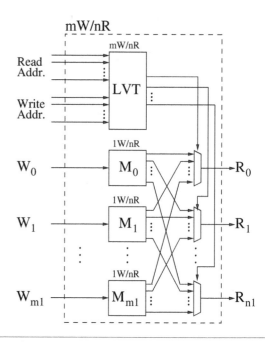

Figure 2: A generalized mW/nR memory implemented using a Live Value Table (LVT)

for the two read ports. At each read port is a multiplexer that is driven by the LVT itself, which selects the most-recently written (i.e., *live*) bank for the read-address. The LVT itself is composed of logic elements (e.g., Altera's ALMs), and itself is a 2W/2R memory, but is only as wide as the log-base-2 of the number of write ports (typically 1-3 bits wide)—and hence can be implemented fairly efficiently.

Figure 2 shows a generalized LVT design. Again, there is a bank of BRAMs per every write port, and each bank is itself a 1W/nR memory composed BRAMs, for a memory with n external read ports. The BRAMs in a bank are arranged using *replication*, where the single write port writes to every BRAM, and the second port for each BRAM is used to provide a read port to the output multiplexers. Hence an LVT design requires a total of $m \cdot n$ BRAMs, plus the logic required to implement the LVT and the multiplexers.

Summary For LVT-based memories, the LVT itself and the output multiplexers together (i) constitute the critical path, and (ii) can require a significant number of logic elements to implement. In the next section we pursue an alternative design that avoids both of these challenges. Later in Section 5 we describe and quantify the ways that we improve on the original LVT design and its multipumped versions to significantly increase its frequency at a small cost in area for shallow designs.

3. AN XOR-BASED DESIGN

In this section we venture to overcome the drawbacks of the LVT design by avoiding the narrow but multi-ported LVT itself required to control the output multiplexers. As introduced earlier, the goal of our design is for each read port to require only the computation of the XOR of values read from BRAMs. In this section we review some of the properties of XOR, build towards a working XOR-based multi-ported memory design, and then discuss how the XOR-based design can be multipumped to trade speed to save area.

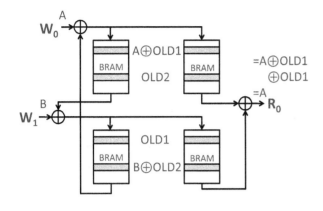

Figure 3: A 2W/1R memory implemented using XOR, **with example data values. Note that only data wires are shown, not address wires.** W_0 **stores the value** A XOR**'ed with the old contents of the other bank** ($OLD1$)**. Similarly,** W_1 **stores the value** B XOR**'ed with the old contents of the other bank** ($OLD2$)**. Reading the location containing** A **computes** $(A \oplus OLD1) \oplus OLD1$ **which returns** A.

3.1 XOR Properties

The bitwise XOR operation is commutative, associative, and has the following properties[1]:

- $A \oplus 0 = A$
- $B \oplus B = 0$
- $A \oplus B \oplus B = A$

The third property, which follows from the first two, implies that we can XOR two values A and B together, and recover A by XORing the result with B. We can exploit this property to allow the XOR of two instances of a location to return the most recent version. For example, suppose location1 contains some OLD value, and then we save a new value A in location2 by XORing it with the OLD value, i.e., by storing $A \oplus OLD$ in location2; explicitly: $location2 = A \oplus location1 = A \oplus OLD$. We can then recover A, i.e., read the most recently-written value, by simply returning the XOR of the two locations, without having to select between them; explicitly: $output = location1 \oplus location2 = (A \oplus OLD) \oplus OLD = A$. While at first this all seems unnecessary, the key is that it allows two write ports to write two separate BRAMS (or banks of BRAMs) simultaneously (like the LVT design), while read ports need only XOR BRAM locations to return a value (unlike the LVT design which requires output multiplexing).

3.2 Simple XOR Designs

We next build on this basic property of XOR to construct a simple 2W/1R memory out of dual-ported BRAMs, as illustrated in Figure 3. Note that the figure shows only data wires and values, not address wires or values. In the design, each write port has its own bank of two BRAMs, and writes for each are copied to both BRAMs—i.e., corresponding locations in all of the BRAMs in a bank always have the same value. When the write port W_0 stores the value A to the upper locations (in grey), it first XORs A with the old value of the same location in W_1's bank ($OLD1$). Similarly,

---

[1]Another interesting use of XOR is to swap the contents of two memory locations A and B without the use of a temporary location in three steps: $A = A \oplus B$; $B = A \oplus B$; $A = A \oplus B$.

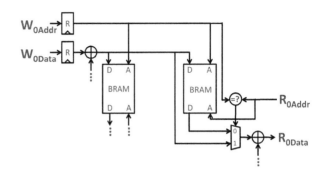

Figure 4: Details of the address wires, registers, and forwarding circuitry used in the XOR design, not shown in other figures for simplicity.

when the write port W_1 stores the value B to the lower locations (also in grey), it first XORs B with the old value of the same location in W_0's bank ($OLD2$).

For the read port, recall that our main design goal is for the read port circuitry to consist solely of an XOR of BRAM outputs, which is achieved. Reading the upper location computes the XOR of both versions of the upper location (from both the upper and lower banks), which results in isolating the value most recently stored to that location by computing $(A \oplus OLD1) \oplus OLD1$, which returns A.

A challenge for the XOR design is that each write requires reading as well, since we must store the write value XOR the old value of that location from the other bank. This increases the number of BRAMs required to implement the design, since extra read ports must be used internally to service writes. This also potentially complicates the design since writes will effectively take two cycles to complete. However, we can keep the XOR design *black-box-compatible* with previous designs and give the illusion that writes effectively take only one cycle with two additions to the design, as illustrated in Figure 4. First, we register the write port addresses and values. Second, we instantiate forwarding circuitry (via Quartus library) that allows the write data value to flow directly to the read data wires in the event that we read a location in the cycle directly after we write that same location. For simplicity these extra registers and forwarding logic are not shown in figures other than Figure 4.

As shown in Figure 5, we next build a slightly more complex memory by adding another read port, constructing a 2W/2R memory. This design functions similarly to the 2W/1R design, except that another column of BRAMs has been added to provide values for the additional read port.

3.3 A Generalized XOR Design

To summarize the XOR design, each time we write the contents of a given location to a particular memory bank, we XOR the new data with the old contents of the same location from all other banks. To read a location we calculate the XOR of the values for that location across all banks, which recovers the latest value written.

In Figure 6 we present a generalized mW/nR XOR design. Each write port has its own bank (row) of BRAMs. To write a value to a location, that value is first XORed with all of the values for that same location from all other banks, and the result of that XOR is then distributed and written to all BRAMs in the current bank. Hence

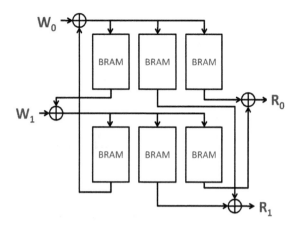

Figure 5: A 2W/2R memory implemented using XOR**. Compared with the 2W/1R memory in Figure 3, an additional column of BRAMs is added to supply values for the additional read port.**

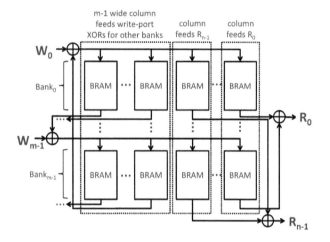

m-1 wide column
feeds write-port
XORs for other banks

column
feeds R_{n-1}

column
feeds R_0

Figure 6: A generalized mW/nR memory implemented using XOR**. Each write port requires a bank (row) of BRAMs. A column of BRAMs is required for each read port, as well as a column for each of one less than the total number of write ports.**

the design requires a column of BRAMs that is as wide as one less than the number of write ports, to provide sufficient internal read ports to support writing. Furthermore, a column of BRAMs is required for each external read port. In summary, the XOR design requires $m * (m - 1 + n)$ BRAMs to provide m writes and n reads.

3.4 Multipumping the XOR Design

For any memory design we can trade speed for area by operating the memory at an internal clock frequency that is a faster multiple of the external clock frequency, giving the illusion of having more ports than are actually supported. We can apply this multipumping [8] to either read or write ports, but not both at once. Doing so interleaves the internal reads and writes and breaks the external appearance of reads logically occurring before writes within the same clock cycle. Note that relaxing this constraint would lead to

a reduction in hardware (Section 6.3), but at the cost of placing the burden of scheduling reads and writes onto the enclosing system.

Multipumping the write ports provides the best speed/area trade-off, resulting in (i) fewer memory banks (Figure 6), which leads to (ii) shallower columns for each read port, and (iii) fewer columns to feed the XOR logic of other write ports. The resulting multipumped XOR design requires $(m/f) * ((m/f) - 1 + n)$ BRAMs, for a given multipumping factor $f \leq m$ (assuming that both f and m are powers of two).

In operation, a multipumped XOR-based memory performs all reads and the first subset of writes on the first *internal* clock cycle, then performs the remainder of the writes afterwards with all operations completing within one *external* clock cycle. For example, with 2x multipumping, all reads and the first half of all writes happen during the first internal cycle, while the second half of the writes happen during the second internal cycle, whose end coincides with the end of a single encompassing external cycle. Since no read happens after a write, a read always returns the current memory contents as opposed to returning a value being written.

Note that in this paper we only consider even factors of multipumping—e.g., 2x and 4x, which means that for every external cycle there are two or four internal cycles. Note also that a design can be multi-pumped to the point where there are as many internal cycles as there are external write ports, meaning that the internal memory requires only one write port, which does not require the XOR mechanism to function—i.e., one write port and N read ports are trivially supported via replication only. Since such designs are not XOR-based (nor LVT-based), we instead call them *fully multipumped*.

4. EXPERIMENTAL FRAMEWORK

In this section we describe our experimental framework. We evaluate the designs on Altera Stratix IV FPGAs, although we expect similar results on comparable quality-grade Xilinx FPGAs. We provide details on Stratix IV BRAMs, the memory designs under study, our CAD flow, and our method for measuring speed and area.

Stratix IV Block RAM (BRAM) Memory Modern FPGAs often implement BRAMs directly on their silicon substrate. These BRAMs typically have two ports that can each function either as a read or a write port. BRAMs use less area and run at a higher frequency than ones created from the FPGA's reconfigurable logic, but do so at the expense of having a fixed storage capacity and number of ports. The Stratix IV FPGAs mostly contain M9K block RAMs[2], which hold nine kilobits of information at various widths and depths. At a width of 32 bits, an M9K holds 256 elements.

CAD Flow We use Altera's Quartus 10.0 to target the Stratix IV EP4SE530H40C2 FPGA, a device of the highest available speed grade and containing 1280 M9K BRAMs. We implement all the designs in generic Verilog-2001 without any Altera-specific modules. We place our circuits inside a synthesis test harness designed to both: (i) register all inputs and outputs to ensure an accurate timing analysis, and (ii) to reduce the number of I/O pins to a minimum as larger circuits will not otherwise fit on the FPGA. The test harness also avoids any loss of circuitry caused by I/O optimization. Shift registers expand single-pin inputs, while

---

[2]These FPGAs also contain M144K and MLAB memories. There are too few M144Ks to fully explore the design space and past work demonstrated that MLABs scale very poorly [8].

registered AND-reducers compact word-wide signals to a single output pin.

We configured the synthesis process to favour speed over area, and enabled all relevant optimizations, including circuit transformations such as register retiming. The impact on area of register retiming varies, depending on the logic found beyond the I/O registers, so the absolute results presented here might not appear in a real system. However, comparing our designs in a real system would yield proportionally similar results. We tested all designs inside identical test harnesses.

We configured the place and route process to make a standard effort at fitting with only two constraints: (i) to avoid I/O pin registers to prevent artificially long paths that would affect the clock frequency, and (ii) to set the target clock frequency to 550MHz, which is the maximum clock frequency specified for M9K BRAMs. Setting a higher target F_{max} does not improve results, and may in fact worsen them if a slower, derived clock exists and thus aims towards an unnecessarily high target frequency, causing competition for fast paths. We assume all clocks to be externally generated and any of their fractions (e.g., half-rate) used in multipumping designs are assumed to be synchronous to the main system clock (i.e., when generated by a PLL).

We report maximum operating frequency (F_{max}) by averaging the results of ten place and route runs, each starting with a different random seed for initial placement. We select the worst-case F_{max} report for the default range of die temperatures of 0 to 85°C. Area does not vary significantly between place and route runs, so we report the first computed result.

Measuring Area When comparing designs as a whole, we report area as the *total equivalent area* (TEA), which estimates the actual silicon area of a design point: we calculate the sum of all the Adaptive Logic Modules (ALMs) used partially or completely, plus the area of the BRAMs *counted as their equivalent area in ALMs*. A Stratix IV ALM contains two Adaptive Lookup Tables (ALUTs), each roughly equivalent to a 6-LUT, two adder and carry-chain stages, and two flip-flops. Wong *et al.* [14] provide the raw layout area data: one M9K has an area equivalent to 28.7 ALMs. This value became known only after publication by LaForest *et al.* [8], which used an estimate.

Designs Considered For simplicity, we consider only the common case of 32-bit-wide memories. We do not consider one-write-one-read (1W/1R) memories as they directly map to a single FPGA BRAM. Similarly, replication trivially enables 1W/nR memories. The challenge lies in creating concurrent multiple *write ports*. We evaluate a representative sample of the range of multi-ported memory configurations with two to eight write ports and four to 16 read ports: 2W/4R, 4W/8R, and 8W/16R. We consider these configurations over memory depths of 32 to 8k entries; some designs with more than 2k entries begin to consume a significant fraction of the large Altera device that we target and hence would likely be impractically large for current-day applications. Although the internal implementations vary, we ensure that all designs are "black-box equivalent" from an outside point of view. Specifically, all ports are unidirectional (read or write only), usable simultaneously within a single external clock cycle, and any latencies between values being written and subsequently readable are equal across designs. Any one design can substitute for another within a system, with clock frequency and resource usage being the only differences. We do not consider memories that may stall (e.g., take multiple cycles to perform a read or write if there exists a resource conflict), although such designs would be compelling future work. Finally, we assume that multiple simultaneous writes to the same

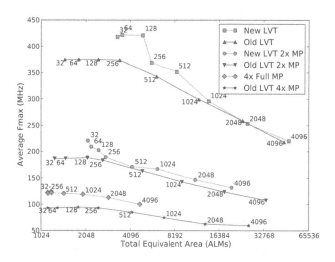

Figure 7: Speed and area of both the original (old) and improved (new) 4W/8R LVT designs. MP means multipumped.

address result in undefined behavior and are thus avoided by the enclosing system.

5. IMPROVEMENTS TO THE LVT DESIGN

For our evaluation and comparison with LVT (later in Section 6), we make two significant improvements to the original LVT design [8]: (i) adding forwarding logic, and (ii) multipumping write ports.

Forwarding Similar to that described earlier for the XOR design (Figure 4), we instantiate forwarding logic around the BRAMs of the LVT design to increase clock frequency for an area cost. Recall that the forwarding logic bypasses the BRAMs such that if a location being written is read during the same cycle, the read will return the new write value (as opposed to the old stored value). To remain compatible with the expected behavior of a one cycle read-after-write latency, we also register write addresses and data. This modified design increases the maximum operating frequency of the BRAMs (including forwarding) from 375MHz to 550 MHz, since there is a frequency cost to the Stratix IV implementation of having a BRAM set to return old data during simultaneous read/write of the same location [2]. Overall this makes the modified LVT designs more competitive.

Multipumping As discussed earlier in Section 3.4, for any multi-ported memory design we can exploit multipumping to trade speed for area savings. Details for how to multipump the LVT design are given by LaForest and Steffan [8]—however their described design multipumps across read ports, while we found it to be an improvement to instead multipump across write ports and hence do so for the LVT implementations in this paper. Furthermore, when the multipumping factor equals the number of external write ports, the number of internal write ports reduces to one, eliminating the need for the LVT circuitry entirely, further saving area and gaining speed—as mentioned previously, we call such designs *fully multipumped*.

Impact Figure 7 shows the speed and area impact of our modifications to the LVT design. For small memories (32-128 entries deep), the new LVT is significantly faster although those are also larger than the corresponding old versions. For the 2x-multipumped memories, all memories are faster than the corresponding original

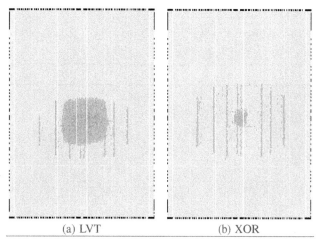

| (a) LVT | (b) XOR |

Figure 8: Circuit layout of 8192-deep 2W/4R memories for (a) LVT and (b) XOR designs, as rendered by Quartus. The thin columns are BRAMs or DSPs (darkened indicates in-use), the dots and dot-clouds are ALMs.

versions, and memories deeper than 256 entries are also smaller. For the 4x-multipumped memories, all memories are both faster and smaller than the original versions. Overall, while there are some trade-offs between area and speed that are apparent for some design points, for simplicity and to focus more on speed than area, we concentrate on the new LVT designs for the remainder of this paper.

6. COMPARING LVT VS XOR

In this section we compare the LVT and XOR approaches to implementing multi-ported memories. We begin by visualizing the resource usage and layout of an example of each design. Next we compare in detail the speed and resource usage of a broad range of memories of varying depths and numbers of ports, and then investigate the impact of multipumping all designs. We summarize the design space by highlighting the designs that minimize delay, ALM usage, or BRAM usage.

6.1 Visualizing Layout

To help visualize the resource diversity of the LVT and XOR approaches, in Figure 8 we present the circuit layout of 8192-deep 2W/4R memories for both XOR and LVT designs, as rendered by Quartus. The thin columns represent BRAMs or DSPs, where darkened areas indicate that the BRAMs are being used, and the dots and dot-clouds represent ALMs. We chose 8192-deep memories because they are large relative to the capacity of the chip and emphasize the differences between the designs. Both designs consume resources in a somewhat circular pattern, due to Quartus' efforts to minimize delay and resource consumption. Looking at BRAM usage, the designs are both the same width, but the XOR design consumes more BRAMs and hence has taller columns of in-use BRAMs. For both designs the ALMs used are clustered in the center, but the LVT design consumes far more ALMs than the XOR design. Considering that each multi-ported memory would be inserted into a larger design, one can visualize how the XOR memory would integrate better with an enclosing design that consumes many ALMs, and the LVT design would integrate better with an enclosing design that demands many BRAMs.

6.2 Varying Depths and Numbers of Ports

In Figure 9 we compare LVT and XOR implementations of 2W/4R, 4W/8R, and 8W/16R memories, with depths varying from

32 entries up to 8192, 4096, and 1024 entries respectively—the memories with more ports exhaust the available BRAMs with fewer entries than the 2W/4R memories. For the figures on the left we plot the average *unrestricted* maximum operating frequency (Fmax)[3] versus area. In particular we report the *total equivalent area* (TEA) in terms of ALMs, that accounts for both ALMs and BRAMs used, as described in Section 4. Note that the x-axis (TEA) is logarithmic.

Fmax vs Area Looking at the results for the 2W/4R memories (Figure 9(a)), we observe first that the LVT designs are superior for both Fmax and TEA for 32 and 64 entries. For designs with more than 64 entries the XOR designs consume less TEA, with the relative savings increasing with the number of entries: e.g., for 8192-entry memories the XOR design is 51% of the area of the LVT design. This TEA difference persists in 4W/8R and 8W/16R memories, with the area advantage going to XOR designs with more than 256 entries.

However, both the XOR and LVT designs trigger anomalies in Quartus (despite averaging these results over 10 seeds as described in Section 4): for example, for LVT 2W/4R designs the 128-entry design is faster than the 32-entry design, and the 1024-entry design is significantly faster than the 512-entry design, contrary to the general trend of Fmax linearly decreasing as memory depth increases. Similarly, the 512-entry LVT 8W/16R design has virtually the same Fmax as the 256-entry design, contrary to expectations since going from 256 to 512 entries doubles the number of BRAMs required. Finally, the smallest 4W/8R XOR designs (< 512 entries) see-saw between higher and lower TEAs and Fmax in a manner seemingly unrelated to the memory depth, as does the Fmax of the 64-entry 8W/16R XOR design.

Disregarding the anomalies, it is apparent that for smaller designs the LVT approach is generally faster or equally-fast as the corresponding XOR designs, while deeper designs save more area and gain more speed from the XOR approach.

BRAMs vs ALMs In the figures on the right of Figure 9, we expand on our TEA metric to view the actual numbers of BRAMs and ALMs consumed by each design—note that both axes are logarithmic. Looking at Figure 9(a) as an example, we first note that the resource diversity of the two designs is clearly evident, with XOR designs consuming far fewer ALMs but more BRAMs than the corresponding LVT designs. *Hence the relative availability of ALMs or BRAMs in a given encompassing design plays a large role in the selection of the best choice of multi-ported memory implementation.* Looking in more detail, we observe that the number of BRAMS used is constant for both designs for 32-256 entries, reflecting the native capacity of each BRAM. The number of ALMs used by XOR memories grows very slowly as memory depth increases, since for the XOR design ALMs are used only to implement the XOR operations and forwarding logic, both of which grow linearly with memory width and only logarithmically with memory depth. In contrast, the number of ALMs used by LVT memories grows with memory depth since (i) they are used to construct the LVT itself, which grows significantly for deep and/or many-ported memories, and (ii) they implement forwarding logic, which grows with the number of BRAMs used. Due to the extra replicated memories required to support the write port XOR operations, XOR designs consume more BRAMS than the corresponding LVT designs—e.g., for the 8192-entry memories the XOR design consumes 25% more BRAMs than the LVT design.

---

[3]Note that a minimum clock pulse width requirement for the BRAMs restricts the actual Fmax to 550MHz on Stratix IV devices.

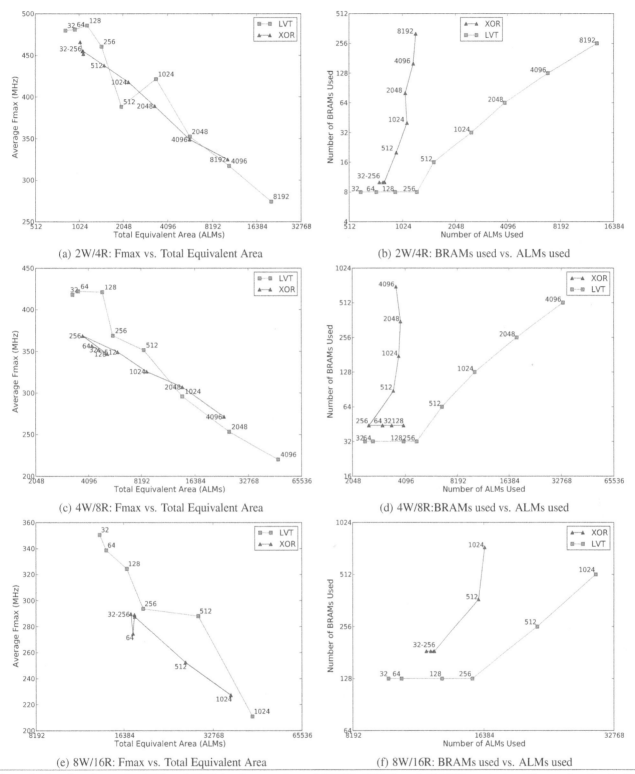

(a) 2W/4R: Fmax vs. Total Equivalent Area

(b) 2W/4R: BRAMs used vs. ALMs used

(c) 4W/8R: Fmax vs. Total Equivalent Area

(d) 4W/8R:BRAMs used vs. ALMs used

(e) 8W/16R: Fmax vs. Total Equivalent Area

(f) 8W/16R: BRAMs used vs. ALMs used

Figure 9: Speed and area for LVT and XOR implementations of 2W/4R, 4W/8R, and 8W/16R memories of increasing depth. For each we show average Fmax vs total equivalent area (on the left), as well as BRAMs vs ALMs (on the right).

	Design that minimizes:		
Depth	Delay	ALMs	BRAMs
32	Equal	LVT	LVT
64	LVT	LVT	LVT
128	LVT	XOR	LVT
256	Equal	XOR	LVT
512	XOR	XOR	LVT
1024	Equal	XOR	LVT
2048	XOR	XOR	LVT
4096	XOR	XOR	LVT
8192	XOR	XOR	LVT

(a) 2W/4R

	Design that minimizes:		
Depth	Delay	ALMs	BRAMs
32	LVT	LVT	LVT
64	LVT	LVT	LVT
128	LVT	Equal	LVT
256	Equal	XOR	LVT
512	Equal	XOR	LVT
1024	XOR	XOR	LVT
2048	XOR	XOR	LVT
4096	XOR	XOR	LVT

(b) 4W/8R

	Design that minimizes:		
Depth	Delay	ALMs	BRAMs
32	LVT	LVT	LVT
64	LVT	LVT	LVT
128	LVT	Equal	LVT
256	Equal	XOR	LVT
512	LVT	XOR	LVT
1024	XOR	XOR	LVT

(c) 8W/16R

Figure 10: For each memory depth, listed is the design that minimizes delay (i.e., has the highest Fmax), the number of ALMs used, or the number of BRAMs used, for (a) 2W/4R, (b) 4W/8R, and (c) 8W/16R memories. Results within 5% are considered "Equal".

Increasing Ports Figures 9(c), (d), (e), and (f) plot the results for 4W/8R and 8W/16R memories, which at a high level show similar trends as the 2W/4R memories, but with some notable differences. As the number of ports increases, so does the frequency gap between LVT and XOR designs, with XOR designs becoming comparatively slower. The TEA advantage of XOR designs also diminishes: e.g., for the 1024-entry 8W/16R memories the XOR design is only 18% smaller. Looking at Figure 9(f), we see that the XOR designs now consume a more significant number of ALMs to support XOR operations and forwarding logic. The area of these functions increases in proportion to the product of the number of read and write ports: e.g., going from 2W/4R to 4W/8R quadruples the number of ALMS required by XOR designs of the same depth.

Navigating the Design Space From the point of view of a system designer, a key question is "*What is the best memory design to use given my constraints?*". To summarize the design space we list in Figure 10 the design that, for each memory depth, minimizes delay (i.e., has the highest Fmax), the number of ALMs used, or the number of BRAMs used, displayed for (a) 2W/4R, (b) 4W/8R, and (c) 8W/16R memories. Note that any results within 5% of each other are considered to be roughly equal due to normal CAD variation and are labeled as such. First, we note that the LVT designs always use the least BRAMs, regardless of depth or number of ports. In terms of ALM use, designs with 64 or fewer entries are most efficiently implemented via the LVT approach, and designs with 256 or greater entries are most efficiently implemented via the XOR approach. For 128 entries the designs have roughly equivalent ALM usage, although for the 128-entry 2W/4R memory the XOR design has a greater than 5% advantage. In terms of maximizing Fmax, the LVT design is generally faster for shallower memories while the XOR design is faster for deeper memories. The crossover point is around 256-1024 entries, depending on the number of ports, and obscured somewhat by the previously-discussed CAD anomalies.

6.3 Impact of Multipumping

Figure 11 shows the average *unrestricted* Fmax vs TEA for LVT and XOR implementations of (a) 2W/4R, (b) 4W/8R, and (c) 8W/16R memories of increasing depth, including the 2x and 4x multipumped (MP) and fully-multipumped (Full MP) designs. Looking first at Figure 11(a) for 2W/4R memories, since these designs have only two write ports, multipumping by 2x results in a fully-multipumped design (rather than an LVT or XOR design). The resulting impact on speed and area is as expected, resulting in memories that are smaller but slower: for example, the 32-entry fully-multipumped memory is 53% of the speed but also 48.7% fewer equivalent ALMs than the 32-entry LVT memory.

The value of considering both LVT and XOR designs is more apparent from Figures 11(b) and (c), where considering the multipumping factor and choice of LVT vs XOR implementation provides designers with a significant range of options in the speed vs area trade-off space. Looking at Figure 11(b) for 4W/8R memories and focusing on the 2x multipumped (2x MP) designs, we observe that LVT designs are faster but significantly larger than their XOR counterparts. At the extreme, the 4096-entry XOR memory consumes 46.4% of the equivalent ALMs of the 2048-entry LVT design, even though the LVT design has half of the entries. The fully-multipumped design (Full MP) further repeats this significant trade-off between speed and area. Multipumping provides a greater relative area savings for XOR designs than for LVT designs: for XOR designs multipumping reduces the number of replicated BRAMs to a greater extent and correspondingly reduces the XOR logic. The range of possibilities provided by these design permutations is significant: for example, for a 4096-entry 4W/8R memory, the design options range from 271MHz using 23,828 equivalent ALMs all the way to 104MHz using 4,875 equivalent ALMs.

For Figure 11(c) for 8W/16R memories we see a similarly-impressive range of design possibilities. Note that we did not include the fully-multipumped implementation for this design as its resulting Fmax is unusably low (i.e., less than 50MHz).

Navigating the Design Space As in the previous section, it is important to have a summary view of the best designs for given constraints. For the multipumped designs, namely the 2x multipumped 4W/8R memories and the 2x and 4x multipumped 8W/16R memories, the result is straightforward: to minimize delay (i.e., maximize Fmax) or to minimize BRAM usage, use the LVT designs; to minimize ALM usage, use the XOR designs.

7. RELATED WORK

Most prior work on multi-ported memories for FPGAs focuses on register files for soft-processors. Simple replication provides the 1W/2R register file required to support a three-operand ISA [1, 5, 6, 10, 15]. Jones *et al.* [7] implement a VLIW soft processor with a multi-ported register file implemented entirely in logic elements, limiting the operating frequency. Saghir *et al.* [12, 13] also implement a multi-ported register file for a VLIW soft-processor, but use replication and banking of BRAMs; however, their compiler must schedule register accesses to avoid conflicting reads and writes. Manjikian aggressively multipumps memories by performing reads and writes on consecutive rising and falling clock edges within a processor cycle [9]—unfortunately, this design forces a multiple-phase clock on the entire system. More recently, Anjam *et al.* [3] successfully use a LVT-based register file for their reconfigurable

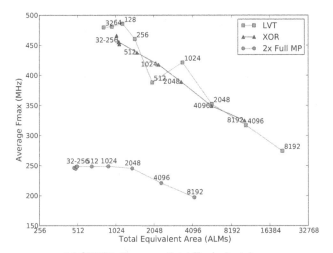

(a) 2W/4R: Fmax vs. Total Equivalent Area

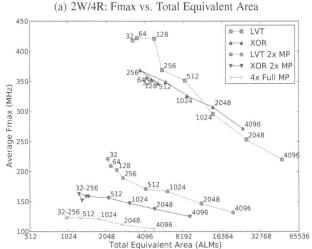

(b) 4W/8R: Fmax vs. Total Equivalent Area

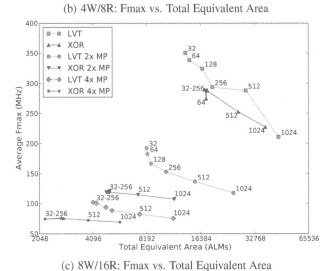

(c) 8W/16R: Fmax vs. Total Equivalent Area

Figure 11: Fmax vs TEA for LVT and XOR implementations of 2W/4R, 4W/8R, and 8W/16R memories of increasing depth, including multipumped designs.

VLIW soft-processor and add one more addressing bit internally to enable splitting a 4W/8R register file into two independent 2W/4R instances. Later work by Anjam *et al.* [4] removes the need for the LVT by avoiding write bank conflicts via compile-time

register renaming, but this solution requires more registers than are architecturally visible.

8. CONCLUSIONS

In this paper we introduced an approach to implementing multi-ported memories—composed of the two-ported block RAMs provided on FPGAs—that exploits the properties of the XOR operation to eliminate both the output multiplexing and logic-based lookup table required by the best prior approach, the Live Value Table (LVT) [8]. Targeting an Altera Stratix IV FPGA, we compared over 100 designs that implement both the LVT and XOR approaches and span a broad range of memory depths and numbers of ports. We found that:

- using forwarding logic improves the maximum speed of LVT designs;

- for both XOR and LVT designs, multipumping the write ports instead of the read ports yields a greater reduction in area;

- the XOR designs use far less logic but more block RAMs than the LVT designs, demonstrating a resource diversity between the two designs that makes them each desirable for different use-cases;

- for shallower designs the LVT approach is generally faster or equally-fast as the corresponding XOR designs, while the XOR designs with more than 64-entries are smaller, and with more than 512 entries are also faster than the corresponding LVT designs;

- exploiting multipumping greatly expands the possible design space with a large range of area and speed trade-offs;

- and multipumped XOR designs are significantly smaller but slower than their LVT counterparts.

To summarize, both the LVT and XOR approaches are valuable and useful in different situations, depending on the constraints and resource utilization (block RAMs vs logic) of the enclosing design. Designers can use the results of this work as a guide when choosing an appropriate design to implement a multi-ported memory.

9. FUTURE WORK

The results of this paper suggest several compelling avenues for future work.

Port to Xilinx FPGA Devices Although the design principles of XOR and LVT memories are generic and applicable to any FPGA device with block RAMs, we have only measured their frequencies and resource usage on Altera FPGAs. These CAD results and hence the trade-off space could be different for Xilinx or other FPGA devices.

Pure Multi-Pumping inside XOR Memory We could reduce the number of BRAM columns and thus the amount of XOR logic required for the XOR design by using the BRAMs in *True-Dual-Port* mode (where each port can perform either a read or a write each cycle), and then multipumping them to appear as a 1W/2R memory each—although True-Dual-Port mode suffers a significant reduction in clock frequency.

Different Ratios of Read/Write Ports For this paper we focused on memories with twice as many read ports than write ports, which are common for many applications such as processor register files. Memories with other ratios of read/write ports should be studied.

Stalling Designs For this paper we also focused on memories that do not stall, meaning that all read and write ports complete their requests every cycle. Designs that trade this restriction for area savings or speed, for example by having reads that may take multiple cycles to service, would also be interesting to investigate.

Power Analysis Compared to XOR memories, LVT memories (i) use fewer BRAMs but more logic, and (ii) access BRAMs with different read/write patterns; hence it would be interesting to determine how the dynamic power power consumption of the two approaches compares.

10. ACKNOWLEDGMENTS

The authors thank Altera and NSERC for financial support. We also thank Vaughn Betz, Jonathan Rose, and Ketan Padalia for help with tuning Quartus and our test harness, and the anonymous reviewers for constructive feedback.

11. REFERENCES

[1] Nios II Processor Reference Handbook. http://www.altera.com/literature/hb/nios2/n2cpu_nii5v1.pdf, March 2009. Version 9.0, Accessed Sept. 2009.

[2] DC and Switching Characteristics for Stratix IV Devices. http://www.altera.com/literature/hb/stratix-iv/stx4_siv54001.pdf, June 2011. Version 5.1, Accessed Aug. 2011.

[3] F. Anjam, M. Nadeem, and S. Wong. A vliw softcore processor with dynamically adjustable issue-slots. In *Field-Programmable Technology (FPT), 2010 International Conference on*, pages 393 –398, dec. 2010.

[4] F. Anjam, S. Wong, and F. Nadeem. A multiported register file with register renaming for configurable softcore vliw processors. In *Field-Programmable Technology (FPT), 2010 International Conference on*, pages 403 –408, dec. 2010.

[5] R. Carli. Flexible MIPS Soft Processor Architecture. Technical report, Massachusetts Institute of Technology, Computer Science and Artificial Intelligence Laboratory, June 2008.

[6] B. Fort, D. Capalija, Z. Vranesic, and S. Brown. A Multithreaded Soft Processor for SoPC Area Reduction. In *IEEE Symposium on Field-Programmable Custom Computing Machines*, pages 131–142, April 2006.

[7] A. K. Jones, R. Hoare, D. Kusic, J. Fazekas, and J. Foster. An FPGA-based VLIW processor with custom hardware execution. In *International Symposium on Field-Programmable Gate Arrays*, 2005.

[8] C. E. LaForest and J. G. Steffan. Efficient Multi-ported Memories for FPGAs. In *Proceedings of the 18th annual ACM/SIGDA international symposium on Field programmable gate arrays*, FPGA '10, pages 41–50, New York, NY, USA, 2010. ACM.

[9] N. Manjikian. Design Issues for Prototype Implementation of a Pipelined Superscalar Processor in Programmable Logic. In *PACRIM 2003: IEEE Pacific Rim Conference on Communications, Computers and Signal Processing*, volume 1, pages 155–158 vol.1, Aug. 2003.

[10] R. Moussali, N. Ghanem, and M. A. R. Saghir. Supporting multithreading in configurable soft processor cores. In *CASES '07: Proceedings of the 2007 international conference on Compilers, Architecture, and Synthesis for Embedded Systems*, pages 155–159, New York, NY, USA, 2007. ACM.

[11] D. A. Patterson, G. Gibson, and R. H. Katz. A case for redundant arrays of inexpensive disks (raid). In *Proceedings of the 1988 ACM SIGMOD international conference on Management of data*, 1988.

[12] M. Saghir and R. Naous. A Configurable Multi-ported Register File Architecture for Soft Processor Cores. In *ARC 2007: Proceedings of the 2007 International Workshop on Applied Reconfigurable Computing*, pages 14–25. Springer-Verlag, March 2007.

[13] M. A. R. Saghir, M. El-Majzoub, and P. Akl. Datapath and ISA Customization for Soft VLIW Processors. In *ReConFig 2006: IEEE International Conference on Reconfigurable Computing and FPGAs*, pages 1–10, Sept. 2006.

[14] H. Wong, V. Betz, and J. Rose. Comparing fpga vs. custom cmos and the impact on processor microarchitecture. In *Proceedings of the 19th ACM/SIGDA international symposium on Field programmable gate arrays*, FPGA '11, pages 5–14, New York, NY, USA, 2011. ACM.

[15] P. Yiannacouras, J. G. Steffan, and J. Rose. Application-specific customization of soft processor microarchitecture. In *FPGA '06: Proceedings of the 2006 ACM/SIGDA 14th international symposium on Field Programmable Gate Arrays*, pages 201–210, New York, NY, USA, 2006. ACM.

Octavo: An FPGA-Centric Processor Family

Charles Eric LaForest and J. Gregory Steffan
Department of Electrical and Computer Engineering
University of Toronto, Canada
{laforest,steffan}@eecg.toronto.edu

ABSTRACT

Overlay processor architectures allow FPGAs to be programmed by non-experts using software, but prior designs have mainly been based on the architecture of their ASIC predecessors. In this paper we develop a new processor architecture that from the beginning accounts for and exploits the predefined widths, depths, maximum operating frequencies, and other discretizations and limits of the underlying FPGA components. The result is Octavo, a ten-pipeline-stage eight-threaded processor that operates at the block RAM maximum of 550MHz on a Stratix IV FPGA. Octavo is highly parameterized, allowing us to explore trade-offs in datapath and memory width, memory depth, and number of supported thread contexts.

Categories and Subject Descriptors

C.1.3 [**Processor Architecture**]: Other Architecture Styles—*Adaptable Architectures*; C.4 [**Performance of Systems**]: Measurement Techniques, Design Studies

General Terms

Design Performance Measurement

Keywords

FPGA, soft processor, multithreading, microarchitecture

1. INTRODUCTION

Making FPGAs easier to program for non-experts is a challenge of increasing interest and importance. One approach is to enable FPGAs to be programmed using software via overlay architectures, for example conventional soft processors such as Altera's NIOS and Xilinx's Microblaze, or more aggressive designs such as soft vector processors [5, 18, 19]. Prior soft processor designs have mainly inherited the architecture of their ASIC-based predecessors with some optimization to better fit the underlying FPGA. However, FPGAs provide a much different substrate than raw transistors, including lookup tables (LUTs), block RAMs (BRAMs), multipliers/DSPs, and various routing resources—all of which have predefined widths, depths, maximum operating frequencies, and

other discretizations and limits [1]. The existence of these artifacts and their characteristics suggests that an FPGA-centric processor architecture, one that is built from the "ground-up" with FPGA capabilities in mind, will differ from a conventional architecture in compelling ways, mainly by using the FPGA resources more efficiently.

1.1 How do FPGAs Want to Compute?

In this work we ask the fundamental question: *How do FPGAs want to compute?* A more exact (but less memorable) phrasing of this question is: *What processor architecture best fits the underlying structures and discretizations of an FPGA?* This question alone is still too broad for the scope of a single research paper, so we narrow our investigation by striving for the following goals for a processor design.

1. To support a highly-threaded data-parallel programming model, similar to OpenCL.
2. To run at the maximum operating frequency allowed by the particular FPGA resources used (e.g.: BRAMs).
3. To have high performance—i.e, not only high-frequency but also reasonable instruction count and processor-cycles-per-instruction.
4. To never stall due to hazards (such as control or data dependences).
5. To strive for simplicity and minimalism, rather than inherit all of the features of an existing processor design/ISA.
6. To match underlying FPGA structures; for example, to discover the most effective width for data elements for both datapaths and storage, as opposed to defaulting to the conventional 32-bit width.

This paper describes the process and results of developing an FPGA-based processor while striving for these goals.

1.2 Octavo

As a starting point, we show the simplest processor design we could imagine in Figure 1, which is composed of at least one multi-ported memory connected to an ALU, supplying its operands and control and receiving its results. We argue that separate data cache and register file storage is unnecessary: on an FPGA both are inevitably implemented using the same BRAMs. We eliminate separate memory and registers, reducing the data and instruction memories and the register file into a single entity directly addressed by the instruction operand fields. For this reason our final architecture is indeed not unlike the simple one pictured, having only a single logical storage component (similar to the scratchpad memory proposed by Chou *et al.* [5]). We demonstrate that this single logical memory eliminates the need for immediate operands and

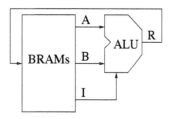

Figure 1: An overview of the architecture of Octavo, composed of a Memory ($BRAMs$) providing operands (A and B) and instructions (I) to an ALU which writes back its results (R) to the same Memory.

load/store operations, but for now requires writing to instruction operands to synthesize indirect memory accesses.

Via the technique of *self-loop characterization*, where we connect a component's outputs to its inputs to take into account the FPGA interconnect, we determine for memories and ALUs the pipelining required to achieve the highest possible operating frequency. This leads us to an overall eight-stage processor design that operates at up to 550MHz on a Stratix IV FPGA, limited by the maximum operating frequency of the BRAMs. To meet the goals of avoiding stalls and maximizing efficiency, we multithread the processor such that an instruction from a different thread resides in each pipeline stage [7, 8, 12, 14, 15], so that all stages are independent with no control or data hazards or result forwarding between them.

We name our processor architecture *Octavo*[1], for nominally having eight thread contexts. However, Octavo is really a processor *family* since it is highly parameterizable in terms of its datapath and memory width, memory depth, and number of supported thread contexts. This parameterization allows us to search for optimal configurations that maximize resource utilization and clock frequency.

1.3 Related Work

Many prior FPGA-based soft processors designs have been proposed, although these have typically inherited the architectures of their ASIC predecessors, and none have approached the clock frequency achieved by Octavo. Examples include soft uniprocessors [3, 17], multithreaded soft processors [6–8, 12, 14, 15], soft VLIW processors [4, 10, 16], and soft vector processors [5, 18, 19]. Jan Gray has studied the optimization of processors specifically for FPGAs [9], where synthesis and technology mapping tricks are applied to all aspects of the design of a processor from the instruction set to the architecture.

1.4 Contributions

In future work we plan to extend Octavo to support SIMD/vector datapaths, multicore interconnection, connection to an OpenCL framework (for its abundance of thread and data parallelism), and evaluation of full applications. In this paper we focus on the architecture of a single Octavo core and provide the following four contributions:

1. we present the design process leading to Octavo, an 8-stage multithreaded processor family that operates at up to 550MHz on a Stratix IV FPGA;

2. we demonstrate the utility of *self-loop characterization* for reasoning about the pipelining requirements of processor components on FPGAs;

3. we present a design for a fast multiplier, consisting of two half-pumped DSP blocks, which overcomes hardware timing and CAD limitations;

4. we present the design space of Octavo configurations of varying datapath and memory widths, memory depths, and number of pipeline stages.

2. EXPERIMENTAL FRAMEWORK

We evaluate Octavo and its components on Altera Stratix IV FPGAs, although we expect proportionate results on other FPGA devices given suitable tuning of the pipeline.

Test Harness We place our circuits inside a synthesis test harness designed to both: (i) register all inputs and outputs to ensure an accurate timing analysis, and (ii) to reduce the number of I/O pins to a minimum as larger circuits will not otherwise fit on the FPGA. The test harness also avoids any loss of circuitry caused by I/O optimization. Shift registers expand single-pin inputs, while registered AND-reducers compact word-wide signals to a single output pin.

Synthesis We use Altera's Quartus 10.1 to target a Stratix IV EP4SE230F29C2 FPGA device of the highest available speed grade. For maximum portability, we implement the design in generic Verilog-2001, with some LPM[2] components. We configure the synthesis process to favor speed over area and enable all relevant optimizations. To confirm the intrinsic performance of a circuit without interference from optimizations—such as register retiming, which can blur the distinction between the circuit under test and the test harness—we constrain a circuit to its own logical design partition and restrict its placement to within a single rectangular area (*LogicLock* area) containing only the circuit under test, excluding the test harness. Any test harness circuitry remains spatially and logically separate from the actual circuit under test.

Place and Route We configure the place and route process to exert maximal effort at fitting with only two constraints: (i) to avoid using I/O pin registers to prevent artificially long paths that would affect the clock frequency, and (ii) to set the target clock frequency to 550MHz, which is the maximum clock frequency specified for M9K BRAMs. Setting a target frequency higher than 550MHz does not improve results and could in fact degrade them: for example, a slower derived clock would aim towards an unnecessarily high target frequency, causing competition for fast routing paths.

Frequency We report the unrestricted maximum operating frequency (F_{max}) by averaging the results of ten place and route runs, each starting with a different random seed for initial placement. We construct the average from the worst-case F_{max} reports over the range of die temperatures between 0 to 85° at a supply voltage of 900mV. Note that minimum clock pulse width limitations in the BRAMs restrict the actual operating frequency to 550MHz, regardless of actual propagation delay. Reported F_{max} in excess of this limit indicates timing slack available to the designer.

Area Area does not vary significantly between place and route runs, so we report the first computed result. We measure area as the count of Adaptive Lookup Tables (ALUTs) in use. We also measure the area efficiency as the percentage of ALUTs actually in use relative to the total number of ALUTs within the rectangular *LogicLock* area which contains the circuit under test, including any BRAMs or DSP Blocks.

---

[1]An *octavo* is a booklet made from a printed page folded three times to produce eight leaves (16 pages).

[2]*Library of Parametrized Modules* (LPM) is used to describe hardware that is too complex to infer automatically from behavioral code.

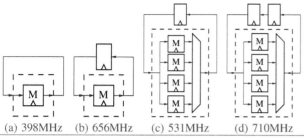

(a) 398MHz   (b) 656MHz   (c) 531MHz   (d) 710MHz

Figure 2: Self-loop characterization of memories reveal that different numbers of pipeline stages absorb the propagation delays depending on their internal configurations. Each of (a)-(d) lists the theoretical maximum frequency of the design, although the BRAM limit of 550MHz is the true limit.

3. STORAGE ARCHITECTURE

We begin our exploration of FPGA-centric architecture by focusing on storage. Since modern mid/high-end FPGAs provide hard block RAMs (BRAMs) as part of the substrate, we assume that the storage system for our architecture will be composed of BRAMs. Since we are striving for a processor design of maximal frequency, we want to know how the inclusion of BRAMs will impact the critical paths of our design. As already introduced, we use the method of *self-loop characterization*, where we simply connect the output of a component under study to its input, to isolate (i) operating frequency limitations and (ii) the impact of additional pipeline stages.

Figure 2 shows four 32-bit-wide memory configurations: 256-word memories using one BRAM with one (2(a)) and two (2(b)) pipeline stages, and 1024-word memories using four BRAMs with two (2(c)) and three (2(d)) pipeline stages. The result for a single BRAM (2(a)) is surprising: without additional pipelining, the F_{max} reaches only 398MHz out of a maximum of 550MHz (limited by the minimum-clock-pulse-width restrictions of the BRAM). This delay stems from a lack of direct connection between BRAMs and the surrounding logic fabric, forcing the use of global routing resources. However, two pipeline stages (2(b)) increases F_{max} to 656MHz, and four pipeline stages (not shown) absorb nearly all delay and increase the achievable F_{max} up to 773MHz. Increasing the memory depth to 1024 words (2(c)) requires 4 BRAMs, additional routing, and some multiplexing logic—and reduces F_{max} to 531MHz. Adding a third pipeline stage (2(d)) absorbs the additional delay and increases F_{max} to 710MHz.

These results indicate that pipelining provides significant timing slack for more complex memory designs. In Octavo we exploit this slack to create a memory unit that collapses the usual register/cache/memory hierarchy into a single entity, maps all I/O as memory operations, and still operates at more than 550MHz. To avoid costly stalls on memory accesses, we organize on-chip memory as a single scratchpad [5] such that access to any external memory must be managed explicitly by software. Furthermore, since an FPGA-based processor typically implements both caches and register files out of BRAMs, we pursue the simplification of merging caches and register file into a single memory entity and address space. Hence Octavo can be viewed as either being (i) registerless, since there is only one memory entity for storage, or (ii) registers-only, since there are no load or store instructions, only operations that directly address the single operand storage.

4. INSTRUCTION SET ARCHITECTURE

The single-storage-unit architecture decided in the previous section led to Octavo's instruction set architecture (ISA) having no

Table 1: Octavo's Instruction Word Format.

Size:	4 bits	a bits	a bits	a bits
Field:	Opcode (OP)	Destination (D)	Source (A)	Source (B)

Table 2: Octavo's Instruction Set and Opcode Encoding.

Mnemonic	Opcode	Action
		Logic Unit
XOR	0000	D = A XOR B
AND	0001	D = A AND B
OR	0010	D = A OR B
SRL	0011	D = A » 1 (zero ext.)
SRA	0100	D = A » 1 (sign ext.)
ADD	0101	D = A + B
SUB	0110	D = A - B
—	0111	*(Unused, for expansion)*
		Multiplier
MLO	1000	D = A * B (Lower Word)
MHI	1001	D = A * B (Upper Word)
		Controller
JMP	1010	PC = D
JZE	1011	if (A == 0) PC = D
JNZ	1100	if (A != 0) PC = D
JPO	1101	if (A >= 0) PC = D
JNE	1110	if (A < 0) PC = D
—	1111	*(Unused, for expansion)*

loads or stores: each operand can address any location in the memory. Immediate values are implemented by placing them in memory and addressing them. Subroutine calls and indirect memory addressing are implemented by synthesizing code, explained in detail later in Section 9. Despite its frugality, we believe that the Octavo ISA can emulate the MIPS ISA.

Table 1 describes Octavo's instruction word format. The four most-significant bits hold the opcode, and the remaining bits encode two source operands (A and B) and a destination operand (D). The operands are all the same size (a address bits), and the width of the operands dictates the amount of memory that Octavo can access. For example, a 36-bit instruction word has a 4-bit opcode, three 10-bit operand fields, and 2 bits unused—allowing for a memory space of 2^{10} (1024) words. Table 2 shows Octavo's instruction set and opcode encoding, with ten opcodes allocated to ALU instructions and the remaining six allocated to control instructions. The Logic Unit opcodes are chosen carefully so that they can be broken into sub-opcodes to minimize decoding in the ALU implementation.

5. MEMORY

Having decided the storage architecture and ISA for Octavo, we next describe the design and implementation of Octavo's memory unit. In particular, we describe the implementation of external I/O, and the composition of the different memory unit components.

I/O Support Having only a single memory/storage and no separate register file eliminates the notion of loads and stores, which normally implement memory-mapped I/O mechanisms. Since significant timing slack exists between the possible and actual F_{max} of FPGA BRAMs, we can use this slack to memory-map I/O mechanisms without impacting our high clock frequency. We map word-wide I/O lines to the uppermost memory locations (typically 2 to 8 locations), making them appear like ordinary memory and thus accessible like any operand. We interpose the I/O ports in

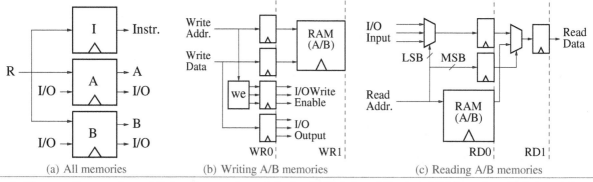

| (a) All memories | (b) Writing A/B memories | (c) Reading A/B memories |

Figure 3: The overall connections of Octavo's memories and the implementation of the A and B Memories with integrated memory-mapped word-wide I/O ports. The RAM component is implemented using BRAMs. Note that both A/B read and writes complete in two cycles, but overlap only for one cycle at RD0/WR1. The I Memory has no I/O and thus reads and writes in a single cycle.

front of the RAM read and write ports: the I/O read ports override the RAM read if the read address is in the I/O address range, while the I/O write ports pass through the write address and data to the RAM. This architecture provides interesting possibilities for future multicore arrangements of Octavo: any instruction can now perform up to two I/O reads and one I/O write simultaneously; also, an instruction can write its result directly to an I/O port and another instruction in another CPU can directly read it as an operand. Similarly, having I/O in instruction memory could enable the PC to point to I/O to execute an external stream of instructions sent from another CPU (although we do not yet support this feature).

Implementation Figure 3 shows the connections of Octavo's memory units and details the construction of the A and B Memories. Each memory behaves as a simple dual-port (one read and one write) memory, receiving a common write value R (the ALU's result), but keeping separate read and I/O ports. The I Memory contains only BRAMs, while the A and B Memories additionally integrate a number of memory-mapped word-wide I/O ports (typically two or four). For the A and B memories, reads or writes take 2 cycles each but overlap for only 1 at $RD0/WR1$. A write (Figure 3(b)) spends its first cycle registering the address and data to RAM, activating one of the I/O write port write-enable lines based on the write address, and registering the write data to all I/O write ports. The data write to the RAM occurs during the second cycle.[3] A read (Figure 3(c)) sends its address to the RAM during the first cycle and simultaneously selects an I/O read port based on the Least-Significant Bits (LSB) of the address. Based on the remaining Most-Significant Bits (MSB) of the address, the second read cycle returns either the data from the RAM or from the selected I/O read port. Our experiments showed that we can add up to about eight I/O ports per RAM read/write port pair before the average operating speed drops below 550MHz.

6. ALU

In this section we describe the development and design of Octavo's ALU components, including the Multiplier, the Adder/Subtractor, the Logic Unit, and their combination to form the ALU.

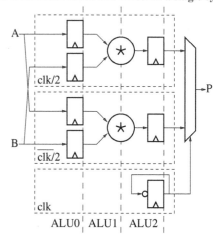

Figure 4: A detailed view of the Multiplier unit, which overcomes the minimum clock pulse width limit of a single multiplier by operating two word-wide multipliers on alternate edges of a half-rate clock $clk/2$, with the correct double-word product P selected by a single state bit driven by the system clock clk.

Multiplier Unit To support multiplication for a high-performance soft processor it is necessary to target the available DSP block multipliers. Although Stratix IV DSP blocks have a sufficiently-low propagation delay to meet our 550MHz target frequency, they have a minimum-clock-pulse-width limitation (similar to BRAMs) restricting their operating frequency to 480MHz for word-widths beyond 18 bits [4].

Figure 4 shows the internal structure of Octavo's Multiplier and our solution to the clocking limitation: we use two word-wide DSP block multipliers[5] in alternation on a synchronous half-rate clock

---

[3]We implemented the RAM using Quartus' auto-generated BRAM write-forwarding circuitry, which immediately forwards the write data to the read port if the addresses match. This configuration yields a higher F_{max} since there is a frequency cost to the Stratix IV implementation of BRAMs set to return old data during simultaneous read/write of the same location [1]. However, since pipelining delays the write to a BRAM by one cycle, a coincident read will return the data currently contained in the BRAM instead of the data being written.

[4]For widths ≤ 18 bits, it might be possible to implement the multiplier with a single DSP block, but current CAD issues prevent getting results consistent with the published specifications [1] for high-frequency implementations.

[5]We implement each multiplier using an LPM instance generated by the Quartus MegaWizard utility. Although the Altera DSP blocks have input, intermediate, and output registers, a designer can only specify the desired number of pipeline stages that begin at the input to the DSP block—hence we cannot specify to use only the input and output registers to absorb the delay of the entire DSP block. We bypass this limitation by instantiating a one-stage-pipelined multiplier and feeding its output into external registers. Later register-retiming optimizations eventually place these external registers into the built-in output registers of the DSP

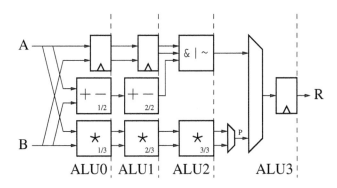

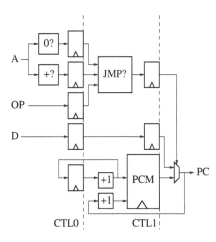

Figure 5: Organization of Octavo's ALU, containing an Adder/Subtractor (+−), a Logic Unit (& | ~), and a Multiplier (∗). The Logic Unit also multiplexes between its results and those of the Adder/Subtractor.

($clk/2$), such that we can perform two independent word-wide multiplications, staggered but in parallel, and produce one double-word product every cycle. In detail, the operands A and B are de-multiplexed into the two half-rate datapaths on alternate edges of the half-rate clock. A single state bit driven by the system clock (clk) selects the correct double-word product (P) at each cycle.

Adder/Subtractor and Logic Unit We also carefully and thoroughly studied adder/subtractors and logic units while building Octavo, again using the method of self-loop characterization described in Section 3. We experimentally found that an unpipelined 32-bit ripple-carry adder/subtractor can reach 506MHz, and that adding 4 pipeline stages increases F_{max} up to 730MHz. An unpipelined carry-select implementation only reaches 509MHz due to the additional multiplexing delay, but requires only two stages to reach 766MHz. Due to the 550MHz limitation imposed by BRAMs, a simple two-stage ripple-carry adder reaching 600MHz is sufficient.

The Logic Unit (& | ~) performs bit-wise XOR, AND, OR, SRL, and SRA operations (Table 2). It also acts as a pass-through for the result of the Adder/Subtractor, which avoids an explicit multiplexer and allows us to separate and control the implementation of the Adder/Subtractor from that of the Logic Unit. The Logic Unit efficiently maps to a single ALUT per word bit: 3 bits for the opcode, plus one bit from the Adder/Subtractor result, and 2 bits for the A and B operands of the bit-wise operations, totaling 6 bits and naturally mapping to a single Stratix IV 6-LUT per output bit.

Combined ALU Design Figure 5 shows the block-level structure of the entire ALU, which combines the Multiplier, Adder/-Subtractor, and Logic Unit. All operations occur simultaneously during each cycle, with the correct result selected by the output multiplexer after four cycles of latency. We optimized each sub-component for speed, then added extra pipeline registers to balance the path lengths. We use the Logic Unit as a pipeline stage and multiplexer to reduce the delay and width of the final ALU result multiplexer. The combined ALU runs at an average of 595MHz for a width of 36 bits.

7. CONTROLLER

Figure 6 shows the design of the Octavo Controller. The Controller provides the current Program Counter (PC) value for each thread of execution and implements flow-control. A Program Counter Memory (PCM) holds the next value of the PC for each

block, yielding a two-stage pipelined multiplier with only input and output registers.

Figure 6: The Controller, which provides the Program Counter (PC) value for each thread of execution and implements flow-control. A Program Counter Memory (PCM) holds the next value of the PC for each thread of execution. Based on the opcode (OP) and the fetched value of operand A (if applicable), the controller may update the PC of a thread with the target address stored in the destination operand D.

thread of execution. We implement the PCM using one MLAB[6] instead of a BRAM, given a typically narrow PC (< 20 bits) and a relatively small number of threads (8 to 16)—this also helps improve the resource-diversity of Octavo and will ease its replication in future multicore designs. A simple incrementer and register pair perform round-robin reads of the PCM, selecting each thread in turn. At each cycle, the current PC of a thread is incremented by one and stored back into the PCM. The current PC is either the next consecutive value from the PCM, or a new jump destination address from the D instruction operand.

The decision to output a new PC in the case of a jump instruction is based on the instruction opcode OP and the fetched value of operand A. A two-cycle pipeline determines if the value of A is zero (0?) or positive (+?), and based on the opcode OP decides whether a jump in flow-control happens (JMP?)—i.e., outputs the new value of the PC from D, instead of the next consecutive value from the PCM. A Controller supporting 10-bit PCs for 8 threads can reach an average speed of 618MHz, though the MLAB implementing the PCM limits F_{max} to 600MHz.

8. COMPLETE OCTAVO HARDWARE

In this section we combine the units described in the previous three sections to build the complete Octavo datapath shown in Figure 7, composed of an instruction Memory (I), two data Memories (A and B), an ALU, and a Controller (CTL).

We begin by describing the Octavo pipeline from left to right. In Stage 0, Memory I is indexed by the current PC and provides the current instruction containing operand addresses D, A, and B, and the opcode OP. Stages 1-3 contain only registers and perform no computation. Their purpose is to separate the BRAMs of Memory I from those of Memories A/B by a suitable number of stages to maintain a 550MHz clock: as shown by the self-loop characterization in Section 3, we must separate groups of BRAMs with at least two stages—having only a single extra stage between the I and A/B memories would yield an F_{max} of only 495MHz for a 36-bit, 1024-word Octavo instance. We insert three stages

---

[6]Memory Logic Array Blocks (MLABs) are small (e.g., 32 bits wide by 20 words deep) memories found in Altera FPGAs.

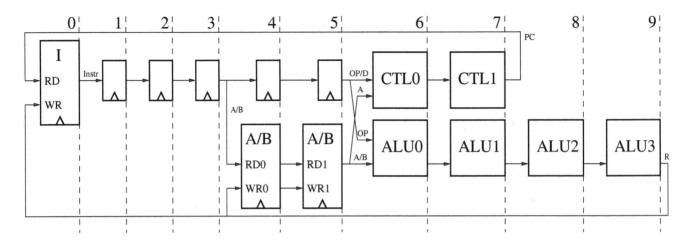

Figure 7: The complete Octavo system: an instruction Memory (I), two data Memories (A and B), an ALU, and a Controller (CTL).

to avoid having an odd total number of stages. Across stages 4 and 5, the A and B memories provide the source operands (of the same name). The ALU spans stages 6-9 and provides the result R which is written back at address D to both Memories A and B (across stages 4 and 5 again), as well as Memory I (at stage 0). The Controller (CTL) spans stages 6 and 7 and writes the new PC back to Memory I in stage 0. The controller contains the PC memory for all threads, and for each thread decides whether (i) to continue with the next consecutive PC value, or (ii) to branch to the new target address D.

There are three main hazards/loops in the Octavo pipeline. The first hazard exists in the control loop that spans stages 0-7 through the controller (CTL)—hence Octavo requires a minimum of eight independent threads to hide this dependence. The second hazard is the potential eight-cycle Read-After-Write (RAW) data hazard between consecutive instructions from the same thread: from operand reads in stages 4-5, through the ALU stages 6-9, and the write-back of the result R through stages 4-5 again (recall that writing memories A/B also takes two stages)—this dependence is also hidden by eight threads. The third hazard also begins at the operand reads in stages 4-5 and goes through the ALU in stages 6-9, but writes-back the result R to Memory I for the purpose of the instruction synthesis introduced in Section 4 and described in detail in the next section. This loop spans ten stages and is thus not covered by only eight threads. Rather than increase thread contexts beyond eight to tolerate this loop, we instead require a *delay slot* instruction between the synthesis of an instruction and its use.

9. OCTAVO SOFTWARE

As described in Section 4, the Octavo ISA supports only register-direct addressing, since all operands are simple memory addresses—hence the implementation of displacement, indirect, or indexed addressing requires two instructions: a first instruction reads the memory location containing the indirect address or the displacement/index, and stores it into the source or destination operand of a second instruction that performs the actual memory access using the modified operand address. The remainder of this section provides examples of indirection implemented using the Octavo ISA, including pointer dereference, arrays, and subroutine calls.

9.1 Pointer Dereference

The C code in Figure 8(a) performs an indirect memory access by dereferencing the pointer b and storing the final value into location a. In the MIPS ISA (Figure 8(b)), this code translates into

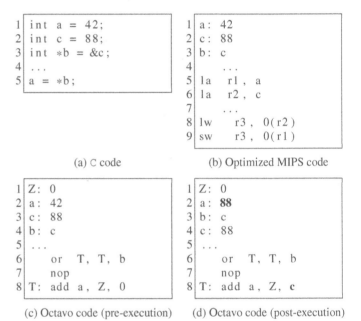

Figure 8: Pointer dereference example.

a pair of address loads (we use the common 'la' assembler macro for brevity) followed by a displacement addressing load/store pair. Since the value of b is known at compile time, we assume that the compiler optimizes-away the dereference and uses the address of c directly.

In the Octavo ISA we synthesize indirect addressing at run-time by placing the address stored in b into a source operand of a later instruction that stores into a the *contents* of the address taken from b. Without load/store operations, we instead use an ADD with "register zero" as one of the operands. Figure 8(c) shows the initial conditions of the Octavo code and begins with a memory location defined as "register zero" (Z) and others containing the same initialized variables (a, b, and c) as the C code. Line 6 contains an instruction that OR's a target instruction T (line 8) with the contents of b (line 3)—note that T's second source operand initially contains zero. A NOP or other independent instruction must exist between the generating instruction and its target due to the 1-cycle RAW hazard when writing to Memory I (Section 8) if executing less than 10 threads. Figure 8(d) shows the result of executing from line 6 onwards, that replaces the zero source operand in T with

```
1 int A[] = { 42, ...};
2 int B[] = { 23, ...};
3 int C[] = { 88, ...};
4 ...
5 *A = *B + *C;
6 A++;
7 B++;
8 C++;
```

(a) C code

```
1  A : 42
2  A': ...
3  B : 23
4  B': ...
5  C : 88
6  C': ...
7     ...
8  la   r1, A
9  la   r2, B
10 la   r3, C
11    ...
12 lw   r5, 0(r2)
13 lw   r6, 0(r3)
14 add  r4, r5, r6
15 sw   r4, 0(r1)
16 addi r1, r1, 1
17 addi r2, r2, 1
18 addi r3, r3, 1
```

(b) MIPS code

```
1  A : 42
2  A': ...
3  B : 23
4  B': ...
5  C : 88
6  C': ...
7  I :    0 1, 1, 1
8     ...
9  T : add A, B, C
10     add T, T, I
```

(c) Octavo code (pre-execution)

```
1  A : 111
2  A': ...
3  B : 23
4  B': ...
5  C : 88
6  C': ...
7  I :    0 1, 1, 1
8     ...
9  T : add A', B', C'
10     add T, T, I
```

(d) Octavo code (post-execution)

Figure 9: Array access example.

the contents of b, and later executes T with the modified operand, storing the contents of c into a. If the compiler knows the value of the pointer b, it can perform these steps at compile-time and synthesize the final instruction—avoiding the run-time overhead. To traverse a linked list or any other pointer-based structure, the target instruction T instead can update the pointer b itself.

9.2 Iterating over Arrays

Despite the apparent inefficiency of needing to synthesize code to perform indirect memory accesses, manipulating the operands of an instruction can also have advantages. For example, the C code in Figure 9(a) describes the core of a loop summing two arrays. Figure 9(b) shows a straightforward translation to MIPS assembly: the same letters as in the C code denote consecutive array locations. After a 3-instruction preamble to load the array addresses into registers $r1$, $r2$, and $r3$, the next four instructions (lines 12-15) load the B and C array element values, sum them, and store them back into the corresponding A element. The last three instructions increment the array pointers.

The equivalent Octavo assembly code in Figure 9(c) works in the same way, but using synthesized code: after directly performing the array element sum at T on line 9, we add 1 to each address operand using a word-wide value I on line 7. This increment value I contains the increment of each array pointer, each shifted to align with the corresponding address field, and a zero value aligned with the opcode field. Adding I to T yields the updated code for the next loop iteration in Figure 9(d). Compared to the MIPS code in

```
1  Z: 0
2  RET1: jmp X, 0, 0
3  RET2: jmp R, 0, 0
4
5  sub:
6        ...
7  E:    jmp X, 0, 0
8
9  caller:
10        ...
11        add E, Z, RET2
12        jmp sub, 0, 0
13 R:     ...
```

(a) pre-execution

```
1  Z: 0
2  RET1: jmp X, 0, 0
3  RET2: jmp R, 0, 0
4
5  sub:
6        ...
7  E:    jmp R, 0, 0
8
9  caller:
10        ...
11        add E, Z, RET2
12        jmp sub, 0, 0
13 R:     ...
```

(b) post-execution

Figure 10: Call/return example, in Octavo code.

Figure 9(b), Octavo requires only two instructions instead of seven to compute the same loop body.

Synthesized code does however increase the size of loop preambles. Octavo's loop preamble overhead could become significant with many short nested loops, but compiler optimizations such as loop coalescing would reduce it. Similarly, induction variable elimination would reduce the amount of synthesized code required for more complex array access patterns.

9.3 Synthesizing Subroutine Calls

Without call stack hardware support, Octavo must synthesize code to implement subroutine linkage using a method previously described by Knuth [11]. While somewhat awkward, having to synthesize CALL and RET instructions saves two scarce opcodes for other uses and enables conditional calls and returns at no extra cost.

Figure 10(a) shows a synthesized CALL and RET pair example. Lines 2 and 3 contain return jumps $RET1$ and $RET2$ that act as the "RET" for specific "CALLs" to sub (lines 5-7). These return jumps get placed by callers at the exit point E of sub, that currently contains a copy of $RET1$ placed there by a previous caller. Before jumping to sub at line 12, the $caller$ will change sub's return jump target from X to R, the return point in the $caller$ at line 13. Figure 10(b) shows the updated code after line 11 executes, with the exit point E updated to return to R. Using JNZ, JZE, JPO, or JNE instead of JMP at line 3 implements a conditional subroutine return. Doing the same at line 12 implements a conditional subroutine call.

This subroutine linkage scheme does not allow re-entrancy: threads cannot intersect in the call graph, including with themselves (i.e., recursive calls must be converted to iterative ones). The compiler must create private copies of the subroutine in such cases.

10. SPEED AND AREA

In this section we examine many varying instances of Octavo as instantiated on a Stratix IV EP4SE230F29C2 device. In particular we measure maximum operating frequency (F_{max}), area usage, and area density over a range of configurations, varying word width, memory depth, and number of pipeline stages. We perform these experiments to confirm that Octavo achieves our stated goals for a processor design (Section 1.1) over a wide range of configurations.

10.1 Maximum Operating Frequency

Our first experiments address whether Octavo's high F_{max} will hold for non-trivial and unconventional word widths and increasing

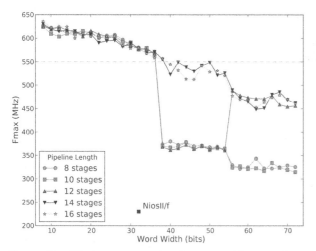

Figure 11: Maximum operating frequency F_{max} vs Octavo word widths ranging from 8 to 72 bits, for Octavo instances with 8 to 16 pipeline stages. In all cases, we limit the memory depth to 256 words to avoid any effect on F_{max}.

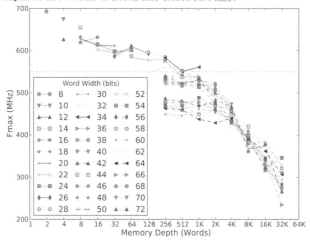

Figure 12: Maximum operating frequency (F_{max}) for a *16-stage* Octavo design over addressable memory depths ranging from 2 to 32,768 words, for word widths from 8 to 72 bits.

memory depths. We find that, over a range of word widths from 8 to 72 bits, F_{max} remains high and degrades smoothly.

Figure 11 shows the maximum operating frequency F_{max} of Octavo for word widths ranging from 8 to 72 bits, and for Octavo instances with 8 to 16 pipeline stages. The dashed line indicates the 550MHz F_{max} upper limit imposed by the BRAMs. As a rough comparison we plot the 32-bit NiosII/f soft processor, reported to be 230MHz for our target FPGA [2]. For this experiment we limited memory depth to a maximum of 256 words so that each memory fits into a single BRAM, avoiding any effect on F_{max} from memory size and layout.

For all pipeline depths, F_{max} degrades slowly from about 625MHz down to 565MHz when varying word width from 8 to 36 bits. For 12 to 16 pipeline stages F_{max} decreases only 28% over a 9x increase in width from 8 to 72 bits, and still reaches just over 450MHz at 72 bits width. Word widths beyond 36 bits exceed the native capacity of the DSP blocks, requiring additional adders (implemented with ALUTs) to tie together multiple DSP blocks into wider multipliers. Adding more pipeline stages to the Multiplier absorbs the delay of these extra adders but increases

total pipeline depth. Increasing pipeline depth by 4 stages up to 12 absorbs the delay of these extra adders.

Unfortunately a CAD anomaly occurs for widths between 38 and 54 bits (inclusive), where Quartus 10.1 cannot fully map the Multiplier onto the DSP blocks, forcing the use of yet more adders implemented in FPGA logic. Increasing the pipelining to 14 stages, again by adding stages in the Multiplier, overcomes the CAD anomaly. Increasing the pipelining to 16 stages has no further effect on Octavo, whose critical path lies inside the Multiplier. The CAD anomaly affects Octavo in two ways: the affected word-widths must pipeline the Multiplier further than normally necessary to overcome the extra adder delay, and also show a discontinuously higher F_{max} than the wider, unaffected word-widths (56 to 72 bits), regardless of the number of pipeline stages. Unfortunately this CAD anomaly hides the actual behavior of Octavo at the interesting transition point at widths of 36 to 38 bits, where the native width of both BRAMs and DSP blocks is exceeded.

Figure 12 shows the maximum operating frequency (F_{max}) for a *16-stage* Octavo design over addressable memory depths ranging from 2 to 32,768 words and plotted for word widths from 8 to 72 bits. We also mark the 550MHz actual F_{max} upper limit imposed by the BRAMs. We use 16 stages instead of 8 to avoid the drop in performance caused by the CAD anomaly.

The previously observed discontinuous F_{max} drop in Figure 11 for Octavo instances with widths of 56 to 72 bits is visible here in the cluster of dashed and dotted lines lying below 500MHz for depths of 256 to 4096 words. Similarly, the cluster of dashed lines above 500MHz spanning 256 to 4096 words depth contains the word widths (38 to 54 bits) affected by the CAD anomaly.

A memory requires twice as many BRAMs to implement widths exceeding the native BRAM maximum width of 36 bits. Unfortunately, the CAD anomaly masks the initial effect on F_{max} of doubling the number of BRAMs for the same depth when exceeding a word width of 36 bits.

For depths up to 256 words, which all fit in a single BRAM, and widths below where the CAD anomaly manifests (8 to 36 bits), F_{max} decreases from 692MHz down to 575MHz, a 16.9% decrease over a 4.5x increase in word width and 128x increase in memory depth (2 to 256 words). For depths greater than 256 words, if we take as example the narrowest width (50 bits) which can address up to 32,768 words, F_{max} decreases 49.8% over a 64x increase in depth (512 to 32,768 words). The decrease changes little as width increases: 42.1% at 72 bits width over the same memory depths. Overall, an increase in memory depth affects F_{max} much more than an increase in width, with the effect becoming noticeable past 1024 words of depth.

Summary We summarize with two main observations: (i) widths > 36 bits require additional logic and pipelining, and (ii) a CAD anomaly forces longer pipelines and hides the actual curves for less than 14 pipeline stages. We also found that at least 12 pipeline stages are necessary for widths greater than 56 bits, modulo the CAD anomaly, and that memory depth has a greater effect on F_{max} than word width, becoming significant beyond 1024 words.

10.2   Area Usage

Our next experiments tests if Octavo's area scales practically as word width and memory depth increase. Figure 13 shows the area used in ALUTs, excluding BRAMs and DSP blocks, over word widths ranging from 8 to 72 bits, for an *8-stage* Octavo design. Where possible, for each width, we plot multiple points each representing an addressable memory depth ranging from 2 to 32,768 words. We also mark the reported 1,110 ALUT area usage of the 32-bit NiosII/f soft processor on the same FPGA family [2].

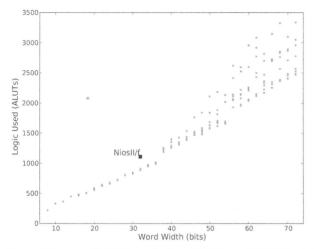

Figure 13: Area (ALUTs, excluding BRAMs and DSP blocks) vs word widths ranging from 8 to 72 bits, for an 8-stage Octavo design. For each width we plot the points representing the possible addressable memory depths, maximally ranging from 2 to 32,768 words.

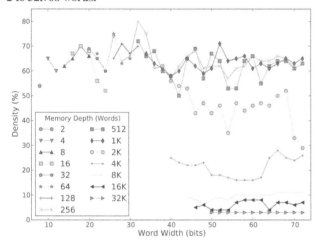

Figure 14: Density, measured as the percentage of ALUTs in actual use within the rectangular area containing an 8-stage Octavo instance, over word widths ranging from 8 to 72 bits and plotted for each addressable memory depth ranging from 2 to 32,768 words. BRAMs and DSP blocks do not count towards ALUT count.

For small memories having less than 256 words, the area used varies roughly linearly, increasing 11.4x in area over a 9x increase in width. The CAD anomaly causes two small discontinuous increases in the ALUT usage: +24.2% while increasing from 36 to 38 bits width, and +16.5% from 54 to 56 bits, both cases for a memory depth of 256 words. Increasing memory depth has little effect on the amount of logic used: at a width of 72 bits, the area increases from 2478 to 3339 ALUTs (+37.5%) when increasing the memory depth from 256 to 32,768 (128x).

Summary We found that area varies roughly linearly with word width, varies little with memory depth, and is also affected by the CAD anomaly.

10.3 Density

Our final experiments seek to find if some Octavo configurations are "denser" than others, leaving fewer ALUTs, BRAMs, or DSP blocks unused within their rectangular area. Figure 14 shows the

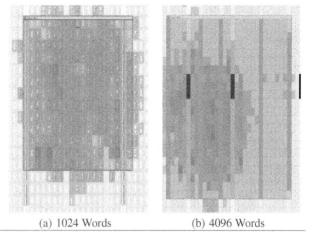

(a) 1024 Words   (b) 4096 Words

Figure 15: Physical layout of an 8-stage, 72-bit wide Octavo instance with (a) 1024 and (b) 4096 memory words. The large shaded rectangular area contains only the ALUTs used by Octavo, any outside ALUTs belong to a test harness and do not count; the darker columns contain the BRAMs implementing the Memory; the pale columns contain DSP blocks implementing the Multiplier and are part of Octavo despite protruding below the rectangular area in one instance; the remaining small blocks denote groups of ALUTs, with shade indicating the relative number of ALUTs used in each group.

density, measured as the percentage of ALUTs in actual use within the rectangular area containing an *8-stage* Octavo instance, over word widths ranging from 8 to 72 bits and plotted for each addressable memory depth ranging from 2 to 32,768 words. BRAMs and DSP block do not count towards ALUT count. Word width has no clear effect, but density drops sharply for depths exceeding 1024 words due to the BRAM columns needing a larger rectangular area to contain them than would compactly contain the processor logic implemented using ALUTs.

Figures 15(a) and 15(b) illustrate the effect of the layout of BRAMs on the density. Each show an 8-stage, 72-bit wide Octavo instance with a memory of 1024 and 4096 words respectively. The large colored rectangular area contains only the ALUTs used by Octavo. Any outside ALUTs belong to a test harness and are ignored. The columns contain the DSP blocks which implement the Multiplier, and the BRAMs for the Memory. The remaining small block denote groups of ALUTs, with shade indicating the relative number of ALUTs in use in each group. When increasing from a 1024 to 4096 word memory, the number of ALUTs used to implement Octavo increases only 15.3%, but the density drops from 65% to 26% due to the unused ALUTs enclosed by the required number of BRAMs.

For memories deeper than 1024 words, we could recover the wasted ALUTs by allowing non-Octavo circuitry to be placed within its enclosing rectangular area, but this choice may negatively affect F_{max} due to increased routing congestion, and prevents the FPGA CAD tools from placing and routing multiple Octavo instances (or other modules) in parallel, lengthening the design cycle. Further work may lead us to create vector/SIMD versions of Octavo to reclaim unused resources.

Summary Our experiments confirm our original intuition that there exists a "sweet spot"—where the number of BRAMs used fits most effectively within the area of the CPU—at approximately 1024 words of memory depth, regardless of word width.

11. CONCLUSIONS

In this paper we presented initial work to answer the question "How do FPGAs want to compute?", resulting in the Octavo FPGA-centric soft-processor architecture family. Octavo is a ten-pipeline-stage, eight-threaded processor that operates at the BRAM maximum of 550MHz on a Stratix IV FPGA, is highly parameterizable, and behaves well under a wide range of datapath and memory width, memory depth, and number of supported thread contexts:

- F_{max} decreases only 28% (625 to 450MHz) over a 9x increase in word width (8 to 72 bits);
- F_{max} decreases 49.8% over a 64x increase in memory depth (512 to 32k words), and almost independently of word width;
- the amount of logic used is almost unaffected by memory depth: at a width of 72 bits, the usage increases from 2478 to 3339 ALUTs (+37.5%) when increasing the memory depth from 256 to 32,768 (128x);
- the amount of logic used varies linearly with word width, increasing 11.4x over a 9x increase in width (8 to 72 bits);
- and the area density is unaffected by word width, but drops sharply for memory depths exceeding 1024 words due to the BRAM columns needing a larger containing rectangular area than that required for the processor logic.

12. FURTHER WORK

Our FPGA-centric architecture approach led us to Octavo, a fast but unconventional architecture. We will next attempt to push more standard processor features back into Octavo to determine whether a high F_{max} can be maintained with more conventional architecture support. For example, we will attempt to provide some support for indirect memory access and possibly eliminate the need for code synthesis and non-re-entrant code. We will also investigate the possibility of allowing fewer threads than pipeline stages via cheap methods for hazard detection and thread scheduling [13]. Beyond a single Octavo datapath, other important avenues of research include scaling Octavo to have multiple datapaths with vector/SIMD support, and to have interconnect, communication, and synchronization between multiple cores. We will also work towards connecting to a data parallel and highly-threaded high-level programming model such as OpenCL. Finally, we hope to explore the applicability of Octavo and its descendants to other FPGA devices.

13. ACKNOWLEDGMENTS

Thanks to Altera and NSERC for financial support, the reviewers for useful feedback and Joel Emer, Intel, for insightful comments.

14. REFERENCES

[1] ALTERA. DC and Switching Characteristics for Stratix IV Devices. http://www.altera.com/literature/hb/stratix-iv/stx4_siv54001.pdf, June 2011. Version 5.1, Accessed Dec. 2011.

[2] ALTERA. Nios II Performance Benchmarks. http://www.altera.com/literature/ds/ds_nios2_perf.pdf, June 2011. Version 7.0.

[3] ALTERA. Nios II Processor. http://www.altera.com/devices/processor/nios2/ni2-index.html, October 2011. Accessed December 2011.

[4] ANJAM, F., NADEEM, M., AND WONG, S. A VLIW softcore processor with dynamically adjustable issue-slots. In *Field-Programmable Technology (FPT), 2010 International Conference on* (dec. 2010).

[5] CHOU, C., SEVERANCE, A., BRANT, A., LIU, Z., SANT, S., AND LEMIEUX, G. VEGAS: Soft Vector Processor with Scratchpad Memory. In *ACM/IEEE International Symposium on Field-Programmable Gate Arrays* (February 2011).

[6] CHUNG, E. S., PAPAMICHAEL, M. K., NURVITADHI, E., HOE, J. C., MAI, K., AND FALSAFI, B. ProtoFlex: Towards Scalable, Full-System Multiprocessor Simulations Using FPGAs. *ACM Trans. Reconfigurable Technol. Syst. 2* (June 2009).

[7] DIMOND, R., MENCER, O., AND LUK, W. CUSTARD - A Customisable Threaded FPGA Soft Processor and Tools . In *International Conference on Field Programmable Logic (FPL)* (August 2005).

[8] FORT, B., CAPALIJA, D., VRANESIC, Z., AND BROWN, S. A Multithreaded Soft Processor for SoPC Area Reduction. In *IEEE Symposium on Field-Programmable Custom Computing Machines* (April 2006).

[9] GRAY, J. Designing a Simple FPGA-Optimized RISC CPU and System-on-a-Chip. http://www.fpgacpu.org/papers/soc-gr0040-paper.pdf, 2000. Accessed December 2011.

[10] JONES, A. K., HOARE, R., KUSIC, D., FAZEKAS, J., AND FOSTER, J. An FPGA-based VLIW processor with custom hardware execution. In *International Symposium on Field-Programmable Gate Arrays* (2005).

[11] KNUTH, D. E. *The Art of Computer Programming, Volume 1 (3rd ed.): Fundamental Algorithms.* 1997.

[12] LABRECQUE, M., AND STEFFAN, J. Improving Pipelined Soft Processors with Multithreading. In *International Conference on Field Programmable Logic and Applications* (Aug. 2007).

[13] LABRECQUE, M., AND STEFFAN, J. G. Fast Critical Sections via Thread Scheduling for FPGA-based Multithreaded Processors. In *in International Conference on Field Programmable Logic and Applications* (2009).

[14] MOUSSALI, R., GHANEM, N., AND SAGHIR, M. Microarchitectural Enhancements for Configurable Multi-Threaded Soft Processors. In *International Conference on Field Programmable Logic and Applications* (Aug. 2007).

[15] MOUSSALI, R., GHANEM, N., AND SAGHIR, M. A. R. Supporting multithreading in configurable soft processor cores. In *CASES '07: Proceedings of the 2007 international conference on Compilers, Architecture, and Synthesis for Embedded Systems* (2007), ACM.

[16] SAGHIR, M. A. R., EL-MAJZOUB, M., AND AKL, P. Datapath and ISA Customization for Soft VLIW Processors. In *ReConFig 2006: IEEE International Conference on Reconfigurable Computing and FPGAs* (Sept. 2006).

[17] XILINX. MicroBlaze Soft Processor. http://www.xilinx.com/microblaze, October 2011. Accessed December 2011.

[18] YIANNACOURAS, P., STEFFAN, J. G., AND ROSE, J. VESPA: portable, scalable, and flexible FPGA-based vector processors. In *Proceedings of the 2008 international conference on Compilers, architectures and synthesis for embedded systems* (2008), CASES '08.

[19] YU, J., EAGLESTON, C., CHOU, C. H.-Y., PERREAULT, M., AND LEMIEUX, G. Vector Processing as a Soft Processor Accelerator. *ACM Trans. Reconfigurable Technol. Syst. 2* (June 2009).

Accelerator Compiler for the VENICE Vector Processor

Zhiduo Liu
zhiduol@ece.ubc.ca
Dept. of ECE, UBC
Vancouver, Canada

Aaron Severance
aaronsev@ece.ubc.ca
Dept. of ECE, UBC
Vancouver, Canada

Satnam Singh
s.singh@acm.org
Google & Univ. of Birmingham
Mountain View, USA

Guy G.F. Lemieux
lemieux@ece.ubc.ca
Dept. of ECE, UBC
Vancouver, Canada

ABSTRACT

This paper describes the compiler design for VENICE, a new soft vector processor (SVP). The compiler is a new back-end target for Microsoft Accelerator, a high-level data parallel library for C++ and C#. This allows us to automatically compile high-level programs into VENICE assembly code, thus avoiding the process of writing assembly code used by previous SVPs. Experimental results show the compiler can generate scalable parallel code with execution times that are comparable to hand-written VENICE assembly code. On data-parallel applications, VENICE at 100MHz on an Altera DE3 platform runs at speeds comparable to one core of a 3.5GHz Intel Xeon W3690 processor, beating it in performance on four of six benchmarks by up to 3.2×.

Categories and Subject Descriptors

C.1.2 [**Multiple Data Stream Architectures (Multiprocessors)**]: Array and vector processors; C.3 [**Special-purpose and Application-based Systems**]: Real-time and Embedded systems

General Terms

Design, Experimentation, Measurement, Performance

Keywords

vector, SIMD, soft processors, scratchpad memory, FPGA

1. INTRODUCTION

FPGAs offer low power operation and great performance potential through massive amounts parallelism. Harnessing the parallelism of FPGAs often requires custom datapath accelerators. C-to-hardware tools assist this process, but still require a lengthy place-and-route and timing closure process every time the software is changed. A soft vector processor

such as VENICE provides an alternative that can accelerate a wide range of tasks that fit the SIMD programming model. To address the programmability issue of previous SVPs, this work presents a vectorizing compiler/back-end code generator based on Microsoft's Accelerator framework.

The VENICE architecture is chosen as a target SVP for this work. It is smaller and faster than all previously published SVPs. For applications that fit the SIMD programming model, VENICE is often fast enough that application-specific accelerators are not needed. For example, running at 100MHz, VENICE can beat the latest 3.5GHz Intel Xeon W3690 processor on data-parallel benchmarks. As well, VENICE achieves speedups up to 370× faster than a Nios II/f running at the same clock speed. The Accelerator compiler described in this paper achieves similar performance levels to manual coding efforts. Compared to past SVPs, the compiler greatly improves the usability of the system.

2. BACKGROUND AND RELATED WORK

Vector processing has been applied to scientific and engineering workloads for decades. It exploits the data-level parallelism readily available in applications by performing the same operation over all elements in a vector or matrix. It is also well-suited for image and multimedia processing.

Vectorizing Compilers. The VIRAM project [7] implemented a vectorizing compiler and achieved good results, auto-detecting over 90% of vector operations [8]. It is based on the PDGCS compiler for Cray supercomputers.

A common concern for soft vector processors is compiler support. Although based on VIRAM, early soft vector processors, VESPA [11, 12] and VIPERS [13, 14], required hand-written inline assembly code and GNU assembler (gasm) support. VESPA researchers investigated the autovectorizing capability of gcc, but have not yet used it successfully [12]. VEGAS [5] uses readable C macros to emit Nios custom instructions, but programmers must still track the eight vector address registers used as operands. This responsibility includes the traditional compiler roles of register allocation and register spilling.

The multi-core Intel SSE3 target of Accelerator [1] is a vector based target with shorter vector length. However, due to a load/store programming model, a fixed number of registers, and load balancing issues, the SSE3 target is entirely different in design than the VENICE target.

Intel's Array Building Blocks (ArBB) [2] is a system for exposing data parallelism. It combines Intel's Ct threaded

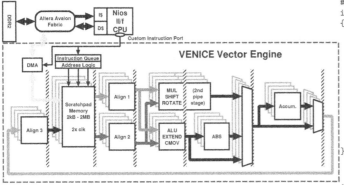

Figure 1: VENICE Architecture

```
#include "vector.h"
int main()
{
    int A[] = {1,2,3,4,5,6,7,8};
    const int data_len = sizeof(A);
    int *va = (int *) vector_malloc( data_len );
    vector_dma_to_vector( va, A, data_len );
    vector_wait_for_dma();
    vector_set_vl( data_len / sizeof(int) );
    vector( SVW, VADD, va, 42, va );  // vector add
    vector_instr_sync();
    vector_dma_to_host( A, va, data_len );
    vector_wait_for_dma();
    vector_free(); // deallocate scratchpad
}
```

Figure 2: VENICE API Adds Scalar to Vector

```
#include "Accelerator.h"
#include "VectorTarget.h"
using namespace ParallelArrays;
using namespace MicrosoftTargets;
int main()
{
    Target *tgtVector = CreateVectorTarget();
    int A[] = {1,2,3,4,5,6,7,8};
    IPA a = IPA( A, sizeof(A)/sizeof(int) );
    IPA d = a + 42;
    tgtVector->ToArray( d, A, sizeof(A)/sizeof(int) );
    tgtVector->Delete();
}
```

Figure 3: Accelerator Code Adds Scalar to Vector

programming model with RapidMind's object system, which provides, similar to Accelerator, C++ libraries for array types and array operations to express data-parallel computation. RapidMind could target CPUs, Cell, and x86 processors; ArBB appears to target only the latter.

VENICE Architecture. A block diagram of VENICE (Vector Extensions to NIOS Implemented Compactly and Elegantly) is shown in Figure 1. Based on VEGAS, VENICE makes the following improvements:

- Instruction-level support for 2D and 3D arrays. These avoid the need for VEGAS auto-increment modes.

- The vector address register file is removed. Hence, there is no need to track and spill the vector address registers. Instead, C pointers are directly used as operands to vector instructions.

- VENICE uses 3 alignment networks in the pipeline. This avoids the performance penalty with VEGAS when operands are misaligned.

- The shared multiplier/shift/rotate structure requires two cycles operational latency, allowing a general absolute value stage to be added after the integer ALU. This is followed by a general accumulator.

The native VENICE application programming interface (API) is similar to inline assembly in C. However, novel C macros simplify programming and make VENICE instructions look like C functions without any run time overhead, e.g. Figure 2 adds the scalar value 42 to a vector.

Each macro dispatches one or more vector assembly instructions to the vector engine. Depending upon the operation, these may be placed in the vector instruction queue, or the DMA transfer queue, or executed immediately.

The VENICE programming model uses a few basic steps: 1) allocate memory in scratchpad, 2) flush data from cache, 3) DMA transfer from main memory to scratchpad, 4) vector setup (e.g., set the vector length), 5) perform vector operations, 6) DMA transfer results from scratchpad to main memory, 7) deallocate memory in scratchpad.

The basic instruction format is vector(VVWU, FUNC, VD, VA, VB). The VVWU specifier refers to 'vector-vector' operation (VV) on integer type data (W) that is unsigned (U). The vector-vector part can instead be scalar-vector (SV), where the first source operand is a scalar value provided by Nios. These may be combined with data sizes of bytes (B), halfwords (H) and words (W). A signed operation is designated by omitting the unsigned specifier (U).

The vector_malloc(num_bytes) call allocates a chunk of scratchpad memory. The vector_free() call frees the entire scratchpad; this reflects the common usage of the scratchpad as a temporary buffer. DMA transfers and instruction synchronization are handled by macros as well. In our experience, DMA transfers can be double-buffered to hide most of the memory latency.

Accelerator. The Accelerator system developed by Microsoft [1, 10] is a domain-specific language aimed at manipulating arrays with multiple back-end targets, including GPUs, multicore Intel CPUs, and VHDL [4]. Accelerator allows easy manipulation of arrays using a rich variety of element-wise operations. The restricted structure of Accelerator programs makes it easy to identify parallelism.

Accelerator data are declared and stored as Parallel Array (PA) objects. Accelerator does a lazy functional evaluation of operations with PA objects. That is, expressions and assignments involving PA objects are not evaluated instantly, but instead they are used to build up an expression tree. At the end of a series of operations, the PA ToArray() method must be called. This results in the expression tree being optimized, translated into native code using a JIT compilation process, and evaluated.

Figure 3 shows code to add the scalar value 42 to a vector in Accelerator. The CreateVectorTarget() function indicates that a subsequent ToArray() call will be evaluated on the vector processor. The IPA type represents an integer parallel array object. The ToArray() call triggers the compiler to generate VENICE-compliant code. Except for creating the proper target, the program is unaware of all hardware-related details, including whether a VENICE processor is being used or its size. To target a different device, one simply renames the CreateXXXTarget() function.

3. VENICE TARGET IMPLEMENTATION

The ability to manipulate arrays is intrinsic to both Accelerator and VENICE. In many cases, a direct translation

```
Front-end:
      input expression_graph;
      convert_to_IR();
      mark_and_add_intermediates();
      move_bounds_to_leaves();
Back-end:
      contant_folding_and_propagation();
      combine_operators();
      eval_ordering_and_ref_counting();
      buffer_counting();
      convert_to_LIR();
      calc_buffer_size();
      assign_buffers_to_input();
      allocate_and_init_memory();
loop:  transfer_data_to_scratchpad();
      set_vector_length();
      write_vector_instructions();
      transfer_result_to_host();
      if( !double_buffering_done ) goto loop;
      output VENICE_C_code;
```

Figure 4: Accelerator Compiler Flow

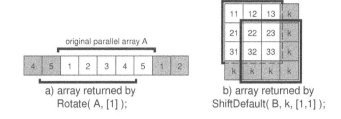

a) array returned by
Rotate(A, [1]);

b) array returned by
ShiftDefault(B, k, [1,1]);

Figure 5: Memory Transform Examples

from Accelerator operators to VENICE instructions is possible. The compiler automatically breaks up large matrices into a series of smaller data transfers that fit in the scratchpad. Also, it uses double-buffering to hide memory latency.

For this work, we do not support JIT. Instead, we use Accelerator as a source-to-source compiler: it writes out another C program annotated with the VENICE APIs, which must be recompiled using gcc.

Figure 4 indicates the sequence of code optimizations and code transformations performed by the compiler. The front-end performs constant folding and common subexpression elimination to produce an optimized intermediate representation (IR). Then, the front-end analyzes all of the memory transforms and array accesses to produce index bounds for each leaf node (input array) in the computation.

Back-end Preliminaries. Next, target-specific optimizations are done before code generation. We found it beneficial to perform our own constant folding in the back-end in addition to existing front-end optimizations. Next, certain short sequences of operators are combined into a single compound VENICE operation, such as a multiply-add sequence or any add/subtract followed by absolute value.

Scratchpad Allocation. To load input data from main memory into the scratchpad, we need to allocate space in scratchpad memory first. The back-end treats the scratchpad as a pseudo-registerfile [6, 9]. This divides the scratchpad into as many equal-size registers (vector data buffers) as needed. However, several techniques are required to limit the number of registers to maximize their size.

To determine the size of these registers, the compiler first counts the total number of registers needed by the program. This is done by first determining an evaluation order for the subexpressions in each tree using a modified Sethi-Ullman algorithm [3]. To re-use registers, the back-end keeps a list of registers acting as input buffers for subsequent calculations, plus the number of remaining references to each of them. Whenever the reference count becomes zero, the register is no longer needed to hold an input array or an intermediate result, allowing the register to be re-used immediately. The total number of registers needed is the sum of registers used to hold leaf (input) data plus temporary intermediate data. After this step, a linear IR (LIR) is generated with references to precise register numbers.

One convenience feature in Accelerator is efficient handling of out-of-bounds array indices coming from memory

transform operations such as Shift() and Rotate(). In the front-end, Accelerator propagates the array bounds back to each leaf node, so the maximum extents are known. The back-end takes this additional information into account and allocates extra memory in the scratchpad for these cases.

All scratchpad memory is freed after each ToArray() call.

Data Initialization and Transfer. The initializing stage copies any user data to the output C file and prepares for memory transforms by padding input arrays with proper values for any out-of-bounds accesses.

Figure 5 demonstrates how input data padding is done. Part a) shows a rotation performed on a 1D array. The original array is white, with the out-of-bounds elements shaded. In this case, the last element 5 appears padded before first element, while the first element 1 appears padded after the last element. The new array formed by Rotate() is indicated by the bold black bar. Part b) shows a shift on a 2D array, up and to the left at the same time. Values past the bounds are initialized with the specified default value of k. The new array formed by ShiftDefault() is highlighted by a bold black box.

DMA transfer instructions are generated after memory allocation and data initialization. In the case where the full array is large, or doesn't entirely fit into scratchpad, the compiler generates code to move data in a double-buffered fashion by pre-fetching. This allows DMA transactions and vector computation to overlap. Overlapping the two can almost completely hide the overhead of the data transfer.

Generation of Vector Instructions. There is nearly a direct mapping of Accelerator operators to VENICE instructions for basic element-wise operations. In a few cases, we have prewritten library code to support Accelerator operators that are not directly supported by VENICE, such as divide, modulo and power.

For memory transforms on PA objects, we discussed in the previous subsection that we handle such operations by initializing the input data with a padded region outside of the normal array bounds. We refer to the examples in Figure 5 again here to demonstrate how memory transforms are executed. With all data properly initialized, extracting partial data from a 1D array is simply done by adding an offset to the starting address in the scratchpad memory. For 2D arrays, the VENICE row stride amounts can be adjusted to step over any padding elements added at both ends.

Implementation Limitations. VENICE does not support floating-point operations, so we are unable to support float, double and quad-float types in Accelerator. The Boolean type uses 32b integers. Most of the Accelerator APIs have been implemented in the VENICE back-end; a few were omitted due to time constraints.

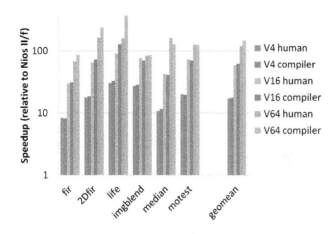

Figure 6: Compiler Speedups

CPU	fir	2Dfir	life	imgblend	median	motest
Xeon W3690	0.07	0.44	0.53	0.12	9.97	0.24
VENICE	0.07	0.29	0.23	0.33	3.11	0.22
Speedup	1.0	1.5	2.3	0.4	3.2	1.1

Table 1: Runtime (seconds) and Speedup

4. RESULTS

Soft processor results were run on an Altera DE3-150 with one DDR2-800 SODIMM. Different VENICE instances use 4, 16, and 64 parallel lanes of 32b ALUs, called V4, V16, and V64, respectively. All processors run at 100MHz, with the DDR2 memory at half rate; this allows easy estimation of runtime using scaled clock rates up to 200MHz.

A set of six benchmarks are used to measure the effectiveness of the compiler at scaling to large size SVPs. All benchmarks use integers because Accelerator does not support byte or short data types. However, smaller data types allow greater performance with VENICE because each 32b ALU can be fractured into four 8b or two 16b ALUs.

Speedups over serial Nios II/f C code for both human and compiler-generated parallel code are shown in Figure 6. The compiler outperforms the human in 11 of the 18 cases; when the human wins, it is only by a small margin, but the compiler often wins by a much larger margin. This is because the compiler puts more effort into the process than a human: 1) it fully unrolls inner loops to reduce overhead; 2) it carefully calculates the maximum buffer size that fits into the scratchpad, rather than conservatively rounding down or guessing; 3) it double-buffers all data transfers; 4) it inlines all function calls. The fastest, life, achieves 370× speedup compared to a Nios II/f. However, humans can sometimes do far better than the compiler; the graph does not show our human-written motion estimation result which is another 1.5× faster because it uses the VENICE accumulator in a way that cannot be expressed in the Accelerator language. Finally, we note that imgblend is memory bandwidth limited at V16, so it does not benefit from more ALUs at V64.

In Table 1, we compare the VENICE compiler (not human) results to a single-core 3.5GHz Intel Xeon W3690 processor compiled with Visual Studio 2010 with -O2. We ran each benchmark 1000 times and measured total runtime. Across the 6 benchmarks, Intel beats VENICE only on imgblend, which is memory bandwidth limited.

5. CONCLUSIONS

This work has shown that compiler-generated results with a soft vector processor can achieve significant speedups on data parallel workloads. Speedups up to 370× versus a Nios II/f, and speedups up to 3.2× versus a 3.5GHz Intel Xeon W3690 are demonstrated. Furthermore, compiler-generated results are comparable to human-coded results.

Currently, Accelerator and VENICE have limited data type support. Accelerator should add support for bitwise operations, plus byte and halfword data types. As well, VENICE should add floating-point data types. In our back-end, some Accelerator APIs are not yet implemented.

6. ACKNOWLEDGMENTS

We thank NSERC for funding, Altera for hardware donations, and the Microsoft Accelerator group for their assistance during this project.

7. REFERENCES

[1] Accelerator. http://research.microsoft.com/en-us/projects/accelerator.

[2] Sophisticated library for vector parallelism. http://software.intel.com/en-us/articles/intel-array-building-blocks/.

[3] A. Appel and K. J. Supowit. Generalizations of the Sethi-Ullman algorithm for register allocation. *Software – Practice and Experience*, 17:417–421, 1987.

[4] B. Bond, K. Hammil, L. Litchev, and S. Singh. FPGA circuit synthesis of accelerator data-parallel programs. In *FCCM*, pages 167–170, Charlotte, North Carolina, USA, 2010.

[5] C. Chou, A. Severance, A. Brant, Z. Liu, S. Sant, and G. Lemieux. VEGAS: Soft vector processor with scratchpad memory. In *FPGA*, pages 15–24, Monterey, California, USA, 2011.

[6] B. Egger, J. Lee, and H. Shin. Scratchpad memory management for portable systems with a memory management unit. In *PACT*, pages 321–330, Seoul, Korea, 2006.

[7] C. Kozyrakis. *Scalable Vector Media Processors for Embedded Systems*. PhD thesis, University of California at Berkeley, May 2002. Technical Report UCB-CSD-02-1183.

[8] C. E. Kozyrakis and D. A. Patterson. Scalable vector processors for embedded systems. *IEEE Micro*, 23(6):36–45, 2003.

[9] L. Li, L. Gao, and J. Xue. Memory coloring: A compiler approach for scratchpad memory management. In *PACT*, pages 329–338, Sydney, Australia, 2005.

[10] D. Tarditi, S. Puri, and J. Oglesby. Accelerator: Using data parallelism to program GPUs for general-purpose uses. In *ASPLOS*, pages 325–355, San Jose, California, USA, 2006.

[11] P. Yiannacouras, J. G. Steffan, and J. Rose. VESPA: portable, scalable, and flexible FPGA-based vector processors. In *CASES*, pages 61–70. ACM, 2008.

[12] P. Yiannacouras, J. G. Steffan, and J. Rose. Data parallel FPGA workloads: Software versus hardware. In *FPL*, pages 51–58, Progue, Czech Republic, 2009.

[13] J. Yu, C. Eagleston, C. Chou, M. Perreault, and G. Lemieux. Vector processing as a soft processor accelerator. *ACM TRETS*, 2(2):1–34, 2009.

[14] J. Yu, G. Lemieux, and C. Eagleston. Vector processing as a soft-core CPU accelerator. In *FPGA*, pages 222–232, Monterey, California, USA, 2008.

FCache: A System For Cache Coherent Processing on FPGAs

Vincent Mirian
Department of Electrical And Computer
Engineering
University Of Toronto
Canada
mirianvi@eecg.toronto.edu

Paul Chow
Department of Electrical And Computer
Engineering
University Of Toronto
Canada
pc@eecg.toronto.edu

ABSTRACT

Much like other computing platforms in the world today, FPGAs are becoming increasingly larger and contain large amounts of reconfigurable logic. This makes FPGAs an acceptable platform for multiprocessor systems. However in today's world of FPGA computing, very limited infrastructure is available to facilitate the creation of cache coherent shared memory systems for FPGAs. This paper introduces FCache, a system for shared memory cache coherent processing on FPGAs. The paper also describes the mapping of the conventional shared bus to FPGAs using two distinct network implemented in FCache. FCache also provides flushing and multithreaded synchronization functionalities, such as locking and unlocking of a mutex variable, which is embedded in its *cache* component. Despite these additional functionalities, results show that FCache has little resource overhead compared to a previous more simplistic cache coherent system that was targeted for FPGAs.

Categories and Subject Descriptors

B.3.2 [**Hardware**]: MEMORY STRUCTURES—*Design Styles*; C.0 [**Computer Systems Organization**]: GENERAL— *System architectures*

General Terms

Design

Keywords

Cache Coherency, FPGA, multiprocessor

1. INTRODUCTION

In software, programming models abstract away the system's architecture from the programmer. These programming models are designed in accordance to two generic memory models: the distributed-memory model and the shared

memory model [1]. The shared memory model is an easier model for accessing memory since the processors share a single view of the memory.

Memory-bound applications in a shared memory system typically suffer from the large off-chip memory latency and congestion. And for this reason, caches are typically found in shared memory systems.

Although designs with shared-memory systems are common in FPGAs, cache-coherent shared-memory systems are not since they are difficult to build. Currently, FPGA vendors do not provide a standard library of intellectual property (IP) that can be used as building blocks to construct shared memory systems with caches. If a coherent shared memory caching system was available for FPGAs, it would ease the programming of heterogeneous multiprocessors that are often implemented on FPGAs.

This paper introduces FCache, a distributed system that simplifies the creation of multiprocessor shared-memory cache-coherent systems on FPGAs.

Section 2 describes an implementation of a cache coherent shared memory system. In Section 3, the paper follows with an introduction to FCache and an explanation of how the FCache implementation surpasses challenges of implementing a shared-memory system on an FPGA. The paper compares the implementation of the FCache with related work in Section 4. The paper continues with a description and analysis of experiments conducted in Section 5. Finally, the paper concludes with remarks in Section 6.

2. SHARED MEMORY SYSTEMS

In the generic shared-memory system model, the shared bus is the medium of communication between the caches and the shared memory. Memory requests, sent from a cache to the shared memory, are transferred over the shared bus. These memory requests are either to read from a memory address or to write to a memory address. With caches implemented with the snoopy protocol, these caches can see the memory request on the shared bus. If a cache contains an up-to-date value associated with the memory address, then the cache would interrupt the memory request on the shared bus and respond with its up-to-date value. This procedure allows coherency to be maintained in the system. The steps involved with interrupting the memory request on the shared bus will hereafter be referred to as the coherency procedure.

One way to achieve consistency amongst the caches is to have ordered access to a memory address. In a shared memory system, the arbiter associated with the shared bus allows

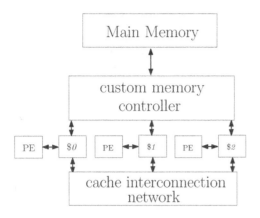

Figure 1: Block diagram of the FCache

at most a single memory request at a time, consequently ordering the memory requests. However, the arbiter and its single point of entry results in the shared bus being a bottleneck in the shared memory system.

3. THE FCACHE

In designing a cache coherent shared memory system for FPGAs, we inherit problems of mapping the shared bus of a shared memory system, and maintaining consistency and coherency throughout the system. The remainder of this section will describe the details of the FCache implementation that address these problems.

3.1 Overview

There are four major components in the FCache: the *custom memory controller*, the *cache interconnection network*, the *cache* and the *processing element*. An instance of the FCache can contain one or more instances of the *cache* and the *processing element*, but only has a single instance of the *custom memory controller* and the *cache interconnection network*. Each instance of the *cache* has a unique integer identifier starting from zero. Figure 1 shows an FCache with three *caches* ($n), where *n* is the unique integer identifier, and three *processing elements* (PE).

In the FCache, each *cache* is connected to the *custom memory controller* and the *cache interconnection network* via a full-duplex FIFO link. In addition, each *cache* is also connected to a *processing element* (PE).

The *cache* has the ability to flush the data contents of the cache, and perform multithreaded synchronization operations such as locking and unlocking a mutex variable. The mutex variables are contained within the global memory space, rendering a mutex variable no different than other variables in an application. The *cache* interface resembles that of a typical BRAM controller [2] with the addition of a flush and mutex signal enabling the use of the embedded functionalities.

3.2 Coherency and Consistency

The FCache uses a distributed approach to manage coherency. On a memory request, a cache will send a coherence message to the *cache interconnection network* that will broadcast the coherence message to the remaining caches connected to the *cache interconnection network*. The coherence message is similar to a memory request, as the co-

herence message contains the memory address and the action the cache wishes to perform. The cache receiving a coherence message will respond to the coherence message with a response message. The response message will contain the state of the cache line mapped to the address from the memory request. Once the *cache* receives a response message from the remaining *caches* in the system, the *cache* will proceed by performing the action dictated by the cache protocol, for example, reading the cache line data from the main memory.

Ordering in a distributed system is generally done with counters maintained at each node. There are two common methods for preserving order in a distributed system: Lamport's timestamps [3] and Vector Clocks [4]. These conventional methods were not used in favour of using the strict ordering method. In contrast to Lamport's timestamps method, the strict ordering method contains less computation; and in contrast to the Vector Clock method, the strict ordering method has a smaller message footprint.

In the strict ordering method, all the *caches* will maintain a logical clock where the value is synchronized and driven by the same clock. To preserve ordering of the coherence message, each coherence message will be timestamped with the logical clock. If two coherence messages map to the same memory address and contain the same timestamp, the priority is arbitrarily given to the coherence message originating from the *cache* with the smallest cache identifier.

The logical clocks have a finite duration. If an application requires more than 2^N clock cycles to complete, where N is the number of bits implementing the logical clock, then the *caches* may become incoherent as the logical clocks will rollover. To prevent this situation from occurring, a logical clock synchronization protocol was developed.

The logical clock synchronization protocol requires a single *cache* node to be identified as the master node and the remainder of the *caches* to be identified as slave nodes. The logical clock values and a summary of the events performed by the master and a slave node are shown in Table 1.

The logical clock synchronization protocol is executed at startup of the FCache or when the logical clocks have reached the maximum value of 2^N - 1. The logical clock synchronization protocol can only be instantiated when there are no pending memory requests from the *caches*. At startup, there are no memory requests and the cache will signal the processing elements that it will not accept any memory requests until the protocol is complete. When the logical clocks have reached the maximum value of 2^N - 1, the *cache* will stop incrementing the logical clock. If the *cache* has a pending memory request, then it must wait until the memory request is satisfied before executing the protocol. Once the memory request is satisfied, the cache will signal the processing elements that it will not accept any memory requests until the protocol is complete. [Table 1 - Step1].

During Step 2 of the protocol, the slave node will send a time message to the master node. Once the master node has received a time message from all the slave nodes, the master node will send: a time message that the *cache interconnection network* will broadcast to the slave nodes, and reset its logical clock [Table 1 - Step3]. During the final Step of the protocol, the slave nodes will reset their logical clocks to a predefined value. The predefined value is equivalent to the number of cycles required for the time message to be sent

Table 1: Summary of the logical clock synchronization protocol

Steps	1	2	3	4
Time	at startup of the FCache or when the logical clocks have reached the maximum value of $2^N - 1$	t-arbitrary value	t-d	t
Summary of Events	cache will signal the processing elements that it will not accept any memory requests	slave node sends time message	master node sends time message	logical clocks are synchronized
master logical clock value	arbitrary value	arbitrary value	0	d
slave logical clock value	arbitrary value	arbitrary value	arbitrary value	t-(t-d)=d

from the master node to the slave node. In Table 1, the predefined value is d.

It is important to note that having no pending memory requests by the *caches* is essential for the protocol to be successful. This factor guarantees no traffic in the *cache interconnection network*, making the calculation of the number of cycles for the predefined value deterministic.

3.3 Mapping the Shared Bus to an FPGA

FPGA buses are typically implemented as multiplexers without provision for snooping capability. The FCache provides snooping capability in the *cache interconnection network*.

Modern FPGAs, in comparison to modern CPU shared memory systems, are handicapped with a much lower achievable clock frequency. In addition, FPGAs have a smaller amount of on-chip memory than CPUs, resulting in a smaller cache size for FPGAs. Therefore, it is very likely that the caches on CPUs would have higher hit rates than the caches implemented on FPGAs. Hence for FPGAs to be competitive, it is more crucial to exploit the parallelism in their design.

The concept of parallelism was applied to the design of the *cache interconnection network*. The *cache interconnection network* is designed to send a packet from a cache message on every clock cycle. Therefore, at any given time, there can be more than one memory request in flight. These memory request are analogous to the coherency procedure described in Section 2. Recall that the maximum number of coherency procedures, at a given time, in the generic shared memory system model is one. Whereas, FCache allows the execution of more than one coherence procedure at any given time. This degree of parallelism will increase performance.

The *cache interconnection network* is responsible for the coherency procedure. It is used only for routing the cache messages, therefore any routing topology can be adopted. However, the only prerequisite on the routing topology used in the *cache interconnection network* is that the predefined value used in the logical clock synchronization protocol can be determined. The fully connected routing topology was adopted in this paper because of the ease of broadcasting messages. Future work consists of analyzing other network topologies that may be better suited for the FPGA architecture.

3.4 Access to Main Memory

In the shared memory system, the caches send memory requests to main memory via the shared bus. In the FCache,

the *caches* communicate with the main memory through the *custom memory controller*. The *custom memory controller* is a multi-ported memory controller accepting memory controller messages from the *caches*. A memory controller message contains a memory request timestamped by the logical clock. The memory requests in the memory controller message instructs the *custom memory controller* to read or write a cache line from main memory. If the memory request is a read command then the *custom memory controller* will respond with the cache line data.

Typical multi-ported memory controllers accept memory requests and the arbitration scheme will decide which memory request to service. The *custom memory controller* differs from typical memory controllers as it services a memory request with the smallest timestamp, representing the earliest memory request. This scheme preserves ordering for a fully coherent and consistent system.

The presence of the *custom memory controller* in the FCache results in two distinct networks, one network for the cache messages sent amongst the *caches* over the *cache interconnection network*, and the other network for the memory controller message sent from the *caches* to the *custom memory controller*. The coherency procedure of a main memory request utilizes the *cache interconnection network*. Whereas the main memory access uses the *custom memory controller*. Having two distinct networks in the FCache provides additional communications parallelism in the design.

4. RELATED WORK

Some work on mapping the shared bus onto an FPGA has been done. In Woods [5], the shared bus is modelled by the central hub. The central hub is responsible for the routing of cache messages, managing coherency and accessing the main memory. These numerous responsibilities results in a resource hungry and large component.

As the number of *caches* attached to the central hub increase, placing and routing of the central hub component on an FPGA becomes increasingly difficult. Since the central hub manages coherency, it cannot be partitioned into smaller components as coherency amongst the caches will be broken.

The *cache interconnection network* is simply a routing network. As the number of *caches* increase, the placing and routing of the *cache interconnection network* on an FPGA can become difficult. However unlike the central hub, the *cache interconnection network* can be split into multiple networks and still preserve coherency and consistency in the FCache. This is possible because the *cache interconnection* network is simply responsible for routing the cache messages, and coherency and consistency are handled by the *cache* nodes (not by a centralized component).

Several infrastructures supporting parallel programming on FPGAs have the mutex operation capabilities. In both Andrews et al. [6] and Yamawaki et al. [7], a separate component is used to store the mutex variables and handle the mutex operations. Their design is similar to the XPS Mutex [8] core provided by Xilinx, which provides a central scheme of control for the mutex variables. This approach is fundamentally different from FCache where the control of the mutex variables is distributed amongst the *caches*.

There are two drawbacks when using a separate component to handle mutex operations. The first is the presence of additional wires required to access the mutex component. As the number of processing elements increase in their system, placing and routing their design onto an FPGA becomes increasingly difficult. In the FCache, a single wire is needed for the mutex signal starting from the *processing elements* and ending at its respective *cache*.

The second drawback is the limitation of the number of mutex variables available to the programmer. In both Andrews et al. and Yamawaki et al., the mutex component is on the FPGA where there is limited storage but the mutex variables are accessed much faster than off-chip memory. In the FCache, the mutex variables are mapped to memory addresses in main memory. But with the benefit of caching and the future implementation of on-chip cache-to-cache data transfer, the accesses to mutex variables in FCache would not usually suffer the off-chip memory latency, since the mutex variable would be on-chip residing in a *cache* and be transferred to another *cache* via the on-chip *cache interconnection network*. This implementation would have comparable access time to that of the mutex component. Furthermore, to the best of the authors' knowledge, the FCache is the first cache coherent system where mutex variables do not require a separate memory component to manage the mutex variables.

5. RESULTS

In this section, the resource utilization of FCache is compared with the system of Woods [5]. Woods is an ideal candidate since its architecture resembles that of FCache and he provides some data on resource utilization. However, Woods only provides the number of Look Up Tables (LUTs) for two components, the *cache* and the central hub.

In both Woods and FCache, the *caches* have the same configuration. The *cache* in Woods uses 613 LUTs, the *cache* in FCache uses 890 LUTs. The additional LUTs in the FCache's *cache* implement the flushing and mutex operations, the distributed functionality of maintaining coherency and consistency, as well as the logical clock synchronization protocol. The central hub in Woods uses 2375 LUTs. The central hub of Woods is analogous to the *cache interconnection network* and some logic in the *custom memory controller*, which utilize 293 and 2073 LUTs respectively, for a total of 2366 LUTs. These utilizations show that FCache does not impose any significant overhead compared to a more simplistic cache implementation.

The maximum frequency of a system with eight caches in both FCache and Woods is 100 MHz. However, Woods clearly states in his thesis that beyond eight caches, the maximum frequency drops. Whereas in FCache, the distributed

nature of its design can scale with much less effect on the maximum frequency, given a network topology with a small wire congestion factor such as the mesh or ring topologies.

6. CONCLUSION

This paper introduces FCache, a distributed system for cache coherent processing on FPGAs. The FCache is currently implemented for shared memory systems. The FCache models the conventional shared bus with two networks on the FPGA, the *cache interconnection network* and the *custom memory controller*. Supporting the coherency procedure via the *cache interconnection network* allows FCache to service more than one main memory request at a time. The *cache* component has embedded functionalities such as flushing the data content of the *cache* and performing multithreaded synchronization operations such as locking and unlocking a mutex variable, with an easy to use interface.

FCache manages coherency with messages and maintains consistency by timestamping these messages with a logical clock. To prevent incoherency and inconsistency by the logical clock values being rolled-over, the logical clocks are synchronized via the logical clock synchronization protocol.

Results show that FCache does not impose any significant overhead compared to a more simplistic cache implementation.

7. ACKNOWLEDGMENTS

We acknowledge Xilinx, CMC and NSERC for the funding and material provided for this project.

8. REFERENCES

[1] David A. Patterson and John L. Hennessy. *Computer Organization and Design, Fourth Edition: The Hardware/Software Interface (The Morgan Kaufmann Series in Computer Architecture and Design)*. Morgan Kaufmann Publishers Inc., San Francisco, CA, USA, 2008.

[2] Xilinx Inc. LMB BRAM Interface Controller (v2.10b), 2009.

[3] Leslie Lamport. Ti clocks, and the ordering of events in a distributed system. *Commun. ACM*, 21:558–565, July 1978.

[4] Friedemann Mattern. Virtual time and global states of distributed systems. In *Parallel and Distributed Algorithms*, pages 215–226. North-Holland, 1989.

[5] David Woods. Coherent Shared Memories for FPGAs. Master's thesis, University of Toronto, Toronto, Canada, September 2009.

[6] David Andrews, Ron Sass, Erik Anderson, Jason Agron, Wesley Peck, Jim Stevens, Fabrice Baijot, and Ed Komp. Achieving Programming Model Abstractions for Reconfigurable Computing. *IEEE Transactions on Very Large Scale Integration (VLSI) Systems*, 16(1):34–44, 2008.

[7] Akira Yamawaki and Masahiko Iwane. An FPGA Implementation of a Snoop Cache With Synchronization for A Multiprocessor System-On-chip. In *ICPADS '07: Proceedings of the 13th International Conference on Parallel and Distributed Systems*, pages 1–8, Washington, DC, USA, 2007. IEEE Computer Society.

[8] Xilinx Inc. XPS Mutex (v1.00c), 2009.

A Lean FPGA Soft Processor Built Using a DSP Block

Hui Yan Cheah [1], Suhaib A. Fahmy [1], Douglas L. Maskell [1], Chidamber Kulkarni [2]

[1] School of Computer Engineering, Nanyang Technological Univesity
Nanyang Avenue, Singapore
hycheah1@e.ntu.edu.sg, {sfahmy, asdouglas}@ntu.edu.sg

[2] Xilinx Inc., San Jose, CA
chidamber.kulkarni@xilinx.com

ABSTRACT

As Field Programmable Gate Arrays (FPGAs) have advanced, the capabilities and variety of embedded resources have increased. In the last decade, signal processing has become one of the main driving applications for FPGA adoption, so FPGA vendors tailored their architectures to such applications. The resulting embedded digital signal processing (DSP) blocks have now advanced to the point of supporting a wide range of operations. In this paper, we explore how these DSP blocks can be applied to general computation. We show that the DSP48E1 blocks in Xilinx Virtex-6 devices support a wide range of standard processor instructions which can be designed into the core of a basic processor with minimal additional logic usage.

Categories and Subject Descriptors

C.1.3 [**Processor Architectures**]: Other Architecture Styles—*Adaptable Architectures*

Keywords

FPGA, DSP blocks, soft processor, hard macro

1. INTRODUCTION

FPGAs have evolved significantly over recent years. From simple, regular arrangements of configurable logic blocks and routing, modern devices now boast increased complexity, in terms of both size, and the variety and capability of primitives offered. Much of this improvement has inevitably been driven by market segments where FPGAs are particularly popular, such as communications and signal processing. This is due to the ease with which such algorithms can be parallelised on FPGAs and the availability of high-level programming techniques that simplify the design process.

Hence, it is not surprising to find that FPGAs have evolved to better suit such applications. The Virtex II brought with it embedded multipliers. A large number of signal processing algorithms make use of multiplications. By embedding

hard multipliers into the silicon, it becomes possible to optimise them for performance while saving the remaining resources for other uses. These later evolved into DSP Blocks: multiply-accumulation units that support the full requirements of a DSP filter tap.

Recently, FPGAs have moved beyond implementation of accelerators for complex algorithms, now housing full systems. Processors are useful when dealing with non-streaming data, in systems with multiple heterogeneous hardware tasks, and for managing complex interfacing. Vendors did previously introduce devices with embedded hard processors such as the PowerPC 405 in the Virtex II Pro, and the PowerPC 440 in the Virtex 4 FX. While these high-end FPGA devices did find an audience, they were out of the budget of many, and so, "soft" processors, implemented using logic resources, have continued to dominate.

In this paper, we connect these two threads. DSP Blocks are indeed highly capable primitives, yet leveraging them outside the DSP domain is extremely difficult, as they were primarily designed to suit such applications. This paper investigates the feasibility of building custom soft-core processors that can allow DSP Blocks to be leveraged beyond their typical target applications, and in a manner accessible to those with minimal FPGA architecture knowledge. The Xilinx DSP48E1 cores included in the most recent Xilinx devices are highly customisable. We aim to build a lean processor around the DSP48E1, with as little extra logic as possible, that supports a full set of standard machine instructions.

The prospects are even more exciting when one considers that modern FPGAs have very many of these blocks; a large Virtex-6 device contains hundreds of such DSP Slices. Hence, such processors could be used to build massively parallel many-core systems. In this paper, we investigate the design of a lean single processor based on the DSP48E1 primitive.

2. RELATED WORK

There is a wide body of work on soft processors for FPGAs. Processors allow systems to be more flexible than would typically be possible in a datapath-only designs. Indeed, soft processors can run complete applications and even operating systems, to better abstract management of embedded systems. Xilinx offers Microblaze, a 32-bit RISC processor, and PicoBlaze, a small 8-bit processor for simple applications. Despite the increased performance of hard processors in the Virtex II Pro and Virtex-4 FX, many have found these soft processors easier to work with, and so they remain very popular.

In research, a number of studies have examined improvements in FPGA soft processor architecture, including improving area consumption, scalability, portability and enabling vector processing. fSE [4] is a soft-processor built using embedded DSP blocks. A comparison between an fSE implementation and a Xilinx Coregen implementation of a 16-point FFT shows that the performance of fSE can surpass that of dedicated hardware. However, good performance can be achieved only if the algorithm fits the multiply-add operating model of the fSE processor. The number of supported instructions is very small, and limited to DSP operations.

In [8, 7], the authors explore the potential of a vector soft-core processor as an accelerator for a scalar main processor. The soft-core is specifically designed to target vector processing with customisable data width and scalable data lanes. Vector processing is demonstrated to offer increased data-level parallelism. In [7], the authors utilised a framework developed in [6] to generate the main scalar processor. However, these processors were designed to be portable, and hence leverage few of the advanced capabilities in modern FPGA devices.

MORA [1] is a multi-core processor made up of an array of small processors called reconfigurable cells (RC). MORA uses a processor-in-memory architecture and each RC has its own internal memory. MORA's matrix-like structure is targeted at multimedia applications and image compression algorithms. The processors are programmed using a specially developed MORA assembly language. While programming in a low-level language yields performance benefits, it increases design complexity for the programmer. Again, MORA does not leverage the capabilities of the DSP Slices on modern FPGAs.

In [2, 3] the authors present a large array of parallel soft processors that use embedded DSP blocks. The processors address the limitations of current soft processors like Microblaze in handling high performance real-time wireless applications. Similar to [4], only a subset of the DSP48E1 arithmetic operations are explored, with implementation limited to just multiply-add and multiply-subtract.

3. A DSP48E1-BASED PROCESSOR

3.1 The DSP48E1 Primitive

The Xilinx DSP48E1 primitive is an embedded hard core present in the Xilinx Virtex 6 and soon to be released 7 Series. It is designed for high-speed DSP computation and is composed of a multiplier and arithmetic logic unit (ALU), along with various registers and multiplexers. A host of configuration inputs allow the functionality of the primitive to be manipulated at both runtime and compile time. Depending on the creativity of the designer, the slice can be configured to support various operations like multiply-add, multiply-accumulate, pattern matching and barrel shift, among others [5].

As shown in Figure 1, the DSP48E1 slice primitive has 4 input ports for data, A, B, C and D, each corresponding to input paths of the multiplier and adder/subtractor or logic unit (ALU). While port D can be used to pre-add a value to input A prior to multiplication, path D is not necessary for basic arithmetic operations, as path A alone is sufficient. In our case, port D is disabled.

The functionality of DSP48E1 is controlled by a combination of dynamic control signals and static parameter attributes. Dynamic control signals allow the slice to run in different configuration modes in each clock cycle. For instance, a designer can change the ALU operation by modifying the control bits of ALUMODE, the ALU input selection by modifying the OPMODE bits, and pre-adder and input pipeline by modifying INMODE. Through manipulation of control signals, the DSP Slice can be dynamically switched between different configurations at runtime. On the other hand, static parameter attributes are specified and programmed at compile time and cannot be modified at runtime. Three different datapath configurations are chosen to demonstrate the flexibility and capability of the DSP48E1 in handling various arithmetic functions. Each of the configurations selected highlights a different functionality and operating mode of the slice primitive.

Multiplication In multiplication, input data is passed through ports A and B as the first and second operand respectively. Three stage registers, A1, B1, M and P are enabled along the multiplication datapath to operate the slice at full-speed.

Addition Addition, in contrast to multiplication, does not require a multiplier, hence it is bypassed and inputs from the A, B and C ports are fed straight into the ALU unit. Since the multiplier is removed from the addition datapath, an extra set of registers has to be enabled to keep the pipelines operating at three stages. This is necessary when designing a processor, so as to have a fixed latency through the DSP Slice, resulting in better controlability; we fix the latency through the primitive at three cycles. To compensate for the stage where register M is bypassed, registers A2 and B2 are enabled.

Compare Similar to addition, the compare operation is configured to follow a non-multiplier datapath with additional pattern detect logic enabled. The pattern detect logic compares the value in register P against a pattern input. If there is a match, an output signal, *patterndetect*, is set to high. The pattern field can be obtained from two sources, a dynamic source from input C or a static parameter field. As we want the pattern detect logic to detect different data patterns, we configure the slice to obtain pattern data from the C input.

3.2 Processor Architecture

Our processor is a scalar processor, loosely based on the MIPS architecture in terms of dataflow and pipeline stages. It executes 32-bit instructions on 16-bit data. Only a single DSP48E1 slice is used, with much of its work being in the ALU. Since the width of input ports A, B and D is less than 32 bits, it made sense to define a data width of 16 bits. There are, in total, 6 stages in the processor execution pipeline with a latency of 1-clock cycle per stage. The full 3-stage pipeline is enabled for the DSP48E1 primitive. The remaining stages are *instruction fetch*, *instruction decode*, and *write-back*. Both instruction decode and operand fetch occur in the same stage. After processing, results from the ALU are written to the register file in the write-back stage. Each DSP48E1 block is located beside a Block RAM (BRAM) slice, separated by a thin column of logic resources in the FPGA fabric. This composition of memory-logic-DSP provides an ideal structure for a fixed instruction size RISC processor architecture. With a BRAM for the instruction

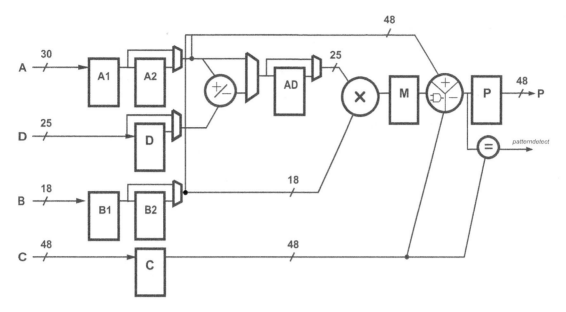

Figure 1: Input paths to pre-adder, multiplier and ALU. Path D can be used as alternative to path A.

	31	27	25		21 20		16 15		11 10		0
Register	opcode	0	Rd		Rs		Rt		00000000000		
Immediate	opcode	1	Rd		Rs		#<imm16>				
Shift	opcode	0	Rd		Rs		#<shift>		00000000000		
Branch	opcode	0	00000		Rs		Rt		00	#<offset>	

Figure 2: Processor instruction format.

and data memory, and simple decoding logic for the RISC instructions, this architecture takes full advantage of available FPGA resources while minimising the use of extra logic at the same time. Furthermore, using the RISC architecture simplifies compiler design and re-targetting as well as program coding.

Instruction Memory Processor instructions are stored in a BRAM memory primitive. Instead of storing instructions externally, off-chip, we take advantage of the abundant on-chip memory resources. A single port ROM block of size 512 x 32 is generated using Xilinx Core Generator. After synthesis, the ROM block is mapped into a single RAMB18E1 primitive. To improve timing, the output of the BRAM is registered.

Register File For the register file, the main issue that needs to be addressed is the requirement for 2 simultaneous reads and 1 write in a single cycle. BRAM primitives support at most 2 operations per cycle and the third operation has to be performed in the next clock cycle. Using BRAM would be interesting, in that this would allow for a large register file, but overcoming the access restrictions would require a more exotic processor architecture; something we plan to investigate in the future. The Xilinx RAM32M primitive is a multi-port distributed RAM designed to implement a register file; this is what we use. This type of RAM is implemented using distributed memory and hence does not consume BRAM resources. Although RAM32M allows 3

reads and 1 write in a single cycle, we only use 2 reads and 1 write at this stage.

Shift LUT In order to shift by n bits (equivalent to multiplying by 2^n), a shift operation requires the second operand (2^n) to be computed before entering the DSP48E1. A shift LUT is constructed to store these values using 2 RAM16X8S primitives, but this affects the processor's timing.

Branch Branching is not evaluated until the end of the execution cycle. In a branch operation, the ALU compares two operands and determines if the operands are equal or otherwise. If the operands are equal, as in the case of BEQ, a status signal *branchsel* is asserted and passed back to the instruction fetch stage along with target address.

3.3 Implementation Results

In this subsection, we analyse the performance of our processor system in terms of frequency, resource usage, latency, and instruction count. The processor is implemented on a Xilinx Virtex 6 XC6VLX240T using Xilinx ISE 13.2. All results are obtained through synthesis of Verilog source code, with all processes run at default settings. The post place and route frequency obtained for speed grades of -3, -2 and -1 are 534 MHz, 470 MHz and 402 MHz respectively, limited by the BRAM access path.

Referring to Table 2, the processor consumes a minimal amount of logic. Since a hardcore primitive is used for the ALU and other functions, only minimal additional circuitry is implemented in the logic fabric. The instruction memory also uses a Block RAM, again freeing logic for other uses. Apart from obvious area savings, this strategy improves the overall timing performance due to the presence of high-speed hardcore primitives in the datapath.

3.4 Code Execution and Analysis

For the analysis of latency and instruction count, the loop kernel of a 3 x 3 median filter is mapped to the processor. Table 3 shows the latency based on idealised conditions with no branch penalty, and realistic conditions with penalty. In idealised conditions, we assume a zero branch

Table 1: Processor Instructions.

Instruction	Operation	Inmode	Opmode	Alumode	Additional Circuitry
nop	none	00000	0000000	0000	none
add	rd = rs + rt	00000	0110011	0000	extra C reg
sub	rd = rt - rs	00000	0110011	0011	extra C reg
mul	rd = rs x rt	10001	0000101	0000	none
muladd	rd = (rs x rt) + ru	10001	0110101	0000	extra C reg
mulsub	rd = (rs x rt) - ru	10001	0110101	0001	extra C reg
mulacc	rd = (rs x rt) + #<feedback>	10001	1000101	0000	none
and	rd = rs and rt	00000	0110011	0000	extra C reg
xor	rd = rs xor rt	00000	0110011	0100	extra C reg
xnr	rd = rs xnr rt	00000	0110011	0101	extra C reg
or	rd = rs or rt	00000	0110011	1100	extra C reg
nor	rd = rs nor rt	00000	0110011	1110	extra C reg
not	rd = rs not rt	00000	0110011	1101	extra C reg
nand	rd = rs nand rt	00000	0110011	1100	extra C reg
lsl	rd = rs << #<shift>	10001	0000101	0000	shift LUT
lsr	rd = rs >> #<shift>	10001	0000101	0000	shift LUT
asr	rd = rs << #<shift>	10001	0000101	0000	shift LUT
mov	rd = rs	00000	0110011	0000	none
beq	(rs == rt) pc = pc + offset	00000	0110011	1100	branch target address
bne	(rs != rt) pc = pc + offset	00000	0110011	1100	branch target address

Table 2: Resource usage on XC6VLX240T.

Resource	Used	Available	Utilization
Slice Registers	238	301,440	< 1%
Slice LUTs	190	150,720	< 1%
DSP	1	768	< 1%
BRAM	1	832	< 1%

penalty. A single inner loop requires 7 instructions and the total number of instructions for the 3 x 3 median filter is 224. Each instruction takes 6 clock cycles to complete and one instruction is fetched every clock cycle. In actual implementation, the branch penalty is as much as 5 instruction cycles. The branching decision is determined by the ALU, and by the time program counter changes, 5 instructions have been fetched.

Table 3: Median filter instruction count and latency.

Loop	Ideal		With penalty	
	Inst count	Latency	Inst count	Latency
Single loop	7	12	12	16
Total	224	227	384	387

4. CONCLUSION

In this paper, we have presented a discussion of the DSP48E1 primitive shown and how it can be harnessed as the core of a general-purpose soft processor. We have developed a processor design that leverages the DSP48E1 to support as many standard assembly instructions as possible, as well as other instructions suited to the primitive's DSP roots, in each case, focussing on using as little extra logic as possible. We have shown that it is possible to build a processor that runs at over 400MHz, using approximately 200 slice LUTs and registers.

5. REFERENCES

[1] S. Chalamalasetti, S. Purohit, M. Margala, and W. Vanderbauwhede. MORA - an architecture and programming model for a resource efficient coarse grained reconfigurable processor. In *NASA/ESA Conf. on Adaptive Hardware and Systems (AHS)*, pages 389–396, 2009.

[2] X. Chu and J. McAllister. FPGA based soft-core SIMD processing: A MIMO-OFDM fixed-complexity sphere decoder case study. In *Proc. Int. Conf. on Field Programmable Technology (FPT)*, pages 479–484, 2010.

[3] X. Chu, J. McAllister, and R. Woods. A pipeline interleaved heterogeneous SIMD soft processor array architecture for MIMO-OFDM detection. In *Proc. Int. Symp. on Applied Reconfigurable Computing (ARC)*, pages 133–144, 2011.

[4] M. Milford and J. McAllister. An ultra-fine processor for FPGA DSP chip multiprocessors. In *Asilomar Conf. on Signals, Systems and Computers*, pages 226 –230, 2009.

[5] Xilinx Inc. *Virtex-6 FPGA DSP48E1 User Guide*, 2011.

[6] P. Yiannacouras, J. Steffan, and J. Rose. Application-specific customization of soft processor microarchitecture. In *Proc. ACM/SIGDA Int. Symp. on Field Programmable Gate Arrays (FPGA)*, pages 201–210, Feb. 2006.

[7] P. Yiannacouras, J. Steffan, and J. Rose. VESPA: Portable, scalable, and flexible FPGA-based vector processors. In *Proc. Int. Conf. on Compilers, Architecture and Synthesis for Embedded Systems (CASES)*, pages 61–70, Oct. 2008.

[8] J. Yu, G. Lemieux, and C. Eagleston. Vector processing as a soft-core CPU accelerator. In *Proc. ACM/SIGDA Int. Symp. on Field Programmable Gate Arrays (FPGA)*, pages 222–232, Feb. 2008.

Functionally Verifying State Saving and Restoration in Dynamically Reconfigurable Systems

Lingkan Gong and Oliver Diessel
School of Computer Science and Engineering
University of New South Wales
Sydney, NSW, Australia 2052
{lingkang,odiessel}@cse.unsw.edu.au

ABSTRACT

Dynamically reconfigurable systems increase design density and flexibility by allowing hardware modules to be swapped at run time. Systems that employ checkpointing, periodic or phased execution, preemptive multitasking and resource defragmentation, may also need to be able to save and restore the state of a module that is being reconfigured. Existing tools verify the functionality of a system that is undergoing reconfiguration. These tools can also be employed if state is accessed using application logic. However, when state is accessed via the configuration port, functional verification is hindered because the FPGA fabric, which mediates the transfer of state between the application logic and the configuration port, is not being simulated. We describe how to efficiently simulate those aspects of the fabric that are used in accessing module state. To the best of our knowledge, this work is the first to allow cycle-accurate simulation of a system partially reconfiguring both its logic and state and a case study shows that our method is effective in detecting device independent design errors.

Categories and Subject Descriptors

B.6.3 [**Logic Design**]: Design Aids—*Verification*; I.6.7 [**Simulation and Modeling**]: Simulation Support Systems

General Terms

Verification

Keywords

FPGA, Dynamically Reconfigurable Systems, Verification, State Saving and Restoration

1. INTRODUCTION

The exponential increase of hardware design costs and risks have driven the electronic industry to use reconfigurable devices such as FPGAs as computing platforms. Com-

pared with customized chips, hardware/software systems implemented on reconfigurable devices achieve more flexibility for potential upgrades or bug fixes over the product life-cycle, while sacrificing acceptable tradeoffs in power, area and performance. Dynamically Reconfigurable Systems (DRS) extend such flexibility by allowing partial reconfiguration of hardware modules at run time. Recent examples of DRS include a networked multiport switch [13], a software-defined radio [10] and a video-based driver assistance system [1].

Apart from reconfiguring module logic, state saving and restoration (SSR) is a common and sometimes essential requirement of DRS. For example, to recover from an error, a module can be restored to a saved checkpoint [7]. For periodic/phased applications, the system configures the required computational module for each period of execution and copies the results across periods [5]. In hardware multitasking, a module can be preempted and later resume execution from the interrupted state [11]. To defragment FPGA resources, modules can be relocated from their original locations to other areas of available resources [6]. In each of these scenarios, module state need to be saved and restored during or after partial reconfiguration.

The architectural flexibility of DRS introduce challenges in functionally verifying the system because conventional verification methods assume that hardware circuits are time-invariant. Although each configuration of a design can be verified using traditional methods, new approaches are required to verify the design *while* it is undergoing reconfiguration, during which modules need to be properly swapped and module state needs to be saved and restored [3].

Register Transfer Level (RTL) simulation is the most common method for verifying hardware design functionality. By visualizing selected signals in a waveform viewer, RTL simulation assists in debugging a design without implementing it. In order to simulate partial reconfiguration of module logic, existing methods use multiplexers to interleave mutually exclusive modules [9], and use simulation-only bitstreams to capture the cycle-accurate behavior of module swapping [4]. This paper extends ReSim [4], our previous work, to support the simulation of a design reconfiguring *both* module logic and module state. In particular, we consider designs that utilize the Configuration Port (CP) to save and restore state (e.g., [6] [11]). The key contributions of this paper are:

- Extending the original use of simulation-only bitstreams (SimB) with state data to enable cycle-accurate simulation of CP-based SSR in DRS designs.

- Providing a case study showing how ReSim assists in simulating and debugging CP-based SSR.

It should be noted that for CP-based SSR, RTL simulation cannot detect device-dependent bugs. Running the DRS on the target device is the only way to accurately verify the system in real-world environments. For example, the locations of the state data on the FPGA fabric can't be checked until they have been determined by implementation. However, it is non-trivial to visualize signals of the implemented design for the sake of debugging because extra effort is required to insert probing logic using vendor tools such as Chipscope. Moreover, the probing logic can only visualize a limited number of signals for a limited period of time. As a result, debugging the design on chip involves costly iterations of probing the relevant signals and reimplementing the design. It is therefore desirable to perform RTL simulation to identify and fix as many *device-independent* bugs as possible in the early stage of the design cycle, and leave the *device-dependent* part of the design to be tested on the target FPGA once the implementation has been completed.

The rest of this paper is organized as follows. Section 2 outlines the background for CP-based SSR. We discuss the architecture of the ReSim library in Section 3. Section 4 illustrates the use of our tool on a case study, while the final section concludes the paper.

2. BACKGROUND AND RELATED WORK

For DRS designs, state data include storage elements such as flip-flops and memory cells such as Block RAM (BRAM) located on the FPGA fabric. To save and restore module state, the designer needs to create a datapath to access the state. There are currently two approaches for creating such a datapath. The first method adds design specific application logic to read/write storage elements (e.g., [7]), whereas the other approach utilizes the configuration port of the FPGA to access flip-flops and memory cells (e.g., [6], [11]). Comparing the two methods, CP-based SSR is more general as the configuration port defines a uniform access method for saving and restoring state anywhere on chip. Therefore, it does not require any changes to module logic, and does not introduce any side effect to the system such as extra resource usage or tighter timing requirements. In contrast, customized state access logic needs to be carefully inserted for each module in the system.

In a typical DRS design (see Figure 1), multiple reconfigurable modules (RM) are mapped to a Reconfigurable Region (RR) and communicate with the static circuitry to perform the required tasks. For CP-based SSR, the static part starts state saving with a synchronization operation, which copies the RM state to the configuration memory of the FPGA fabric. The user design then reads the state data by sending frame addresses, i.e., the addresses of the configuration memory, to the configuration port and sampling the returned data. In this process, requests from the user design and responses from the fabric are all delivered in the form of bitstreams, a device-specific, packet-like data structure containing commands (e.g., synchronize, read) and parameters (e.g., frame addresses). The restoration process reverses the state saving procedure. Please refer to the configuration guide (e.g., [14]) for further details of CP-based SSR.

The primary difficulty of simulating and verifying CP-based SSR is the conflict between simulation accuracy and

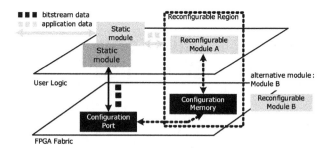

Figure 1: State saving and restoration in DRS

productivity. Because the state access datapath (i.e., the configuration memory and the configuration port depicted as black boxes in Figure 1) is part of the FPGA fabric, the most straightforward way to simulate CP-based SSR involves modeling the fabric. Unfortunately, vendor tools do not provide a simulation model for the fabric and even if such a model were available, simulating the design by modeling the fabric is performed at too low a level. For the sake of productivity, it is desirable that functional simulation should be independent of the FPGA fabric. This means it is better simulating user logic via HDL signals than simulating the bit settings in the configuration memory. It is also better to avoid the time-consuming implementation step in the iterative simulation-design cycle.

Alternatively, if accuracy can be sacrificed, CP-based SSR can be simulated without dependence on the FPGA fabric. The Dynamic Circuit Switch method modifies the RTL description of all storage elements in simulation so as to save and restore state [9]. By extending SystemC, ReChannel use call-back functions to access arbitrary variables in the modeled design [8]. However, the functionality added to the storage elements and the call-back functions do not exist on the target device and the simulated design therefore doesn't replicate what is implemented. Furthermore, these methods fail to model the interaction between user logic and the configuration port and thus only offer limited assistance in verifying CP-based SSR.

Although ReSim facilitates cycle-accurate simulation of configuration port accesses, it only supports the verification of a DRS reconfiguring its logic [4]. The simulation-only bitstream concept introduced in ReSim raises simulation accuracy by linking accesses to the simulated configuration port with module swapping. This work extends the application of the simulation-only bitstream by transferring module state via its contents. With this extension, ReSim is the first work to support cycle-accurate yet physically independent simulation of CP-based SSR.

3. RESIM LIBRARY

The core idea of ReSim is to use a simulation-only layer to emulate the physical fabric of FPGAs so as to achieve the desired balance between accuracy and physical independence. Figure 2 redraws Figure 1 with all the physically dependent blocks (solid black boxes) replaced by the components of our simulation-only layer (open black boxes). These components are known as simulation-only artifacts or artifacts. It should be noted that the artifacts only model the aspects of the fabric that are essential to partial reconfiguration. In particular, the configuration bitstreams are replaced by simulation-only bitstreams, possible configuration ports are represented by an ICAP artifact, and the part of configuration memory to which each reconfigurable region (RR) is mapped is substituted by an Extended Portal.

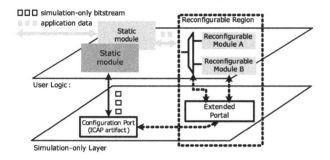

Figure 2: Using the simulation-only layer

To simulate partial reconfiguration of module logic, ReSim connects all RMs in parallel in the same way as existing approaches. However in ReSim, the selection of the active module is controlled by the Extended Portal, which is in turn triggered by writing a SimB to the ICAP artifact. SimBs are also used to simulate CP-based SSR. In particular, the static module transfers a SimB instead of a real bitstream to the ICAP artifact. By parsing the SimB, the ICAP artifact extracts the readback parameters, and controls the Extended Portal. The Extended Portal probes the RM state (typically modeled by HDL signals), and returns the state data to the ICAP artifact. Finally, the ICAP artifact returns the retrieved state data to the user design as a readback SimB. Restoring the state of a RM also utilizes the artifacts and SimBs to mimic the behavior of the FPGA fabric. However, a different SimB from the one for state saving is used, and the state data are copied back to the HDL signals. Using the SimB, ReSim mimics the behavior of the FPGA fabric during partial reconfiguration, and significantly improves the accuracy of simulating the user design.

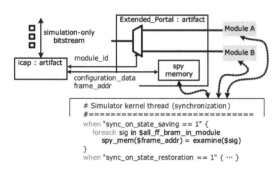

Figure 3: Extended Portal

Figure 3 illustrates the architecture of the Extended Portal. Mimicing the FPGA fabric, the Extended Portal instantiates a spy memory as a substitute for the configuration memory to buffer the state data. To simulate state saving, for example, a Simulator-kernel thread (SKT) component probes RM signals when a synchronization operation is requested by the static module. The SKT uses simulator commands (e.g. ModelSim `examine`, `force` commands) to extract values of arbitrary HDL signals. The extracted signal values are then buffered to the spy memory, and are

returned to the static module on each read from the ICAP artifact.

As the simulation-only layer doesn't rely on any physical details of the target device, the simulation of CP-based SSR is physically independent. The ReSim library is built upon the existing SystemVerilog language and Open Verification Methodology (OVM) [2], an open source SystemVerilog class library. As a result, ReSim is fully compatible with existing and mainstream EDA tools, although the current implementation of ReSim only supports ModelSim.

In order to avoid requiring the designer to create artifacts from scratch, ReSim automatically generates all artifacts based parameters defined in a script. The parameters include the name of each RM, the IO signals for each RR, the target FPGA family, etc. When executed, the script then creates parameterized artifacts by calling ReSim APIs.

4. A CASE STUDY

The hardware architecture of our case study is similar to a reference design from Xilinx [12] (see Figure 4). Although the original design only partially reconfigures the logic, we modified the software running on the microprocessor to support state saving and restoration as well. Our case study runs a periodic application. In each period of execution, the `xps_math` module is dynamically reconfigured with either an `adder` core or a `maximum` core as two alternative RMs. Apart from computation, each core maintains a `statistic` register, and its value is copied across configuration periods.

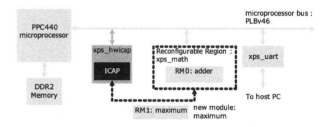

Figure 4: CP-based SSR case study, after [12]

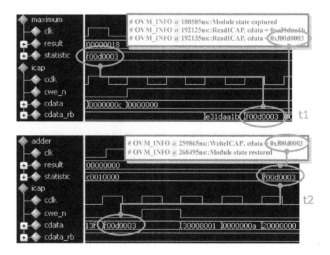

Figure 5: Waveform example

Figure 5 shows two waveform segments obtained by simulating the design using ModelSim 6.5g. Here, an old `maximum` core is reconfigured to a new `adder` core, and the value of the `statistic` register is copied across the two cores.

- **@t1**: Before reconfiguration, A readback SimB is written to the ICAP artifact requesting to save the value of the `statistic` register (0xf00d0003). The ICAP artifact is switched to read mode and returns the retrieved state data (see the `cdata_rb` signal).

- **@t2**: After reconfiguration, the `statistic` register of the new `maximum` core is initially zero. Then a restoration SimB containing the saved value is written to the ICAP artifact (see the `cdata` signal), after which the `statistic` register is restored to the desired value.

We detected dozens of bugs in our design using the above cycle-accurate simulation of CP-based SSR. For example, the restoration of the `statistic` register should have been skipped during the first round of execution as there was no previous execution period. The `statistic` register saved by the `xps_hwicap` driver was not properly returned to the periodic application, and was destroyed at the end of the function call. These bugs were detected as a consequence of incorrect values being restored to the `statistic` register. In the debugging process, it took less than a minute to complete one iteration of compiling and simulating the design on a Windows XP, Intel 2.53G Dual Core machine.

The simulated design was subsequently tested on an ML507 board with a Virtex-5 FX70T FPGA and we detected two device dependent bugs. As one example, the saved data were wrong because we failed to invert each bit of the saved state data as required by Virtex-5 FPGAs. In this cases, ReSim failed to mimic the exact behavior of the target device. Although this bug was trivial and would not have occurred with experienced designers, we used 3 iterations to trace the cause of this bug. Each iteration involved inserting new probing logic and re-implementing the design, and took 53 min to complete on the same machine used for simulation.

In this study, the workload for integrating ReSim with the testbench included: 50 lines of Tcl script for generating the artifacts, 10 lines of Verilog code to instantiate the artifacts, and 4 ModelSim commands to start the SKT. In contrast, the workload for modifying the `xps_hwicap` driver to support CP-based SSR included 1300 lines of C code. The simulation overhead of ReSim is proportional to the number of registers to be saved and restored. The overhead is also proportional to the frequency of SSR in a simulation run as each saving and restoration triggers a synchronization between the spy memory and the HDL signals. For our case study, the simulation overhead was negligible.

5. CONCLUSIONS AND FUTURE WORK

Functional verification has become a significant challenge in DRS designs. It is therefore essential to perform RTL simulation of the reconfiguration process, including the saving and restoration of state, as part of a full system simulation. This paper proposes extending ReSim [4] to simulate CP-based SSR. In particular, we use simulation-only bitstreams to correlate the accesses of a simulated configuration port with the state of a simulated module.

As ReSim abstracts out the details of the FPGA fabric, the simulation-only layer can be regarded as a vendor-independent device. Simulation can be thought of as functionally verifying a design on such a vendor-independent device. Although RTL simulation is only an approximation to the target device, our case study demonstrated that with negligible development and simulation overhead, simulation using ReSim captured a reasonable number of bugs and avoided most of the costly iterations of using Chipscope. Meanwhile, as ReSim only offers limited help in detecting device dependent bugs, it is desirable that a DRS be separated into a device dependent part and a device independent part. ReSim can then assist in verifying the device independent part, and the rest of the design should be tested on the target device.

6. REFERENCES

[1] CLAUS, C., ZEPPENFELD, J., MULLER, F., AND STECHELE, W. Using Partial-Run-Time Reconfigurable Hardware to accelerate Video Processing in Driver Assistance System. In *Design, Autom. Test in Europe* (2007), pp. 1–6.

[2] GLASSER, M. *Open Verification Methodology Cookbook.* Mentor Graphics Corporation, 2009.

[3] GONG, L., AND DIESSEL, O. Modeling Dynamically Reconfigurable Systems for Simulation-based Functional Verification. In *Field-Prog. Cust. Comput. Machines* (2011), pp. 9–16.

[4] GONG, L., AND DIESSEL, O. ReSim: A Reusable Library for RTL Simulation of Dynamic Partial Reconfiguration. In *Field-Prog. Tech.* (2011).

[5] JIANG, Y.-C., AND WANG, J.-F. Temporal Partitioning Data Flow Graphs for Dynamically Reconfigurable Computing. *IEEE Trans. VLSI Syst. 15*, 12 (2007), 1351–1361.

[6] KALTE, H., AND PORRMANN, M. Context Saving and Restoring for Multitasking in Reconfigurable Systems. In *Field-Prog. Logic and App.* (2005), pp. 223–228.

[7] KOCH, D., HAUBELT, C., AND TEICH, J. Efficient Hardware Check pointing: Concepts, Overhead Analysis, and Implementation. In *Field-Prog. Gate Arrays* (2007), pp. 188–196.

[8] RAABE, A., HARTMANN, P. A., AND ANLAUF, J. K. ReChannel: Describing and Simulating Reconfigurable Hardware in SystemC. *ACM Trans. Design Autom. Electr. Syst. 13*, 1 (2008), 15.

[9] ROBERTSON, I., AND IRVINE, J. A Design Flow for Partially Reconfigurable Hardware. *ACM Trans. Embedded Comput. Syst. 3*, 2 (2004), 257–283.

[10] SEDCOLE, P., BLODGET, B., ANDERSON, J., LYSAGHT, P., AND BECKER, T. Modular Partial Reconfiguration in Virtex FPGAs. In *Field-Prog. Logic and App.* (2005), pp. 211–216.

[11] SIMMLER, H., LEVINSON, L., AND MANNER, R. Multitasking on FPGA Coprocessors. In *Field-Prog. Logic and App.* (2000), pp. 121–130.

[12] XILINX INC. *Partial Reconfiguration of a Processor Peripheral (UG744)*, 2009.

[13] XILINX INC. *Partial Reconfiguration User Guide (UG702)*, 2010.

[14] XILINX INC. *Virtex-5 FPGA Configuration User Guide (UG191)*, 2010.

A Configurable Architecture to Limit Wakeup Current in Dynamically-Controlled Power-Gated FPGAs

Assem A. M. Bsoul and Steven J. E. Wilton
Department of Electrical and Computer Engineering
University of British Columbia
Vancouver, British Columbia, Canada
absoul@ece.ubc.ca   stevew@ece.ubc.ca

ABSTRACT

A dynamically-controlled power-gated (DCPG) FPGA architecture has recently been proposed to reduce static energy dissipation during idle periods. During a power mode transition from an off state to on state, the wakeup current drawn from power supplies causes a voltage droop on the power distribution network of a device. If not handled appropriately, this current and the associated voltage droop could cause malfunction of the design and/or the device. In DCPG FPGAs, the amount of wakeup current is not known beforehand as the structures of power-gated modules are application dependent; thus, a configurable solution is required to handle wakeup current. In this paper we propose a programmable wakeup architecture for DCPG FPGAs. The proposed solution has two levels: a fixed *intra-region* level and a configurable *inter-region* level. The architecture ensures that a power-gated module can be turned on such that the wakeup current constraints are not violated. We study the area and power overheads of the proposed solution. Our results show that the area overhead of the proposed inrush current limiting architecture is less than 2% for a power gating region of size 3x3 or 4x4 tiles, and the leakage power saved is more than 85% in a region of size 4x4 tiles.

Categories and Subject Descriptors

B.7.1 [**Integrated Circuits**]: Types and Design Styles— *gate arrays, VLSI (very large scale integration)*

General Terms

Design

Keywords

FPGA, Power Gating, Architecture, Inrush Current, Leakage Power

1. INTRODUCTION

Static power dissipation is a major component of the total power consumption in field-programmable gate array (FPGA) devices based on sub-90 nm CMOS technology nodes [2]. A recent white paper from Xilinx shows that even with improved process technology, static power could be as large as dynamic power for 28 nm technology node [11]. This matches the prediction that the effect of static current will increase with continuous technology scaling [21]. The operation of some low-power applications, such as mobile and hand-held devices, is dominated by idle periods with small bursts of activity; this may cause the leakage energy consumed due to static power to surpass that dissipated during activity periods.

Recently, a *dynamically-controlled* power-gated (DCPG) FPGA architecture has been proposed as a way to reduce idle periods' power consumption [5]. During their idle periods, functional blocks in this architecture could be powered down by using power gating, thus reducing their static power dissipation. Unlike *statically-controlled* power gating (SCPG) in which the states (on or off) of the different parts in an FPGA device are set at configuration time [18, 8, 23], DCPG enables run-time control of the power state based on the application's behavior.

An important issue in DCPG architectures is the amount of current that is drawn from the power supply during the wakeup phase, known as the *wakeup* or *inrush current*. This current can be large and can cause a temporary IR drop across the power rails. This temporary IR drop, called *voltage droop*, may cause functional errors due to reduced noise margins and degraded performance by corrupting data storage elements, generating incorrect combinational logic output, or resulting in violations of the timing constraints [13]. Figure 1 shows an illustration of inrush current and the associated voltage droop. As the amount of inrush current

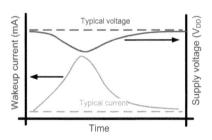

Figure 1: Hypothetical inrush current and the associated voltage droop

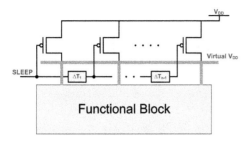

Figure 2: Illustration of a typical wakeup sequencing technique

increases, the droop on the power grid will increase, resulting in a violation of power integrity constraints if the circuit is not designed appropriately.

In application-specific integrated circuits (ASICs) that employ power gating, the inrush current problem is well understood, and many solutions have been proposed in the literature. These solutions revolve around staggering the wake up phase of the logic to guarantee that voltage droop constraints are not violated [10, 7, 6]. This can be done by chaining the power-on signal to turn on power gates in a specific timing sequence by using appropriate delays. Figure 2 shows an illustration of a typical power-on sequencing solution. As can be seen in the figure, the delay to turn on the next stage in the parallel sleep transistor (ST) chain allows the virtual V_{DD} node to be partially charged, such that when the next ST is turned on, the amount of current will remain within a specified constraint.

In a DCPG FPGA architecture, the problem of handling inrush current is different from that in ASICs. FPGAs are flexible in order to implement a wide range of applications. This means that the structures of power-gated modules that can be mapped to a DCPG FPGA is not known beforehand. The DCPG regions in an FPGA that will be used to implement a specific power-gated module and the amount of wakeup current are not known at fabrication time; therefore, it is not feasible to implement a fixed inrush current handling circuitry in such architecture. Instead, it must be flexible enough to support a variety of scenarios. To the best of our knowledge, there has been no published work that considers the problem of inrush current in a power-gated FPGA architecture.

In this paper we propose a configurable inrush current handling architecture suitable for DCPG FPGAs. Our proposed architecture contains delay elements to sequence the wakeup phase of a power-gated module implemented in a DCPG FPGA architecture. Although the experienced designer could handle inrush current manually in a DCPG FPGA (as suggested in [5]), this increases the design overhead, and has disadvantages that our proposed technique overcomes as we discuss in the next section.

Our proposed architecture has two levels. The lower level, which we refer to as the *intra-region* level, consists of circuitry to wake up a single power gating region (PGR). This circuitry sequences the enabling of a series of sleep transistors, and ensures that the inrush current resulting from powering up a single PGR does not violate the constraints set by the power grid. This level is not configurable; the required delays must be determined when the chip is fabricated. The upper level, which we refer to as the *inter-region* level, consists of circuitry to sequence the turning on of re-

gions in the same power-gated module to ensure voltage drop constraints are not violated. Since it is not known at fabrication time how big power-gated modules will be or where they will be on the chip, this level must be *configurable* using static RAM bits that are set when the chip is configured. As described above, this combination of configurable and static wakeup circuitry is unique to FPGAs.

The paper is organized as follows. Section 2 provides background on the sources of voltage drop during the turn-on phase of a power gating architecture, discusses previous works related to handling wakeup current, and provides a background on the DCPG FPGA architecture framework assumed in this paper. Section 3 discusses the power grid model used in this paper, and provides analysis of the effect of inrush current in the DCPG FPGA architecture. Section 4 shows the proposed circuits for handling inrush current and discusses the associated architectural tradeoffs. Section 5 shows the experimental setup and the results for our study. Finally, we conclude the paper in Section 6, and point out directions for future work.

2. BACKGROUND

2.1 Inrush Current in Power-Gated Designs

In power-gated designs, sleep transistors are used as power switches that can be turned off when a functional block is idle to disconnect it from the power supplies. This significantly reduces leakage power dissipation [9]. Assuming a header PMOS switch, as in the DCPG FPGA architecture used in this paper, if a functional block remains in sleep mode for a long time, all internal devices and the virtual V_{DD} node of the block will gradually discharge.

As the functional block is turned on, a sudden charging of all floating nodes in the block will result in a large current to be drawn from the power supplies that flows through the sleep transistors. This large surge current causes voltage drop on the power grid due to IR and $L\frac{\partial i}{\partial t}$ drops, leading to functional errors and degraded performance, and may cause reliability problems due to electromigration [12].

IR drop is the voltage drop on the power network metal lines due to their resistance. Usually, the IR drop is analyzed in static power grid analysis techniques, where all metal segments of the power grid are replaced by their equivalent resistances, and the functional blocks are modeled as current sources that draw the maximum average current that can be estimated by power analysis techniques. In this paper, we focus on the IR component of the power grid voltage drop.

The second source of the droop is the inductive voltage drop ($L\frac{\partial i}{\partial t}$). This voltage drop mechanism occurs due to current transients. Typically, inductive voltage drop has a significant effect at the package level of the power distribution network.

2.2 Previous Work

There have been some previous works that have proposed the use of power gating in FPGAs.

Bsoul and Wilton suggested handling inrush current manually in their DCPG FPGA architecture [5]. In their approach, the designer creates the power controller such that separate signals are used to wake up every region in a power-gated module; the power controller must be designed to guarantee that a small amount of the logic in a power-gated module is turned on at a time in order to limit the maximum

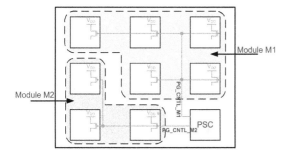

Figure 3: Using a power controller to control the power state of the different functional blocks

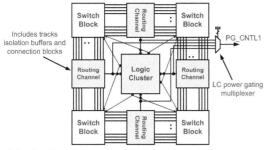

(a) The basic dynamic power gating architecture

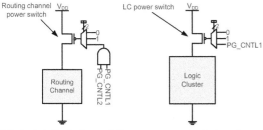

(b) A detailed view of the power control signal connections

Figure 4: Dynamic power gating architecture for a logic cluster and its routing channels

inrush current. In addition to the complexity that a designer may face in this approach, which is not desirable in FPGAs, the power dissipation of the power controller can increase, offsetting any leakage-energy saving opportunities that may exist in the application. Moreover, the additional signals generated from a power controller will compete with other circuit signals for the FPGA's routing resources, which may negatively affect routability and timing performance, thus affecting energy saving opportunities.

The works in [8, 4, 23, 17] discussed the use of dynamic power modes in FPGAs without addressing the effects of inrush current, and how it can be handled.

Kim et al. proposed reducing glitches on the ground and power rails due to inrush current by dynamically controlling the gate-to-source voltage (V_{gs}) of sleep transistors [14]. They also proposed daisy-chaining the wakeup of the sleep transistors in a chain of size-increasing sleep transistors.

Howard and Shi proposed reducing inrush current by splitting the chip into logic rows, each powered up by one or few sleep transistors [10], with a controller to stagger their turn on. They also proposed a two-stage power-on method as an

alternative solution. A trickle chain of sleep transistors is turned on to slowly charge the floating nodes, followed by the turn on of the main chain to fully charge the nodes.

Shi and Li proposed a programmable power gating unit [22]. The unit is composed of multiple daisy chains of sleep transistors, and can be configured to select which chains are turned on first to trickle charge the design, and which chains are activated later to fully charge the virtual nodes to V_{DD}.

Calimera et al. presented a power gating reactivation technique based on modulating the size of sleep transistors with delay elements in order to limit the wakeup current [6]. The authors presented an algorithm that can find the optimal sizes of the sleep transistors (STs) in the delay chain for a specific standard cell library.

The above mentioned solutions in [14, 10, 22, 6] are suitable for handling inrush current for designs where the functional blocks are known beforehand. Unlike ASICs, FPGAs are configurable, and they need a solution that is suitable to a wide range of applications.

2.3 Dynamically-Controlled Power Gating

The DCPG FPGA architecture [5] is illustrated in Figure 3; the figure shows an FPGA that has an application composed of two *power-gated modules*, M1 and M2. A power state controller (PSC) is synthesized from the information that describes the behavior of the application, such as the data flow graph (DFG) of the application [4] or any other suitable description. This controller could exploit the idle periods that the modules in the application may experience by turning off the logic in the idle modules, and turning them back on when they exit their idle periods. This requires routing *power control signals* from the controller to the logic blocks that support power gating.

Figure 4 shows an example of the basic power gating architecture [5]. In this figure, a logic cluster has four input pins and four connection blocks, distributed uniformly on its four sides. Each of the connection blocks can be used either to route an endpoint of a connection to the corresponding input pin, or to route a power control signal that controls the sleep transistor (ST) of the cluster. This scheme allows power control signals to be routed from, say, an on-chip power controller to the target logic clusters in the same way that conventional signals are routed in the original architecture. If a power control signal is to be routed through an input pin, then that input pin is not used as an input to the logic implemented by the cluster. The power state can be either statically set (on/off) or dynamically controlled using the power control signal. This is achieved by proper configuration for the 3:1 multiplexer that drives the ST.

The outputs of the connection blocks are fed as inputs to the *power gating multiplexer* of the logic cluster. This multiplexer selects the input pin that will be used as the power control signal for the cluster and the bordering connection blocks; this signal is labeled *PG_CNTL1* in the figure. As shown in Figure 4(b), *PG_CNTL1* could drive the gate of the sleep transistor to turn it off for low-leakage mode, or to turn it on for normal circuit activity.

The track isolation buffers in a routing channel are shared between the connection blocks of the two neighboring logic clusters; therefore, it is important to not turn the routing channel off if either of the neighboring logic clusters is on. This is ensured in this architecture by ANDing the power

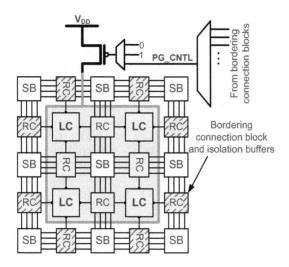

Figure 5: An example region of the dynamic power gating architecture

control signals of the two logic clusters, namely *PG_CNTL1* and *PG_CNTL2*, as shown in Figure 4(b).

This architecture can be extended to larger regions that can be turned off as a unit, thus reducing the area overhead of the required power gating circuitry [5]. In this case, a group of logic clusters and routing channels that are spatially close to each other could be power-gated by the same power control signal. Figure 5 illustrates a region of power gating of size 2 (a region size of R means the region has R^2 tiles). The bordering routing channels can be used as access points for the power control signals. Obviously, other variations of this architecture are possible where a subset of the connection boxes could be used to provide inputs to the power gating multiplexer instead of using all connection boxes.

The DCPG architecture in [5] does not turn off SRAM configuration bits or flip-flops inside the logic clusters in order to retain their values. Moreover, the architecture assumes that switch boxes are always turned on. We follow the same assumptions in this paper. Note that although switch boxes are not turned off, power gating of a logic cluster and its routing channels reduces the leakage power of a tile by more than 40% [5].

3. EFFECT OF INRUSH CURRENT

In this section, we will show that the inrush current seen by the baseline DCPG architecture can cause a large voltage droop, motivating our design in Section 4. Our estimation methodology is as follows. We first model a power grid based on estimates of the current drawn from the power supply during normal operation of an FPGA. We then model the impact of the additional current drawn when a region of the DCPG FPGA architecture is turned on, and show that, with the same size power grid, an unacceptable voltage droop occurs. Finally, we show that if we were to increase the size of the power grid to supply the required additional current, the area required by the power grid increases significantly, motivating our alternative approach.

3.1 Baseline Power Grid Model

Our model of the power grid is similar to that in [15, 16]. We assume a mesh-like power grid structure [19], in which each metal layer has alternating V_{DD} and GND lines. The

number of lines in each layer is determined based on the width and spacing between the lines. Vias are provided at the intersection of V_{DD} (GND) lines in adjacent layers. This builds a large V_{DD} (GND) net that can provide power for the transistors. Clean V_{DD} (GND) sources (power supplies) are positioned, and evenly distributed in the top metal layer of the power grid.

In order to determine the width of the power and ground lines, and the number of clean power supplies at the top level, we must first estimate the expected current requirements of the device during normal operation of the DCPG FPGA architecture (at times other than when regions of the chip are turning on or off). To do this, we first mapped the twenty largest MCNC circuits to an FPGA with parameters $N = 6$, $I = 16$, $W = 90$, $F_s = 3$, $F_{c,in} = 0.2$, and $F_{c,out} = 0.1$. We then used an enhanced version of the Poon power model [20] (modified to better model leakage power based on curve fitting of HSPICE simulation results) to estimate the power dissipated by each design *per FPGA tile*, averaged over all tiles in the design. We then used the maximum such power per tile across all designs; this quantity is denoted P_{max_avg}. For the architecture parameters above, we found that $P_{max\_avg} = 400\,\mu$W (for 45 nm technology).

We then created a parameterized HSPICE model of the power grid. Each metal segment and via in the power grid is modeled as a resistance. FPGA tiles are modeled as independent current sources. These current sources correspond to the current drawn by each FPGA tile during the normal operation (which depends on P_{max_avg}).

We assume the connections between the power grid and *each* tile are distributed in an array of size $n \times n$ where n represents the granularity at which we model the power grid at the lowest metal layers. A large value of n would lead to more accurate results at the cost of longer simulation times; in our experiments, we found that $n = 2$ gives adequate accuracy with reasonable simulation time. This is shown on the right side of Figure 6, where there are 2x2 current sources per tile. The amount of current drawn by each current source is $I_{src} = P_{max\_avg}/(V_{DD} \times n \times n)$.

We then performed HSPICE simulations, and iteratively adjusted the width of each metal line in the power grid and the number of clean power sources until the IR drop across the power grid was less than 5% of V_{DD} (in our case, this corresponds to 50 mV). This represents the size of the power grid that can supply current to the DCPG architecture during normal operation. Note that this approximation is both pessimistic and optimistic. It is pessimistic because we have assumed the largest current seen by all of our benchmark circuits, and assumed this current is drawn from each tile. It is unlikely that during normal operation, all tiles would draw this maximum current all the time. It is optimistic, because in a real FPGA, the power grid would likely be *over-designed* (i.e., provisioned to supply more current than the benchmarks might predict). We consider the impact of over-provisioning the power network in Subsection 3.3.

3.2 Voltage Droop Estimation

The power grid described above was sized to provide adequate current during normal operation. When a region is turned on, there will be additional inrush current, which will cause voltage droop on the power rails.

To estimate the magnitude of the voltage droop, we created a detailed transistor-level model of a single region.

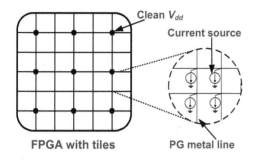

Clean V_{dd}

Current source

FPGA with tiles

PG metal line

Figure 6: Mesh power grid structure

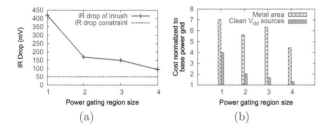

(a)   (b)

Figure 7: (a) IR drop and (b) the cost of power grid sizing to handle region's inrush current, for different region sizes

We then used HSPICE to determine the amount of current drawn per tile when a region is turned on. We then used the original power grid model from Subection 3.1, replacing the normal operating current I_{src} with this new (larger) current. Using HSPICE, we then measured the maximum voltage droop that occurs on the voltage rails.

Figure 7(a) shows the voltage droop as a function of region size (a region size of R means each region has R^2 tiles). As can be seen in the figure, for all region sizes, the droop is more than 100 mV, which is twice our target of 50 mV (5% of V_{DD}). This could lead to incorrect operation of the FPGA.

It is interesting to note that the amount of IR drop reduces as the region size increases. This is because the per tile inrush current is smaller for larger region sizes. This phenomenon happens because as the region size increases, the number of the bordering routing channels increases, i.e., the number of RCs pulled out to the region's borders increases. These bordering RCs have their own STs that turn on faster than the region's ST, resulting in a small overlap between their inrush current and the inrush current of the region's ST. Thus, the per tile floating nodes that need to be charged when turning on the region's sleep transistor decreases as the region size increases, resulting in smaller inrush current per tile.

3.3 Over-Provisioning the Power Grid

One naive solution to the large voltage droop problem is to over-provision the power grid. To investigate this, we again iteratively adjusted the metal width and the number of clean V_{DD} supplies until HSPICE predictions showed that the voltage droop due to the inrush current is below our 50 mV target. Using an area model, we were able to estimate the area impact of doing this; Figure 7(b) shows the ratio of the area required by the modified power grid to the area of the original power grid, as a function of region

size. Clearly, the overhead in doing this is significant. This result motivates the more intelligent inrush current limiting architecture described in the next section.

4. PROPOSED ARCHITECTURE

In this section, we describe our new power-gating architecture for handing inrush current in a DCPG FPGA. The architecture consists of strategically placed configurable and non-configurable delay elements and sleep transistors that can be used to ensure that inrush current does not violate the constraints set by the power grid. The proposed architecture has two levels: a fixed *intra-region* level and a configurable *inter-region* level. Each are described below.

4.1 Intra-Region Level

As described in Section 2, a *power gating region* (PGR) consists of one or more *tiles* (a tile is a logic block surrounded by routing), and is the smallest unit of granularity that can be turned on or off. The purpose of the *intra-region* power gating architecture is to limit the amount of current drawn by a single region when it is turned on. By limiting the current drawn by a single region, it becomes possible to turn on multiple regions simultaneously; this tradeoff will be revisited in Subsection 4.3.

The top portion of Figure 8 shows the intra-region power gating architecture. Instead of using a single sleep transistor (ST) for the whole region, a set of parallel STs is used for the region. At a wakeup event, the STs are turned on sequentially in order to limit the inrush current to a set value (I_{max_tile}). Clearly, I_{max_tile} must be small enough that voltage droop does not occur on the power rail when a single region is turned on; as will be described in Subsection 4.3, there are tradeoffs that may motivate a significantly smaller value of I_{max_tile}.

The value of I_{max_tile} is fixed, and is determined at fabrication time. As a result, the delay elements and the sleep transistors (STs) do not need to be configurable. For a given I_{max_tile}, the sizes of the delay elements and the STs can be found using the sizing algorithm proposed in [6]. This algorithm takes the current constraint, a minimum delay value

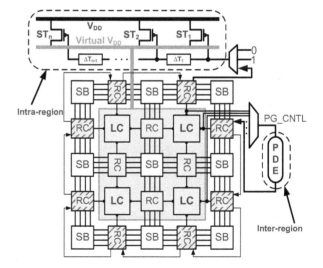

Figure 8: Power gating region with the inrush current limiting circuitry

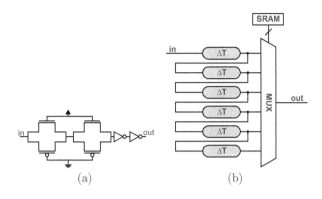

Figure 9: (a) Fixed and (b) programmable delay elements used in the proposed architecture

supported by delay elements, and an HSPICE representation of the circuitry that makes up the region, and produces a list of STs and their sizes and the delays between the turn on of the STs.

Unlike [6], which assumes that all delays are created by a set of identical delay elements, we use a library of delay elements. The smallest delay element in our library has a delay of 100 ps. Figure 9(a) shows an example delay element that consists of a series of transmission gates. Larger delays can be realized by increasing the chain length or by increasing the length of the gate of individual devices.

Note that it is also necessary to turn on the STs for the surrounding routing channels (RCs are connection blocks and track isolation buffers) of the region since they have their own virtual V_{DD} nodes (see Subection 2.3). As shown in Figure 8, the output signal of the power gating multiplexer is chained through all RCs to wakeup STs that are in sleep mode. Figure 10 shows an example of the first two RCs.

The design of the wakeup circuit in the bordering RCs of a power gating region (PGR) is similar to that for the internal part of the PGR ([6] is used to size sleep transistors and to insert delay elements). The only difference is the additional 2:1 multiplexer at the output of the last delay element in an RC's power gating circuit to select among the power control signal that was used for the RC, or the delayed output from the last delay element. This multiplexer will ensure that the correct wakeup signal is routed to the next RC in the chain even if the current RC is statically powered on/off.

Although the authors in [6] suggest that their sizing algorithm is optimal, we believe that manual optimizations can be further achieved in our architecture because we have control over the design of delay elements, unlike in [6] where they assume a specific library of cells. However, we leave the exploration of further optimizations as future work.

4.2 Inter-Region Level

The circuitry described in the previous subsection ensures that a single region can be turned on without violating the current constraint of the power grid. However, in practice there might be excess current caused by other activity, such as unrelated signals passing through the region (for switch boxes that are not powered down), or from transient signal changes while turning on a region. Furthermore, it is unlikely that the part of a user circuit that will be turned on will be confined to a single architectural region. If multiple regions are turned on simultaneously, then, even with the

architecture described in the previous subsection, current constraints might be violated causing a large voltage droop. Since the pattern of which regions will turn on together is specific to the user circuit, it is impossible to design a fixed architecture that is suitable for all application circuits. This is different than the more traditional problem of designing wakeup circuitry for a fixed-function chip (such as an ASIC); our architecture must be flexible enough to work for a wide variety of user circuit scenarios.

Our approach is to provide a Programmable Delay Element (PDE) for each region on the chip. As shown in Figure 8, the PDE is inserted just after the region gating multiplexer that selects which of the region's inputs is used as a power gating signal. Figure 9(b) shows our implementation of each PDE. Each of the blocks labeled ΔT represent a delay element of the minimum amount of delay required to wakeup a PGR.

The proposed architecture allows the CAD tool to configurably delay the turn on of individual regions, to limit the number of regions that turn on simultaneously. The maximum number of regions that can turn on at a time is dictated by the architecture, and will be denoted R_{concur}. If a user's circuit contains a power-gated functional block that occupies R_{block} regions, then the CAD tool can logically divide the block into R_{block}/R_{concur} parts. Each part would be configured with a different delay, ensuring that when a wakeup event occurs, no more than R_{concur} regions are turned on at a time. The value of R_{concur} is dictated by the architecture, and depends on the design of the power grid as well as the value of I_{max_tile}; these tradeoffs will be revisited in Subsection 4.3.

Note that the size of the multiplexer in the PDE dictates the maximum number of regions that can be turned on due to one wakeup event; if more regions are to be turned on, the CAD tool can create a multi-cycle wakeup circuit out of the general-purpose FPGA logic.

4.3 Architectural Tradeoffs

There is a complex set of tradeoffs between the architectural parameters in our wakeup circuitry and the area of the architecture and the power and wakeup time of applications. The value of I_{max_tile}, selected when designing the intra-region architecture, determines the sizes of the delay elements. While it may seem desirable to make this as large as possible to minimize the area of the intra-region level circuitry, and to minimize the time to turn on a single region, doing so reduces the achievable value of R_{concur}, since each region draws more instantaneous current. A lower value of R_{concur} means that more "steps" are required to turn on a large functional block. In addition, a lower value of R_{concur} would imply more fingers in the PDE multiplexer, increas-

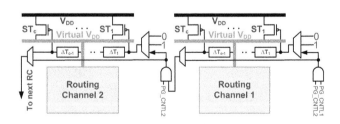

Figure 10: Example for details inside the power gating of RCs with inrush current limiting circuitry

Table 1: Terms used in finding t_{idle}

Term	Meaning
R_{block}	Number of PGRs a block occupies
R_{concur}	Number of PGRs that can be turned on concurrently
t_{idle}	The idle period for a functional block
P_{on_leak}	Average leakage power for an inactive region if not in sleep mode
P_{off_leak}	Average leakage power for a region in sleep mode
t_{turn_on}	Time to turn on up to R_{concur} regions simultaneously
t_{turn_off}	Time to turn off a region
E_{turn_on}	Energy consumed to turn on a region
E_{turn_off}	Energy consumed to turn off a region

ing its area and leakage power. The size of the power grid also affects these parameters; a larger power grid would be able to supply more instantaneous current with an acceptable voltage droop, relaxing the requirements of our power gating architecture. The optimization of all of these parameters is a complex problem; in Section 5, we experimentally investigate some of these tradeoffs.

4.4 Conditions for Energy Savings

Power-gating a functional block is only beneficial if the idle time is longer than a certain threshold. To establish the mathematical model for this constraint, we use the terms in Table 1 that are related to the DCPG FPGA architecture.

If the functional block is not placed in sleep mode during its idle period, then its power consumption during idle period using the power gating architecture is:

$$E_{idle} = t_{idle} \times R_{block} \times P_{on\_leak} \qquad (1)$$

On the other hand, if the block is placed in sleep mode during its idle period, then the energy consumption during sleep mode and to enter and exit sleep mode is:

$$E_{sleep} = t_{idle} \times R_{block} \times P_{off\_leak} + R_{block} \\ \times (E_{turn\_on} + E_{turn\_off}) + E_{DPT} \qquad (2)$$

where E_{DPT} is the energy consumed *during power transitions*; that is, during the time of the turn-on phase and during the time of the turn-off phase.

During the period of turning off a power-gated module, only R_{concur} regions can be turned off simultaneously. Therefore, some of the regions will be dissipating leakage energy until the whole power-gated module is powered down. Similarly, during the power up phase, some of the regions will dissipate leakage energy until the whole power-gated module is turned on. The leakage energy dissipated for these regions during both the turn-on and turn-off phases of a power-gated module is accounted for in 3.

$$E_{DPT} = (t_{turn\_on} + t_{turn\_off}) \times (P_{off\_leak} + P_{on\_leak}) \\ \times \sum_{i=1}^{m-1} (R_{block} - i \times R_{concur}) \qquad (3)$$

where $m = \lceil R_{block}/R_{concur} \rceil$.

Therefore, putting a functional block in sleep mode is more energy efficient than not doing so when $E_{idle} > E_{sleep}$.

This can also be written in terms of the block's idle time (t_{idle}) as:

$$t_{idle} > \frac{R_{block} \times (E_{turn\_on} + E_{turn\_off}) + E_{DPT}}{R_{block} \times (P_{on\_leak} - P_{off\_leak})} \qquad (4)$$

Note that the analysis above does not consider other parameters that play a role on deciding whether using sleep mode is feasible or not, such as the energy consumed by a power controller. However, such terms can be integrated when dealing with higher level energy models that describe the energy for a complete system.

5. EXPERIMENTAL SETUP AND RESULTS

In this section we evaluate the proposed inrush current limiting architecture, and investigate some of the architectural tradeoffs identified in Section 4.3.

5.1 Experimental Setup

Analysis Settings

We assume 45 nm [24] with $V_{DD} = 1$ V. Power consumption and duration of power transitions were measured assuming the worst case temperature of 85°C. In the algorithm that finds the sizes of the sleep transistors in the intra-region architecture, however, we assume a temperature of 25°C. This is because it is possible that an on transition happens after a long idle period, in which the temperature has gone down. At this temperature, a sleep transistor can deliver current that is larger than at higher temperatures.

Power Grid

We use the methodology explained in Section 3 to build the power grid used in this study. The power grid is assumed for a chip of 3×3 mm size (57×57 tiles). The power grid has been synthesized manually using M1-M4, with a pitch of 30 µm for the top metal layers. Clean V_{DD} and GND sources were distributed at a spacing of 400 µm.

For the lowest metal layer, which has the current sources that represent the tiles' circuits, we used a pitch that is equivalent to the physical width of a tile divided by the number of sources ($n = 2$) in each metal segment passing over a tile. Thus, there are $n \times n$ current sources connected to the lowest metal layer lines over each tile. The physical length of a tile was found by mapping the number minimum-width transistor areas (MWTA) of a tile to the physical dimensions of the tile, assuming square tiles [3].

We assume that the maximum allowed voltage drop at a virtual V_{DD} node is 100 mV (10% of V_{DD}). The source of this drop is 50 mV from the IR drop on the power grid, and 50 mV drop on sleep transistors.

Power Gating Architecture

To size the sleep transistors, we first used HSPICE simulations of the transistor-level model of a power gating region to determine the total effective width of the sleep transistors, and then used the algorithm from [6] to break this into individual sleep transistors as discussed in Section 4. In choosing the total effective width, we assumed a maximum allowable voltage drop of 50 mV on the sleep transistor of a power gating region during normal operation, and a worst case temperature of 85°C. Our sizing method requires an estimate of the activity of the nodes in a region; rather than

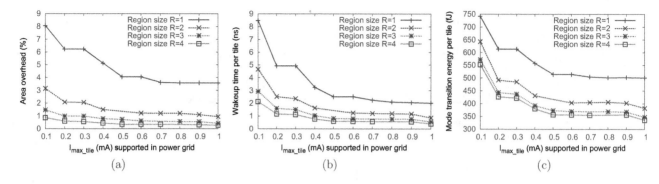

Figure 11: Intra-region level (a) area overhead, (b) wakeup time, and (c) mode transition energy results

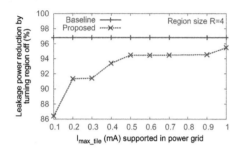

Figure 12: Leakage power reduction in off-state for baseline [5] and proposed intra-region architectures

performing extensive power analysis using the Poon power model, we assume an activity of 30% for these nodes. We found that this approximation over-estimates the power and results in larger than necessary sleep transistors, however, the overall impact on the results is small. Similar to the architecture in [5], we assumed that switch boxes and storage elements are not turned off during sleep mode.

FPGA Architecture and Area Model

We use similar FPGA architecture parameters as the ones used in Section 3: $N = 6$, $I = 16$, $W = 90$, $F_s = 3$, $F_{c,in} = 0.2$, and $F_{c,out} = 0.1$. To calculate areas, we used the MWTA model from [3]. Note that in some cases, as in delay elements, we had to increase the resistance of transistors (to achieve larger delays) by increasing the gate length of transistors. To account for this in the area model, we first calculated the area for the minimum sized transistor in the 45 nm technology node, assuming MOSIS scalable CMOS design rules [1]. Then we calculated the area for the transistor as we increase the gate length. We found that increasing the gate length by a factor of l results in $0.125 \times l$ increase in the area of the transistor. We used the same scaling factor to scale results from the MWTA area model for transistors that have larger-than-minimum gate length.

5.2 Results and Discussion

Intra-Region Level

In order to understand the area, timing, and energy overheads of the proposed architecture, we varied the maximum supported current by the power grid in each tile location; this corresponds to different power grid area costs.

Figure 11(a) shows the area overhead of the delay ele-

ments and 2:1 multiplexers in the intra-region level for different PGR sizes compared to the area of a tile that has no power gating circuitry. The area of a tile's switch box is not included. Figure 11(b) shows the wakeup time. We can see that as I_{max_tile} increases, the area overhead and the wakeup time decrease. This is due to a reduction in the number of stages of delay elements as well as a reduction in the size of each delay element. It is clear from the figures that larger PGR sizes have smaller area overhead and a smaller wakeup time per tile.

Figure 11(c) shows the energy consumption during power mode transitions (sum of both during turning on and during turning off) for different region sizes. This energy is due to delay elements as well as inrush current. As I_{max_tile} increases, the energy due to power mode transition decreases. It is interesting to note that the energy due to the inrush current dominates the total energy. For example, for $I_{max\_tile} = 400\ \mu A$, the transition energy due to inrush current is about 86% of the total energy (graphs not included due to space constraints). Although increasing the wakeup time leads to decreased instantaneous power due to smaller inrush current, the overall energy is increased.

Figure 12 shows the reduction in leakage power of a region achieved by turning off that region for both the baseline architecture (from [5]) and the proposed architecture described in Section 4, both compared to the leakage in a region in the baseline architecture. A region size $R = 4$ (4x4 tiles) is assumed in this figure. As can be seen, turning off a region has a dramatic effect on leakage power, however, the savings are smaller for the proposed architecture. This is due to the leakage energy of the delay elements. We believe, however, that circuit-level optimizations can reduce the leakage power, such as using larger gate length for devices in buffers that drive sleep transistors and combining the wakeup of multiple routing channels simultaneously. These optimizations will be investigated as future work.

The extra leakage overhead in the proposed architecture also occurs when the region is in its on-state, as shown in Figure 13. Again, we anticipate that circuit level optimizations could reduce this overhead somewhat.

Inter-Region Level

In the architecture assumed in this section, the power grid is sized sufficiently such that as long as each region draws no more than I_{max_tile}, it can supply enough current to turn on all regions on the chip simultaneously assuming there is no other activity. However, in practice, there will be other

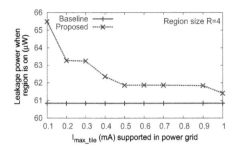

Figure 13: Leakage power in on-state for baseline [5] and proposed intra-region architectures

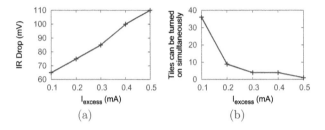

Figure 14: Implications of per tile excess current (additional to inrush current) during turn on

activity, either from unrelated signals passing through the region (switch blocks are not powered down) or from transient signal changes while turning on a region. In the presence of these extra transitions, there is a limit to the number of regions that can be turned on simultaneously.

Figure 14(a) shows the IR drop in the power rail due to this extra current, which we denote I_{excess}, assuming the worst case in which all power gating regions in a chip are turned on simultaneously. Clearly, the IR drop violates our constraint of 50 mV. Figure 14(b) shows the impact this has on the number of regions that can be turned on simultaneously, R_{concur}. As can be seen, as I_{excess} increases, the number of tiles that can be turned on simultaneously drops significantly.

As described in Section 4.3, there is a relationship between the number of regions that can be turned on at a time, R_{concur}, and the size of the PDE in each tile. As R_{concur} increases, fewer "steps" are required to turn on a large power-gated functional block, meaning each PDE can be smaller. Figure 15(a) shows this relationship for two example functional block sizes: a small one (ex5p, 109 tiles) and a large one (clma, 775 tiles). These functional block sizes represent what architects may use as a target when designing the intra- and inter-region circuits (i.e., the largest sizes of power-gated functional blocks that can be turned on using only one turn-on event). As R_{concur} increases, the required PDE size decreases, which leads to smaller area/power overheads of the inter-region level (Figure 15(b)), but at the cost of power grid sizing or the area/power overheads of the intra-region level.

Minimum T_{idle} to Achieve Energy Saving

Figure 16 shows the minimum idle time (t_{idle}) that is required for a functional block in order to achieve energy savings when turned off using the proposed architecture. We used the relation in (4) to obtain the results in this figure.

The results are reported assuming a region with $R = 4$ and $I_{max\_tile} = 400\ \mu A$. As R_{concur} increases, the required PDE size (number of supported delays) decreases, which leads to better area/power overheads of the inter-region level; this leads to smaller minimum t_{idle}.

As can be seen in Figure 16, in the worst case (when $R_{concur} = 1$), the minimum t_{idle} that is required in order to achieve energy savings is about 200-900 ns. These times are relatively small when compared to idle times that applications, such as mobile devices, experience in real life. For example, for an applications that runs at 500 MHz, a functional block only needs to be idle for about 450 cycles in order to achieve energy savings!

6. CONCLUSION AND FUTURE WORK

Wakeup inrush current can cause large voltage droop on the power distribution network in a chip, leading to a malfunction of the design or the device. In DCPG FPGAs, the problem is different from that in ASICs as the structure of applications and the power-gated modules is not known at fabrication time; thus, a configurable architecture is required to solve this problem.

In this paper, we presented a configurable architecture to limit the wakeup current during turn-on in dynamically-controlled power-gated FPGAs. Our approach has a *fixed* intra-region level, and a *configurable* inter-region level. Appropriate design of the intra-region level ensures that voltage droop constraints are not violated in a power gating region. By combining the intra-region and inter-region levels, it is possible to provide design-time configurability for the turn-on of multiple regions in a power-gated application.

We investigated different tradeoffs associated with the two levels of the inrush current limiting architecture. We found

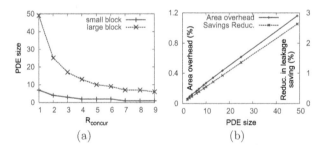

Figure 15: (a) Required PDE size for different R_{concur} and (b) implications of PDE size on inter-region level for region size $R = 4$

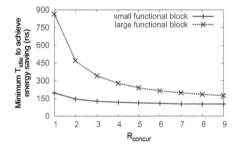

Figure 16: Minimum idle time of a functional block to achieve energy saving assuming region size $R = 4$

that the overhead of the proposed intra- and inter-region architecture is small in terms of its area and power.

As future work, we hope to investigate other tradeoffs associated with the proposed architecture, such as the best combination of power grid sizing, intra-region level over-design, and inter-region level flexibility that can achieve best power/area results. Clearly, this is a complex multi-objective optimization problem that depends not only on the architecture parameters, but also on the target application domain.

Another interesting area of future work is the circuit-level optimization that can be performed in order to reduce the power/area overheads and to shorten turn on times. Circuit-level optimizations of delay elements and the buffers that are inserted between delay elements and sleep transistors, and the possibility to turn on multiple routing channels simultaneously could lead to smaller area/power overheads and shorter wakeup times, resulting in more energy savings.

Another interesting avenue for future work is the development of CAD tools that can automatically detect opportunities for power gating functional blocks. Today, ASIC designers identify these blocks manually through design intent files, and initially, this is the way we would expect this architecture to be used. However, in the long term, automating this process would significantly simplify system design.

7. REFERENCES

[1] MOSIS Scalable CMOS Design Rules. http://www.mosis.com/Technical/Designrules/scmos/scmos-main.html.

[2] Stratix II vs. Virtex-4 Power Comparison & Estimation Accuracy. Altera Corp. white paper WP-S20805-01 (v1.0), Aug. 2005.

[3] V. Betz, J. Rose, and A. Marquardt, editors. *Architecture and CAD for Deep-Submicron FPGAs*. Kluwer Academic Publishers, Norwell, MA, USA, 1999.

[4] R. P. Bharadwaj, R. Konar, P. T. Balsara, and D. Bhatia. Exploiting Temporal Idleness to Reduce Leakage Power in Programmable Architectures. In *ASPDAC'05*, pages 651–656, 2005.

[5] A. Bsoul and S. Wilton. An FPGA Architecture Supporting Dynamically Controlled Power Gating. In *Field-Programmable Technology, 2010. (FPT). Proceedings. 2010 IEEE International Conference on*, pages 1–8, Dec. 2010.

[6] A. Calimera, L. Benini, A. Macii, E. Macii, and M. Poncino. Design of a Flexible Reactivation Cell for Safe Power-Mode Transition in Power-Gated Circuits. *TCSI*, 56(9):1979 –1993, Sept. 2009.

[7] A. Calimera, L. Benini, and E. Macii. Optimal MTCMOS Reactivation Under Power Supply Noise and Performance Constraints. In *DATE'08*, pages 973–978, 2008.

[8] A. Gayasen, Y. Tsai, N. Vijaykrishnan, M. Kandemir, M. J. Irwin, and T. Tuan. Reducing Leakage Energy in FPGAs Using Region-Constrained Placement. In *FPGA'04*, pages 51–58, 2004.

[9] S. Henzler. *Power Management of Digital Circuits in Deep Sub-Micron CMOS Technologies (Springer Series in Advanced Microelectronics)*. Springer-Verlag New York, Inc., Secaucus, NJ, USA, 2007.

[10] D. Howard and K. Shi. Power-On Current Control In Sleep Transistor Implementations. In *VDAT'06*, pages 1–4, 2006.

[11] J. Hussein, M. Klein, and M. Hart. Lowering Power at 28 nm with Xilinx 7 Series FPGAs. Xilinx, Inc. white paper WP389 (v1.1), Jun. 2011.

[12] Y.-M. J. Kaijian Shi, Zhian Lin and L. Yuan. Simultaneous Sleep Transistor Insertion and Power Network Synthesis for Industrial Power Gating Designs. *JCP*, 3(3):6–13, Mar. 2008.

[13] M. Keating, D. Flynn, R. Aitken, A. Gibbons, and K. Shi. *Low Power Methodology Manual: For System-on-Chip Design*. Springer Publishing Company, Incorporated, 2007.

[14] S. Kim, S. V. Kosonocky, and D. R. Knebel. Understanding and Minimizing Ground Bounce During Mode Transition of Power Gating Structures. In *ISLPED'03*, pages 22–25, 2003.

[15] A. Kumar and M. Anis. IR-Drop Management CAD Techniques in FPGAs for Power Grid Reliability. In *ISQED'10*, pages 746–752, 2009.

[16] A. Kumar and M. Anis. IR-Drop Aware Clustering Technique for Robust Power Grid in FPGAs. *Systems, IEEE Trans. on VLSI*, 19(7):1181–1191, July 2011.

[17] C. Li, Y. Dong, and T. Watanabe. New Power-Aware Placement for Region-Based FPGA Architecture Combined with Dynamic Power Gating by PCHM. In *ISLPED'11*, pages 223–228, Piscataway, NJ, USA, 2011. IEEE Press.

[18] Y. Lin, F. Li, and L. He. Routing Track Duplication with Fine-Grained Power-Gating for FPGA Interconnect Power Reduction. In *ASPDAC'05*, pages 645–650, 2005.

[19] S. R. Nassif. Power Grid Analysis Benchmarks. In *ASPDAC'08*, pages 376–381, 2008.

[20] K. K. W. Poon, S. J. E. Wilton, and A. Yan. A Detailed Power Model for Field-Programmable Gate Arrays. *TODAES*, 10(2):279–302, 2005.

[21] K. Roy, S. Mukhopadhyay, and H. Mahmoodi-Meimand. Leakage Current Mechanisms and Leakage Reduction Techniques in Deep-Submicrometer CMOS Circuits. *Proce. of the IEEE*, 91(2):305–327, Feb. 2003.

[22] K. Shi and J. Li. A Wakeup Rush Current and Charge-up Time Analysis Method for Programmable Power-Gating Designs. In *SOCC'07*, pages 163–165, 2007.

[23] T. Tuan, S. Kao, A. Rahman, S. Das, and S. Trimberger. A 90 nm Low-Power FPGA for Battery-Powered Applications. In *FPGA'06*, pages 3–11, 2006.

[24] W. Zhao and Y. Cao. New Generation of Predictive Technology Model for Sub-45nm Design Exploration. *ISQED'06*, 0:585–590, 2006.

Reducing the Cost of Floating-Point Mantissa Alignment and Normalization in FPGAs

Yehdhih Ould Mohammed Moctar[1]   Nithin George[2]

Hadi Parandeh-Afshar[2]   Paolo Ienne[2]   Guy G. F. Lemieux[3]   Philip Brisk[1]

[1]Department of Computer Science
and Engineering
University of California, Riverside
{moctar, philip}@cs.ucr.edu

[2]School of Computer and
Communication Sciences
École Polytechnqiue Fédérale de
Lausanne (EPFL)

[3]Department of Electrical and
Computer Engineering
University of British Columbia
lemieux@ece.ubc.ca

{nithin.george, hadi.parandehafshar, paolo.ienne}@epfl.ch

ABSTRACT

In floating-point datapaths synthesized on FPGAs, the shifters that perform mantissa alignment and normalization consume a disproportionate number of LUTs. Shifters are implemented using several rows of small multiplexers; unfortunately, multiplexer-based logic structures map poorly onto LUTs. FPGAs, meanwhile, contain a large number of multiplexers in the programmable routing network; these multiplexer are placed under static control of the FPGA's configuration bitstream. In this work, we modify some of the routing multiplexers in the intra-cluster routing network of a CLB in an FPGA to implement shifters for floating-point mantissa alignment and normalization; the number of CLBs required for these operations is reduced by 67%. If shifting is not required, the routing multiplexers that have been modified can be configured to operate as normal routing multiplexers, so no functionality is sacrificed. The area overhead incurred by these modifications is small, and there is no need to modify every routing multiplexer in the FPGA. Experiments show that there is no negative impact in terms of clock frequency or routability for benchmarks that do not use the dynamic multiplexers.

Categories and Subject Descriptors

B.7.1 [**Integrated Circuits**]: Types and Design Styles – *gate arrays.*

General Terms

Design, Performance.

Keywords

Field Programmable Gate Array (FPGA), Floating-point, Mantissa Alignment, Normalization.

1. INTRODUCTION

There is considerable interest in using FPGAs to accelerate scientific applications that are dominated by floating-point computations. As FPGAs have abundant spatial parallelism, the best strategy to optimize a floating-point datapath for an FPGA is to minimize the size of each operator, as doing so maximizes the number of operators that can be synthesized onto a device of fixed size; this, in turn, maximizes throughput. Recent work on floating-point datapath compilation for FPGAs has identified the wide shifts required for mantissa alignment and normalization as significant sources of area overhead [16, 17]. For IEEE single-precision floating-point addition, mantissa alignment requires a right shift from 0-24 bits on a 24-bit mantissa (which includes the "hidden '1'" in the most significant position); and normalization requires a left shift from 0-27 bits on a 27-bit mantissa that has been extended with three additional bits ("guard," "round," and "sticky,") which are used for rounding.

Shifters are generally implemented using several layers of multiplexers. In this paper, we consider FPGAs with 6-input lookup tables (6-LUTs), which can implement a 4:1 multiplexer in a single layer of logic level; since a 4:1 multiplexer has two control bits, it maps perfectly onto a 6-LUT. The 24- and 27-bit shifters required for mantissa alignment and normalization can be implemented with three layers of 6-LUTs: two layers of 4:1 multiplexers and one layer of 2:1 multiplexers. The 24- and 27-bit shifters require 72 and 81 6-LUTs respectively.

A typical single-precision floating-point adder requires 350 to 550 6-LUTs, depending on various implementation choices (e.g., support for all IEEE rounding modes, vs. support for just one). Regardless, of the details, the two shifters consume 153 6-LUTs, which is a significant portion of the overall area of the operator.

1.1 Static vs. Dynamic Multiplexing in FPGAs

Multiplexer-based logic maps inefficiently onto LUTs [24], and shifters are no exception. With that in mind, it is interesting to note that FPGAs contain many multiplexers in their routing networks; however, these multiplexers are not accessible to the user, as they are placed under static control of the FPGA's configuration bitstream. In principle, we would like to leverage some of these multiplexers to implement the shifters required for mantissa alignment and normalization in floating-point addition; this paper describes architectural mechanisms to accomplish this goal, along with the supporting CAD tools.

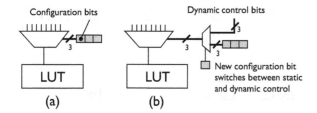

(a)   (b)

Figure 1. A multiplexer in a traditional FPGA's routing network is placed under static control of the configuration bitstream (a); an extension that allows the user to configure the multiplexer to have static or dynamic control (b).

As a motivating example, consider the 8:1 multiplexer shown in Figure 1(a), which drives one input of a LUT; the other LUT inputs are driven by similar multiplexers, which are not shown. Three FPGA configuration bits drive the multiplexer's selection inputs. The purpose of this multiplexer is to provide some flexibility to the FPGA CAD tools—in particular, the router—when synthesizing a circuit onto the FPGA. In this case, there are 8 physical wires within the FPGA that can connect to this LUT input, via the multiplexer. One signal must route to that particular LUT input, and the router is given 8 possible wires to use. Once the route is complete, the configuration bits are set to select the chosen wire. This configuration is *static,* i.e., it does not change until the FPGA is reprogrammed. As there is no possibility to dynamically drive the selection inputs of this multiplexer, there is no possibility for the user to utilize it as an actual 8:1 multiplexer. As it is not architecturally visible, the typical user—who is not an FPGA architect—will be completely unaware of its existence.

Figure 1(b) illustrates a *Static-Dynamic Multiplexer (SD-MUX),* which can be configured for either static or dynamic control. A 2:1 multiplexer now drives the configuration inputs of the 8:1 multiplexer. The 2:1 multiplexer can select either the control bits or a set of wires that are available to the user to provide dynamic control. An extra configuration bit drives the selection input of the 2:1 multiplexer, thus allowing the user to configure the 8:1 multiplexer to provide either static or dynamic control. This basic idea easily generalizes to a multiplexer with any number of inputs, as long as a sufficient number of control bits are provided.

When the SD-MUX is configured to provide static control, one signal can be routed to any of the 8 multiplexer inputs, and the configuration bits are set accordingly, as shown in Figure 2(a); as noted earlier, this provides flexibility to the router, as there is fierce competition for routing resources. When the SD-MUX is configured as a dynamic multiplexer, as shown in Figure 2(b), 8 signals are routed to the 8 multiplexer inputs in pre-specified order; e.g., if the user logic expects the multiplexer to select signal x when the selection bits are 010, then x must be routed to multiplexer input 010 in order to preserve this functionality; thus, the flexibility afforded to the router in the static case is sacrificed.

If we assume that the multiplexers in the routing network are 27:1 or larger, then 24 of them can implement mantissa alignment, and 27 can implement normalization. If we ignore the other LUT inputs, and configure the LUT to implement the identity function, then these two shifters can be implemented using 51 LUTs: a savings of 66.7% over the LUT-based implementation. This paper provides a solution that realizes this best-case savings.

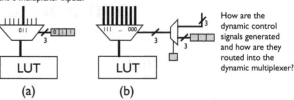

(a)   (b)

Figure 2. A static multiplexer provides flexibility to the router: in this case, 8 inputs are available to route one signal (a); a dynamic multiplexer offers no flexibility to the router, as all multiplexer inputs are used, and routes must be computed for the dynamic control signals as well.

1.2 Fundamental Challenges

The benefit of the SD-MUX is evident for circuits that require a significant amount of multiplexing; however, the introduction of dynamic multiplexers into an FPGA fabric comes at a non-negligible cost, and creates new challenges for physical design tools. The following issues must be addressed in order to justify the inclusion of dynamic multiplexers in an FPGA fabric:

(1) Given that FPGA routing networks consume as much as 90% of on-chip area [7], is the area overhead of replacing static multiplexers with SD-MUXes justifiable?

(2) When the SD-MUX is configured for dynamic control, how can the router overcome the lack of flexibility arising from the fact that 8 input signals must be routed to 8 multiplexer inputs in a pre-specified order, as shown in Figure 2(b)? How is routability achieved in the general case?

(3) How are the dynamic control bits generated, and how are they routed into SD-MUX, as noted in Figure 2(b)?

The answer to question (1) is that SD-MUXes can be introduced sparingly. Realistic user circuits may contain a significant amount of multiplexer-based logic that benefits from the presence of dynamic multiplexers; however, they also contain other logic that maps better onto existing FPGA logic and arithmetic resources, such as LUTs, carry chains, and DSP blocks. As an example, floating-point operators require a large number of LUTs for shifters, but also include components that would not benefit from dynamic multiplexers, such as fixed-point adders and multipliers and leading zero counters. Thus, there is no need to replace more than a handful of static multiplexers with SD-muxes.

The answer to questions (2) and (3) is solved through CAD algorithms. As shown in Figure 2(b), SD-MUXes configured as dynamic multiplexers impose significantly more constraints on the router than static multiplexers. To handle these constraints, the CAD tools extract *macro-cells,* which are subcircuits comprised of the user logic that will use the dynamic multiplexers, plus the immediately preceding logic layer as well. The macro-cells are placed-and-routed separately from the remainder of the circuit. This ensures that the router can satisfy all of the constraints imposed by the dynamic multiplexers without having the macro-cells compete with the remainder of the circuit for limited routing resources in congested areas. Placement then proceeds as normal, with some additional provisions to handle the macro-cells: the placer can move the entire macro-cell around within the FPGA, but cannot change the placement within the macro-cell. Once placement completes, routing resources within each macro-cell are reserved; unused routing resources within the perimeter of the macro-cell are not reserved, as their usage does not affect the macro-cell's functionality. The remainder of the circuit is then routed as normal, with the restriction that the reserved routing resources within each macro-cell are not perturbed, thereby ensuring its correct functionality.

1.3 Paper Goals and Organization

The remainder of this paper focuses on the architectural design choices to enable the SD-MUXes, the supporting CAD toolflow, and the experimental evaluation thereof. In particular, it is of great importance to ensure than the introduction of SD-MUXes does not significantly impair the performance of industrial-scale circuits that do not use the dynamic multiplexing functionality; fortunately, no adverse affects were observed in our experiments.

Section 2 describes the architectural modification that are necessary to integrate SD-MUXes into an FPGA, including the key design choices and trade-offs involved. Section 3 describes the CAD flow in greater detail. Section 4 presents our experimental evaluation, including an analysis of the benefits that SD-MUXes offer floating-point datapath circuits. Section 5 summarizes related work, and Section 6 concludes the paper.

2. ARCHITECTURAL MODIFICATIONS TO INTEGRATE SD-MUXES INTO FPGAS

This section describes the necessary architectural enhancements to FPGAs to integrate SD-MUXes into the routing fabric. Starting with an overview a typical FPGA architecture (Section 2.1), we consider two locations in the FPGA to introduce the SD-MUXes (Sections 2.2 and 2.3). Lastly, we introduce the macro-cell and describe how a standard FPGA CAD flow can be modified to achieve routability (Section 2.4).

2.1 FPGA Architecture Overview

This paper targets an FPGA architecture based on the *Versatile Place and Route (VPR)* tool, which is publicly available from the University of Toronto [20, 21]. The user specifies several architectural parameters in a configuration file. VPR generates an FPGA architecture based on these parameters. For each LUT, VPR generates a *Basic Logic Element (BLE)*, depicted in Figure 3(a), which includes a bypassable flip-flop. The BLE can be configured to implement either combinational or sequential logic.

BLEs are grouped into clusters called *Configurable Logic Blocks (CLBs)*, with fast *Intra-Cluster Routing*, as shown in Figure 3(b). A *Connection Block (C Block)* interfaces the CLB with the global routing segments on either side. Figure 3(c) shows a floorplan of an *island-style FPGA*, which includes *Switch Blocks (S Blocks)* at the intersection points between horizontal and vertical routing channels, as well as I/O pads.

The user specifies several parameters that VPR uses to generate the logic and routing architecture:

K: the LUT size (i.e., a K-LUT);

N: the number of LUTs per CLB;

I: the number of CLB inputs;

W: the number of segments per routing channel; and

Fc_{in} and *Fc_{out}:* C Block connectivity parameters

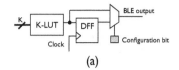

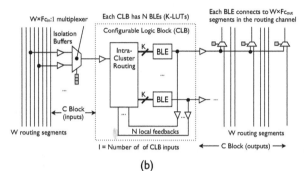

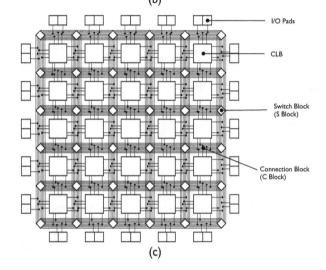

Figure 3. A Basic Logic Element (BLE) (a); a Configurable Logic Block (CLB) contains several BLEs, with fast intra-cluster routing; the Connection Block (C Block) interfaces the intra-cluster routing inside the CLB with the global routing network (b); the floorplan of a generic island-style FPGA (c).

As shown in Figure 3(b), each input multiplexer in the C Block selects one of $W \times Fc_{in}$ wires in the adjacent routing channel on the left, and each BLE drives $W \times Fc_{out}$ segments in the adjacent routing on the right. Most FPGAs use single driver routing [18], so the C Block output is a conceptual description of the routing topology (i.e., which BLE outputs drive which segments in the channel). The multiplexers shown on the right-hand side of Figure 3(b) are actually implemented in the S Blocks, which are shown (without detail) in Figure 3(c). Figure 3(b) depicts inputs coming in from the left hand side of the CLB and outputs leaving to the right; in actuality, inputs and outputs may enter and exit from all four sides of the CLB, and the user may specify the percentage of inputs and outputs that enter and exit on each side.

Given all of these architectural parameters, VPR generates the interconnect topology algorithmically. A number of architectural parameters, such as those that describe the switch box, have been omitted from this discussion for brevity.

The intra-cluster routing depicted in Figure 3(b) is a crossbar that connects I inputs and N local feedbacks to the $K \times N$ LUT inputs in the CLB. VPR 5.0 implements intra-cluster routing as a full crossbar, which provides a connection between every CLB input and LUT input. Full crossbars are costly in terms of area and power, but guarantee routability: i.e., any combination of signals routed to CLB inputs can be routed to any desired combination of LUT inputs. Highly routable sparse crossbar topologies for intra-cluster routing have also been investigated in recent years [7, 19].

Ahmed and Rose determined that the ideal number of CLB inputs is $I = K \times (N+1)/2$, which is less than the total number of LUT inputs, $K \times N$. This suffices because many signals fan-out to multiple LUT inputs within a CLB after the FPGA has been configured. As each CLB input (other than LUT feedbacks) is driven by a $W \times Fc_{in}:1$ multiplexer, reducing the number of CLB inputs reduces the overall cost of the C Block, at the expense of some flexibility. In other words, N independent K-input logic functions cannot be packed into a CLB due to I/O limitations, despite the fact that the CLB has sufficient LUT capacity.

Recall that the goal of this paper is to replace static multiplexers in the routing network with SD-MUXes. In Figure 3(b), there are two locations where this is possible: the C Block (input), and intra-cluster routing, as discussed in the next two subsections.

2.2 Integrating SD-MUXes into the C Block

Example 1. To illustrate the integration of an SD-MUX into a C Block, let us consider a conditional swap, which has three inputs, I_0, I_1 and c, and two outputs, J_0 and J_1. The operation is:

$$J_0 = c \,?\, I_1 : I_0 \quad , \quad J_1 = c \,?\, I_0 : I_1 \qquad (1)$$

Figure 4 depicts a portion of the C Block that has been modified with two SD-MUXes to implement the conditional swap. A static multiplexer provides the control bit, while two SD-MUXES compute J_0 and J_1. In this particular example, $W = 8$ segments per channel and $Fc_{in} = 0.5$, i.e., each C Block multiplexer connects to 4 wires in the channel. Each of the three 4:1 multiplexers in the C Block are implemented using three 2:1 multiplexers; two of the 2:1 multiplexers have been replaced with SD-MUXes in Figure 4.

The C Block in Figure 4 imposes routing constraints that must be satisfied in order to deliver input signals I_0 and I_1 in the correct order to the SD-MUX inputs. In particular, I_0 and I_1 must be routed on routing segments w_6 and w_7; the order is irrelevant, i.e., either I_0 can be routed on w_6 and I_1 on w_7, or vice-versa; similarly, I_0 and I_1 must be routed on w_4 and w_5 as well. The condition bit, c, has greater flexibility: it can be routed on w_0, w_1, w_2, or w_3.

The placer and router must satisfy these constraints. Let F_0 and F_1 be K-input logic functions that compute conditional swap inputs I_0 and I_1. F_0 and F_1 must be synthesized on LUTs whose outputs collectively drive a subset of the wires that satisfy the aforementioned constraints. Moreover, this assumes that such a combination of LUTs actually exists. Although it may be possible to satisfy this constraint for a 3-input conditional swap operation, it will be much more difficult to satisfy for a 24- or 27-bit shifter.

Example 2. Consider a 4-bit left shift with rotation. The inputs are $I_0...I_3$ and the outputs are $J_0...J_3$; two control bits c_0 and c_1 specify the shift amount (0-3 bit positions). Once again, we assume that $W = 8$ and $Fc_{in} = 0.5$, and the C Block contains 4:1 multiplexers. As the shifter has four data inputs rather than two, each of the four data inputs, $I_0 ... I_3$ must connect to *exactly one* input of each C Block multiplexer in a pre-determined pattern.

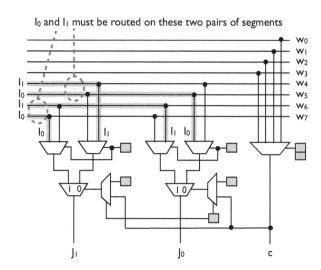

Figure 4. A C Block modified to implement a conditional swap by introducing two SD-MUXes. The requirements to deliver the correct signals to the SD-MUX inputs impose stringent routing constraints.

Figure 5(a) depicts a portion of the C Block that produces the lower-order data outputs, J_0 and J_1; however, the interconnection topology does not allow the design to be satisfied due to conflicts on the routing segments. For example, the multiplexer that produces J_0 requires I_0 to be routed on segment w_4, while the multiplexer that produces J_1 requires I_1 to be routed on the same segment concurrently. Similar conflicts occur on segments w_5, w_6, and w_7. In contrast, Figure 5(b) depicts an interconnect topology that eliminates the routing conflicts; of course, this topology *only* satisfies a 4-bit left shift with rotation, and would not necessarily be helpful for some other type of multiplexer-based circuit.

As mentioned earlier, dynamic multiplexers impose strict ordering constraints on the signals that are connected to their inputs. The examples shown in Figures 4 and 5 demonstrate that these constraints are propagated into the global routing network when SD-MUXes are integrated into the C Block. The ability to implement very simple multiplexing circuits, as shown in Figures 4 and 5, is dependent on the interconnect topology between the CLBs and the routing network; this interconnect topology depends on parameters W, Fc_{in} and Fc_{out}, and the algorithm [4] that generates the routing network from these parameters.

Although it may be possible to modify the routing network generation algorithm to favor certain interconnect topologies, it is difficult to determine whether the basic idea will generalize to larger structures. For example, in a 27-bit shifter, input bit I_0 will fan out to 27 outputs, $J_0 ... J_{26}$. This means that a single routing segment or a subset of segments driven by the same BLE must individually or collectively fan-out to 27 pre-specified C Block SD-MUX inputs, all in close quarters. Similarly I_1 will need to fan-out to 26 pre-specified C Block SD-MUX inputs, etc. Moreover, this must be done with parameters that are representative of commercial FPGAs, e.g., $N = 10$, $W = 300$, $Fc_{in} = 0.15$, and $Fc_{out} = 1/N = 0.1$. The likelihood of success in this case is too low to be considered realistic; consequently, we conclude that the C Block is not a particularly promising location to integrate SD-MUXes into the routing fabric.

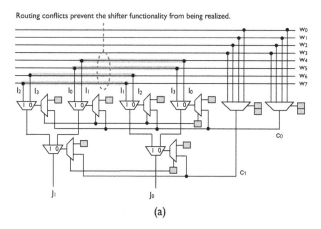

Routing conflicts prevent the shifter functionality from being realized.

(a)

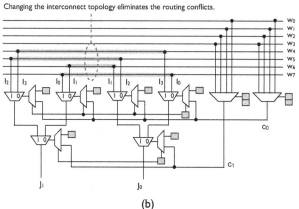

Changing the interconnect topology eliminates the routing conflicts.

(b)

Figure 5. A conflict in the interconnect topology makes it impossible to implement a 4-bit rotator using SD-MUXes in the C Block (a); changing the interconnect topology can eliminate the conflict (b).

2.3 Integrating SD-MUXes into Intra-Cluster Routing

Alternatively, we can introduce SD-MUXes into the intra-cluster routing instead of the C Block. The primary advantage of this approach is that it eliminates the routing constraints that arise due to the interconnection topology between the routing segments and the C Block. Instead, the interconnection topology constraints are internal within the intra-cluster routing. Signals that drive a specific SD-MUX input for dynamic multiplexing, as shown in Figure 6, are routed to pre-selected CLB inputs. Each pre-selected CLB input connects to one of the SD-MUX inputs: some to the data inputs, and others to the selection inputs.

Figure 6 shows an example of a 4:1 SD-MUX integrated into the intra-cluster routing; a significant portion of the intra-cluster routing is omitted from Figure 6 to conserve space. Two CLB inputs provide dynamic control (they may also drive other multiplexers, which are not depicted in the figure); control signals c_0 and c_1 must be routed to these two inputs. The other four CLB inputs drive the data inputs of the SD-MUX. The input signals are routed to these four CLB inputs in a specific order, e.g., the SD-MUX selects input I_0 if $c_1 c_0 = 00$. Thus, the connection topology between CLB inputs and SD-MUX inputs determines which signals must be routed to each pre-selected CLB input.

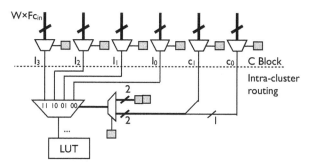

Figure 6. Integrating an SD-MUX into intra-cluster routing imposes a strict ordering on the signals that are routed to the CLB inputs.

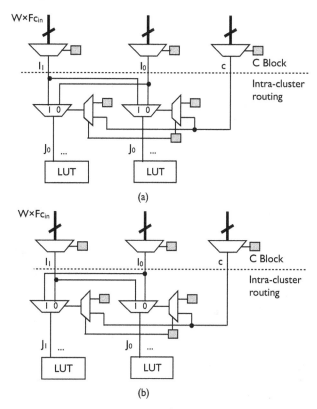

(a)

(b)

Figure 7. Integration of SD-MUXes into the intra-cluster routing. The interconnect topology may force both SD-MUXes to implement the same logic function when configured to implement dynamic control (a); rearranging the topology enables the SD-MUXes to implement different functions (b).

In Figure 6, any 4-input multiplexer can be realized by permuting either the control or the data bits; however, additional restrictions are imposed when we consider multiple-output functions because each CLB input may connect to multiple SD-MUX inputs.

Example 3. Let us reconsider the conditional swap operation from Example 1; this time, we want to implement it using SD-MUXes in the intra-cluster routing rather than the C Block. Figure 7(a) shows an initial attempt. Due to the interconnection topology within the intra-cluster routing, both SD-MUXes conditionally select the same input bit, i.e., they both compute logic function J_0.

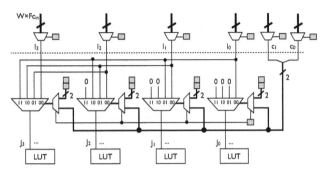

Figure 8. Intra-cluster routing with SD-MUXes modified to support a 4-bit left shift.

By swapping the order of I_0 and I_1 at the CLB inputs, then this intra-cluster routing topology would compute J_1, rather than J_0; however, by changing the topology, as shown in Figure 7(b), the two SD-MUXes compute logic functions J_0 and J_1, respectively. In this case, swapping the order of I_0 and I_1 at the inputs would likewise swap the order of J_0 and J_1 at the outputs.

Example 4: Figure 8 illustrates intra-cluster routing with SD-MUXes that can implement a 4-bit left shifter (bits shifted in are set to zero). The basic interconnection pattern shown here easily generalizes to a larger shifter sizes. In this case, the SD-MUXes implement all of the shifting functionality; the LUTs are configured to pass the SD-MUX outputs through unmodified.

In Figure 8, many of the SD-MUX inputs in are '0' bits. It is not immediately clear how these bits should be handled. It would be unrealistic to extend some of the SD-MUXes to account for a large number of '0' bits, e.g., in a K-bit shifter, the SD-MUX that computes least significant output bit J_0 requires $K-1$ '0' bits, the SD-MUX that computes J_1 requires $K-2$ '0' bits, etc.; the area overhead required to support larger shifters would be prohibitive.

Another issue is that the routing network may invert signals en route. The LUT sink is usually reprogrammed to compensate if some of its inputs arrive with the wrong polarity. SD-MUXes, however, are not programmable in this respect. One possibility is to add programmable inversion at the shifter inputs; however, this incurs significant area overhead. Another option is to reprogram the previous layer of LUTs that generate the shifter inputs to compute the complement of its logic function; however, this logic layer may have a large fan-out, where some fan-out bits are inverted and others are not, rendering this approach ineffective.

We can solve both the '0' SD-MUX input bit problem and the inversion problem by using LUTs in conjunction with the SD-MUXes. Each SD-MUX output drives a LUT input; we can then route the control bits to the remaining LUT inputs. The LUT is then programmed to invert the SD-MUX output if the selected input arrives in inverted form. The LUT is also programmed to output a '0' for the appropriate control bit combinations (e.g., $c_1c_0 = \{01, 10, 11\}$ for J_0 in Figure 8), which eliminates the need to route '0' bits to the SD-MUX inputs.

The CLBs in modern high performance FPGAs contain 6-LUTs; this limits the number of control bits that can be supported using this approach to 5 or less, which, in turn, limits the maximize SD-MUX size to 32:1. This suffices for the 24- and 27-bit shifters used for single-precision floating-point mantissa alignment and normalization. Figure 9 illustrates the preceding discussion for the second least significant bit, J_1 of a 27-bit shifter.

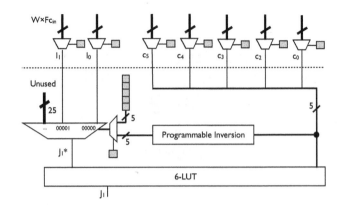

I_0 Not Inverted I_1 Not Inverted							I_0 Inverted I_1 Not Inverted							I_0 Not Inverted I_1 Inverted							I_0 Inverted I_1 Inverted						
c_4	c_3	c_2	c_1	c_0	J_1^*	J_1	c_4	c_3	c_2	c_1	c_0	J_1^*	J_1	c_4	c_3	c_2	c_1	c_0	J_1^*	J_1	c_4	c_3	c_2	c_1	c_0	J_1^*	J_1
0	0	0	0	0	0	0	0	0	0	0	0	0	1	0	0	0	0	0	0	0	0	0	0	0	0	0	1
0	0	0	0	1	1	1	0	0	0	0	1	0	0	0	0	0	0	1	1	1	0	0	0	0	1	0	0
0	0	0	1	0	0	0	0	0	0	1	0	0	1	0	0	0	1	0	1	1	0	0	0	1	0	1	1
0	0	0	1	1	1	1	0	0	0	1	1	1	1	0	0	0	1	1	0	0	0	0	0	1	1	0	0
--	--	--	--	--	--	0	--	--	--	--	--	--	0	--	--	--	--	--	--	0	--	--	--	--	--	--	0

Figure 9. A LUT in conjunction with an SD-MUX solves the problems of inverted input bits and generates '0' outputs when appropriate. This example is the second least significant output bit, J_1, of a 27-bit left shifter. Four truth tables are possible, depending on whether I_0 and I_1 are inverted. Programmable inversion is necessary for the five control bits.

CLB parameters also limit the size of the SD-MUXes that can be introduced. The intra cluster routing has a total of $I+N$ inputs and $N \times K$ outputs. The inputs are the I CLB inputs provided by the C Block plus N LUT feedbacks from within the CLB. The cluster contains N K-LUTs; each LUT input is an output of the intra-cluster routing. A typical modern high-performance FPGA has $N = 8$, $K = 6$, and $I = K \times (N+1)/2 = 27$, using the formula provided by Ahmed and Rose [1]. To support a 27-bit shifter, we need to increase I to 32, to account for the control signals.

In VPR 5.0, the intra-cluster routing is a full crossbar. Given these parameter values, the intra-cluster routing would be composed of 48 40:1 multiplexers. Modern FPGAs, however, use sparsely populated crossbars [7, 19]. Depending on the population density of the sparse crossbar, the multiplexers may be smaller than 27:1. In this case, we would either need to limit the shift amount in accordance with the multiplexer size, or introduce SD-MUXes that are larger than the pre-existing static multiplexers; this latter option is unfavorable, because it introduces asymmetry in terms of delays: i.e., the delay through a statically configured SD-MUX is greater than the delay through a standard static multiplexer, which could affect performance and complicate routing.

The interconnection topology (i.e., which CLB inputs connect to exactly which SD-MUX inputs) has a significant impact on our ability to implement shifters in the intra-cluster routing; this was illustrated quite clearly by Figure 7. Figure 8 illustrates the general interconnect topology pattern required for a left shifter (which easily generalizes to more than 4 inputs), and a left shifter can implement a right shifter by reversing the order of the inputs. Shifters that perform rotation (e.g., Figure 5) require a different topology as they do not shift-in zeroes. To summarize, the topology must account for ordering constraints on SD-MUX inputs in order to ensure correctness.

Lastly, we do not advocate the introduction of SD-MUXes into every CLB, as the vast majority of CLBs in a given FPGA will not be configured to implement dynamic multiplexing circuits in most realistic designs. CLBs containing SD-MUXes are a new form of heterogeneity, similar in principle to the introduction of DSP blocks and block RAMs in past FPGAs. As a rough estimate, we suggest at most 10% of the CLBs in an FPGA should be enhanced with SD-MUXes, and that those that are enhanced should be laid out in columns within the FPGA; the column-based layout echoes the way that DSP blocks and block RAMs are currently laid out in FPGAs, and therefore makes intuitive sense.

2.4 Ensuring Routability with Macro-Cells

Consider a 27-bit shifter implemented with SD-MUXes integrated into the intra-cluster routing. In accordance with prior notation, let $I_0 \ldots I_{26}$ and $J_0 \ldots J_{26}$ denote the shifter inputs and outputs, and let $c_0 \ldots c_4$ denote the control bits. This is a total of 32 inputs (including control bits) and 27 outputs. Eight CLBs, $CLB_0 \ldots CLB_7$ realize the shifter. LUT L_i computes shifter input I_i, LUT S_i computes shifter output J_i, and LUT C_i computes control bit c_i.

Figure 10 depicts the interconnection pattern for the 27-bit shifter. The structure depicted in Figure 10 is called a *macro-cell*, because the LUTs and CLBs are pre-placed and routed. Without loss of generality, if the placer (generally an iterative improvement algorithm) randomly moves L_5 to a new CLB, the likelihood is quite small that a legal route will be found that delivers shifter input I_5 to the pre-specified CLB inputs in CLB_5, CLB_6, and CLB_7. By fixing the locations of the LUTs relative to one another in the macro-cell, routability is achieved.

3. CAD SUPPORT FOR MACRO-CELLS

We used VPR 5.0 [20] for architectural simulation, placement, and routing, T-VPack for packing [22], and ABC for logic synthesis and technology mapping [3].

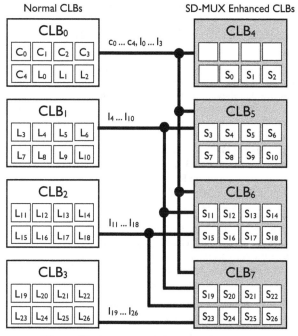

Figure 10. A macro-cell for a 27-bit shifter.

3.1 Programming Model, Assumptions, and Technology Mapping

We assume that the programmer will add annotations to the HDL code to specify when to configure the programmable macro-cell as a shifter, similar to how DSP blocks and carry chains are used. The technology mapper explicitly binds the annotated shifters to macro-cells rather than mapping them to LUTs. Large shifters are decomposed into smaller ones if macro-cell capacity is exceeded. Next, we extract the layer of LUTs that precedes each shifter, e.g., LUTs $L_0 \ldots L_{23}$ in Figure 10. The structure of the macro-cell effectively pre-packs, pre-places, and pre-routes these subcircuits.

3.2 Macro-cell Placement and Routing

VPR's router, which is based on PathFinder [23], assumes that CLB intra-cluster routing is a full crossbar. Any path from the source to a CLB input can route a net: the crossbar connects all CLB inputs to all LUT inputs. We modified VPR to allow the user to specify specific CLB inputs pins as targets for certain sinks. VPR can find a legal route for a macro-cell, establishing a path from the LUT source that computes each net to all of its pre-specified inputs in the second macro-cell layer. VPR successfully routed 24- and 27-bit shifters in macro-cells using this approach.

Macro-cells are placed-and-routed offline, prior to the rest of the circuit. Placement of the shifter onto SD-MUXes within the macro-cell is deterministic. Placement of the LUT layer preceding the shifter is more flexible: any placement that successfully routes all nets within the macro-cell suffices. We try to pack the LUTs tightly into a small number of CLBs in the vicinity of the shifter.

3.3 Global Placement and Routing

Extensive modifications were made to VPR's placer [21] in order to handle macro-cells. The input is a netlist, which may or may not contain macro-cells, and an architectural description of the FPGA in which certain columns have been annotated to indicate CLBs that have been enhanced with SD-MUXes.

Each shifter in the netlist, along with the layer of preceding LUTs, is placed onto a macro-cell. Each macro-cell is routed up-front. SD-MUXes in the remaining (unused) macro-cells are configured as normal CLBs, similar to how shadow clusters are used [11]. The placer considers all other CLBs to be functionally equivalent.

VPR's placer uses simulated annealing. We implemented two placement strategies. In the first, we place shifters onto macro-cells and fix their placement; the placer moves normal soft logic clusters around, but does not perturb the placement of the shifters onto macro-cells. The second option relaxes this constraint, and moves both soft logic clusters around the FPGA and may also move any shifter onto an unused macro-cell.

Macro-cells configured as shifters are similar to DSP blocks from the perspective of the CAD tools. The difference is that unused logic and routing resources within each macro-cell, after it has been placed-and-routed, remain available to the global placer and router and can be used by the rest of the circuit.

4. EXPERIMENTAL RESULTS

Our experimental goals are twofold. Firstly, we wish to quantify reduction in LUT count that can be achieved by synthesizing shifters onto SD-MUX enabled macro-cells. Secondly, we wish to ensure that the inclusion of macro-cells does not adversely affect routability for industrial-scale benchmarks.

4.1 Floating-Point Operators

We consider a set of single-precision multi-operand floating-point adders that have already been optimized for area. These operators are similar to those produced by Altera's floating-point datapath compiler [16, 17], which removes redundant normalizations. We used designs published by Verma et al. [25], which were slightly smaller than those produced by Altera's compiler. We used the smallest design approach, which implemented the internal fixed-point multi-operand addition using a tree of 3-input adders.

For a *K*-input adder, we denormalize *K-1* mantissas using shifters; the mantissa corresponding the largest exponent is not shifted. Normalization is only applied once, at the output of the operator.

For 2, 4, 8, and 16-input adders, Figure 11 reports that the area savings (in terms of Altera's ALMs) obtained by the macro-cell range from 25% to 32%. Assuming that the number of LUTs and CLBs in an FPGA are fixed, this means that 33-40% more operators can be packed into an area of fixed size when macro-cells are used to implement shifters.

We did not measure the effect of the macro-cells on critical path delay or pipeline depth of the adders. The throughput of floating-point data paths is driven mostly by spatial parallelism; reducing the area of an operator increases the number of operators that can be synthesized on a fixed area device. The area savings reported in Figure 11, thus, translate indirectly into increased throughput for parallel floating-point data paths that use these operators.

4.2 Experimental Setup: VPR

We modeled an FPGA enhanced with macro-cells using VPR 5.0 [20]. We did not use VPR 6.0, which is now part of the Verilog-To-Routing (VTR) flow, for these studies because it did not have timing models in-place at the time this work was performed. As our baseline, we took one of the VPR architecture files from the iFAR repository [14, 15]. Table 1 lists the baseline parameters for our architecture. CLB inputs and outputs are evenly distributed around all four sides of the CLB.

VPR explicitly models a C Block, but does not model the intra-cluster routing; as it is a full crossbar, only its delay is modeled. We do not model SD-MUXes explicitly. Our experiments strive to show that macro-cells, which reserve a non-trivial quantity of routing resources in localized areas, do not adversely affect the ability to route large-scale circuits that contain shifters.

Macro-cells are organized as vertical columns when they are introduced into the FPGA. The motivation is to mimic the layout of modern FPGAs. For example, logic clusters are generally laid out as columns; so are DSP blocks, block RAMs, etc. Only a small proportion of CLBs in the FPGA contain SD-MUXes.

For each experiment, we placed each benchmark once and routed it three times using different random number seeds. The delay for each benchmark is the average delay of the three runs. This reduces the noise in our delay results as different random number seeds can yield significantly different routing results.

4.3 Benchmarks

We selected the ten largest IWLS 2005 benchmarks [10], which are described in Table 2, to evaluate the impact of the macro-cell on large-scale applications. Using VPR, we synthetically added macro-cells (shifters) to these benchmarks; our goal is to ascertain whether these shifters, when pre-placed and routed onto macr-cells, adversely affect area, delay, and routability.

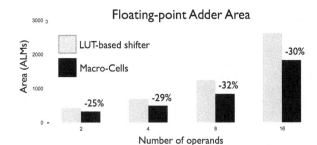

Figure 11. The area savings obtained by macro-cell based implementations of 24- and 27-bit shifters used for mantissa alignment and normalization on four optimized multi-operand single-precision floating-point adders.

Table 1. FPGA architectural parameters

Parameter	Value	Parameter	Value
LUT Size	6	Fc input	0.15
Cluster Size	8	Fc output	0.1
Channel Width	96	Technology*	65nm CMOS
Cluster Inputs	36	Tile Area**	18940

* Berkeley predictive models   ** Min-width transistors

Table 2. Ten largest IWLS 2005 benchmarks

Benchmark	Description
ac97_ctrl	Interface to external AC 97 audio codec
aes_core	Advanced Encryption Standard (AES)
des_perf	16-cycle pipelined DES/3-DES Core
ethernet	10/100 Mbps IEEE 802.3/802.3u MAC
mem_ctrl	Embedded memory controller
pci_bridge32	Bridge interface to PCI local bus
systemcaes	Area-optimized AES implementation
usb_func	USB 2.0 compliant core
vga_lcd	Embedded VGA/LCD controller
wb_conmax	Wishbone Interconnect Matrix IP Core

We modified each benchmark's netlist to include 20, 40, 60, 80, and 100 shifters, which are connected at arbitrarily chosen points to ensure that they are not completely disjoint from the remaining logic. In VPR, we pre-allocated macro-cell columns and pre-placed the shifters and preceding layers of logic onto them. We pre-routed the macro-cells, and marked the routing resources used as unavailable. Our primary concern was that locking down these resources up-front would adversely affect the quality of the routes obtained for the remainder of the circuit; fortunately, practically no degradation was observed.

VPR generates a custom FPGA for each benchmark, based on its demand for logic and routing resources. Each benchmark is repeatedly placed and routed, varying the channel width each time; VPR converges onto the minimum channel width (W_{min}) for which a legal route can be found. The average W_{min} obtained by VPR across all benchmarks (with no macro-cells) here is 84.4.

Figure 12 reports the area of each benchmark with a varying number of shifters. The area is reported in terms of minimum-width transistors; this accounts for the fact that CLBs that have been augmented with SD-MUXes are larger than regular CLBs.

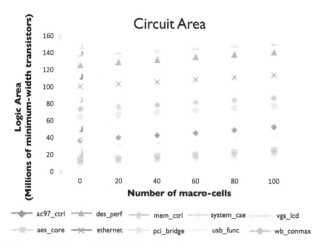

Figure 12. Area of the 10 circuits synthesized using VPR 5.0.

4.4 Routability

Figure. 13 reports the critical path delay of the IWLS benchmarks with a varying number of macro-cells; we observe practically no impact on critical path delay from the inclusion of as many as 100 shifters per benchmark. As noted in Section 3.3, we considered two different placement strategies: a *constrained* strategy in which the logic placed onto macro-cells is fixed a-priori, and an *unconstrained* strategy in which the placer can move the macro-cell logic (the shifter, and logic layer preceding it) onto any macro-cell. The results reported in Figure. 13 are for the constrained strategy; we observed that the unconstrained strategy produced essentially identical results, where the differences in delays for each data point are in the range of tens of pico-seconds.

Figure. 14 shows that introducing macro-cells may adversely affect W_{min}, as each macro-cell requires some routing resources. For many benchmarks, W_{min} steadily increases when the number of macro-cells ranges from 20 to 80, but decreases rapidly from 80 to 100. The reason for this observation is that VPR automatically generates an FPGA that is sized to a specific application; based on the number of CLBs used and I/O pads required, VPR generates the smallest square FPGA that can provide sufficient resources. VPR then repeatedly places and routes the circuit to determine W_{min}.

Many of the IWLS benchmarks are I/O bound, so CLB utilization is relatively sparse, and there is relatively little congestion in the routing network. Each macro-cell that is added increases CLB utilization, and introduces congestion, which increases W_{min}. If we assume a fixed-size FPGA, eventually, the inclusion of more macro-cells will cause utilization to exceed 100%. VPR then generates a larger FPGA, with much lower utilization; consequently, the benchmark circuit routes much easier, and W_{min} is reduced. This is precisely what occurred, for example, for benchmarks aes_core and des_perf (and a few others) between 80 and 100 macro-cells in Figure 14. It is important to recall that these benchmarks are synthetic. A floating-point operator, in contrast, would contain shifters and use the available macro-cells. Moreover, W_{min} as reported in Figure 14 is much smaller than the routing channel width of commercially available FPGAs.

These experiments demonstrate that macro-cells are quite useful for benchmarks that contain shifters, while their presence will not adversely affect other benchmarks that do not contain shifters.

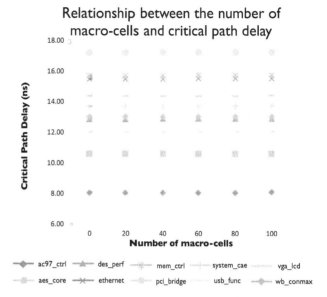

Figure 13. Introducing as many as 100 macro-cells into the benchmarks does not increase the critical path delay.

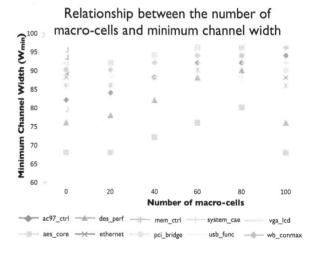

Figure 14. Introducing macro-cells into the benchmarks does have some affect on channel width; however, the dimensions of the FPGA change as well, as the number of macro-cells changes each benchmark's demand for logic resources.

5. RELATED WORK

The goal of this work is to reduce the cost of mantissa alignment and normalization in floating-point operations. One alternative is to integrate floating-point units as hard blocks [2, 5, 9]; however, applications that are not floating-point intensive will be unable to use these blocks. To date, FPGA vendors do not sell device families with dedicated blocks for floating-point applications. Beauchamp et al. [2] advocate integrating hard shifters or 4:1 multiplexors in parallel with FPGA logic; however, when the shifters are not used, the nearby routing resource are wasted; and when the 4:1 multiplexors are used, significant routing resources are still required to form large shifters.

Shifters and multiplexers can be synthesized onto multipliers in the DSP blocks [8, 12], and Xilinx has added 17-bit barrel shifters to their DSP48E1 blocks [26]; however, a DSP block used for shifting, cannot perform other operations. Benchmarks that require multiplication and shifting can still benefit from FPGAs containing DSP blocks and macro-cells.

Floating-point datapath compilers use arithmetic transformations to synthesize floating-point operations efficiently on FPGAs [6, 16, 17]; reducing the cost of normalization is one of their goals. These compilers achieve better performance and logic density than using 2-input operators, but they sacrifice IEEE compliance. Our approach is amenable to IEEE-compliant operators.

A patent by Kaviani (*Xilinx*) [13] exposes the selection bits of C block multiplexers to the programmer; the idea is similar to Xilinx Virtex FPGAs, which do not have intra-cluster routing. No CAD tools are described, so the affect on routability is unknown.

6. CONCLUSION

The macro-cells introduced in this paper can implement 27-bit shifters for single-precision floating-point mantissa alignment and normalization. The macro-cells reduce the area of floating-point addition clusters by up to 32%, which increases the number of operators that can be synthesized into a fixed-area device. This aligns well with the strategy employed by Altera's floating-point datapath compiler [16, 17]. Our experiments show that macro-cells do not adversely affect routability for benchmarks that do not contain shifters. Future work will look to integrate macro-cells with FPGAs that contain sparse intra-cluster routing, and to see whether it is possible to extend them into the C Block.

REFERENCES

[1] Ahmed, E., and Rose, J. The effect of LUT and cluster size on deep-submicron FPGA performance and density. *IEEE Trans. VLSI*, vol. 12, no. 3, March, 2003, pp. 288-298. DOI= http://dx.doi.org/10.1109/TVLSI.2004.824300

[2] Beauchamp, M. J., Hauck, S., Underwood, K. D., and Hemmert, K. S. Architectural modifications to enhance the floating-point performance of FPGAs. *IEEE Trans. VLSI*, vol. 16, no. 2, Feb. 2008, pp. 177-187. DOI= http://dx.doi.org/10.1109/TVLSI.2007.912041

[3] Berkeley Logic Synthesis and Verification Group. "ABC: A system for sequential synthesis and verification.: December 2005 release. URL= http://www.eecs.berkeley.edu/~alanmi/abc

[4] Betz, V., and Rose, J., "Automatic generation of FPGA routing architectures from high-level descriptions," ACM/SIGDA Int. Symp. FPGAs (FPGA '00), pp. 175-184, Feb. 10-11, 2000, DOI= http://doi.acm.org/10.1145/329166.329203

[5] Chong, Y. and Parameswaran, S., "Flexible multi-mode embedded floating-point unit for field programmable gate arrays," ACM/SIGDA Int. Symp. FPGAs (FPGA '09), pp. 171-180, Feb. 22-24, 2009, DOI= http://doi.acm.org/10.1145/1508128.1508155

[6] de Dinechin, F., Klein, C., and Pasca, B., "Generating high-performance custom floating-point pipelines," Int. Conf. Field Programmable Logic and Applications (FPL '09), Aug. 31- Sept. 2, 2009. DOI=http://dx.doi.org/10.1109/FPL.2009.527255/

[7] Feng, W. and Kaptanoglu, S. Designing Efficient Input Interconnect Blocks for LUT Clusters Using Counting and Entropy. *ACM Trans. Reconfigurable Technol. Syst.*, vol. 1, no. 1, Mar. 2008, pp. 1-28. DOI= http://doi.acm.org/10.1145/1331897.1331902

[8] Gigliotti, P., "Implementing barrel shifters using multipliers," XAPP – Application Note: Virtex II Family, pp. 1-4, Aug., 2004. URL= http://www.xilinx.com/support/documentation/application_notes/xapp195.pdf

[9] Ho, C. H., et al., Floating-point FPGA: architecture and modeling. *IEEE Trans. VLSI*, vol. 17, no. 12, Dec. 2009, pp. 1709-1718. DOI= http://dx.doi.org/10.1109/TVLSI.2008.2006616

[10] IWLS 2005 benchmarks. URL= http://iwls.org/iwls2005/benchmarks.html

[11] Jamieson, P., and Rose, J., "Enhancing the area-efficiency of FPGAs with hard circuits using shadow clusters," *IEEE Trans. CAD*, vol. 18, no. 12, Dec. 2010, pp. 1696-1709. DOI = http://dx.doi.org/10.1109/TVLSI.2009.2026651

[12] Jamieson, P., and Rose, J., "Mapping multiplexers onto hard multipliers in FPGAs," 3rd Int. IEEE Northeast Workshop on Circuits & Systems (IEEE-NEWCAS '05), pp. 323-326, June 19-22, 2005. DOI= http://dx.doi.org/10.1109/NEWCAS.2005.1496692

[13] Kaviani, A., FPGA with improved structure for implementing large multiplexors. U.S. patent, no. US 6,556,042 B1, Apr. 29, 2003.

[14] I. Kuon and J. Rose, "Area and delay trade-offs in the circuit and architecture design of FPGAs," ACM/SIGDA Int. Symp. FPGAs (FPGA '08), pp. 149-158, Feb. 24-26, 2008, DOI= http://doi.acm.org/10.1145/1344671.1344695

[15] I. Kuon and J. Rose, "Automated transistor sizing for FPGA architecture exploration," ACM/IEEE Design Automation Conference (DAC '08), pp. 792-795, June 8-13, 2008, DOI= http://doi.acm.org/10.1145/1391469.1391671

[16] Langhammer, M., "Floating point datapath synthesis for FPGAs," Int. Conf. Field Programmable Logic and Applications, (FPL '08), pp.355-360, Sept. 8-10, 2008. DOI= http://dx.doi.org/10.1109/FPL.2008.4629963

[17] Langhammer, M., and Vancourt, T., "FPGA floating point datapath compiler," IEEE Symp. 17th IEEE Symp. Field-programamble Custom Computing Machines (FCCM '09), April 5-7, 2009. DOI = http://dx.doi.org/10.1109/FCCM.2009.54

[18] Lemieux, G. Lee, E. Tom, M., and Yu, A. "Directional and single-driver wires in FPGA interconnect," IEEE International Conference on Field-Programmable Technology (FPT '04), pp. 41-48, Dec. 6-8, 2004, DOI: http://dx.doi.org/10.1109/FPT.2004.1393249

[19] Lemieux, G, and Lewis, D. "Using sparse crossbars within LUT clusters," ACM/SIGDA Int. Symp. FPGAs (FPGA '01), pp. 59-68, Feb. 11-13, 2001, DOI= http://doi.acm.org/10.1145/360276.360299

[20] Luu, J., Kuon, I., Jamieson, P., Campbell, T., Ye, A., Fang, W. M., and Rose, J. "VPR 5.0: FPGA CAD and architecture exploration tools with single-driver routing, heterogeneity and process scaling," ACM/SIGDA Int. Symp. FPGAs (FPGA '09), pp. 133-142, Feb. 22-24, 2009, DOI= http://doi.acm.org/10.1145/1508128.1508150

[21] Marquardt, A., Betz, V., and Rose, J. "Timing-driven placement for FPGAs," ACM/SIGDA Int. Symp. FPGAs (FPGA '00), pp. 203-213, Feb. 10-11, 2000, DOI= http://doi.acm.org/10.1145/329166.329208

[22] Marquardt, A., Betz, V., and Rose, J. "Using cluster-based logic blocks and timing-driven packing to improve FPGA speed and density," ACM/SIGDA Int. Symp. FPGAs (FPGA '99), pp. 37-46, Feb. 21-23, 1999, DOI= http://doi.acm.org/10.1145/296399.296426

[23] McMurchie, L., and Ebeling, C. "PathFinder: a negotiation-based performance-driven router for FPGAs," ACM/SIGDA Int. Symp. FPGAs (FPGA '95), pp. 111-117, Feb. 12-14, 1995, DOI= http://doi.acm.org/10.1145/201310.201328

[24] Metzgen, P., and Nancekievill, D. Multiplexer restructuring for FPGA implementation cost reduction. Design Automation Conf. (DAC '05) pp. 421-426, June 13-17, 2005, DOI= http://doi.acm.org/10.1145/1065579.1065692

[25] Verma, A., et al. "Synthesis of floating-point addition clusters on FPGAs using carry-save arithmetic," Int. Conf. Field Programmable Logic and Applications (FPL '10), pp. 19-24, Aug. 31- Sep. 2, 2010.

[26] Xilinx Corporation. Virtex-6 FPGA DSP48E1 Slice User Guide UG369 (v1.2), September 16, 2009. URL= http://www.xilinx.com/support/documentation/user_guides/ug369.pdf

Poster Session 1

Accelerating Short Read Mapping on an FPGA

Yupeng Chen, *Nanyang Technological University*
Bertil Schmidt, *Johannes Gutenberg University Mainz*
Douglas Leslie Maskell, *Nanyang Technological University*
Contact: ASDouglas@ntu.edu.sg

The explosive growth of short read datasets produced by high throughput DNA sequencing technologies poses a challenge to the mapping of short reads to a reference genome in terms of sensitivity and execution speed. Existing methods often use a restrictive error model for computing the alignments to improve speed, whereas more flexible error models are generally too slow for large-scale applications. Although a number of short read mapping software tools have been proposed, designs based on hardware are relatively rare. In this paper, we present a hybrid system for short read mapping utilizing both software and field programmable gate array (FPGA)-based hardware. The compute intensive semi-global alignment operation is accelerated on the FPGA. The proposed FPGA aligner is implemented with a parallel block structure to gain computational efficiency. We also propose a block-wise alignment algorithm to approximate the score of the conventional dynamic programming algorithm. Our performance comparison shows that the FPGA achieves an average speedup of 38 for the alignment operation on a Xilinx Virtex5 FPGA compared to the GASSST software implementation. For the overall execution time, our hybrid system achieves an average speedup of 2.4 compared to GASSST at comparable sensitivity and an average speedup of 1.8 compared to the popular BWA software at a significantly better sensitivity.

ACM Categories & Descriptors: C.1.3 Other Architecture Styles: Heterogeneous (hybrid) systems; **B.5.1** Design: Styles (e.g., parallel, pipeline, special-purpose)

Keywords: Short read alignment, Semi-global alignment, parallel processing, hybrid system, FPGA

The Masala Machine: Accelerating Thread-Intensive and Explicit Memory Management Programs with Dynamically Reconfigurable FPGAs

Mei Wen, Nan Wu, Qianming Yang, Chunyuan Zhang, Liang Zhao, *National University of Defense Technology*
Contact: yqm21249@163.com

A uniform FPGA-based architecture, an efficient programming model and a simple mapping method are paramount for PPGA technology to be more widely accepted. This paper presents MASALA, a dynamically reconfigurable FPGA-based accelerator specifically for parallel programs written in thread-intensive and explicit memory management (TEMM) programming models. The system uses TEMM programming model to parallelize the demanding application, including decomposing the application into separate thread blocks, decoupling compute and data load/store etc. Hardware engines are included into the MASALA by using partial dynamic reconfigure modules, each of which encapsulates Thread Process Engine implementing the thread functionality in hardware. A data dispatching scheme is also included in MASALA to enable the explicit communication among multiple memory hierarchies such as between inter-hardware engines, the host processor and hardware engines. At last, the paper illustrates a Multi-FPGA prototype system of the presented architecture: MASALA-SX. A large synthetic aperture radar (SAR) image formatting experiment shows that the MASALA architecture facilitates the construction of a TEMM program accelerator by providing it with greater performance and less power consumption than current CPU platforms, but without sacrificing programmability, flexibility and scalability.

ACM Categories & Descriptors: C.1.3 Other Architecture Styles: Adaptable architectures; **D.1.3** Concurrent Programming: Parallel programming

Keywords: FPGA, MASALA, Accelerator, SAR, Explicit Memory Management, Thread-intensive

Timing Yield Improvement of FPGAs Utilizing Enhanced Architectures and Multiple Configurations Under Process Variation

Fatemeh Sadat Pourhashemi, Morteza Saheb Zamani, *Amirkabir University of Technology*
Contact: fpoorhashemi@aut.ac.ir

Designing with field-programmable gate arrays (FPGAs) can face with difficulties due to process variations. Some techniques use reconfigurability of FPGAs to reduce the effects of process variations in these chips. Furthermore, FPGA architecture enhancement is an effective way to degrade the impact of variation. In this paper, various FPGA architectures are examined to identify which architecture can achieve larger parametric yield improvement utilizing multiple configurations as opposed to single configuration. Experimental results show that by increasing cluster size from 4 to 10, yield improvement increases from 2.82X to 4.48X. However, changing look-up table (LUT) size from 4 to 7 results in yield improvement degradation from 2.82X to 1.45X, using 10 configurations compared to single configuration over 20 MCNC benchmark circuits. These results indicate that multi-configuration technique causes larger timing yield improvement in FPGAs with larger cluster size and smaller LUT size..

ACM Categories & Descriptors: B.8.2 Performance Analysis and Design Aids

Keywords: FPGA Architecture, Multiple configurations, Process Variation, Timing Yield

A Field Programmable Array Core for Image Processing

Declan Walsh, Piotr Dudek, *The University of Manchester*
Contact: declan.walsh@postgrad.manchester.ac.uk

Massively parallel processor arrays have been shown to be an effective and suitable choice for image processing tasks [1]. More recently, some of the state of the art processor arrays have been used for real-time machine vision tasks such as intelligent transport system applications [2] or video processing on mobile applications [3] providing a much more powerful solution than a conventional processor. A number of Single Instruction Multiple Data (SIMD) processor arrays have been implemented on FPGAs [4]-[6], which are particularly suited to implementing such processor architectures because of their similarities of both being arrays of fine grained logic elements. In this work, we propose an FPGA implementation of a processor array where the processing elements (PEs) are as small as possible, while providing local memory sufficient for processing greyscale images. The PE is then replicated to form an array. A 32×32 PE array is implemented on a Xilinx Virtex 5 XC5VLX50 FPGA using the four-neighbour connectivity with the possibility to scale up using a larger FPGA. The processor array operates at a frequency of 96 MHz and executes a peak of 98.3 giga operations per second (GOPS) (bit-serial operations). A binary edge detection algorithm is performed in 52.08 ns. Uploading and downloading a binary image in a 32×32 array takes an extra 687.5 ns. Sobel edge detection of an 8-bit greyscale image is performed in 5.33 μs. Uploading and downloading an 8-bit greyscale image in a 32×32 array takes 5.36 μs. With larger FPGAs being available in the future, the array sizes comparable to state of the art custom designed ICs can be implemented on these FPGAs.

ACM Categories & Descriptors: C.1.2 Multiple Data Stream Architectures (Multiprocessors): Single-instruction-stream, multiple-data-stream processors (SIMD); **I.4.0** General

Keywords: FPGA, SIMD, architecture, image processor array

EmPower: FPGA Based Emulation of Dynamic Power Management Algorithms for Multi-Core Systems on Chip

Sundaram Ananthanarayanan, *Anna University*
Chirag Ravishankar, Siddharth Garg, Andrew Kennings, *University of Waterloo*
Contact: siddharth.garg@uwaterloo.ca

Dynamic power management for multi-core system on chip (MPSoC) platforms has become an increasingly critical design problem. In this paper, we present EmPower, an FPGA based emulation, validation and prototyping framework for dynamic power management research targeted at MPSoC platforms. EmPower supports two advanced power management features – per-core dynamic frequency scaling and clock gating, and power-aware thread migration. We also provide two fully-functional parallel applications for benchmarking–video encoding and software-defined radio. Our experimental results indicate that EmPower provides up to 36X improvement in run-time compared to cycle-accurate software simulations, and enables accurate and efficient exploration of the design space of power management algorithms.

ACM Categories & Descriptors: C.5.0 General: **B.5.0** General

Keywords: FPGA Implementation, Frequency Scaling, Power Management, Thread Migration

Poster Session 2

Adaptive FPGA-Based Robotics State Machine Architecture Derived with Genetic Algorithms

Jesus Savage, Rodrigo Savage, *Universidad Nacional Autonoma de Mexico*

Marco Morales-Aguirre, Angel Kuri-Morales, *Instituto Tecnologico Autonomo de Mexico*

Contact: robotssavage@gmail.com

This paper discusses how to generate mobile robots' behaviors using genetic algorithms, GA. The behaviors are built using state machines implemented in a programmable logic device, an FPGA, and they are encoded in such a way that a state machine architecture executes them, controlling the overall operation of a small mobile robot. The behaviors generated by the GA are evaluated according to a fitness function that grades their performance. Basically, the fitness function evaluates the robot's performance when it goes from an origin to a destination. In our approach each individuals' chromosome represents, given a set of inputs coming from the sensors and the current state, the next state and outputs that controls the robot's movements. For each generation the GA needs to evaluate population's individuals, doing this with the real robot it would required to much time, that would be impossible to do. Thus, the GA needs a simulator, as close as it can be to the real robot and its environment. The simulator gets the individuals' chromosomes and executes the algorithm state machine represented by them, it simulates the movements of the robot depending of the output generated in the present state and the simulated robot's sensors. Our objective was to prove that GA is a good option as a method for finding behaviors for mobile robots' navigation and also that these behaviors can be implemented in FPGAs.

ACM Categories & Descriptors: B.6.1 Design Styles: Sequential circuits

Keywords: Genetic Algorithms, Robots Behaviors

A Novel Full Coverage Test Method for CLBs in FPGA

Yong Fu, Chi Wang, Liguang Chen, Jinmei Lai, *Fudan University*

Contact: 10212020011@fudan.edu.cn

FPGA's configurability makes it difficult for FPGA's manufacturers to fully test it. In this paper, a full coverage test method for FPGA's Configurable Logic Blocks (CLBs) is proposed, through which all basic logics of FPGA's every CLB can be fully tested. Innovative test circuits are configured to build repeatable logic arrays for look-up tables, distributed random access memories, configurable registers and other logics. The programmable interconnects needed to connect CLBs in these test circuits are also repeatable, making the configuration process much easier and the test speed much faster. The test method is implemented on different scales of Xilinx Virtex chips, where 19 test configuration circuits are needed to achieve 100% coverage for all CLBs. Besides, the method is transplantable and independent of FPGA's array size. To evaluate the test method reliably and guide the process of test vectors generation, a fault simulator - Turbofault is used to simulate FPGA's test coverage.

ACM Categories & Descriptors: B.8.1 Reliability, Testing, and Fault-Tolerance

Keywords: CLB, Fault Coverage, FPGA, Test Circuits Design

Constraint-Driven Automatic Generation of Interconnect for Partially Reconfigurable Architectures

Andre Seffrin, *Center for Advanced Security Research Darmstadt*

Sorin A. Huss, *TU Darmstadt*

Contact: andre.seffrin@cased.de

Dynamic partial reconfiguration allows the exchange of hardware configurations on FPGAs at run-time. Within a reconfigurable system that supports several different modules, resource requirements for interconnect between these modules may be considerably high. Enabling communication via a crossbar may require too many resources. State-of-the-art modelling methods for partial dynamic reconfiguration already support the fine-grained description of interaction between the partial modules. We propose both an online and an offline method for automatically generating interconnect according to such communication constraints, aiming at a low resource usage. The online algorithm determines an appropriate port assignment for the partial modules by means of a greedy approach and exploits port overlaps. The offline algorithm employs simulated annealing in order to find a proper port assignment and also incorporates the scheme for exploiting port overlaps. Constraint-generated interconnect requires significantly less resources than a crossbar, even if only a random port assignment is used. Proper port assignment by the online method reduces these requirements by an additional 10%, and using the offline method reduces them by an additional 30% on average. Online port assignment is faster than the offline method by several orders of magnitude. The interconnect generation tool introduced in this work takes textual input of communication constraints and automatically generates a corresponding hardware description in VHDL.

ACM Categories & Descriptors: B.5.2 Design Aids: Optimization

Keywords: Interconnect, partial dynamic reconfiguration, port assignment, crossbar, constraint-based design

Thermal-Aware Logic Block Placement for 3D FPGAs Considering Lateral Heat Dissipation

Juinn-Dar Huang, Ya-Shih Huang, Mi-Yu Hsu, Han-Yuan Chang, *National Chiao Tung University*
Contact: sali.ee95g@nctu.edu.tw

Three-dimensional (3D) integration is an attractive and promising technology to keep Moore's Law alive, whereas the thermal issue also presents a critical challenge for 3D integrated circuits. Meanwhile, accurate thermal analysis is very time-consuming and thus can hardly be incorporated into most of placement algorithms generally performing numerous iterative refinement steps. As a consequence, in this paper, we first present a fine-grained grid-based thermal model for the 3D regular FPGA architecture and also highlight that lateral heat dissipation paths can no longer be assumed negligible. Then we propose two fast thermal-aware placement algorithms for 3D FPGAs, Standard Deviation (SD) and MineSweeper (MS), in which rapid thermal evaluation instead of slow detailed analysis is utilized. Moreover, both take the lateral heat dissipation into consideration and focus on distributing heat sources more evenly within a layer in a 3D FPGA to avoid creating hotspots. Experimental results show that SD and MS achieve 12.1%/7.6% reduction in maximum temperature and 82%/56% improvement in temperature deviation compared with a classical thermal-unaware placement method only at the cost of minor increase in wirelength and delay. Moreover, MS merely consumes 4% more runtime for producing thermal-aware placement solutions.

ACM Categories & Descriptors: B.7.2 Design Aids: Placement and routing

Keywords: Three-dimensional integration, 3D FPGAs, thermal-aware placement, logic block placement

Power-Aware FPGA Technology Mapping for Programmable-VT Architectures

Wei Ting Loke, *Xilinx Inc. & National University of Singapore*
Yajun Ha, *National University of Singapore*
Contact: weitingl@xilinx.com

In this paper, we present a framework for leakage power reduction in FPGAs with programmable-VT architectures, with focus on dual-VT technology mapping. The use of Reverse Back Bias (RBB) circuit techniques is recognized as one of the possible strategies in mitigating leakage power, a critical problem in circuits deploying deep submicron process technologies. FPGAs with the ability to tune LUT VT via RBB oer the potential of reducing leakage power with no sacrifice to circuit speed. Today, Altera☐s Stratix line of FPGAs oer some levels of VT programmability, but with optimizations limited to the post-P&R stage. We present a novel technology mapper (RBBMap), logic block packer (RBBPack) and placement-and-routing tool (RBBVPR) that together demonstrate the advantages in moving RBB optimizations upwards to the technology mapping level.

Compared to an existing power-optimized technology mapping tool Emap, our framework oers an average of 44.41% savings in average logic block leakage power and 30.88% savings in average total energy consumption. We also illustrate why our work is potentially superior to another comparable work DVMap-2 that utilizes a dual-VDD approach.

ACM Categories & Descriptors: B.6.3 Design Aids: Optimization; **B.7.1** Types and Design Styles: Gate arrays

Keywords: Programmable-VT, Dual-VT, Multiple Voltage, Reverse Back Bias, Technology Mapping, FPGA

FPGA-RR: An Enhanced FPGA Architecture with RRAM-Based Reconfigurable Interconnects

Jason Cong, Bingjun Xiao, *University of California, Los Angeles*
Contact: xiao@cs.ucla.edu

In this study, we explore the use of Resistive RAMs (RRAMs) as candidates for programmable interconnects in FPGAs. An RRAM cell can be programmed between high resistance state and low resistance state, with an on/off ratio close to MOSFET. It provides an opportunity to use an RRAM as a routing switch at a much smaller area cost than its CMOS counterpart. RRAMs can be fabricated over CMOS circuits using CMOS-compatible processes to have a more compact gate array. Our recent work (presented in NanoArch'2011) demonstrated significant potential of area, delay, and power reduction from using RRAMs in FPGAs. But some design problems remain open. The programming of RRAM switches integrated in interconnects is one important problem. We show that the high-level architecture of programming circuits for RRAM switches should be modified to avoid potential logic hazard. Also the programming cells used in previous works have an area overhead even larger than RRAM itself. We manage to reduce this overhead significantly with utilization of the non arbitrary pattern of RRAM integration in FPGA interconnects. In addition we suggest a novel buffering solution for FPGA interconnects in light of the low area cost of RRAM-based routing switch. We propose on-demand buffer insertion, where buffers can be connected to interconnects via RRAMs to dynamically reflect the demand of the netlist to map onto FPGA. Compared to conventional buffering solution which are pre-determined during fabrication and can only be optimized for general case, our solution shows further area savings and performance improvement. The resulting FPGA architecture using RRAM for programmable interconnects is named FPGA-RR. We provide a complete CAD flow for FPGA-RR.

ACM Categories & Descriptors: B.7.1 Types and Design Styles: Advanced technologies

Keywords: RRAM FPGA, Resistive RAM, programming, buffer

Poster Session 3

Efficient In-System RTL Verification and Debugging Using FPGAs

Proshanta Saha, Chuck Haymes, Ralph Bellofatto, Bernard Brezzo, Mohit Kapur, Sameh Asaad, *IBM T.J. Watson Research Center*
Contact: asaad@us.ibm.com

FPGAs have become indispensible in processor design, bring-up and debug. Traditionally FPGAs have been used in prototyping, allowing end-users to emulate functionality of a specific component of a processor. However, as the complexity of processors grows, another aspect of processor design, RTL verification, has become a prime target for acceleration using FPGAs. Software-only RTL simulation and verification tools are no longer sufficient for many verification tasks as they often incur long execution time penalties. Software simulation time for a basic Linux kernel bring-up on a BlueGene/Q [1] processor, with 16 user PowerPC A2 cores, for example, could easily exceed several years.

An important feature of RTL verification acceleration using FPGAs is its fast debugging capabilities. The ability to quickly and accurately pinpoint the location of an anomaly in an RTL source is highly desirable. This paper proposes efficient in-system debugging techniques on FPGAs for RTL verification. We show how a network of over 45 Virtex 5 LX330 FPGAs can be efficiently used to read out state information of the BlueGene/Q processor. We also demonstrate how the new in-system debugging technique is 250x faster than comparable methods.

ACM Categories & Descriptors: B.6.3 Design Aids, **C.1.4** Parallel Architectures

Keywords: FPGA debugging, FPGA-based acceleration, processor simulation

Parallel FPGA Placement Based on Individual LUT Placement

Chris C. Wang, Guy G. F. Lemieux, *University of British Columbia*
Contact: chrisw@ece.ubc.ca

This work describes a novel approach to FPGA placement. Most conventional FPGA CAD flow clusters circuits into CLBs prior to placing it. We show that is it possible to achieve 28% and 21% improvement in wirelength and minimum channel width respectively, while suffering only 1.8 % in critical path delay by placing individual LUTs directly. By utilizing a good parallel placer, the novel approach can achieve speedups over the conventional uni-processor placers as well.

ACM Categories & Descriptors: B.7.2 Design Aids; Placement and routing

Keywords: parallel placement

Dataflow-Driven Execution Control in a Coarse-Grained Reconfigurable Array

Robin Panda, Scott Hauck, *University of Washington*
Contact: robin@ee.washington.edu

Coarse Grained Reconfigurable Arrays (CGRAs) are a promising class of architectures for accelerating applications using a large number of parallel execution units for high throughput. While they are typically very efficient for a single task, all functional units are required to perform in lock step; this makes some classes of applications more difficult to program and efficiently use resources. Other architectures like Massively Parallel Processor Arrays (MPPAs) are better suited for these applications and excel at executing unrelated tasks simultaneously, but the amount of resources dedicated to a single task is limited.

We are developing a new architecture with the design flexibility of an MPPA and the throughput of a CGRA. A key to the flexibility of MPPAs is the ability for subtasks to execute independently instead of in lock step with all other tasks on the array. Adding this capability requires special control circuitry for architectural support in a CGRA. We decribe the modifications required and our solutions. Additionally, we also describe the CAD tool modification and application developer concerns for utilizing the resulting hybrid CGRA/MPPA architecture.

ACM Categories & Descriptors: C.1.2 Multiple Data Stream Architectures (Multiprocessors); Multiple-instruction-stream, multiple-data-stream processors (MIMD), **B.2.1** Design Styles; Parallel

Keywords: CGRA, MPPA, architecture

OpenCL Memory Infrastructure for FPGAs

S. Alexander Chin, Paul Chow, *University of Toronto*
Contact: xan@eecg.toronto.edu

Programming models assist developers in creating high performance computing systems by forming a higher level abstraction of the target platform. OpenCL has emerged as a standard programming model for heterogeneous systems and there has been recent activity combining OpenCL and FPGAs. This work introduces memory infrastructure for FPGAs and is designed for OpenCL style computation, complementing previous work. An Aggregating Memory Controller is implemented in hardware and aims to maximize bandwidth to external, large, high-latency, high-bandwidth memories by finding the minimal number of external memory burst requests from a vector of requests. A template processing array with soft-processor and hand-coded hardware elements was also designed to drive the memory controller. The Aggregating Memory Controller is described in terms of operation and future scalability and the created processing array is described as a flexible structure that can support many types of processing solutions. A hardware prototype of the memory controller and processing array was implemented on a Virtex-5 LX110T FPGA. Two micro-benchmarks were run on both the soft-processor elements and the hand-coded hardware cores to exercise the memory controller. Results for effective memory bandwidth within the system show that the high-latency can be hidden using the Aggregating Memory Controller by increasing the number of threads within the processing array.

ACM Categories & Descriptors: B.6.1 Design Styles

Keywords: FPGA, OpenCL, memory coalescing, memory aggregation, DRAM

Operation Scheduling and Architecture Co-Synthesis for Energy-Efficient Dataflow Computations on FPGAs

Colin Yu Lin, Ngai Wong, Hayden Kwok-Hay So, *The University of Hong Kong*
Contact: hso@eee.hku.hk

Compiling high-level user applications for execution on FPGAs often involves synthesizing dataflow graphs beyond the size of the available on-chip computational resources. One way to address this is by folding the execution of the given dataflow graphs onto an array of directly connected simple configurable processing elements (CPEs). Under this scenario, the performance and energy-efficiency of the resulting system depends not only on the mapping schedule of the compute operations on the CPEs, but also on the topology of the interconnect array that connects the CPEs. This paper presents a framework in which the operation scheduler and the underlying CPE interconnect network topology are co-optimized on a per-application basis for energy-efficient FPGA computation. Given the same application, more than 2.5x difference in energy-efficiency was achievable by the use of different common regular array topologies to connect the CPEs. Moreover, by using irregular application-specific interconnect topologies derived from a genetic algorithm, up to 50% improvement in energy-delay-product was achievable when compared to the use of even the best regular topology. The use of such framework is anticipated to serve as part of a rapid high-level FPGA application compiler since minimum hardware place-and-route is needed to generate the optimal schedule and topology.

ACM Categories & Descriptors: B.7.2 Design Aids

Keywords: Architecture Synthesis, Dataflow Computation, Energy-Efficient, FPGA, Operation Scheduling

Poster Session 4

Post-Silicon Debugging Targeting Electrical Errors with Patchable Controllers

Masahiro Fujita, Hiroaki Yoshida, *University of Tokyo*
Contact: fujita@ee.t.u-tokyo.ac.jp

Due to continuous increase of design complexity in SoC development, the time required for post-silicon verification and debugging keeps increasing especially for electrical errors and subtle corner case bugs, and it is now understood that some sort of programmability in silicon is essential to reduce the time for post-silicon verification and debugging. Although an easiest way to achieve this is to use FPGA for entire circuits, performance especially in terms of power efficiency compared with pure hardwired logic may be significantly inferior. Here, we discuss partial use of such in-field programmability in control parts of circuits for post-silicon debugging processes for electrical errors and corner case logical bugs. Our method deals with RTL designs in FSMD (Finite State Machine with Datapath) by adding partially in-field programmability, called 'patch logic', in their control parts. With our patch logic we can dynamically change the behaviors of circuits in such a way to trace state transition sequences as well as values of internal values periodically. Our patch logic can also check if there is any electrical error or not periodically. Assuming that electrical errors occur very infrequently, an error can be detected by comparing the equivalence on the results of duplicated computations. Through experiments we discuss the area, timing, and power overhead due to the patch logic and also show results on electrical error detection with duplicated computations..

ACM Categories & Descriptors: **C.3** Special-Purpose and Application-Based Systems: Real-time and embedded systems

Keywords: Post-silicon debug, Formal analysis, Hardware patch

Algorithm and Architecture Optimization for Large Size Two-Dimensional Discrete Fourier Transform

Berkin Akin, Peter A. Milder, Franz Franchetti, James C. Hoe, *Carnegie Mellon University*
Contact: bakin@ece.cmu.edu

We present a poster showcasing our FPGA implementations of two-dimensional discrete Fourier transform (2D-DFT) on large datasets that must reside off-chip in DRAM. These memory-bound large 2D-DFT computations are at the heart of important scientific computing and image processing applications. The central challenge in creating high-performance implementations is in the carefully orchestrated use of the available off-chip memory bandwidth and on-chip temporary storage. Our implementations derive their efficiency from a combined attention to both the algorithm design to enable efficient DRAM access patterns and datapath design to extract the maximum compute throughput at a given level of memory bandwidth. The poster reports results including a 1024x1024 double-precision 2D-DFT implementation on an Altera DE4 platform (based on a Stratix IV EP4SGX530 with 12 GB/s DRAM bandwidth) that reached over 16 Gflop/s, achieving a much higher ratio of performance-to-memory-bandwidth than both state-of-the-art CPU and GPU implementations.

ACM Categories & Descriptors: **C.3** Special-Purpose and Application-Based Systems: Signal processing systems

Keywords: 2D-DFT, Bandwidth, DRAM, FFT, FPGA

Early Timing Estimation for System-Level Design Using FPGAs

Hugo Andrade, Arkadeb Ghosal, Rhishikesh Limaye, Sadia Malik, Newton Petersen, Kaushik Ravindran, Trung Tran, Guoqiang Wang, Guang Yang, *National Instruments Corp.*
Contact: gerald.wang@ni.com

FPGA devices provide flexible, fast, and low-cost prototyping and production solutions for system design. However, as the design complexity continues to rise, the design and synthesis iterations become a labor intensive and time consuming ordeal. Consequently, it becomes imperative to raise the level of abstraction for FPGA designs, while providing insight into performance metrics early in the design process. In particular, an important design time problem is to determine the maximum clock frequency that a circuit can achieve on a specific FPGA target before full synthesis and implementation. This early quantification can greatly help evaluate key design characteristics without reverting to tedious runs of the full implementation flow. In this work, we focus on the predictability of timing delay of circuits composed of high-level blocks on an FPGA. We are well aware of difficulties in tackling uncertainties in early timing estimation, e.g., an inherent gap between a high-level representation and gates/wires; extremely difficult delay estimation due to the randomness in physical design tools, etc. We show that the estimation uncertainties can be mitigated through a carefully characterized timing database of primitive building blocks and refined timing analysis models. We primarily focus on applications composed of data-intensive word-level arithmetic computations from the DSP domain and specified using static dataflow models. Our experiments indicate that for these applications, timing estimates can be obtained reliably within a good error margin on average and in the worst case. As future work, we plan to fine tune the timing database by modeling resource utilization effects and inter-primitive/actor routing delay via variants of Rent's rule and related efforts. We are also interested in exploring dynamic sub-cycle timing characterization.

ACM Categories & Descriptors: **D.2.2** Design Tools and Techniques, **D.1.2** Automatic Programming

Keywords: FPGA, early timing analysis, system-level design

Scalable Architecture for 135 GBPS IPV6 Lookup on FPGA

Yi-Hua E. Yang, *Futurewei Technologies*
Oguzhan Erdem, *Middle-East Technical University*
Viktor K. Prasanna, *University of Southern California*
Contact: YiHua.Yang@huawei.com

High-speed IP lookup remains a challenging problem in next generation routers due to the ever increasing line rate and routing table size. In addition, the evolution towards IPv6 also requires long prefix length, sparse prefix distribution, and potentially very large routing tables. Previous solutions have relied on complex trie compression as well as Bloom filters to achieve high forwarding rate for IPv6.In this paper, we propose a novel Combined Length-Infix Pipelined Search (CLIPS) architecture for IPv6 routing table lookup on FPGA. CLIPS solves the longest prefix match (LPM) problem by combining both prefix length and infix pattern search. Binary search in prefix length is performed on the 64-bit routing prefix of IPv6 down to an 8-bit length range in $\log\left(64/8\right)=3$ phases; each phase performs a fully-pipelined infix pattern search with only one external memory access. A fourth and the last phase then finds the LPM (if any) within the 8-bit length range in a compressed multi-bit trie.

We describe the algorithms and data structures used for the CLIPS construction, run-time operation, dynamic update and false-positive avoidance. The proposed solution improves the on-chip memory efficiency on FPGA and maximizes the external SRAM utilization; additional properties for ensuring the practicality of our scheme include the modular construction, easy dynamic update, and simple resource allocation. Using a state-of-the-art FPGA, our CLIPS prototype supports up to 2.7 millioin IPv6 prefixes when employing 33 Mbits of BRAM and 4 channels of external SRAM. The prototype achieves a sustained throughput of 264 million IPv6 lookups per second, or 135 Gbps with minimum size (64-byte) packets.

ACM Categories & Descriptors: H.4.3 Communications Applications; Internet

Keywords: IP lookup, packet forwarding, longest prefix match, binary search tree, search trie, Tree Bitmap

FPGA'12, February 22–24, 2012, Monterey, California, USA.
ACM 978-1-4503-1155-7/12/02.

Author Index

www.ingramcontent.com/pod-product-compliance
Lightning Source LLC
Chambersburg PA
CBHW080357060326
40689CB00019B/4046